Intelligent Design and Religion as a Natural Phenomenon

The International Library of Essays on Evolutionary Thought
Series Editor: Neil Levy

Titles in the Series:

Philosophy of Evolutionary Biology: Volume I
Stefan Linquist

Evolutionary Psychology: Volume II
Stefan Linquist and Neil Levy

Evolutionary Ethics: Volume III
Neil Levy

The Evolution of Culture: Volume IV
Stefan Linquist

Intelligent Design and Religion as a Natural Phenomenon: Volume V
John S. Wilkins

Intelligent Design and Religion as a Natural Phenomenon
Volume V

Edited by

John S. Wilkins

Bond University and the University of Sydney, Australia

LONDON AND NEW YORK

First published 2010 by Ashgate Publishing

Published 2016 by Routledge
2 Park Square, Milton Park, Abingdon, Oxon OX14 4RN
711 Third Avenue, New York, NY 10017, USA

Routledge is an imprint of the Taylor & Francis Group, an informa business

British Library Cataloguing in Publication Data
Intelligent design and religion as a natural phenomenon.
Volume 5. – (International library of essays on
evolutionary thought)
1. Psychology, Religious. 2. Genetic psychology.
3. Intelligent design (Teleology)
I. Series II. Wilkins, John S., 1955
200.1'9-dc22

Library of Congress Control Number: 2009938973

ISBN 9780754627630 (hbk)

Contents

Acknowledgements

The editor and publishers wish to thank the following for permission to use copyright material.

ABC-CLIO Inc. for the essay: Allen D. MacNeill (2006), 'The Capacity for Religious Experience is an Evolutionary Adaptation to Warfare', in F. A. Stout (ed.), *The Psychology of Resolving Global Conflicts: From War to Peace*, Westport, CT: Praeger Security International, pp. 257–84. Copyright © 2006 ABC-CLIO Inc.

American Anthropological Association for the essay: Pierre Liénard and Pascal Boyer (2006), 'Whence Collective Rituals? A Cultural Selection Model of Ritualized Behavior', *American Anthropologist*, **108**, pp. 814–27. Copyright © 2006 American Anthropological Association.

Brill Academic Publishers for the essays: Joseph Bulbulia (2005), 'Are There Any Religions? An Evolutionary Exploration', *Method & Theory in the Study of Religion*, **17**, pp. 71–100. Copyright © 2005 Brill Academic Publishers; James W. Dow (2006), 'The Evolution of Religion: Three Anthropological Approaches', *Method and Theory in the Study of Religion*, **18**, pp. 67–91. Copyright © 2006 Brill Academic Publishers; Justin L. Barrett and Brian Malley (2007), 'A Cognitive Typology of Religious Actions', *Journal of Cognition and Culture*, **7**, pp. 201–11. Copyright © 2007 Brill Academic Publishers; Pascal Boyer (1992), 'Explaining Religious Ideas: Elements of a Cognitive Approach', *Numen*, **39**, pp. 27–57. Copyright © 1992 Brill Academic Publishers; Jesse M. Bering and Dominic D.P. Johnson (2005), '"O Lord … You Perceive my Thoughts from Afar": Recursiveness and the Evolution of Supernatural Agency', *Journal of Cognition and Culture*, **5**, pp. 118–42. Copyright © 2005 Brill Academic Publishers.

John Wiley and Sons for the essays: Loyal Rue (2000), 'Religion Generalized and Naturalized', *Zygon*, **35**, pp. 587–602. Copyright © 2000 John Wiley and Sons; Ursula Goodenough (2000), 'Exploring Resources of Naturalism: Religiopoiesis', *Zygon*, **35**, pp. 561–66. Copyright © 2000 John Wiley and Sons; Lyle B. Steadman and Craig T. Palmer (1997), 'Myths as Instructions from Ancestors: The Example of Oedipus', *Zygon*, **32**, pp. 341–50. Copyright © 1997 John Wiley and Sons; Deborah Kelemen (2004), 'Are Children "Intuitive Theists"? Reasoning about Purpose and Design in Nature', *Psychological Science*, **15**, pp. 295–301. Copyright © 2004 John Wiley and Sons; Gregory R. Peterson (2000), 'God, Genes, and Cognizing Agents', *Zygon*, **35**, pp. 469–80. Copyright © 2000 John Wiley and Sons; Azim F. Shariff and Ara Norenzayan (2007), 'God Is Watching You: Priming God Concepts Increases Prosocial Behavior in an Anonymous Economic Game', *Psychological Science*, **18**, pp. 803–809. Copyright © 2007 John Wiley and Sons; Gregory R. Peterson (2002), 'The Intelligent-Design Movement: Science or Ideology?', *Zygon*, **37**, pp. 7–23. Copyright © 2002 John Wiley and Sons; William Grey (1987), 'Evolution and the Meaning of Life', *Zygon*, **22**, pp. 479–96. Copyright © 1987 John Wiley and Sons; Barbara Forrest (2000), 'The Possibility of Meaning in Human Evolution', *Zygon*, **35**, pp. 861–80. Copyright © 2000 John Wiley and Sons.

The Royal Society for the essay: Maurice Bloch (2008), 'Why Religion is Nothing Special But is Central', *Philosophical Translations of the Royal Society B*, **363**, pp. 2055–61. Copyright © 2008 The Royal Society.

Springer for the essays: Joseph Bulbulia (2004), 'The Cognitive and Evolutionary Psychology of Religion', *Biology and Philosophy*, **19**, pp. 655–86. Copyright © 2004 Springer; David Sloan Wilson (2005), 'Testing Major Evolutionary Hypotheses about Religion with a Random Sample', *Human Nature*, **16**, pp. 382–409. Copyright © 2005 Springer. Candace S. Alcorta and Richard Sosis (2005), 'Ritual, Emotion, and Sacred Symbols: The Evolution of Religion as an Adaptive Complex', *Human Nature*, **16**, pp. 323–59; Copyright © 2005 Springer; Elliott Sober (2002), 'Intelligent Design and Probability Reasoning', *International Journal for Philosophy of Religion*, **52**, pp. 65–80. Copyright © 2002 Springer; John S. Wilkins and Wesley R. Elsberry, (2001), 'The Advantages of Theft over Toil: The Design Inference and Arguing from Ignorance', *Biology and Philosophy*, **16**, pp. 711–24. Copyright © 2001 Springer.

Taylor & Francis for the essays: Francisco J. Ayala (2003), 'Intelligent Design: The Original Version', *Theology and Science*, **1**, pp. 9–32. Copyright © 2003 Taylor & Francis; Justin L. Barrett (2007), 'Is the Spell Really Broken? Bio-psychological Explanations of Religion and Theistic Belief', *Theology and Science*, **5**, pp. 57–72. Copyright © 2007 Taylor & Francis; Howard J. Van Till (2008), 'How Firm a Foundation? A Response to Justin L. Barrett's "Is the Spell Really Broken?"', *Theology and Science*, **6**, pp. 341–49. Copyright © 2008 Taylor & Francis.

Every effort has been made to trace all the copyright holders, but if any have been inadvertently overlooked the publishers will be pleased to make the necessary arrangement at the first opportunity.

Series Preface

The theory of evolution is one of science's great achievements. Though to those outside science, it may seem that the theory is controversial, within science there is no controversy at all about its basic form. Moreover, the theory of evolution plays a pivotal role in guiding new research. 'Nothing in biology makes sense except in the light of evolution', Theodosius Dobzhansky famously wrote; the theory of evolution unifies disparate subfields of biology and generates testable predictions for each. The success of the theory and its explanatory fecundity for biology cannot be doubted. But might the theory also be capable of illuminating phenomena outside the direct purview of biology?

The volumes in this series are dedicated to exploring this question. They bring together some of the best writings of the past two decades which explore the relevance of evolution and evolutionarily-inspired thought to arenas of human life beyond the merely biological. Volumes focus on whether it is productive and illuminating to attempt to understand our most distinctive achievements and our most intimate features as evolved phenomena. Is the content of moral systems explained by evolution? To what extent are the processes of selection and reproduction that explain changes in gene frequencies also at work in explaining the reproduction of ideas? Can evolution shed light on why we think as we do, perceive as we do, even feel as we do? Might even our idea of God – and perhaps with it the perennial temptation to reject evolution in the name of religion – be explained by evolutionary thought?

Answering these questions requires not only a detailed grasp of the phenomena we aim to explain – the contours of religious thought, the features of morality, and so on – but also an understanding of the theory we aim to apply to the field. Though the theory of evolution is not itself controversial within science, there are lively controversies about its details. One volume of this theory is devoted to writings which illuminate these controversies and deepen our understanding of the mechanisms of evolution. It is only if we have an appreciation of how evolution works that we can begin to assess attempts to extend its reach to culture, to the mind, to morality and to religion.

The volumes are edited by experts in the philosophy of biology and include sensitive and thoughtful discussions of the material they contain. Naturally, in selecting the papers for inclusion, and given the large amount of high quality thought on the philosophy of biology, and on each of the topics covered by these volumes, it was necessary to make some hard choices. Each editor has chosen to focus on particular controversies within the field covered by their volume; on each topic, a range of views is canvassed (including the views of those who deny that evolution can contribute much to the understanding of non-biological features of human beings).

Evolution is our story; in coming to understand it, we come to understand ourselves. Readers of these volumes should be left with a deepened appreciation for the power and ambition of evolutionary thought, and with a greater understanding of what it means to be an evolved being.

NEIL LEVY

Florey Neuroscience Institutes, Australia and University of Oxford, UK

Introduction

Daniel Dennett notes that in the opening lines to *The Natural History of Religion* David Hume distinguished two questions: 'As every enquiry, which regards religion, is of the utmost importance, there are two questions in particular, which challenge our attention, to wit, that concerning *its foundation in reason*, and that concerning its *origin in human nature*' (Dennett, 2006, p. 26; Hume, Green and Grose, 1875, vol. 2, p. 309) Whether science challenges religion's 'foundation in reason' is a familiar question of the philosophy of religion. Both the general epistemological stance of the natural sciences and particular claims that have scientific warrant have often been presented as challenges to religion. The focus of this volume, however, is on Hume's second question, which has received far less attention from philosophers despite its impressive intellectual lineage and its prominence in contemporary science and popular culture.

Hume's second question is whether religious behaviour, institutions and experiences can be explained in natural terms. Attempts to 'naturalize' religion seek to locate it alongside war, trade and the manufacture of pottery as simply one more characteristic human activity.[1] This is an old project, arguably going back to the classical world. But it has recently become an active focus of research in evolutionary biology. The results of this research have become a hot topic for both scientific technical writing and popular science writing.

Over the last decade, a strident public debate has arisen about the nature and origin of religions. The core question is the adaptive nature of religion – whether it is socially beneficial or a malign influence that 'infects' people, as has been recently claimed by such authors as Richard Dawkins (2006), Sam Harris (2004), Daniel Dennett (2006) and Christopher Hitchens (2007). Controversies include how exactly religion evolved, whether by individual or group selection, if it is adaptive, and if not, whether and how it is a side effect of evolution. The topicality of this debate can be seen from its role in 'culture wars' in the US and increasingly in the rest of the world, and in attacks on evolution by religious figures, both in the US and elsewhere. Evolutionary biology has been identified as one aspect of a 'naturalistic world-view' endemic among scientists and essentially incompatible with religious commitments (Plantinga, 1993; Johnson, 1995; Ruse, 2001).

Historically there have been three main approaches to explaining religion naturalistically, each of which can be placed in the broader context of a theory of evolution. The first, which was very popular in the late nineteenth century, explains religious sentiment as a universal human psychological trait (Baring-Gould, 1892; James, 1902; Wells, 1918). Edward Burnett Tylor (1871) offered an influential account of this kind based closely on positivist concepts of the stages of civilization. The second approach, derived from the work of Émile

---

[1] See Grinde (1998); Wunn (2000, 2002); Boyer (2001); Wilson (2002); Hopkins (2003); Atran and Norenzayan (2005); Bulkeley (2005); Kelemen (2005); Landau, Greenberg and Solomon (2005); Martin (2005); Pyysiäinen (2005); Sosis and Alcorta (2005); Bulbulia, Chapter 1, this volume; Dow, Chapter 5, this volume.

Durkheim (Durkheim and Lukes,1982), sees religion as a socioeconomic phenomenon which either maintains or challenges the status quo, depending on the politico-social interests of its adherents. The third approach sees religion as a distinct cultural evolutionary process, decoupled from biological evolution (Dennett, 2006).

The reason this has become a hot-button issue stems from the current prominence of scientific accounts of the origin of religion. Religion is a focus of research in a cluster of scientific disciplines focused on evolutionary biology and its applications to human behaviour. The resultant prospect of 'naturalizing' religion is a key battleground in the ongoing culture war between religious advocates who are critical of some aspects of the scientific world-view and those who take it upon themselves to defend science in the public arena (Johnson, 1995; Ruse, 2001; Shanks, 2004; McGrath, 2005; Dawkins, 2006), including the media stars commonly labelled the 'new atheists' (Harris, 2004, 2006; Dawkins, 2006; Dennett, 2006). Hence, it is crucially important at this point to critically analyse current efforts to naturalize religion, to understand how cultural, psychological, and biological forms of naturalization relate to one another, and to assess claims about the implications of the science for religion itself and for its role in our society.

The main approaches to naturalizing religion today draw their explanatory resources from evolutionary biology, or from quasi-Darwinian accounts of cultural evolution, or from some combination of these two. Early Darwinian approaches to explaining religious conformity and morality in evolutionary terms appealed to the benefits they confer on social groups, following a comment of Darwin's in *The Descent of Man* in which he explained the evolution of ethical behaviour in terms of the differential success of 'tribes' (1874: 162). Ethically inclined groups, Darwin argued, would do better on average than purely egoistical groups (note that then and now naturalistic explanations of religion are often intertwined with naturalistic explanations of morality). Darwin's account seems to require a group selectionist view of social evolution, and since the 1960s both biologists and philosophers of science have argued against this on theoretical grounds (Bradley, 1999; Dow, Chapter 5, this volume). However, in recent years group selectionist accounts have had a dramatic resurgence, following the revival of group selection models more generally (Wilson, 1997b, 1997a; Wilson and Sober, 1998; Wilson and Wilson, 2007).

An alternative kind of explanation relies on the ideas of the population geneticist William Hamilton (1964a, 1964b) and his followers, particularly those concerning 'kin selection'. On this account, altruistic behaviours such as self-sacrifice are explicable, because they tend to increase not the individual's fitness, but their 'inclusive fitness', which includes the fitness of the copies of their genes that exist in their relatives. Hence insofar as religious behaviours contribute to the flourishing of kin, they can be explained without recourse to selection on groups as such. Apparently fitness-reducing behaviours like celibacy and other-regarding behaviours that are often associated with religious rules become explicable when viewed in the light of Hamilton's evolutionary models (Wright, 1994). Explanations of this kind are popular in contemporary 'evolutionary psychology' (Barkow, Cosmides and Tooby, 1992; Badcock, 2000).

Methodological and philosophical problems arise with each of these evolutionary accounts of social institutions and behaviours. There is continuing debate over the scientific legitimacy of sociobiology and its recent incarnation, evolutionary psychology.[2] This question – essentially whether human behaviour can be explained in the same way as animal behaviour and human physiology – turns on several of the basic issues in the philosophy of the life sciences, such as the levels of selection,[3] the limits of 'adaptationism',[4] the nature of mind[5] and whether there is any such thing as human nature.[6] Evolutionary explanations of collective, institutional phenomena like religion raise specific issues of their own. Several theorists allege that this makes cultural evolution the more relevant framework for naturalizing religion (Grinde, 1998; Dennett, 2006; Dow, Chapter 5, this volume). Still more specific issues facing evolutionary accounts of religion, whether biological or cultural, concern what religion is an adaptation to, or if not an adaptation itself, what the traits it is a by-product of are adaptations to, and in general what is the social and ecological 'selective environment' of religion (Jacobson, 1993; Reynolds and Tanner, 1995).

Recent research on primates indicates that there are some shared behavioural and psychological traits between them and humans that may shed light on human religious behaviour.[7] This approach would seem to have some unique methodological advantages. In applying the evidence of social organization and rule-following from primate studies the unwarranted analogies made by early sociobiologists, who analogized behaviour in distant species like gulls or gazelles (Morris and Morris, 1966; Morris, 1967; Tiger and Fox, 1989), can be avoided (Griffiths, 1996). Moreover, evolutionary explanations of behaviour are on sounder ground when conducted in conjunction with neurobiological and developmental studies (Griffiths, 2007).

A key part of the topic is to locate current naturalistic theories of religion within the overall problems such theories address. These can be presented as a series of issues on which any theorist must make choices.

The first issue is how directly the proposed explanation relates to religion. In the broadest terms we can distinguish *functionalism*, or the view that religion fulfils some social or biological task, from *consequentialism*, in which religion is a consequence of some other process that is functional. In the evolutionary context, this becomes the distinction between *adaptationist* accounts and *by-product* or '*spandrel*'[8] accounts. A 'spandrel' is a by-product of selection for some function other than the one that the spandrel serves.

---

2 See Wilson (1975); Boyd and Richerson (1985, 2005); Durham (1991); Barkow, Cosmides and Tooby (1992); Wright (1994); Weingart (1997); Cronk (1999); Badcock (2000); Segerstråle (2000); Dupré (2001); Rose and Rose (2001); Griffiths (2004); Buller (2005); Richerson and Boyd (2005); Wilson and Wilson (2007).

3 See Brandon (1988); Wilson (1989); Bradley (1999); Keller (1999); Leigh (1999).

4 See Sober (1994); Orzack and Sober (1996); Andrews, Gangestad and Matthews (2002); Lewens (2004); Scher (2004).

5 See Fodor (1983, 2000); Bennett (1990); Pinker (1994, 1997); Dretske (1995); Scholl and Leslie (1999); Carruthers and Chamberlain (2000); Buller (2005); Carruthers (2006).

6 See Hull (1986); Griffiths (2002); Pinker (2002).

7 See McGrew (1998); de Waal (2001); Bekoff, Allen and Burghardt (2002); Russon and Begun (2004); de Waal *et al.* (2006); Cheney and Seyfarth (2007); Jensen, Call and Tomasello (2007); Pollick and de Waal (2007); Premack (2007).

8 See Gould and Lewontin (1979); Rose and Lauder (1996); Gould (1997); Pigliucci and Kaplan (2000).

A second issue is which theory will provide the needed explanatory resources. The choices include (1) individual explanations based on the adaptiveness of the underlying biology (Dow, Chapter 5, this volume); (2) group selectionist explanations based on the benefit of the trait to whole groups, or to groups formed by kin (Wilson, 2002); and (3) cultural evolutionary accounts in which religions are simply traditions that evolve along with the rest of culture.

Cultural evolutionary explanations themselves are of several kinds: (1) *memetic accounts* (Dennett, 2006), according to which religions are parasitic cultural replicators – a variant of this is that religions are autonomous cultural replicators but not necessarily disadvantageous to the 'host'; (2) *social ecology accounts*, in which religion adapts to the cultural, economic and social needs of the believers' environment (Reynolds and Tanner, 1995); (3) *costly signalling of commitment*, in which religions serve to identify commitment to the group in virtue of the high cost of adherence (Sosis and Bressler, 2003; Dow, Chapter 5, this volume); and (4) classical anthropological '*cultural evolution*' which tended to have a quite non-Darwinian flavour, being orthogenetic or directed, and was once considered to be rationally driven (Carneiro, 2003).

These different sources of explanatory resources stand in complex relations of coherence, compatibility and mutual support. The costly commitment account is compatible, for example, with a cultural evolution account and also a group selectionist account. These and many other possible combinations exist in the literature.

Another choice that theorists must make is between targeting the genesis of *religion* as a putatively universal human trait or the evolution of particular *religions* or religious traditions. To explain the latter, one must appeal to the properties of the particular religion in its social context, and whether these properties are culturally adaptive (that is, functional) or not. To explain the former, one has to give an account of how the tendency to religiosity might have developed over time, and whether it is biologically adaptive. For instance, one might say that religiosity evolved from cognitive and emotional capacities still seen in other species. A biological or cognitive explanation will not explain directly why, for example, Christianity has displaced traditional animisms in Africa.

The definition of 'religion' itself is highly contested. Some hold that a world-view is not a religion unless it posits supernatural agents or afterlife, in which case some religions turn out not to be religions at all – perhaps Zen Buddhism or Confucianism, although popular forms of these religions, such as Mahayana and Vajrayana, usually involve ancestor spirits, supernatural demons and devas, and even divine beings borrowed from other religions, such as Brahmin. Moreover, there is a crucial distinction between *elite* and *popular* forms of a religion that must be held in mind. Elite forms are often more sophisticated and philosophical, but that doesn't mean the popular forms will be as well, and the phenomenon to be explained by theories of religion is usually the popular form, not the 'Spinozan' form (see, for example, Dawkins, 2006).

Others hold that this sort of definition relies too heavily on the familiar religions of the Western tradition, particularly Christianity, and attempt a more general definition that includes overall metaphysical world-views, in which case Marxism-Leninism may turn out to be a religion. It can be argued that in single tribal societies, such as semi-nomadic societies and village or small town societies before the rise of large-scale sedentary agrarian civilizations, there is little traction to be had from distinguishing ordinary social rituals from 'religious' rituals (Liénard and Boyer, Chapter 13, this volume), although there is one ritual site that predates agriculture and may have been involved in its evolution (Heun, Haldorsen and Vollan, 2008). The choices a theorist makes on this issue will constrain and be constrained by the choices made on the issues listed above. For example, it may be that cultural evolutionary accounts of religion necessarily explain only the development of *particular* religions and not the overall phenomenon of religion. The definition of religion implicit in a theoretical approach also interacts with whether the explanation offered from that approach has the effect of undermining the religious commitments it seeks to explain.

Historical Considerations

Attempts to explain religion naturally go back a long way. Cicero, in his *On the Nature of the Gods* (*De Natura Deorum*), attempted to deflate popular religion in favour of an Epicurean notion of gods as distant and unconcerned with human lives, and therefore presented arguments that religion was more like

> the dreams of madmen than the considered opinions of philosophers. For they are little less absurd than the outpourings of the poets, harmful as these have been owing to the mere charm of their style. The poets have represented the gods as inflamed by anger and maddened by lust, and have displayed to our gaze their wars and battles, their fights and wounds, their hatreds, enmities and quarrels, their births and deaths, their complaints and lamentations, the utter and unbridled license of their passions, their adulteries and imprisonments, their unions with human beings and the birth of mortal progeny from an immortal parent. With the errors of the poets may be classed the monstrous doctrines of the magi and the insane mythology of Egypt, and also the popular beliefs, which are a mere mass of inconsistencies sprung from ignorance. (Book I)

In the fourth century BCE, Euhemerus, working at the court of Cassander of Macedon in a work now lost, argued that the gods had been historical individuals, kings, heroes, benefactors, who had been venerated and turned into mythological figures. His ideas were influential on early Christians, who used Euhemerus' views to deflate the pagan deities in favour of their universal god. Snorri Sturlsson, a twelfth-century Icelandic historian and Christian, also did this for the Norse pantheon in the *Prose Edda* (Brown, 1946). Hume himself suggested a similar account for mythical figures like Hercules, Theseus and Bacchus (Hume, Green and Grose, 1875, vol. 2, p. 313).

After the Enlightenment began, various authors also invoked a kind of euhemerism, including David Friedrich Strauss. In the post-evolutionary era, however, naturalizing religion became popular, with the *Religionsgeschichteschule* (History of Religion school) developing out of the new 'higher criticism' approach to biblical texts, arguing that religion as we now see it is a sociological and historical outcome of previous cultural developments, such as the monotheism of Akhneton (Hinnells, 2005).

Evolutionary accounts of religion begin with Auguste Comte, who held that societies and civilizations pass through three stages: from the *theological*, through the *metaphysical*, to the *positive* state. This typology was at the same time both formal and ahistorical, allowing classification of societies at a synchronic point, and historical, indicating how the society in question had developed and would evolve in future. Theological stages were, in effect, the default state of humans before they had advanced. Herbert Spencer, a friend of Darwin's, presented a similar view. But the best representatives are Caird (1894) and Martineau (1888), who offered historical and psychological accounts. Edward Burnett Tylor (1871) offered an evolutionary account based closely on Comtean stages. The default or ancestral religion was, he said, *animism*, a term he coined, in which humans projected an *anima* or motivating soul on ordinary objects in the world. Subsequent evolution of religion was cognitive and conceptual, as cultures acquired more and better knowledge of the world.

Historical and psychological accounts of the origins of religion have dominated discussions since (for example Hopkins, 1923). Anthropological studies also add the notion of ritual as a social cohesion force. All of these rely in some way or another on natural properties of humans, whether they call it human nature or not. Sociology provides economic explanations. Max Weber (1963) suggested that individuals with 'charisma' held psychological sway over others, and that worship derived from abstracting this property to the natural world. He famously held that different ideas affect social development and that, in particular, the Protestant ethic was responsible for the capitalist revolution in the West (Weber, 1930). Émile Durkheim, in contrast, held that religions were out-workings of the socioeconomic conditions of the believers and that religion acted to cohere society by affecting the beliefs of individuals (Durkheim, 1915; Durkheim and Pickering, 1975). He held that the primordial religious distinction was between the sacred and the profane, and that the original religion was totemism, in which a clan or tribe is watched over by a totem or sacred object or entity.

The psychological approach is exemplified by William James' classic study, *The Varieties of Religious Experience* (1902), in which he argued that all religious dispositions were at base biological and evolved. Psychoanalysts also took this up, and Karl Jaspers, an existentialist psychiatrist, held that we are disposed towards the transcendent by nature. Jaspers also coined the term 'Axial Age' for the period roughly from 800 to 200 BCE in which most of the modern world religions came into being. Jaspers thought that this simultaneous development was unconnected in different regions and cultures, although few would now hold that (Jaspers, 1951: 99–102 cf. Armstrong, 2007). Freud, on the other hand, thought religion to be a kind of infantilism that the mature healthy individual grew out of, a view shared by some modern psychological researchers (Bering, McLeod and Shackelford, 2005; Bering and Shackelford, 2005; Bering, 2006a, 2006b).

In the period following the development of the modern evolutionary synthesis, so-called, in the period from 1930 to 1950, researchers began to consider what role biology played in the behaviour of organisms, and out of the study of animal behaviour, or ethology, developed several strands of research. One came to be known as 'sociobiology' (Wilson, 1975; Wilson and Wilson, 2007), and such things as moral behaviour and religious conformity were given evolutionary explanations in terms of the benefit to groups, following Darwin's comment in *The Descent of Man*, discussed earlier. Ethically inclined groups would, on average, do better than the purely egoist groups. This seems to require a group selectionist view of social evolution, and many have argued against that on theoretical grounds.

Another strand relied on the ideas of William Hamilton (1964a, 1964b) in which behaviours that benefited kin were selectively favoured. Hence, on this account, altruistic behaviours such as self-sacrifice were explicable, because they tended to increase not the individual's fitness, but their 'inclusive fitness', which was the fitness of the same genes they had in all their relatives. So insofar as religious behaviours contribute to the flourishing of kin, they can be explained without recourse to selection of groups as such.

Others rejected this interpretation as being unduly adaptationist (Rose and Rose, 2001). And so a dispute arose between those who held that evolution was usually if not always adaptive and those who thought that much of evolution was the result of chance or a by-product of other things that were adaptive. The by-product theory is sometimes called 'spandrelism', after *spandrel*, a term of Gould and Lewontin's (1979) taken from architecture where it denotes the triangular space formed when arches form a rectangular support for a dome. The space is necessary, but not useful as a support, and so it was used, for example, in churches to paint images of saints on. Religion has been claimed as a by-product of various cognitive and neurological features: of our social capacities, of epilepsy, of our empathy with others and so on. Of course, if something *began* as a spandrel, it does not follow that it will remain one. Once some feature enters the population, it is subjected to selective pressures like any other, and so it can become adaptive very rapidly, even if it *began* as a spandrel. The origin and subsequent history of adaptive features needs to be distinguished.

Disciplinary Approaches

Anthropological accounts of religion focus heavily on the role of ritual in social organization. Some, such as Richard Sosis and colleagues (Sosis, 2000; Sosis and Bressler, 2003; Sosis and Alcorta, 2005), argue that religion is a form of ritual behaviour which has the function of enforcing collective action and social cohesion. Defining religion in terms of ritual behaviour, however, means that religion is not a unique trait, as ritual behaviour applies in most social institutions (consider sport). Loyal Rue (Chapter 6, this volume), on the other hand takes this conclusion seriously: religion is not about gods or the supernatural at all; it is about *us* and our society. Atran and Norenzayan (2005: 713) use a hybrid definition: 'passionate communal displays of costly commitments to counterintuitive worlds governed by supernatural agents'. In this account, religion is a set of rituals and beliefs that are hard to fake, and signal commitment to the community. The 'honest advertising' account here is both evolutionary and anthropological. A similar account has been given by Sanderson (2008), in which religion is a substitute for kin attachments in urbanized environments.

The use of primate studies (*primatology*) has recently entered the discussion again, after a long hiatus. In the 1970s, sociobiologists like Tiger and Fox (1989) argued by analogy with other animals that humans were motivated by social bonds. More recently, Frans de Waal and his colleagues (de Waal, 1982, 1996, 2001, 2005; de Waal and Tyack, 2003; de Waal *et al.*, 2006) have drawn a number of conclusions about human nature based on our shared evolutionary heritage with apes (or, rather, *other* apes, as we ourselves are members of the ape clade) as a way into the much more complex topic of purely human traits, overlaid as they are with cultural and social biases of observers. In other words, sociobiological accounts should start with apes in order to avoid projecting our prejudices onto the biology. Barbara King (2007), a specialist in ape and baboon social communication, has argued that religion is a side effect of ordinary ape empathy and social grooming played out in complex symbolic societies.

Modern sociobiologists appeal to various kinds of theories of social adaptation. Kin selection accounts are based on the work of Trivers (1971) and Hamilton (1964a, 1964b), in which one's fitness is increased by engaging in acts that are reciprocally beneficial, a process Trivers called 'reciprocal altruism' and which is encapsulated by people saying 'you scratch my back,

I'll scratch yours'. In situations where the members of your group are kin, this increases your inclusive fitness by aiding those who carry similar genes to you. But in an increasingly urbanized society, as Sanderson (2008) notes, these kin relationships are correspondingly disrupted, and so more abstract deities may play a role in uniting people in the new forms of social structure that appeared around the middle of the Axial Age, by encouraging attachments. On this account, monotheisms are a way of ensuring that individuals are not isolated, and that mutual love, respect and mercy are practised against genetic interests of xenophobia.

Cognitive psychology often appeals to the view that humans evolved to see agency in things, as our cognitive abilities had to make rapid inferences on the reasons for the actions of other humans (Atran and Norenzayan, 2005; Barrett and Malley, Chapter 10, this volume; but see Sperber, 2004 for a dissenting view). Consequently we have a strong 'agency detector', which Barrett (2000) dubbed the Hypersensitive Agency Detector Device or HADD. This is very similar to Dennett's 'Intentional Stance' (1987), which is the native approach humans take to causal reasoning, ascribing intentions to things (and echoes the view presented by Tylor in the nineteenth century). As a result, we find agency in natural processes such as thunderstorms, disease, natural disasters and crop failures, and infer that an agent caused them. The widespread religious ritual of propitiating or satisfying these supernatural agents with sacrifices or duties is a result. In many pre-agrarian societies, the form of religion known as *shamanism* often involves making ancestors happy in this way, and ancestor worship is widespread in many cultures.

Particular kinds of religious experiences are also explained as psychological by-products, and have been since James' *The Varieties of Religious Experience* (1902). James distinguished between the institutional or public forms of religion and the personal and mystical forms, which he focused upon.

How well these arguments stand up is a matter of future research. At this point, as Richerson and Newson (2008) note, we simply do not know enough to be sure if religion in general, or particular types of religions, are adaptive or side effects.

Intelligent Design

Intelligent design (ID) in the modern period is more than a revival of the natural theology that began with Harvey and Ray and culminated with Paley and the Bridgewater Treatises in the early part of the nineteenth century (Bowler, 2007). Nor is it a continuation of the ancient argument that design indicates the universe is the product of intelligence (Sedley, 2007). Instead, modern intelligent design is an outgrowth of the creation science movement (Shanks, 2004) that had its beginnings in the 1960s.

The basic difference between the classical and modern forms is that the classical theologians were attempting to discern God's nature or plan from the order of the universe (and in particular the living world). The modern form is an attempt to establish that there is a God based on the complexities of biological functions and structures (Forrest and Gross, 2004). As such it falls into the long tradition of arguments for the existence of God and is a special case of the cosmological proof.

But it does not end there. ID also sees that naturalistic explanations of biological function or apparent design compete with the basic premises of ID, and so its proponents have targeted 'naturalism', and in particular 'Darwinism', as being false or at least disputable, inventing a 'teach the controversy' strategy for inserting ID into American, and to a lesser extent some European, curricula at the secondary level. A later development of this is 'academic freedom', in which teachers in high schools are to be able to question scientific views they do not like.

These are political aspects of ID, and some argue the entirety of it, but a number of philosophers have challenged the epistemology and overall philosophical foundations of ID. The two senses of 'naturalism' that, for example, Alvin Plantinga has specified – 'philosophical' naturalism and 'methodological' naturalism – are in his work defined in terms of excluding God's existence for the former, and His relevance to investigation and explanation for the latter (Plantinga, 1996, 1993). Hence both topics are of philosophical interest in the wider sense when approaching the topic of religion and evolution. It should be noted that when philosophers talk about explanations in terms solely of physical properties, they tend to refer to 'physicalism'. And a major reason why scientific methodology or epistemology does not refer to the actions of God is that any event can be explained by divine intervention, and hence its explanatory power reduces, as Darwin said, to a mere restatement of the phenomena.

Outcomes of Naturalizing Religion

If evolution explains our religious, moral and other values in natural terms, how does this leave us with respect to meaning in life? If religion is natural, what does this mean about the truth of that religion? Can we still accept a religion that is a natural process?

In many ways this is a parallel question to naturalizing ethics and epistemology. If our brains and epistemic capacities evolved, can we trust our reasonings and results in science? And if ethics evolved, does this mean that there are no moral values?

The essays in the final section of this volume address these and similar matters. Chapter 23, William Grey's essay from 1987, shows how little such matters had been discussed in the philosophical literature a mere twenty or so years ago, but since then there has been a substantial increase in discussions, in part due to the influence of ID. The so-called 'new atheists', Dawkins, Dennett, Hitchens and others, have raised the question whether it is even possible to believe in religion now. Oddly, this is a topic that resurfaces in different generations, one of the first being Hume's discussions on natural religion (1947), and in the late nineteenth century the debate was in full flight. Some historical perspective is very useful to cool the blood.

Darwin himself said that such matters were beyond his abilities, and that he mistrusted the conclusions of a 'modified monkey brain'; but his ideas continue to have a lasting impression on this debate.

References

Andrews, Paul W., Gangestad, Steven W. and Matthews, Dan (2002), 'Adaptationism – How to Carry Out an Exaptationist Program', *Behavioral and Brain Sciences*, **25**, pp. 489–504.

Armstrong, Karen (2007), *The Great Transformation: The Beginning of our Religious Traditions*, New York: Anchor Books.

Atran, Scott and Norenzayan, Ara (2005), 'Religion's Evolutionary Landscape: Counterintuition, Commitment, Compassion, Communion', *Behavioral and Brain Sciences*, **27**(06), pp. 713–30.

Badcock, C.R. (2000), *Evolutionary Psychology: A Critical Introduction*, Cambridge: Polity; Malden, MA: Blackwell.

Baring-Gould, Sabine (1892), *The Origin and Development of Religious Belief*, London: Longmans, Green.

Barkow, Jerome H., Cosmides, Leda and Tooby, John (eds) (1992), *The Adapted Mind: Evolutionary Psychology and the Generation of Culture*, Oxford and New York: Oxford University Press.

Barrett, Justin L. (2000), 'Exploring the Natural Foundations of Religion', *Trends in Cognitive Sciences*, **4**, pp. 29–34.

Bekoff, Marc, Allen, Colin, and Burghardt, Gordon M. (2002), *The Cognitive Animal: Empirical and Theoretical Perspectives on Animal Cognition*, Cambridge, MA: MIT Press.

Bennett, Laura J. (1990), 'Modularity of Mind Revisited', *British Journal for the Philosophy of Science*, pp. 429–36.

Bering, Jesse M. (2006a), 'The Cognitive Psychology of Belief in the Supernatural', *American Scientist*, **94**, pp. 142–49.

Bering, Jesse M. (2006b), 'The Folk Psychology of Souls', *Behavioral and Brain Sciences*, **29**, pp. 453–93.

Bering, Jesse M., McLeod, K. and Shackelford, T. (2005), 'Reasoning about Dead Agents Reveals Possible Adaptive Trends', *Human Nature*, **16**, pp. 360–81.

Bering, Jesse M. and Shackelford, Todd K. (2005), 'Supernatural Agents May Have Provided Adaptive Social Information', *Behavioral and Brain Sciences*, **27**(06), pp. 732–33.

Bowler, Peter J. (2007), *Monkey Trials and Gorilla Sermons: Evolution and Christianity from Darwin to Intelligent Design*, New Histories of Science, Technology, and Medicine, Cambridge, MA and London: Harvard University Press.

Boyd, Robert and Richerson, Peter J. (1985), *Culture and the Evolutionary Process*, Chicago: University of Chicago Press.

Boyd, Robert and Richerson, Peter J. (2005), *The Origin and Evolution of Cultures*, Evolution and Cognition Series, New York: Oxford University Press.

Boyer, Pascal (2001), *And Man Creates God: Religion Explained*, New York: Basic Books.

Bradley, B.J. (1999), 'Levels of Selection, Altruism, and Primate Behavior', *Quarterly Review of Biology*, **74**(2), pp. 171–94.

Brandon, Robert N. (1988), 'The Levels of Selection: A Hierarchy of Interactors', in H. Plotkin (ed.), *The Role of Behavior in Evolution*, Cambridge, MA: MIT Press, pp. 51–71.

Brown, Truesdell S. (1946), 'Euhemerus and the Historians', *Harvard Theological Review*, **39**(4), pp. 259–74.

Bulkeley, Kelly (2005), 'Future Research in Cognitive Science and Religion', *Behavioral and Brain Sciences*, **27** (06), pp. 733–34.

Buller, David J. (2005), *Adapting Minds: Evolutionary Psychology and the Persistent Quest for Human Nature*, Cambridge, MA: Bradford Book/MIT Press.

Caird, Edward (1894), *The Evolution of Religion: The Gifford Lectures Delivered before the University of St. Andrews in Sessions 1890–91 and 1891–92* (2nd edn), 2 vols, Glasgow: James Maclehose.

Carneiro, Robert L. (2003), *Evolutionism in Cultural Anthropology: A Critical History*, Boulder, CO: Westview Press.

Carruthers, Peter (2006), *The Architecture of the Mind: Massive Modularity and the Flexibility of Thought*, Oxford, UK: Clarendon Press.

Carruthers, Peter and Chamberlain, Andrew (2000), *Evolution and the Human Mind: Modularity, Language, and Meta-cognition*, Cambridge, UK and New York: Cambridge University.

Cheney, Dorothy L. and Seyfarth, Robert M. (2007), *Baboon Metaphysics: The Evolution of a Social Mind*, Chicago: University of Chicago Press.

Cronk, Lee (1999), *That Complex Whole: Culture and the Evolution of Human Behavior*, Boulder, CO: Westview Press.
Darwin, Charles Robert (1874), *The Descent of Man and Selection in Relation to Sex* (2nd edn), London: John Murray.
Dawkins, Richard (2006), *The God Delusion*, London: Bantam Press.
de Waal, Frans (1982), *Chimpanzee Politics: Power and Sex Among Apes*, London: Cape.
de Waal, Frans (1996), *Good Natured: The Origins of Right and Wrong in Humans and Other Animals*, Cambridge, MA: Harvard University Press.
de Waal, Frans (2001), *The Ape and the Sushi Master: Cultural Reflections by a Primatologist*, New York: Basic Books.
de Waal, Frans (2005), *Our Inner Ape: A Leading Primatologist Explains Why We Are Who We Are*, New York: Riverhead Books.
de Waal, Frans, Macedo, Stephen, Ober, Josiah and Korsgaard, Christine M. (2006), *Primates and Philosophers: How Morality Evolved*, Princeton, NJ: Princeton University Press.
de Waal, Frans and Tyack, Peter L. (2003), *Animal Social Complexity: Intelligence, Culture, and Individualized Societies*, Cambridge, MA: Harvard University Press.
Dennett, Daniel C. (1987), *The Intentional Stance*, Cambridge, MA: MIT Press.
Dennett, Daniel Clement (2006), *Breaking the Spell: Religion as a Natural Phenomenon*, London: Allen Lane.
Dretske, Fred I. (1995), *Naturalizing the Mind*, Cambridge, MA: MIT Press.
Dupré, John (2001), Human Nature and the Limits of Science, New York: Oxford University Press.
Durham, William H. (1991), *Coevolution: Genes, Culture, and Human Diversity*, (Stanford, CA: Stanford University Press.
Durkheim, Émile (1915), *The Elementary Forms of the Religious Life*, London: Allen and Unwin.
Durkheim, Émile and Lukes, Steven (1982), *The Rules of Sociological Method*, New York: Free Press.
Durkheim, Émile and Pickering, W.S.F. (1975), *Durkheim on Religion: A Selection of Readings with Bibliographies*, London and Boston: Routledge and Kegan Paul.
Fodor, Jerry A. (1983), *The Modularity of Mind*, Cambridge MA: MIT Press.
Fodor, Jerry A. (2000), *The Mind Doesn't Work That Way: The Scope and Limits of Computational Psychology*, Cambridge, MA: MIT Press.
Forrest, Barbara and Gross, Paul R. (2004), *Creationism's Trojan Horse: The Wedge of Intelligent Design*, Oxford and New York: Oxford University Press.
Gould, Stephen Jay and Lewontin, Richard C. (1979), 'The Spandrels of San Marco and the Panglossian Paradigm: A Critique of the Adaptationist Programme', *Proceedings of the Royal Society London B*, **205**, pp. 581–98.
Gould, Stephen Jay (1997), 'The Exaptive Excellence of Spandrels as Term and Prototype', *Proceedings of the National Academy of Sciences of the United States of America*, **94**, pp. 10750–55.
Griffiths, Paul E. (1996), 'The Historical Turn in the Study of Adaptation', *British Journal for the Philosophy of Science*, **47**(4), pp. 511–32.
Griffiths, Paul E. (2002), 'What is innateness?', *The Monist*, **85** (1), pp. 70–85.
Griffiths, Paul E. (2004), 'Emotions as Natural and Normative Kinds', *Philosophy of Science*, **71** (5), 901–11.
Griffiths, Paul E. (2007), 'Evo-Devo Meets the Mind: Towards a Developmental Evolutionary Psychology', in R. Sansom and R.N. Brandon (eds), *Integrating Development and Evolution*, Cambridge, MA: MIT Press.
Grinde, Bjorn (1998), 'The Biology of Religion: A Darwinian Gospel', *Journal of Social and Evolutionary Systems*, **21** (1), pp. 19–28.
Hamilton, W.D. (1964a), 'The Genetical Evolution of Social Behaviour. I', *Journal of Theoretical Biology*, 7(1), pp. 1–16.

Hamilton, W.D. (1964b), 'The Genetical Evolution of Social Behaviour. II', *Journal of Theoretical Biology*, **7**(1), pp. 17–52.

Harris, Sam (2004), *The End of Faith: Religion, Terror, and the Future of Reason*, (1st edn.), New York: W.W. Norton & Co.

Harris, Sam (2006), *Letter to a Christian Nation*, (1st edn.), New York: Knopf.

Heun, M., Haldorsen, S. and Vollan, K. (2008), 'Reassessing Domestication Events in the Near East: Einkorn and Triticum Urartu', *Genome*, **51**(6), pp. 444–51.

Hinnells, John R. (2005), *The Routledge Companion to the Study of Religion*, London and New York: Routledge.

Hitchens, Christopher (2007), *God is Not Great: How Religion Poisons Everything*, New York: Twelve.

Hopkins, E. Washburn (1923), *Origin and Evolution of Religion*, New Haven, CT: Yale University Press.

Hopkins, E. Washburn (2003), *Origin and Evolution of Religion*, Whitefish, MT: Kessinger Publishing.

Hull, David L. (1986), 'On Human Nature', in A. Fine and P.K. Machamer (eds), PSA 1986: *Proceedings of the 1986 Biennial Meeting of the Philosophy of Science Association* (2; East Lansing, MI: Philosophy of Science Association, pp. 3–13.

Hume, David (1947), *Dialogues Concerning Natural Religion* (2nd edn with supplement), London: Nelson.

Hume, David, Green, T.H. and Grose, T.H. (eds) (1875), *Essays, Moral, Political, and Literary*, 2 vols, London: Longmans, Green and Co.

Jacobson, Esther (1993), *The Deer Goddess of Ancient Siberia: A Study in the Ecology of Belief*, Leiden; New York: E.J. Brill.

James, William (1902), *The Varieties of Religious Experience: A Study in Human Nature*, New York and London: Longmans, Green.

Jaspers, Karl (1951), *Way to Wisdom: An Introduction to Philosophy*, London: Gollancz.

Jensen, Keith, Call, Josep, and Tomasello, Michael (2007), 'Chimpanzees are Rational Maximizers in an Ultimatum Game', *Science*, **318** (5847), pp. 107–09.

Johnson, Phillip E. (1995), *Reason in the Balance: The Case Against Naturalism in Science, Law & Education*, Downers Grove, IL: InterVarsity Press.

Kelemen, Deborah (2005), 'Counterintuition, Existential Anxiety, and Religion as a By-product of the Designing Mind', *Behavioral and Brain Sciences*, **27** (06), pp. 739–40.

Keller, Laurent (ed.), (1999), *Levels of Selection in Evolution*, Princeton NJ: Princeton University Press.

King, Barbara J. (2007), *Evolving God: A Provocative View of the Origins of Religion*, New York and London: Doubleday.

Landau, Mark Jordan, Greenberg, Jeff, and Solomon, Sheldon (2005), 'The Motivational Underpinnings of Religion', *Behavioral and Brain Sciences*, **27** (06), pp. 743–44.

Leigh, Egbert Giles (1999), 'Levels of Selection, Potential Conflicts, and their Resolution: The Role of the "Common Good"', in Laurent Keller (ed.), *Levels of Selection in Evolution*, Princeton NJ: Princeton University Press, pp. 15–30.

Lewens, Tim (2004), *Organisms and Artifacts: Design in Nature and Elsewhere*, Cambridge, MA: The MIT Press.

Martin, Luther H. (2005), 'Toward a New Scientific Study of Religion', *Behavioral and Brain Sciences*, **27** (06), pp. 744–45.

Martineau, James (1888), *A Study of Religion, Its Sources and Contents*, 2 vols, Oxford: Clarendon Press.

McGrath, Alister E. (2005), *Dawkins' God: Genes, Memes, and the Meaning of Life*, Oxford: Blackwell Publishing.

McGrew, W.C. (1998), 'Culture in Nonhuman Primates?', *Annual Review of Anthropology*, 1998, pp. 301–28.

Morris, Desmond (1967), *The Naked Ape: A Zoologist's Study of the Human Animal*, London: Cape.

Morris, Ramona and Morris, Desmond (1966), *Men and Apes*, London: Hutchinson.

Orzack, S.H. and Sober, H. (1996), 'How to Formulate and Test Adaptationism', *American Naturalist*, **148**(1), pp. 202–10.

Pigliucci, Massimo and Kaplan, Jonathon (2000), 'The Fall and Rise of Dr Pangloss: Adaptationism and the Spandrels Paper 20 Years Later', *Trends in Ecology & Evolution*, **15**(2), pp. 66–70.

Pinker, Steven (1994), *The Language Instinct: The New Science of Language and Mind*, London: Allen Lane.

Pinker, Steven (1997), *How the Mind Works*, New York: Norton.

Pinker, Steven (2002), *The Blank Slate: The Modern Denial of Human Nature*, London: Allen Lane.

Plantinga, Alvin (1993), *Warrant and Proper Function*, New York: Oxford University Press.

Plantinga, Alvin (1996), 'Science: Augustian or Duhemian?', *Faith and Philosophy*, **13**(3), pp. 368–94.

Pollick, Amy S. and de Waal, Frans B.M. (2007), 'Ape Gestures and Language Evolution', *PNAS*, 0702624104.

Premack, David (2007), 'Human and Animal Cognition: Continuity and Discontinuity', *Proceedings of the National Academy of Sciences*, **104**(35), pp. 13861–67.

Pyysiäinen, Ilkka (2005), 'God: A Brief History with a Cognitive Explanation of the Concept', *Temenos*, **41**(1), pp. 77–128.

Reynolds, Vernon and Tanner, R.E.S. (1995), *The Social Ecology of Religion*, New York: Oxford University Press.

Richerson, Peter J. and Boyd, Robert (2005), *Not by Genes Alone: How Culture Transformed Human Evolution*, Chicago: University of Chicago Press.

Richerson, Peter J. and Newson, Lesley (2008), 'Is Religion Adaptive? Yes, No, Neutral, But Mostly, We Don't Know', in J. Bulbulia, R. Sosis, E. Harris, R. Genet, C. Genet and K. Wyman (eds), *The Evolution of Religion: Studies, Theories and Critiques*, Santa Margarita, CA: Collins Foundation Press.

Rose, Michael R. and Lauder, George V. (1996), 'Post-spandrel Adaptationism', in Michael R. Rose and George V. Lauder (eds), *Adaptation* (San Diego, CA and London, UK: Academic Press, pp. 1–8.

Rose, Hilary and Rose, Steven P.R. (2001), *Alas, Poor Darwin: Arguments Against Evolutionary Psychology*, London: Vintage.

Ruse, Michael (2001), *Can a Darwinian be a Christian? The Relationship between Science and Religion*, Cambridge and New York: Cambridge University Press.

Russon, Anne E. and Begun, David R. (2004), *The Evolution of Thought: Evolutionary Origins of Great Ape Intelligence*, Cambridge, UK and New York: Cambridge University Press.

Sanderson, Stephen K. (2008), 'Religious Attachment Theory and the Biosocial Evolution of the Major World Religions', in J. Bulbulia, R. Sosis, E. Harris, R. Genet, C. Genet and K. Wyman (eds), *The Evolution of Religion: Studies, Theories and Critiques*, Santa Margarita, CA: Collins Foundation Press.

Scher, J. Steven (2004), 'A Lego Model of the Modularity of the Mind', *Journal of Cultural and Evolutionary Psychology*, **2**(3), pp. 249–59.

Scholl, Brian J. and Leslie, Alan M. (1999), 'Modularity, Development and "Theory of Mind"', *Mind and Language*, **14**(1), pp. 131–53.

Sedley, David N. (2007), *Creationism and Its Critics in Antiquity*, Sather Classical Lectures, Berkeley and London: University of California Press.

Segerstråle, Ullica (2000), *Defenders of the Truth: The Sociobiology Debate*, Oxford: Oxford University Press.

Shanks, Niall (2004), *God, the Devil, and Darwin: A Critique of Intelligent Design Theory*, New York: Oxford University Press.

Sober, Elliott (1994), *From a Biological Point of View: Essays in Evolutionary Philosophy*, Cambridge Studies in Philosophy and Biology, Cambridge, UK and New York: Cambridge University Press.

Sosis, Richard (2000), 'Religion and Intragroup Cooperation: Preliminary Results of a Comparative Analysis of Utopian Communities', *Cross-Cultural Research*, **34**(1), pp. 70–87.

Sosis, Richard and Alcorta, Candace (2005), 'Is Religion Adaptive?', *Behavioral and Brain Sciences*, **27**(06), pp. 749–50.

Sosis, Richard and Bressler, Eric R. (2003), 'Cooperation and Commune Longevity: A Test of the Costly Signaling Theory of Religion', *Cross-Cultural Research*, **37**(2), pp. 211–39.

Sperber, Dan (2004), 'Agency, Religion, and Magic', *Behavioral and Brain Sciences*, **27**, pp. 750–51.

Tiger, Lionel and Fox, Robin (1989), *The Imperial Animal* (2nd edn), New York: Henry Holt. First published 1971.

Trivers, Robert L. (1971), 'The Evolution of Reciprocal Altruism', *Quarterly Review of Biology*, **46**(1), pp. 35–57.

Tylor, Edward Burnett (1871), *Primitive Culture: Researches into the Development of Mythology, Philosophy, Religion, Language, Art and Custom*, London: Murray.

Weber, Max (1930), *The Protestant Ethic and the Spirit of Capitalism*, New York: Scribner.

Weber, Max (1963), *The Sociology of Religion*, Boston: Beacon Press.

Weingart, Peter (1997), *Human by Nature: Between Biology and the Social Sciences*, Mahwah, NJ: Lawrence Erlbaum Associates.

Wells, Wesley Raymond (1918), 'The Biological Value of Religious Belief', *The American Journal of Psychology*, **29**(4), pp. 383–92.

Wilson, David Sloan (1989), 'Levels of Selection: An Alternative to Individualism in Biology and the Human Sciences', *Social Networks*, **11**, pp. 257–72.

Wilson, David Sloan (2002), *Darwin's Cathedral: Evolution, Religion, and the Nature of Society*, Chicago: University of Chicago Press.

Wilson, David Sloan and Sober, Elliot (1998), 'Multilevel Selection and the Return of Group Level Functionalism – Response', *Behavioral & Brain Sciences*, **2**, pp. 305–06.

Wilson, David Sloan and Wilson, Edward O. (2007), 'Rethinking the Theoretical Foundation of Sociobiology', *Quarterly Review of Biology*, **82**(4), pp. 327–48.

Wilson, Edward O. (1975), *Sociobiology: The New Synthesis*, Cambridge, MA: Belknap Press of Harvard University Press.

Wright, Robert (1994), *The Moral Animal: Evolutionary Psychology and Everyday Life*, New York: Pantheon Books.

Wunn, Ina (2000), 'Beginning of Religion', *Numen*, **47**(4), pp. 417–52.

Wunn, Ina (2002), 'Die Evolution der Religionen', Universität Hannover.

Part I
Adaptation

[1]

The cognitive and evolutionary psychology of religion

JOSEPH BULBULIA

Religious Studies, Victoria University P.O. Box 600, Wellington, New Zealand (e-mail: joseph.bulbulia@vuw.ac.nz)

Key words: cognitive psychology, costly signalling, evolution, evolutionary psychology, God, religion, ritual

Abstract. The following reviews recent developments in the cognitive and evolutionary psychology of religion, and argues for an adaptationist stance.

Introduction

Religious cognition presents significant explanatory questions to those interested in the evolutionary biology of our species.[1] Suppose the function of cognition, in the widest sense, is to help an organism deal, in the widest sense, with environmental complexity (Godfrey-Smith 2002). It is easy to appreciate how the ability to construct mental maps or for colour vision emerged in complex organisms given the enhancements to reproduction these bring. However, a functional explanation for religious cognition is less obvious. Assume that gods do not figure as genuine aspects of environmental complexity. Given the costs of religious cognition – misperceiving reality as phantom infested, frequent prostrations before icons, the sacrifice of livestock, repetitive terrifying or painful rituals, investment in costly objects and architecture, celibacy, religious violence and non-reciprocal altruism, to name a few – it seems selection should have weeded out any religious tendency. But religious conviction and practice is extremely commonplace. It is universal among hunter gathers and emerges in all modern societies (Rappaport 1999). Archaeologists trace religion back to our earliest Sapiens progenitors (Trinkhaus and Shipman 1993; Mithen 1999). Atheism seems to be a relatively recent and rare phenomenon, and though secular pundits have long predicted the demise of religion, it continues to flourish. It seems the human mind is especially prone to religion, in spite of the associated costs. Why?

656

To a crude approximation, there are two dominant research strands in the naturalistic study of religion.[2] In one camp are those that see religious cognition as a by-product of the evolved mind. For these spandrelists, religion has no adaptive value per se. The psychological architecture that produces god-related thought and activity has evolved for other purposes, and religion falls out of it as relatively harmless noise. On the other side are the adaptationists, who view religion as exquisitely functional, an elegant mechanism best explained as the target of natural selection, and best discovered by reverse-engineering its design. Below I highlight recent developments in both camps and suggest (i) why I think the evidence is stronger on the adaptationist side, and (ii) how I think adaptationism matters to the study of religion.

Strand 1: Spandrel explanations

Given the universality of religion, its strong motivational aspects, and behavioural consequences, venturing a functionalist explanation may seem irresistible. Viewing our species as one among many, an alien scientist might compare our strong and elaborate religious tendencies to the migratory instincts, territorial defence rituals, and intricate sexual displays of other animals [compare (Laughlin and McManus 1979; Smith 1979)]. Noticing a discrepancy between the outlay of nature on the one side, and how religious persons understand and interact with their world on the other, the scientist might conclude that selection outfitted our species with internal god-projectors – systems that distort experience to generate supernatural conviction, emotion, and behaviour. Here the poverty of stimulus could not be more extreme, nor could religious responses be more robust. Consider adolescent Khoisa males in Southern Africa who endure excruciating ritual circumcision only to live in exile in a desert environment without any food or water until they heal. The initiates risk infection, dehydration, exposure, and willingly submit to certain agony. The Khoisa claim the gods demand this ordeal of them. But how can chopping bits of genitals before the heavens improve survival?

Traditional theories of religion provide a suite of candidate functions – enhanced solidarity and co-ordination among the faithful, an answer book to life's riddles, an existential purpose generator, a means for providing hope and solace to the suffering, an adaptation for inter-group warfare, or for morality, and various combinations thereof (Preus 1987). In an effort to understand the god-projector, what it does beyond warping the outlay of reality, the alien naturalist might look to how these distortions enable the religious to relate to and manipulate their world, and other people, in ways that bolster reproduction. Though supernatural beings cannot improve survival – they don't exist – perhaps through religion we somehow do.

But need religion enhance reproduction to evolve? Interestingly, Darwin didn't think so. In the *The Descent of Man*, Darwin devoted only several paragraphs to the subject of human religious tendencies, amazingly little given the place of religion in human life (Darwin 1871/1981). Darwin concluded that our religious inclinations are best explained as spandrels of consciousness. He noticed that:

(i) religious cognition isn't a natural or psychological kind, but rather a composite of many distinct and overlapping elements: "the feeling of religious devotion is a highly complex one, consisting of love, complete submission to an exalted and mysterious superior, a strong sense of dependence, fear, reverence, gratitude, hope for the future, and perhaps other elements" (p. 68).

(ii) these elements yield to apparent cultural variation and many cultures lack any concept of "God" known to the Abrahamic faiths: "there is ample evidence, derived ... from men who have long resided with savages, that numerous races have existed and still exist, who have no idea of one or more gods, and who have no words in their languages to express such an idea" (p. 64). Darwin suggested that varieties of religious thought and behaviour materialize through the influence of social and institutional structures, by way of nurture's effects on common human nature. The broad spectrum of religiosity – savage through noble (to use Darwin's categories) – suggests that the best explanation of religion comes through an understanding of how culture assembles religious elements.

(iii) religious elements are not localised to our species:

> The tendency in savages to imagine that natural objects and agencies are animated by spiritual or living essences, is perhaps illustrated by a little fact which I once noticed: my dog, a full-grown and very sensible animal, was lying on the lawn during a hot and still day; but at a little distance a slight breeze occasionally moved an open parasol, which would have been wholly disregarded by the dog, had any one stood near it. As it was, every time that the parasol slightly moved, the dog growled fiercely and barked. He must, I think, have reasoned to himself in a rapid and unconscious manner that movement without any apparent cause indicated the presence of some strange living agent, and no stranger had a right to be on his territory (p. 67).

Contemplating blood rituals, trials by poison and fire, witchcraft and other "superstitions" Darwin summed up his spandrelist view: "These miserable and indirect consequences of our highest faculties may be compared with the incidental and occasional mistakes of the instincts of the lower animals" (p. 69). For Darwin, the elemental strands of religiosity can be seen in other animals very clearly as the by-products of ordinary cognition. Given that environmental complexity really is complex, religious "mistakes" should not be surprising. Different cultures generate distinctive religious doctrines, practices, and institutions because the inhabitants of those cultures are prone to supernatural errors.

In Darwin's classic statement religion serves no adaptive function. But if Darwin wasn't tempted to Darwinize religion, why should we?

Starting in the 1990s cognitive psychologists began to seriously explore specific features of religious cognition. Following in Darwin's footsteps, they argued that the aspects of religious cognition are most fruitfully understood not as parts to a globally adaptive system but as spandrels of other systems. Once we understand how these other integrated, modular, information processors work, we'll understand how they wind up accidentally generating supernatural thought as noise.

One of the first theorists to apply cognitive psychology to religion was Stewart Guthrie who in his 1993 monograph *Faces in the Clouds* argued that religion is mainly a by-product of agency detection systems (Guthrie 1993). Guthrie understood that the cornerstone of any religious life is religious experience. You can't throw a brick at a church or temple without hitting someone who has had a powerful religion-affirming encounter with the supernatural. Otherwise, without evidence, why commit to the gods? What makes religion plausible, for Guthrie, is our experience of the world as filled with animate beings.[3]

Contrary to the methodological assumptions of late 20th century anthropology, Guthrie didn't think that anthropomorphic tendencies could be explained solely as products of local culture and context. You don't "learn" to read gods into the fabric of reality. Rather god-mongering is a panhuman tendency; even secularists do it, for example when we perceive faces in clouds or a man on the moon. Guthrie explained anthropomorphism as resulting from perceptual hypersensitivity to persons. We animate the world with human life because we need to find other people whenever they are there, and faced with vague reality, perception gambles cautiously. In doing so, we lose little if they are not there and gain much if they are. Religious experience emerges from a hyperactive agent detection device, what the Justin Barrett calls "HADD" (Barrett 2000).

Assume that selection could have enhanced accuracy in the perception of persons. For Guthrie the payoff for enhanced accuracy did not warrant the costs of more a discriminating detection system. Because selection conserves HADD, religious beings spring from our minds like jack-in-the-boxes. They are projected everywhere because our brains overcompensate when facing vague reality. Like Darwin's dog, our cognitive organisation leads us to

Chart 1. Guthrie's wager

Seeing person when it is unlikely that a person is there (odds = 0.1), but where payoffs for perception are high (+10,000 utiles). Assume a cost of false perception is –10 utiles. Selection will favour HADD, if evolving perfect perception is difficult or more costly than to an organism than running HADD.

Perception	Reality: person there	Reality: person not there
Benefit of HADD = 901	+10,000 (0.1) or +1000	–10 (0.9) or –9
Opportunity cost of no HADD = –1000	–10,000 (0.1) or –1000 (opportunity cost)	0 (0.9) or 0

attribute natural effects to intentional causes, and so to project human-like beings into the world. We do this in our rapid unconscious inferences to the best explanation for what we perceive. Because these inferences are to human-like beings, we get worked up in ways that activate the social mind.

Notice that on this approach, evolution didn't design us to be religious any more than it designed us to love cinema or fast food. Once HADD is in place, religiosity falls out as an innocuous after-effect.

But why should we perceive only human-like beings and not also dangerous predators, food, potential mates and other reproductively important distractions (crouching with fear before clouds, exhibiting Pavlovian responses to the moon, or erotic responses to shadows?) And "human-like" needs to be disambiguated. In important ways, the gods are not at all human-like. They possess *supernatural* traits and powers. Given the "human" in "HADD," why are the gods conceived – always – as *not human*? And how does HADD explain religious rituals and institutions? While it is understandable how Darwin's dog could have responded to a moving gate by barking, it is not obvious why we would respond to vague reality with a Sistine Chapel, or a Mecca, or with painful rituals? – "I detect an agent, therefore, off with my foreskin."

Around the time Guthrie was publishing *Faces in the Clouds*, Pascal Boyer, a young French anthropologist then at Cambridge, began what was arguably a more rigorous application of cognitive psychology to religion. Like Guthrie, Boyer was impressed with the pervasiveness of religious thought and behaviour and unimpressed with standard anthropological explanations that we learn religion from culture. For Boyer, anthropological locutions about "learning" fail to elucidate the processes by which we acquire religious representations, obscuring this extremely perplexing dimension of human nature. Boyer wanted to better understand, at the level of cognitive architecture, just how the mechanisms for the acquisition and dispersion of religious understandings and practices work. His early research and subsequent career has been grounded in the view that religious ideas are attractive and spread because they activate multiple features of the intuitive inferences systems that govern our natural understandings of the world. Very generally these systems are panhuman, aspects of biological rather than cultural inheritance. They also function largely implicitly. Apparently, no one teaches us folk physics and psychology. They develop along predictable schedules that resemble the growth of organs. When psychological architecture encounters specific conceptual information supplied by the environment, it triggers rich understandings whose intricacy far outstrips external factors. On Boyer's view, religious concepts function like pin-balls falling into spring loaded psychological pockets, which, when they strike them just right, trigger an ornate array of largely implicit mental representations and psychological responses. Regularities in the system that generates these responses allow for psychological generalizations across cultures. Critically, there is no single system like HADD that is responsible for religious cognition. Like Darwin, Boyer doesn't think religious thought emerges from only one cognitive domain. And because religion is complex, its explanation is likely complex: there are no explanatory "magic bullets" in the naturalistic study of religion (Boyer 2001).

A critical property of any religious concept, on Boyer's view, is its minimally counterintuitive structure (Boyer 1994; Boyer and Ramble 2001). Boyer hypothesised that religious concepts violate *a few but not many* of our intuitive expectations for the relevant natural kinds. This is what makes them interesting and memorable.

To understand Boyer's reasoning imagine how religious concepts located at each extreme of the conceptual spectrum will function. Consider the case where religious concepts are ordinary and so do not violate intuitive expectations. Plainly, ordinary concepts will not be felt as arresting or considered worth talking about. Few would care if "god" referred to a car dealer who lives in Toledo Ohio, or to a phase in the development of turnips. On the other hand a being with an absolute power to create and destroy is memorable. Were you to meet the THE CREATOR OF THE UNIVERSE at a cocktail party you'd likely remember his name, tell others about it, and try to get on the Deity's good side. Given that counterintuitive concepts startle us, we are more likely to discuss them, which is why they spread in populations.[4]

Now imagine a concept that violates innumerably many of our intuitive expectations for the relevant kind. Will such a concept not startle us even more? Not if we can neither represent nor remember its properties. Consider the God of St. Thomas's five volume Summa Theologica. Thomas's ultra complicated Deus gives even the brightest seminary students mental cramps. Thomas's theology – though clearly written – is ambiguous enough to have fuelled over seven centuries of theological debate. His "God" concept remains too detailed and counterintuitive for common or garden believers to employ. Boyer suggests that as violations of intuitive psychology pile up beyond the attention-grabbing threshold, we become less able to understand and retain theological meanings. The gods must be strange, but once they become too exotic we lose our mental grip on them.

By the 1990s important experimental evidence began to emerge supporting Boyer's theory. The cognitive psychologists Justin Barrett and Frank Keil prompted religious devotees in American and India represented their gods in ways that made them far more anthropomorphic than the theologically explicit representations that these believers of each tradition consciously assent to in explicit doctrines and creeds (Barrett and Keil 1996). God or Shiva knows all, but you still need to pray if you want to communicate your intentions. This discrepancy between explicit theology and implicit religion has been duplicated in numerous experiments, revealing the gods of living religion to depart from the officially sanctioned versions theologians describe (Barrett and Keil 1998; Boyer 1998; Barrett 2000; Boyer and Ramble 2001).

Interestingly, Boyer and Barrett's line on minimally counterintuitive agents patches an oversight in Guthrie's HADD based explanation. Clearly, the supernatural is *never* conceived as an ordinary agent (Boyer 2003). There is always some conceptual twist. Satan is a talking serpent, not a serpent. Shiva has eight arms, not two. Ganesh doesn't just have a big nose; he's endowed with an elephant's trunk. If religion could be explained by HADD then we'd come to believe in ordinary persons animating the world. There are few absolutely universal rules in human culture. That religious thought *always* centres on non-natural or supernatural entities is one of them. The violation of natural expectation is what generates the distinctively sacred quality of supernatural conviction. It is what causes one's neck hairs to stand on end.

Of course we are not just attracted to the gods, but also to god-centred celebrations. No account of religion could be complete without an account of religious ritual. As Khoisa initiates demonstrate, religion doesn't just concern exotic belief; religious commitment may involve disfiguration, fire walks, starvation, awkward bodily postures and manipulations, the recitations of scriptures, psychedelic drugs, dance, and countless other special behaviours and practices. Around the same time that Boyer was developing his minimal counterintuitive violation theory of supernatural concept dispersion, the philosophers Tom Lawson and Robert McCauley began applying cognitive principles they learned from linguistics to religious ritual (Lawson and McCauley 1990). The result was a massively ambitious and complex theory – along the lines of generative grammar – aimed at elucidating and predicting structured religious behaviour. The latest refinement of the theory explains rituals in virtue of the (largely implicit) cognitive processes unleashed through either dramatic or repetitive practices (McCauley and Lawson 2002). Dramatic religious rituals produce what Harvey Whitehouse calls "flashbulb effects" – such rituals burn religious representations into our memories, rendering them salient and vivid (Whitehouse 2000). You would forget your mother's name before forgetting the day a knife was taken to your genitalia. On the other hand ritual repetition – doing something over and over and over again – screws religious representations firmly into cognitive place. Repeat: "The Lord is my Shepard" often enough, and do not be surprised that the phrase becomes believable and normative. [See (Whitehouse 2000) for a similar view].

In an impressive book, Scott Atran attempts to draw these various spandrelist threads together to provide an overview of cognitive research (Atran 2002). Like Boyer, Atran believes that we should not expect a simple explanation religious thought and action. Atran furthermore addresses two key problems that Boyer's earlier work exposes but did not (at that time) clearly answer.

First, there is a difference between remembering a minimally counterintuitive representation and becoming ontological committed to it. Many classicists maintain extremely detailed understandings of Greek religion, but few end up worshipers of Zeus. Atran noticed that Boyer and other cognitive psychologists face the "Mickey Mouse Problem." We can represent and easily recall fictional characters like Mickey Mouse (or Zeus) yet few adults come to believe Mickey Mouse actual exists outside the fiction. Compare "Mickey Mouse" with any religious conviction, say "Lord Jesus", and notice that the religious concept possesses something approximating mathematical certainty. It may also carry moralistic overtones. You will make no friends of religious persons suggesting their gods are as fictional as Donald Duck. Atran thinks we need to attend more carefully to how religious information grinds through diverse psychological systems to produce ontological and moral commitments. For example, the systems that generate existential meaning and purpose are susceptible to supernatural concepts capable of answering existential questions and providing hope and solace [compare (Bering 2003)]. Mickey Mouse is cold comfort on the cancer ward. Yet a Loving Sky Father who will carry me through the pain and separation of death helps me to make better sense of my life, and so I am attracted to the idea. Atran observes that the systems generating strong solidarity and that produce signals of commitment are susceptible to supernatural concepts that police social contracts. That is why there are religions of gods but not of sports teams or musical groups, which may also serve as rallying points. Michael Jordon is only metaphorically a god worthy of worship. He cannot bring rewards commensurate with moral goodness to all. But Yahweh or Allah can. And so individuals uphold and support these concepts to secure co-operation and morality with co-religionists.

Thus for Atran, in any instance where a religious concept flourishes, the precise brand of supernatural causation it supplies is attractive because it activates specific (and diverse) psychological systems. There is sense in which the particular form a religious *concept* takes owes to *its* adaptive features, and so the expression of a religious concept is the result of a selection process. This is true of many cultural products, from automobiles to videogames. However, the underlying psychological systems that accommodate religious concepts were not designed to process them for reproductive advantage. Like Boyer, Atran thinks religious information merely excites systems evolved for other purposes: "Religion has no evolutionary function per se. It is rather that moral sentiments and existential anxieties constitute – by virtue of evolution – ineluctable elements of the human condition, and that the cognitive invention, cultural selection and historical survival of religious beliefs owes, in part, to success in accommodating these elements" (p. 279).

It seems to me that Boyer and Atran provide something like the following methodological agenda to bring cognitive insights to Darwin's spandrelist intuitions about religious cognition:

Step 1: Take a feature of religious thought (like perception of a supernatural agents, categorization of supernatural entities, ritual understanding, etc.)

Step 2: Notice how this feature is a spandrel of some adaptive psychological mechanism or collection of psychological mechanisms (like agent detection, categorization of natural entities, sensory vividness and memory, etc.)

Step 3: Notice how these adaptive psychological mechanisms, even subtracting costly religious spandrels, enhance reproduction.

Step 4: Use the cognitive and developmental analysis of these adaptive psychological mechanisms to shed light on the religious spandrel in question.

Repeat 1–4 for other features of religious cognition.

664

Here we have a powerful methodological agenda for developing a biologically grounded theory of religion that is not adaptive. Whenever a new feature of religious cognition is noticed or discovered it can be linked to an adaptive psychological system as a non-selected after-effect.

I think the rationale behind spandrel accountancy is plausible. The strategy simplifies explanation by minimising the complexity ascribed to cognitive design. There is no need to view religious cognition as engineered; no need to find specific dedicated psychological architecture to explain why we believe in gods. Spandrelists understand that a highly specialized brain will be prone to cognitive hiccups. As long as religion doesn't kill or maim too much, selection may preserve religious tendencies because it preserves the more broadly functional design that produces them. [For general overviews see (Barrett 2000; Andresen 2001; Boyer 2001; Pyysiainen 2001; Boyer 2003)].

But there is a way of looking at religious waste as itself an aspect of exquisite design.

Explanatory strand 2: Adaptationism

Consider how religious coalitions are more effective than secular coalitions. Let's suppose Barney and Fred want to undertake reciprocal exchange. They live in the Pleistocene and so can't rely on the police or the courts to enforce any of their agreements. Both stand to benefit from mutual aid, but as is so often true with reciprocity, both stand to benefit even more from defecting, receiving but not giving, the stuff of prisoner's dilemmas. We know that the potential for defection poses no insurmountable barriers to reciprocity. Organisms have evolved a suite of elaborate devices to secure and enhance co-operation. Dunbar presents plausible evidence that the evolutionary driver of our big brain was the presence of other people, for detecting and dealing with friend and foe in large social groups (Dunbar 1998). If we view religion as an aspect of the social mind, we can begin to understand how the excessive costs associated with it may actually be exquisite adaptations that selection targeted to enhance.

Amotz Zahavi's costly signalling theory holds the key (Zahavi 1977; Grafen 1990; Zahavi and Zahavi 1997). Take the prisoner's dilemma that obtains between predatory lions and gazelle in the African savannah. Both predator and prey wants the other to die willingly: the gazelle want the cats to starve without ever chasing them, and the lions want the gazelle to jump into their hungry mouths. But there is an opportunity for co-operation, even here where reproductive interests appear to completely diverge. Neither the lions nor the gazelle want a chase at each encounter. Mainly, gazelle are faster than lions; so constant chasing will leave both exhausted to no end.

But were fit gazelle able to accurately signal their speed, a pointless waste of resources could be avoided on both sides. Such a signal/detection system has in fact evolved to mediate the relation between predator and prey. Gazelle communicate their fitness by stotting – vigorously leaping up and down in place. This is exactly the opposite of what you would expect an animal to do to avoid the jaws of a fast and hungry predator. Stotting makes a gazelle more visible, and flushes its muscles with lactic acid. Why not run or hide? It is precisely because only fit gazelle can afford the costs associated with stotting – can do all that, and still run away and not get caught – that lions have learned to assess the information to avoid senseless pursuits. Through costly signalling a prisoner's dilemma has been averted. Zahavi describes many such examples. Cost may authenticate a resource or intention because waste is the luxury only the resource rich or predictably committed can afford.

In understanding how costly religious displays convey information that solves prisoner's dilemmas, we can better understand why the objects of religious beliefs are agents with very specific supernatural properties. Boyer notes that the objects of religious devotion and piety are typically, "full access strategic agents." The gods or ancestor spirits, or impersonal forces like "Karma" and "Grace," are beings and forces with the power to scrutinizing the morality of earth dwellers. God's eye sees everything, and in particular every virtue and transgression.[5] Boyer explains this feature as a spandrel of the social mind. The moral record of others provides critical strategic information. Barney is more likely to trust Fred if Fred has a proven track record of honour, less if Fred's dealings remain unknown. So for Boyer, when we're confronted with full-access strategic agents concepts, our passion for surprise and wonder is activated, and we take a special interest in them (Boyer 2000).

Yet here again, taking an interest is one thing, belief quite another. Santa Clause knows who's been naughty and nice, but I do not worship him or seek a religion of Clausanity. Precisely what makes Clause interesting makes him, to me and everyone else over the age of 9, absurd. The concept is propagated, and the festival of Clause is celebrated each December, but only as a child's game. Bona fide religious conviction, however, is a deadly serious affair. (I will return to Santa below.)

Consider another common feature of the gods. Not only do they observe moral behaviour, they are empowered to reward and punish it. In no living religion are the gods inert. Much of religious practice is an attempt to communicate and exchange with the gods, to receive benefits, appease, or bring merit to oneself and the community of the faithful. The gods can dish out hurt – eternal damnation in hellfire, reincarnation as a garden shrub, bus terminal-like purgatories, and so on. But they frequently bring fortune to the good and righteous – lusty heavens, reincarnation as an emperor, release from

666

Chart 2. Predator/Prey prisoner's dilemma for fit prey encounter

Some arbitrary (but not wild) assumptions:

- Stotting is a costly signal that only fit Gazelle can perform.
- The energy of a chase is 1000Kj for Lion and Gazelle.
- The energy of stotting is 50Kj for Gazelle.
- Prize for Gazelle kill is 5000Kj.
- Chance of a lion catching a fit gazelle = 0.
- Chance of a lion catching an unfit gazelle = 0.9.
- Gazelle is lion's only food source.

Chase game where fit gazelle cannot signal speed: *Chase dominates Stay*

Action	Fit gazelle chases	Fit gazelle stays	Unfit gazelle chases	Unfit gazelle stays
Lion chases	−1000 −1000	Death 5000	(0.9) Death −1000 4000 (0.9)	Death 5000
Lion stays	Death	0 Death	−1000 Death	0 Death

Chase Game where fit gazelle reliably signal speed through stotting and Lion can detect this as signal of fitness: Stot/Stay Dominates Chase for fit gazelle/lion interactions, and Chase dominates unfit gazelle/lion interactions.

Action	Fit gazelle stots and chases	Fit gazelle stots and stays if lion doesn't chase otherwise chase	Unfit gazelle chases	Unfit gazelle stays
Lion chases	−1050 −1000	−1050 −1000	(0.9) Death −1000 4000 (0.9)	Death 5000
Lion stays	−1050 Death	−50 Death	−1000 Death	0 Death
Lion chases only if gazelle doesn't stot	−1050 0	−50 0	(0.9) Death −1000 4000 (0.9)	Death 5000

Chart 3. Ordinary prisoner's dilemma: defection dominates co-operation

Action	Barney co-operates	Barney defects
Fred co-operates	Prison = 1 year Prison = 1 year	Prison = 0 years Prison = 25 years
Fred defects	Prison = 25 years Prison = 0 years	Prison = 15 years Prison = 15 years

Chart 4. Prisoner's dilemma when both players act on the belief in a perfect supernatural reward and punishment regime. Heaven = maximally desirable. Hellfire = maximally undesirable. Assumption. Chart 3 gives actual outcomes. Chart 4 gives perceived outcomes. Co-operation dominates defection

Action	Barney co-operates	Barney defects
Fred co-operates	Prison = 1 year + heaven Prison = 1 year + heaven	Prison = 0 years + hellfire Prison = 25 years + Heaven
Fred defects	Prison = 25 years + heaven Prison = 0 years + hellfire	Prison = 15 years + Heaven Prison = 15 years + heaven

the cycle of birth and re-birth, profound insight and protection from harms way. Even the unrewarding gods of ancient Greece, Asia, and some African religious traditions tend to punish those who act badly, and in the case of Greek religions, their descendents [see (Hunter 1994)]. A god that harms everyone, but the righteous less, can still police social exchange. The key to the theory of religion as an adaptation for social exchange is that all-seeing gods impinge on our lives to hold us morally accountable. The supernatural causation represented through religious conviction is one capable of solving prisoner's dilemmas between those who share similar religious outlooks. In an ordinary prisoner's dilemma, economic rationality favours defection. But religious persons views the world as bound by supernatural causation, one that alters the relevant payouts for exchange. Even gods that bring few rewards and mainly punish are capable of policing exchange.[6]

Moreover extremely pro-social behaviour – suicide for one's group and its socially acceptable equivalent, celibacy, may be viewed as desirable, given supernatural causation.

668

Chart 5. Punishing Gods game: Zeus who punishes all, but the righteous less. Hates = Hell

	Barney defects	Barney co-operates
Fred defects	Hates Hates	–1000 Hates
Fred co-operates	Hates –1000	–1000 –1000

Chart 6. Celibacy game with supernatural rewards

Action	Perceived outcome
Celibate priest	Sexual frustration + Heaven
Carnal priest	Sexual Gratification + Hellfire

Clearly policing costs are substantially reduced in communities of prudent individuals who believe their transactions are perfectly policed by supernatural beings. All things equal then, members of religious coalitions are at an advantage over non-religious coalitions. They pay less to secure reciprocity. Solitary organisms would have no use for religious illusions, but in social species the illusions, when shared, benefit all who exchange under their spell [For discussion see (Atran 2002; Bulbulia 2004; Johnson and Kruger (in press).]

Notice however that while religious individuals living among religious cohorts extract the full benefits available to co-operating groups, irreligious invaders will be even more handsomely rewarded, deriving all the benefits of exchange but paying no cost for reciprocity. Clearly religious communities are open to invasion.[7] An individual who sees only natural causation will flourish among moral supernaturalists, and over time, naturalistic inclinations in her offspring will come to dominate mixed communities of religious altruists and irreligious defectors.[8] It seems that secularist ballast over the long haul sinks religious reciprocity.

Religious signalling, however, may keep the gods afloat. By producing and detecting hard-to-fake signals of religious commitment, the god-fearing can certify authentic exchange partners, sifting impious outlaws from the devout. The model predicts that religious deeds function as discriminatory screens through which religious groups preserve their integrity, an intuition echoed in the Christian Gospel: "They profess to know God but by their deeds they deny Him" (Tit 1: 16).

But what constitutes a religious deed? Answer: a costly signal capable of authenticating religious commitment.

The anthropologist William Irons is the first theorist I know of to describe religious behaviour as a commitment device (Irons 1996). On Iron's view, costly religious expressions are hard-to-fake signals that authenticate commitment to moralistic supernatural agency. For example, among Yomut Turkmen of northern Iran, a version of highly conspicuous Islamic practice infuses life. This is especially true among travelling Yomut whose ostentatious display signals to strangers commitment to a common morality, thus enhancing solidarity and trust (Irons 2001).

Importantly then, religious individuals do pay for invisible police forces. The cost of policing is the price of hard-to-fake religious display. But where these costs are lower than secular policing, the distortions that produce religious cognition, it seems, will be favoured by selection.

Evidence of supernatural security applies to many types of prisoner's dilemma. Irons and his colleagues Lee Cronk and Shannon Steadman found that among the people of Utila (a Bay Island of Honduras), men prefer to marry women who frequently attend church. This preference, however, is not reversed: women are not as interested in expressions of religious piety. Irons and his associates note that men in Honduras spend months away from home in maritime work. Because a woman knows the mother of her children, and a father cannot be certain, husbands should be especially interested in the sexual virtue of their wives. Given Utilan work regimes, the threat posed by infidelity is especially high. Irons concludes that hard-to-fake religious signally is favoured among Utilan women, who are far more religious than Utilan men, because the religious causation secures virtue (Irons 2001).

University of Connecticut anthropologist Richard Sosis has done more than any other researcher to test and develop the costly signalling theory of ritual. In a comparative study of two hundred religious and secular communes in the 19th century, Sosis determined that the religious communities were far more likely to outlast their non-religious counterparts – four times as likely in any given year (Sosis 2000). In a subsequent study, Sosis and Bressler determined that compared to secular communes, religious communes imposed over twice as many costly requirements on their members, and that the number of costly requirements was positively correlated with group lifespan. Interestingly a similar effect did not hold for secular communes, where costly requirements did not correlate with secular commune lifespan (Sosis and Bressler 2003). Adding further support to the costly signalling theory

670

of ritual, Sosis and Ruffle have shown that religious ritual influences co-operation in contemporary religious kibbutzim. Using common-pool resource games, the authors found that religious males were significantly more altruistic in their play than were religious and secular females, and secular males. The authors discovered no sex differences in co-operation among the secular kibbutz members, eliminating the possibility that there were differences in the ways males and females play the game. Noting that only orthodox men are expected to participate in communal prayer three times a day, the authors conclude that costly ritual participation (rather than any inherent differences between the sexes) accounts for the discrepancy (Sosis and Ruffle 2003).

It may be, of course, that altruists are more inclined to partake of ritual, rather than vice versa. But consider nature's economy. While it is possible to explain costly religious behaviour as accidents, costly signalling theory enables us to view these costs as adaptations. Given the enhancements to individual life that comes through co-operation, it should be unsurprising that selection has outfitted us with dedicated cognitive equipment to secure it. Moreover the success of religious communes over their secular counterparts is evidence for religious altruism as a special form of social glue.

Generalising, it is possible to view aspects of ritual activity described by cognitive psychologists in a different light. One reason rituals exhibiting "flashbulb" effects may be dramatic is that rituals frequently inflict punishment and ordeal to assess commitment. The drama comes from either enduring an ordeal or scrutinizing it. But the theory can explain repetitive religious rituals as well. Only a committed Christian will endure a boring sermon week after week; a ritual to which many atheists would prefer the stimulation of dental surgery. Fasting on Yom Kippur or during Ramadan is an entrance requirement for many Jewish or Islamic communities, here again deprivation proving commitment. Buddhists must sit still for hours and do nothing – pure torture for those not interested in Buddhist liberation. These rituals screen by imposing sensory deficits and extreme opportunity costs on those who partake of them.

It is important to be precise about how subjecting ritual goers to specific traumas and ordeals actually tests the presence and strength of altruistic commitments. Such costs assess devotion by rendering expected utilities explicit in ways directly related to supernatural belief. Ritual trials need to be arranged so that only those actually committed to the relevant gods would be willing to subject themselves to the trials.

Consider Bruce the believer deciding whether to partake in the strenuous or tedious rituals of his religion. The costs of participating in a ritual times their frequency must be discounted by the conditional probability that the gods will bring about some better outcome outweighing the costs. If Bruce genuinely accords a high probability to future supernatural reward then Bruce perceives that:

> **The cost of ritual participation $\times$ frequency $<$ conditional probability of value from pleasing gods.**

Now consider Sally the defecting atheist. Sally would like to receive the spoils of defection from social exchange, but must discount those benefits from the costs of ritual participation multiplied by their frequency. Sally expects zero future returns to make up for these costs. Rather she anticipates only more ritual pain or drudgery. Beyond this expense, there is the real possibility that Sally will be caught out as defector – given this is her plan – and hence the requirement to factor in additional risk. It is easy to see that how the expected utility from costly ritual action may well exceed the likelihood of any advantage from cheating the devout.

Sally finds religious ritual hard because she perceives:

> **Conditional probability of value from cheating the devout $<$ costs of ritual participation $\times$ frequency**

Notice that ritual costs are not arbitrary. For ritual to be an effective test, it must accurately measure religious commitment. It must reliably reflect a commitment to a system of supernatural causation capable of altering outcomes favourable to those who believe in it (and so act altruistically towards others similarly committed.) The logic is simple: if Sally doesn't believe the gods will repay her ritual sacrifice, then it is unlikely she will believe they will repay her altruistic sacrifice. Whatever Sally may *say* about her conviction, rituals assess whether she is willing to put her money where her mouth is [for discussion see (Sosis 2003; Sosis and Alcorta 2003; Bulbulia 2004).]

Of course, defectors can still invade, where the payoffs from defection are extremely high. In such cases, the expectation if for ritual costs and frequency to move upward [see discussion of (Chen 2004) below]. Importantly, beyond the actual expense and opportunity cost of participating in religious ritual, religious rituals are structured to prompt public displays of god centred *emotions*. Emotions are notoriously hard to fake (Ekman 1975; Frank 1988; Ramachandran and Blakeslee 1998). Most of us cannot convincingly produce on demand an expression of love and devotion to say Zeus, or some other god we don't believe exists. Rituals that elicit god-centred emotions in public can serve to hold such religious commitment open to public scrutiny, insulating religious groups from defectors.

Sosis theorizes that ritual performance actually impacts belief (Sosis 2003). On his view, repeated ritual performance lowers the perceived costs of ritual action (or increase the perceived benefits) by generating conviction in the supernatural outlook that imbues ritual with meaning. Thus ritual not only serves as a forum for signalling godly commitments, it inculcates religion by generating affirming religious experiences. Sosis and Alcorta hypothesize that religious ritual alters affective centres in the brain, educating the mind to feel religious wonder and passion toward the relevant religious symbols and mythology of a group (Sosis and Alcorta in preparation), a view echoing sociologist Emile Durkheim's work theory of "collective effervescence" a century ago (Durkheim 1964 [1915]).

The model makes some surprising predictions. In cases of extreme hardship, where common-pool resource problems abound and the threats of defection are high, the expectation is for the cost outlays to intensify and to become more frequent. When the chips are down the religious will produce *more* effort and expend *more* resources proving their faith. Interestingly Daniel Chen, a graduate student in economics at MIT has shown that during the Indonesian financial crisis of 1996–1997, Muslim participation in religious rituals became both more frequent and more intense (Chen 2004) see also (Johnson 2003). Given lower resource availability, both the benefits of social exchange and the threats posed to it by defection would have increased, thereby increasing the threshold standards for discriminatory religious signalling. This is only one case, and it will be interesting to see whether this prediction of costly signalling theory generalises.

Religion and group selection

It may be that the exchange-based understanding of religion is founded on too narrow a conception of reciprocity. Selection produces design through the differential success of replicating entities. Implicit in the adaptationist approaches I have been considering is the idea that selection operates on gene lineages through the differential reproductive success of religious individuals (that is, their ephemeral phenotypes) who propagate them. But selection may act at any replicating entity, given certain constraints (Sterelny 2000). David Sloan Wilson has recently argued that the religious groups may function as adaptive units (Wilson 2002). The benefits uniquely available to social species do not just flow directly from the mutual aid-giving of co-operating individuals, but through highly indirect channels opened through group-level structures, which those engagements create and maintain. Where resources can only be acquired through the integrated action of several individuals, the functional organization of groups relative to competing groups may generate adaptive features at the group level [See also (Hardin 1995)]. For Wilson, the best explanation for religious thought and behaviour is that it facilitates strongly integrated and functionally adaptive groups [for evidence see (Wuthnow 1994)]. Functionally adaptive groups purdure through time by adapting themselves to variable local circumstances. This may go some way to explaining why the Christianity of Canada is relatively benign when compared to the more violent strains of Christianity in Northern Ireland or the Balkans. In the latter cases, limited resources lead to severe group competition. To access these resources, individuals must sink their individuality in the identity of a group whose fate they share, one for all. David Wilson suggests that religious cognition in particular has been selected because, "Supernatural agents and events that never happened can provide blue-prints for action that far surpass factual accounts of the natural world in clarity and motivating power." (p. 42)

From the analysis of supernatural causation above we can see how this works. The gods punish and reward in ways that defy imperfect natural justice, and so bolster morality more thoroughly than secular alternatives. In extreme conditions of inter-group competition, it may be that psychological mechanisms are favoured that (i) track distinctively group affirming religious information and (ii) integrate this information with social behaviour.

Sosis argues that if religion emerged to promote group well-being then there would be little need for doctrines stressing human frailty (Sosis 2003). We'd all be good religious citizens, naturally. Over time, group selected moralities would tend to weed out psychological tendencies to religious thought and practice. We wouldn't need supernatural police forces if our natural inclinations were to help each other. Strong group instincts similar to parental love could facilitate reciprocity without the resource wastes of religious practice. But Wilson is careful to point out that groups are subject to internal competition among members. Group level adaptations may include internal checks and balances to these contests. In fact, from the vantage point of group selection theory, ritual signalling could be a group level adaptation that enables functionally co-ordinated units to emerge in situations where individual optimizers face prisoner's dilemmas. Groups that better facilitate these dilemmas compete better against groups at war with themselves. Moreover that costly signalling makes the join-defect-leave strategy less appealing accords well with the idea that religion is group level adaptation: groups that pray together stay together, and so flourish against other groups.

One avenue to testing Wilson's theory would be to study the forms of altruism that develop in religious groups. Forms of apparent non-reciprocal altruism – practices of Christian agape (universal love), Hindu and Buddhist non-violence and vegetarianism, Muslim and Jewish charities, and other forms of sacrifice may prove to be group supporting activities rather than mere Zahavian signals of religious commitment. The trick is to isolate signalling variables from group-sacrifice variables and see whether extreme

674

pro-social sacrifice can be generated where there is no chance it will be detected. Missionary practices in which entire families risk life and limb to spread a faith may prove a fertile source for empirical testing. It would be interesting to see whether missionaries sacrifice when the probable inclusive benefits are extremely low. If religious altruism doesn't fit a model of inclusive fitness maximization, then costly signalling and reciprocity may be inadequate to explain all the varieties of religious sacrifice.

The group selection theory of religion faces some difficult conceptual problems. For example, the theory requires an empirically adequate definition of a religious group. Consider two Presbyterian churches that view themselves as members of a single overarching religious sect. Imagine they compete with each other for members and funds. Is this one group or two (or both)? At any rate, it is probably unhelpful to think of religious groups as mere collections of individuals. To think so obscures the ways in which the physical, legal, and theological products of collective activity endure over time. Clearly these environmental legacies influence the fitness of those (individuals and/or groups of individuals) who inherit them. Of course, ambiguity over the unit of selection is a problem that faces any group selectionist theory, not merely those crafted to explain religion.

Beyond coming to definitional grips with "religious group," it seems to me that the exact psychological pathways that lead individuals to generate group level adaptations need to be articulated with some clarity. Whether the idea that selection can produce group level adaptations is capable of generating robust understandings of intricate psychological architecture needs to be demonstrated, not assumed. In his book, Wilson admirably describes numerous features of certain religious groups that promote their survival, but how individual psychological design reliably generates these group level patterns remains somewhat obscure. I think Wilson would acknowledge this, and argue for the urgency of more empirical work, for example along the lines of Sosis's excellent kibbutzim studies.

The correct approach: Darwin's or Darwinian?

Consider how Wilson's group-selectionist argument can be taken in spandrelist directions. Earlier I played up religion's generality. But it seems not everyone is committed to supernatural causation. I take as evidence myself. In some places, like Western Europe, Scandinavia, New Zealand and Australia, explicit religious commitments are rapidly declining. And clearly not all group competition falls along religious lines. Often ethnicity, sex, or political affiliation provides the relevant ties that bind groups. Further many people require no God to back morality. They feel justice is its own reward. Importantly, this is so with many religious persons. Baston et al. have shown that "quest oriented" religious persons view faith as a method for personal or social transformation, and are generally more tolerant and pro-social than the "extrinsically" religious, who act with the expectation of supernatural profits (Batson and Schoenrade et al. 1993).[9] If Wilson is correct, and we assume variable resource distributions over our evolutionary history, then it would be unsurprising if group-markers proved to be flexible: religious when religion binds – ethnic, sexist, nationalistic, democratic or otherwise secular when these ties secure more powerful and effective alliances. The strategy "one-for-all" clearly does not always reduce to "all-for-god(s)." Given the prominence of secular communal organisation, it seems difficult to rule out the hypothesis that religion is a spandrel of a more basic group-oriented sociability, as Atran argues. [Though we would need some account for the success of religious communes over secular alternatives observed in (Sosis 2000).]

Critical to the spandrel/adaptationist debate is how individuals, especially children come to acquire their religions. If religion were part of genetic endowment, we would expect certain features to spring from internal architecture, perhaps according to developmental schedules, in ways that generalize across cultures. Recently there has been increasing experimental attention devoted to the religiosity of children, though gaps in our understanding of childhood religion remain large.

Barrett et al. have argued that children before the age of five reason easily about god-like beings, before they have developed robust folk psychological powers (Barrett and Richert 2003). The authors presented children between the ages of 3–8 with nonhuman agent puppets: a kitty cat, a monkey, and a little girl named "Maggie." Children were told that of these puppets, the kitty cat has a special power for seeing things in the dark. Children were then shown a darkened box and asked to report what they saw inside. All children reported they could see nothing. After, illumination of the box revealed it to contain a red block. The children were then asked what each of the puppets would see if they were to look inside the darkened box. Preschoolers reported that Maggie would be able to see the contents of the box. At age 5, children reasoned correctly that Maggie as well as the monkey would not be able to see the block. Older children however reasoned that only the kitty cat could see the block. This is consistent with developmental studies suggesting that recursive mental state attributions (X doesn't know that Y knows that P) do not appear until after pre-school. However, children *at all ages* reported that God would see the block. This supports the idea that children do not reason about God as just another person but rather as a different type of agent with nonhuman properties. On the author's view "children may be better prepared to conceptualize the properties of God than for understanding

676

humans" (p. 60). Importantly, the authors do not ascribe adaptive significance to this developmental feature: gods are easier to conceptualize because god-concepts require less computational power than ordinary agent concepts do (like ascriptions of false belief.) Nevertheless, it is startling that the first agent concepts children apparently acquire are of gods, which they then apply to persons, whom they endow with familiar god-like powers such as omniscience and omnipotence. Knight et al. have come to similar conclusions in a cross-cultural study (Knight and Sousa et al. forthcoming).

These experiments suggest that very early in cognitive development, children easy grasp with rich inferential understandings the meaning of supernatural agent concepts. But there may be far more to the developmental story. Deb Kelemen has recently argued that beyond mere facility with supernatural agent concepts, children are "intuitive theists" who explain much of their world through supernatural concepts (Kelemen in press). Kelemen points to converging bodies of research suggesting children are biased to reason about the natural world in terms of intentions and purpose, as well as to view natural phenomena as intentionally created by non-human agents. With respect to promiscuous teleological intuitions, when prompted to respond to the question "what is this for?" American 4–5 year olds find the question appropriate not only to body parts and artefacts, but also to living things like lions ("to go in the zoo") and non-biological natural kinds like clouds ("for raining"). Moreover when asked whether raining is what a cloud "does" or what it is "made for," pre-schoolers favour explanations that natural entities are "made for something" and that is their reason for being here (Kelemen 1999). Kelemen has shown that rampant teleology endures well into elementary school, especially with respect to object properties, with the teleological biases only beginning to moderate at age 9–10. When asked to perform a "science" task and decide whether ancient rocks were pointy because "bits of stuff piled up for a long period of time" (i.e. a physical process) children preferred "self-survival" functions like "so that animals wouldn't sit on them and smash them" and "artefact" functions like "so that animals could scratch on them when they got itchy" (Kelemen 1999). This bias to favour teleological explanation for non-living as well as living natural object properties persists even when children are told that adults employ physical explanations, a pattern also observed in British children (Kelemen 2003). Given that parents actually prefer non-teleological explanations, the child's promiscuous teleology remains difficult to explicate in terms of social acquisition.

It appears moreover that children view non-artefact items and events of the world as caused by supernatural agents. With respect to childhood explanatory biases for nature's origins, Evans found that regardless of the religiosity in their homes, children exhibit a bias for intentional accounts (Evans 2000;

Evans 2001). For example when asked: "how do you think the very first sun-bear got here on earth?" 8–10 year olds from both fundamentalist and non-fundamentalist American homes favour creationist accounts, namely that "god made it" over other teleological accounts "a person made it" or non-teleological accounts, "it just happened." 5–7 year olds exhibit the same explanatory bias for both animate and inanimate objects. It is only among 11–13 year old non-fundamentalist children that patterns of divergence emerge. Gelman and Kremer have found that children prefer to explain the existence of remote natural items (like oceans) as made by God (Gelman and Kremer 1991). A similar result has been observed among British children (Petrovich 1997). It appears that while children identify people as the designing agents of artefacts, they distinguish god as the designing agent of nature (Kelemen and DiYanni 2002). Additionally, Bering has shown that over the age of 5, though critically not before, children explain random events in nature as caused by invisible agents (Bering in press). This result is especially interesting because it shows that recursive theory of mind abilities actually enhance the domain of supernatural explanation.

Though in its infancy (so to speak), the developmental literature suggests that children's pervasive teleological ideas about things and events of the world are closely linked to their endorsements of intentional design by a supernatural agency – leading them to distinguish supernatural beings as the designing agents of nature. Moreover current evidence suggests the systems that generate these beliefs emerge without any specific or robust cultural input (Kelemen in press). It may well be that a child's default theory of the world includes an "intuitive theism." Speculating further, it may be as with language, that children are endowed with all possible religions, acquiring their religious idiolect largely by forgetting [for discussion see (Bulbulia in preparation).] If so then perhaps the Clausanity of children at Christmas is best explained as just such a proto-religion.

Reverse engineering religion beyond intuitive theism, we'd expect the developmental pattern to generate specifically moralizing gods by the age that hunter-gatherer children begin to exchange with non-relatives, roughly mid-adolescence. In a startling experiment, Bering has shown that children before the age of three use supernatural concepts to orient moral behaviour – this pattern emerging long before they acquire robust folk psychological capacities, let alone transact with strangers (Bering in press). In a series of experiments, Bering primed very young children with a supernatural agent concept, "Princess Alice," who is described as invisible, present, and who "really likes good boys and girls – I bet she really like you!" The children are then led to a game in which they are asked to guess the contents of a box, for which correct answers are rewarded. Before starting the game the experimenter says she must leave the room, and instructs the child to stay in the room until she returns, "but don't worry, you won't be alone, because Princess Alice will be in the room with you." A second experimenter, watching the child from another room, flashes a light on and off if the child attempts to "cheat" by removing the box lid. The child is then observed to see whether she will continue "cheating." Bering writes, "preliminary evidence suggests that even 2.5 year-olds display the inhibitive response after encountering the unexpected event in the midst of their cheating. Even these youngest children act as if they have been 'caught red-handed'. ... Some of these children display behavioural signs of dejection and fear. Moreover many children respond on the unexpected event as soon as the experimenter returns to the room. To quote one very excited [girl] '*Princess Alice is real!*' " (Bering in press). Bering takes these experiments as initial evidence that a child's intuitive theism has a moral component.

So does intuitive theism, particularly if it is susceptible to moralizing varieties, seal the adaptationist case? Not yet. It may be that the child's predilection for intentional explanation emerges from cognitive features dedicated to explaining human agency, which in their undeveloped form prefer simpler supernatural agent concepts, as Barrett suggests (though this would leave unexplained the retention of intuitive theism to late childhood). Moralistic notions may come on-line early because moral restraint is important to survival, and acute morality + simple agent concepts = young children's moralizing religion. It cannot be ruled out that the moralized religion of adolescents and adults is disassociated from these early developmental processes, resulting for example from the sort of informational cuing Atran postulates. That is, it may be that information relevant to solidarity and exchange is sometimes religious information. If so then there is no need for dedicated religious features to explain its uptake and use.

I have been assuming that religion polices morality, but is this case? There is some preliminary evidence suggesting that the moral causation critical to adaptationist theories of religion may have appeared only recently. Roes and Raymond have shown that moralizing gods are favoured in larger differentiated groups where religious elites draw disproportionately large slices of the resource pie (Roes and Raymond 2003). In larger groups as total resource pies grow proportionately larger through enhanced efficiencies, lower castes can still benefit even as fat casts on top grow fatter. In less stratified hunter-gather communities the case is different. While a class of religious elites, shamans, are commonplace, these religious experts do not seem to draw greater resources. In fact some of the most egalitarian societies we know of are hunter-gatherer. Interestingly, the gods don't seem to matter as much in them. It may be that moralistic religions only surface as societies became more differentiated. An echelon of religious elites use the moral absolutisms of religion to promote both pie and personal-slice growing agendas [though see (Cronk 1994)]. If this story proves roughly correct, then the developments that brought moralizing religion into the world would have occurred too late in our evolutionary history for human psychological architecture to be substantially effected. Instead the best explanation for such tendencies may simply be that individuals target information that builds functionally adaptive groups. We get moralizing religion where strong socially differentiated groups are more likely to prevail over rival groups. If so, then no special purpose mechanisms for religion needs to be postulated. Moralizing religion, like the wheel or farming, may be a cultural invention, independently discovered in multiple places because the moralizing religious concepts help us (individuals or groups of individuals, depending on your evolutionary story) to flourish.

However, in my view, the evidence against moralising religion relies heavily on how anthropologists describe specific cases. Consider the !Kung! people of the Kalahari Desert. Their creator god is envisioned as a stupid and lazy sky being with long hair and a horse, who takes little interest in human affairs (Katz 1984). In !Kung! culture, the dominant supernatural power is a healing energy called "num." Ritual life centres on healing dances that occur several times a month. In these dances, the community gathers in all-night festivals, in which !Kung! healers lay their hands on sick persons healing them with num and releasing these salubrious energies to all present (Katz 1997). It is not obvious that these rituals support a conception of reality in which the righteous are rewarded and defectors are punished. Nevertheless, the rituals provide ample scope for enhanced solidarity. The close physical proximity of participants, the touching and stroking, the intense focus on those who are unwell, the holding of each other through dance, strike me as clear hard-to-fake demonstrations of social commitment. Here we find social grooming writ large. Moreover to challenge the num-centred conception of reality would be to undermine the sacred underpinning of !Kung! social life. Rather than conceiving of !Kung! religion as non-moralizing, it may be more accurate to expand the conception of supernatural enforcement at the centre of solidarity theories to embrace the multifarious ways in which num-centred illusions (and other apparently non-moral religious understandings more generally) bind individuals together. There may be multiple forms of supernatural glue, apart from clearly moralistic gods who wield heavens and hells. At any rate, it seems to me that spandrelists need to take care in their selection of counter-examples to moralizing religion.

Let me lay my cards on the table. I am an adaptationist. In my view, the strongest evidence for adaptive account of religion comes from the precise

680

way in which religious distortions mediate a believer's relationship to the world and to other people. We should not forget that the self-deception involved in religious cognition operates on a massive scale: for innumerably many people, powerful and dramatic religious understandings and dramas are thickly draped over an impoverished secular reality. For religion to happen at all, there must be an active distorting and biasing of experience strong enough to erect cathedrals and to bring people to their knees. Notice that these tremendous deceptions, though motivating and normative in various ways, are nevertheless encapsulated to prevent people from seriously harming themselves: generally, the Cartesian certainty of religious conviction does not leave the exigencies of life up to gods. The faithful believe the gods will provide, but still till fields, provide for children, arm themselves against attack, and seek medicine when ill. Actually committing to supernatural causation without constraint is a recipe for disaster. But in fact religious cognition mainly enhances health [for example see (Ellison 1991; Hummer and Rogers et al. 1999; McClenon 2002; Sosis and Alcorta 2003), though see (Livingston 2002)]. Looking at religion from the vantage point of our alien naturalist, it strikes me that this functionality combined with the overarching concinnity of the system that produces religious thought, its modularity and adaptation of parts to whole, is best explained as the target of natural selection. Again, Irons and Sosis have shown us that even the harms that religion brings are not inefficiencies when viewed as signals authenticating religious commitment. And David Wilson has taught us that even beyond religious signalling, religious illusions may support pure sacrifice and organisational commitments to group welfare, where individual fates are anchored to collective futures.

In my view, a broader understanding of supernatural causation able to capture Atran's insights into the place of religion outside of social exchange (as a provider of hope and solace, and an impetus to heath) will need to be developed to make sense of full spectrum of religious conviction and practice on the ground. Very basically, if believing in supernatural causation helps us to recover from illness or meet the terrors of life, then tendencies to fall into such deceptions will be conserved and more intricately articulated (McClenon 1997; Bulbulia 2003; Sosis and Alcorta 2003). If so, the relationship between "moralizing supernatural belief" and "existential supernatural belief" may prove interesting. An optimal psychological design would shift between conceptions of just gods and conceptions of loving gods (and other conceptions) to suit circumstances. In some instances obliging Barney may think a *vengeful god* will punish him for cheating Fred. This distortion secures co-operation, where accuracy would favour defection. In other instances, defecting Barney may seek a *loving god's* assistance to free him from a punishing Fred's torture chamber. This distortion sustains courage and hope, where accuracy would warrant despondency. On average the optimism that supernatural illusions warrant may prove helpful to individuals like Barney whose actual prospects are dim.

Speculating about optimal design raises a critical issue. In my view, the main (and maybe the only) reason the spandrel/adaptationist debate should be interesting to naturalists is methodological. If religion is a product of natural selection, then reverse engineering techniques may help unravel features of its internal architecture. Time and again biases and distortions have yielded their secrets to functionalist reasoning (Trivers 2001). If adaptationists are correct, then we can use ideal designs to formulate testable hypotheses about information flow and behavioural outputs. We can then work to discern and unravel intricacies in the systems that regulate supernatural thought. It may even turn out, surprisingly, that the apparent demise of religion in some places is illusory. Perhaps secular conceptions of reality are nevertheless informed by quasi-supernatural understandings, artefacts of our ancestral way of coming to grips with the world and each other. Bering observes that even atheists are prone to supernatural commitments, "even the atheist's God seems to bite through its muzzle from time to time" (Bering in press) see also (McCauley 2000). Or it may be that distinctively supernatural understandings can be genuinely suppressed in secular communities – as seems to happen with racists, sexists, and homophobic biases. Clearly there remain a lot of maybes surrounding the naturalistic study of religion.

To sum up, the past ten years have witnessed a renaissance in the psychology of religion. We have learned a great deal about how supernatural thought develops and spreads, and how it mediates social relationships. But much remains obscure, and the hard work lies ahead of us.

Acknowledgements

Many thanks to Nick Agar, Jesse Bering, Rich Sosis, and Kim Sterelny for very helpful comments on earlier drafts.

Notes

[1] A variety of definitions of "religion" and "religious cognition" circulate in the naturalistic study of religion. Loose conceptions, for example counting just about anything that involves ritual and non-mechanical causal beliefs create a pseudo-universal of "religion" similar to "human practice" and "imagination" inquiry into which demands the kind of theory-of-everything that science cannot supply [for a related stance see Boyer, P. (2001). *Religion*

682

Explained: the evolutionary origins of religious thought. New York, Basic Books. I am interested in motivating beliefs and practice relative to supernatural agents and powers. Departing somewhat from ordinary language, I call these supernatural agents and powers "gods."

[2] For the most part I will not discuss developments in the social psychology of religion, a more descriptive than explanatory field (though clearly these descriptions provide evidence for those interested in explanation). There is a large literature inspired by the writings of Freud, Jung, and more recent European philosophers, which I will ignore, because they do not hold themselves accountable to standards of naturalistic inquiry. I do not consider meme theoretic accounts according to which religion is explained in virtue of the adaptive properties of religious ideas in accommodating themselves to human minds, which support them. Here religion is an adaptation, but not our adaptation. I ignore this approach because it has yet to inspire any experimentally fruitful psychological research programme.

[3] Guthrie started out as an anthropologist of Japanese culture. Japan's indigenous religious tradition, Shinto, holds that the world is brimming with "kami." These are god-like nature deities who infuse nature. Many contemporary Japanese, while claiming not to be religious, still believe in and pay tribute to the kami. And new religious movements largely based on kami-like animisms proliferate at an extremely high rate. In Japan, the bullet train, camera cell phone, and pocket-sized supercomputer co-exist with a thoroughly deified conception of nature: see Reader, I. and G. J. J. Tanabe (1998). *Practically Religious: Worldly Benefits and the Common Religion of Japan*. Honolulu, University of Hawaii Press.

[4] Here Boyer's work echoes David Hume's theory of religion. In the essay "On Miracles," Hume credits our "passion for surprise and wonder" (p. 71) and "inclination to the marvellous" (p. 73) for the popularity of supernatural stories – what Hume calls "pious frauds" (p. 83). Hume thought we believe in miracles precisely *because* they are outlandish. Boyer seeks to bring to Hume's observation the precision of 21st century cognitive psychology, to elaborate the cognitive apparatus that leave us susceptible to religious wonder.

[5] Boyer observes the gods are not imagined as knowing the contents of your refrigerator or the correct way to change motor oil, though these inference follow from their imagined powers Boyer, P. (2001). *Religion Explained: the evolutionary origins of religious thought*. New York, Basic Books.

[6] Clearly this and otherworldly desires are in part influenced by culture and a more detailed account of religion would need to take account of how particular reward/punishments schemes become desirable. For example many of us would prefer a life of karma to the annihilation of nirvana.

[7] For simplicity, I ignore redundant cheater detector and punishment systems.

[8] For simplicity, I ignore what Skyrms calls "viscous communities" of related exchange partners Skyrms, B. (1996). *Evolution of the social contract*. New York, Cambridge University Press. In such communities co-operation evolves more easily than in communities of largely unrelated exchange partners.

[9] Though it may be that "extrinsically religious" define their groups more narrowly – morality is always the morality of a group "our people" – with indifference or moralistic aggression to those outside, depending on resource distributions and scarcities. See Hartung, J. (1996). "Love They Neighbor: The Evolution of In-Group Morality." *Skeptic* **3**(4): 86–99.

References

Andresen, J. (ed.): 2001, *Religion in Mind*, Cambridge University Press, Cambridge.

Atran, S.: 2002, *In Gods We Trust: The Evolutionary Landscape of Religion*, Oxford University Press, New York.

Barrett, J. and Keil, F.: 1996, 'Conceptualizing a Nonnatural Entity', *Cognitive Psychology* **31**, 219–247.

Barrett, J.L.: 2000, 'Exploring the Natural Foundations of Religion', *Trends in Cognitive Sciences* **4**(1) (January).

Barrett, J.L. and Keil, F.C.: 1998, 'Cognitive Constraints on Hindu Concepts of the Divine', *Journal for the Scientific Study of Religion* **37**, 608–619.

Barrett, J.L. and Richert, R.A.: 2003, 'Anthropomorphism or Preparedness? Exploring Children's God Concepts', *Review of Religious Research* **44**(3), 300–312.

Batson, C.D. and Schoenrade, P. et al.: 1993, *Religion and the Individual: A Social-Psychological Perspective*, Oxford University Press, New York.

Bering, J.: 2003, 'Towards a Cognitive Theory of Existential Meaning', *New Ideas in Psychology* **21**, 101–120.

Bering, J.M.: in press, 'The Evolutionary History of an Illusion: Religious Causal Beliefs in Children and Adults', in B. Ellis and D. F. Bjorklund (eds.), *Origins of the Social Mind: Evolutionary Psychology and Child Development*, Guilford Press, New York.

Boyer, P.: 1994, *The Naturalness of Religious Ideas: A Cognitive Theory of Religion*, University of California Press, Berkeley, CA.

Boyer, P.: 1998, 'Cognitive Aspects of Religious Ontologies: How Brain Processes Constrain Religious Concepts', in T. Alhback (ed.), *Theories and Method in the Study of Religion*, Donner Institute.

Boyer, P.: 2000, 'Functional Origins of Religious Concepts: Ontological and Strategic Selection in Evolved Minds', *Journal of the Royal Anthropological Institute* **6**, 195–214.

Boyer, P.: 2001, *Religion Explained: The Evolutionary Origins of Religious Thought*, Basic Books, New York.

Boyer, P.: 2003, 'Religious Thought and Behaviour as By-products of Brain Function', *Trends in Cognitive Sciences* **7**(3), 119–124.

Boyer, P. and Ramble, C.: 2001, 'Cognitive Templates for Religious Concepts: Cross-Cultural Evidence for Recall of Counter-Intuitive Representations', *Cognitive Science* **25**, 535–564.

Bulbulia, J.: 2003, 'Review of James McClenon, Wondrous Healing: Shamanism, Human Evolution and the Origin of Religion', *Method and Theory in the Study of Religion* **15**(1).

Bulbulia, J.: 2004, 'Religious Costs as Adaptations that Signal Altruistic Intention', *Evolution and Cognition* **10**(1).

Bulbulia, J.: in preparation, 'Are There any Religions?'.

Chen, D.: 2004, 'Economic Distress and Religious Intensity: Evidence from Islamic Resurgence During the Indonesian Financial Crisis', *Under Review*.

Cronk, L.: 1994, 'Evolutionary Theories of Morality and the Manipulative Use of Signals', *Zygon* **29**(1), 81–101.

Darwin, C.: 1871/1981, *The Descent of Man and Selection in Relation to Sex*, Princeton University Press, Princeton.

Dunbar, R.I.: 1998, 'The Social Brain Hypothesis', *Evolutionary Anthropology* **6**, 178–190.

Durkheim, E.: 1964 [1915], *The Elementary Forms of the Religious Life*, George Allen & Unwin Ltd., London.

Ekman, P.: 1975, *Unmasking The Face*, Prentice-Hall, Englewood Cliffs, NJ.

Ellison, C.: 1991, 'Religious Involvement and Subjective Well-Being', *Journal of Health and Social Behavior* **32**, 80–89.

Evans, E.M.: 2000, 'The Emergence of Beliefs about the Origin of Species in School-Age Children', *Merrill Palmer Quarterly* **46**, 221–254.

Evans, E.M.: 2001, 'Cognitive and Contextual Factors in the Emergence of Diverse Belief Systems: Creation versus Evolution', *Cognitive Psychology* **42**, 217–266.

Frank, R.: 1988, *Passions Within Reason: The Strategic Role of The Emotions*, Norton and Company, New York.

Gelman, S. and Kremer, K.: 1991, 'Understanding Natural Cause: Children's Explanations of How Objects and their Properties Originate', *Child Development* **62**, 396–414.

Godfrey-Smith, P.: 2002, 'Environmental Complexity, Signal Detection, and the Evolution of Cognition', in M. Bekoff, C. Allen and G. Burghardt (eds.), *The Cognitive Animal: Empirical and Theoretical Perspectives on Animal Cognition*, The MIT Press, Cambridge MA, pp. 135–142.

Grafen, A.: 1990, 'Biological Signals as Handicaps', *Journal of Theoretical Biology* (144), 517–546.

Guthrie, S.: 1993, *Faces in the Clouds: A New Theory of Religion*, Oxford University Press, New York.

Hardin, R.: 1995, *One for All: The Logic of Group Conflict*, Princeton University Press, Princeton, NJ.

Hartung, J.: 1996, 'Love They Neighbor: The Evolution of In-Group Morality', *Skeptic* **3**(4), 86–99.

Hummer, R.A. and Rogers, R.G. et al.: 1999, 'Religious Involvement and U.S. Adult Mortality', *Demography* **36**, 273–285.

Hunter, V.: 1994, *Policing Athens: Social Control in the Attic Lawsuits, 420–320 B.C.*, Princeton University Press, Princeton, NJ.

Irons, W.: 1996, 'Morality as an Evolved Adaptation', in J. P. Hurd (ed.), *Investigating the Biological Foundations of Morality*, Edwin Mellon Press, Lewiston, pp. 1–34.

Irons, W.: 2001, 'Religion as Hard-to-Fake Sign of Commitment', in R. Nesse (ed.), *Evolution and the Capacity for Commitment*, Russell Sage Foundation, New York.

Johnson, C.: 2003, 'During Economic Turmoil, Religion is 'Insurance', *Science and Theology News* **4**.

Johnson, D. and Kruger, O.: in press, 'The Good of Wrath: Supernatural Punishment and the Evolution of Cooperation', *Political Theology*.

Katz, R.: 1984, *Boiling Energy: Community Healing among the Kalahari Kung*, Harvard University Press, Cambridge, MA.

Katz, R.: 1997, *Healing Makes Our Hearts Happy: Spirituality and Cultural Transformation among the Kalahari Ju/ 'Hoansi*, Inner Traditions International Ltd., New York.

Kelemen, D.: 1999, 'The Scope of Teleological Thinking in Preschool Children', *Cognition* **70**, 241–272.

Kelemen, D.: 1999, 'Why are Rocks Pointy? Children's Preference for Teleological Explanations of the Natural World', *Developmental Psychology* **35**, 1440–1453.

Kelemen, D.: 2003, 'British and American Children's Preference for Teleo-Functional Explanations of the Natural World', *Cognition* **88**, 201–222.

Kelemen, D.: in press, 'Are Children "Intuitive Theists"?: Reasoning about Purpose and Design in Nature', *Psychological Science*.

Kelemen, D. and DiYanni, C.: 2002, 'Children's Ideas about Nature: The Role of Purpose and Intelligent Design', *Manuscript in Submission*.

Knight, N. and Sousa, P. et al.: forthcoming, 'Children's Attributions of Beliefs to Humans and God: Cross-Cultural Evidence.'

Laughlin, C. and McManus, J.: 1979, 'Mammalian Ritual', in E. D'Aquili, C. Laughlin and J. McManus (eds.), *The Spectrum of Ritual*, Columbia University Press, New York, pp. 80–116.

Lawson, E.T. and McCauley, R.N.: 1990, *Rethinking Religion: Connecting Cognition and Culture*, Cambridge University Press, Cambridge.

Livingston, K.: 2002, 'Reason, Faith, and The Good Life: Does Strong Doubt Permeate Good Health?' *Free Inquiry* (Winter 2001/02).

McCauley, R.N.: 2000, 'The Naturalness of Religion and the Unnaturalness of Science', in F.C. Keil and R.A. Wilson (eds.), *Explanation and Cognition*, Cambridge University Press, Cambridge, pp. 61–85.

McCauley, R.N. and Lawson, E.T.: 2002, *Bringing Ritual to Mind*, Cambridge University Press, New York.

McClenon, J.: 1997, 'Shamanic Healing, Human Evolution, and the Origin of Religion', *Journal for the Scientific Study of Religion* **36**(3), 345–354.

McClenon, J.: 2002, *Wondrous Healing: Shamanism, Human Evolution, and the Origin of Religion*, Northern Illinois University Press, DeKalb, Illinois.

Mithen, S.: 1999, 'Symbolism and the Supernatural', in R.I. Dunbar, C. Knight and C. Power (eds.), *The Evolution of Culture*, Rutgers University Press, New Brunswick, NJ, pp. 147–171.

Petrovich, O.: 1997, 'Understanding of Non-natural Causality in Children and Adults: A Case Against Artificialism', *Psyche and Geloof* **8**, 151–165.

Preus, J.S.: 1987, *Explaining Religion: Criticism and Theory from Bodin to Freud*, Yale University Press, New Haven.

Pyysiainen, I.: 2001, *How Religion Works: Towards a New Cognitive Science of Religion*, Brill, Leiden.

Ramachandran, V.S. and Blakeslee, S.: 1998, *Phantoms in the Brain: Probing the Mysteries of the Human Mind*, Quill William Morrow, New York.

Rappaport, R.A.: 1999, *Ritual and Religion in the Making of Humanity*, Cambridge University Press, Cambridge, England.

Reader, I. and Tanabe, G.J.J.: 1998, *Practically Religious: Worldly Benefits and the Common Religion of Japan*, University of Hawaii Press, Honolulu.

Roes, F.L. and Raymond, M.: 2003, 'Belief in Moralizing Gods', *Evolution and Human Behavior* **24**, 126–135.

Skyrms, B.: 1996, *Evolution of the Social Contract*, Cambridge University Press, New York.

Smith, W.J.: 1979, 'Ritual and the Ethology of Communicating', in E. D'Aquili, C. Laughlin and J. McManus (eds.), *The Spectrum of Ritual*, Columbia University Press, New York, pp. 51–79.

Sosis, R.: 2000, 'Religion and Intragroup Cooperation: Preliminary Results of a Comparative Analysis of Utopian Communities', *Cross-Cultural Research* **34**(1), 77–88.

Sosis, R.: 2003, 'Book Review: Darwin's Cathedral: Evolution, Religion, and the Nature of Society', *Evolution and Human Behavior* **24**, 137–143. Costly Signalling Theory, Religion, God.

Sosis, R.: 2003, 'Why Aren't We All Hutterites?' *Human Nature* **14**(2), 91–127.

Sosis, R. and Alcorta, C.: 2003, 'Signalling, Solidarity, and The Sacred: The Evolution of Religious Behavior', *Evolutionary Anthropology* **12**, 264–274.

Sosis, R. and Alcorta, C.: in preparation, 'Religion, Emotion, and Symbolic Ritual: The Evolution of an Adaptive Complex.'

686

Sosis, R. and Bressler, E.: 2003, 'Co-operation and Commune Longevity: A Test of the Costly Signaling Theory of Religion', *Cross-Cultural Research* **37**(2), 11–39.

Sosis, R. and Ruffle, B.: 2003, 'Religious Ritual and Cooperation: Testing for a Relationship on Israeli Religious and Secular Kibbutzim', *Current Anthropology* **44**(5), 713–722.

Sterelny, K.: 2000, 'Niche Construction, Developmental Systems and the Extended Replicator', in R. Gray, P. Griffiths and S. Oyama (eds.), *Cycles of Contingency*, MIT Press, Cambridge.

Trinkhaus, E. and Shipman, P.: 1993, *The Neandertals: Changing the Image of Mankind*, Knopf, New York.

Trivers, R.: 2001, 'Self-Deception in Service of Deceit', in R. Trivers (ed.), *Natural Selection and Social Theory*, Oxford University Press, New York.

Whitehouse, H.: 2000, *Arguments and Icons*, Oxford University Press, Oxford.

Wilson, D.S.: 2002, *Darwin's Cathedral: Evolution, Religion, and the Nature of Society*, University of Chicago Press, Chicago.

Wuthnow, R.: 1994, *Producing the Sacred: An Essay on Public Religion*, University of Illinois Press, Champaign, IL.

Zahavi, A.: 1977, 'The Testing of The Bond', *Animal Behavior* **25**, 246–247.

Zahavi, A. and Zahavi, A.: 1997, *The Handicap Principle: A Missing Piece of Darwin's Puzzle*, Oxford University Press, New York.

[2]

A Scientific Definition of Religion

by

James W. Dow

Abstract

Religion is a collection of behavior that is only unified in our Western conception of it. It need not have a natural unity. There is no reason to assume, and good reason not to assume, that all religious behavior evolved together at the same time in response to a single shift in the environment. This article does not look at the religion as a unified entity and seek a definition of its essence. Instead, it looks at what science needs to know in order to discover how and why religion came into existence as a human behavior. What does science need to know about religion, or how should religion be defined so that science can look at it? A definition that refers to observable behavior is required. Then, a preliminary hypothesis to orient observations is proposed. I suggest a preliminary hypothesis consisting of three stages in the evolution of religion: (1) a cognizer of unobservable agents, (2) a sacred category classifier, and (3) a motivator for public sacrifice. Each one of these stages is a nucleus of modern anthropological theorizing. Although they all come together in the Western folk concept of religion, this article proposes that they are independent evolutionary complexes that should not be lumped together, but should be investigated as separate types of religious behavior.

Understanding Religion

Religion is a human activity that can be easily accepted only within the framework of reality that it creates for itself. If you accept the existence of whatever myth, god, spirit, or supernatural force that a religion proposes, then you can see the logic of all that follows. However, most of the entities, gods or whatever, that are the basis of religious thought and action cannot have their existence validated by direct observation. How do non-believers understand religion? Simply saying that the believers are crazy or living in a different world will not suffice. The believers are also normal human beings. They are no crazier than anyone else.

There is another way to look at religion, through science. Science has provided human culture with an excellent understanding of the natural world and human behavior. However, for the scientist, the logic of religious behavior is not simple. The scientist must understand religion as the complex workings of a human brain that is not responding directly to observable reality. The cause of religious behavior for the scientist does not lie in myth but in an understanding of why human beings do and think what they do. Among other explanations, science has found that they do what they do because they have been made that way by evolution. Evolution is one key to understanding of religion from a scientific point of view.

Is religious behavior rational? Is it the mobilization of available means to achieve certain ends? The sociologists Stark and Fink (2000) argue that religious behavior is actually rational in an economic sense in spite of the fact that the believers work with unobservable actors and magical processes. The rationality is economic and can be seen in the social and material rewards that flow from participation in religious groups. When there is a market place for different faiths, individuals usually choose, consciously or unconsciously, the faith that brings them the most rewards. The rationality in this case is apparent when one measures the rewards that flow from different religious activities. So, despite its apparent irrationally, religious activity can have a latent economic rationality. However, economic rationality is the surface manifestation of underlying trophic tendencies built into the mammalian brain. This is seen is the optimal foraging behavior of most species. Thus, evolution can cause economic rationality.

Is religion an exclusively human behavior? Religion has some particularly human characteristics and some pre-human ones as well. It depends on the unique human ability to communicate with language. Religion, as we know it, needs language, but that does not mean that it has freed itself from pre-human behavior found among primates, mammals, and even reptiles. Religion has rituals and non-human animals have rituals. Birds have rituals, reptiles have rituals, and they communicate symbolically with other members of their species. They just do not use the same linguistic structures that humans use.

Does human psychology explain religion? Religion can be examined by psychological science. Psychological explanations satisfy many social scientists (eg. Hinde 1999; Kirkpatrick 2005). Religion has obvious psychological functions. It takes care of: the need for a comforting parent figure, the need to explain difficult things, the need to fight depression, the need to deny mortality, etc. However, psychology does not explain how humans got to be religious. Although psychological explanations tell us why people do religious things, they do not tell us how religion got started and why it continues. They do not tell us about religion's evolutionary past or future. They tell us how religion works in the mind, but they do not tell us how the mind got that way. The mind is a product of evolution, not its cause.

Evolutionary Science

Because the evolution of living forms takes place slowly, it cannot be easily observed. We understand it primarily through the traces that it leaves. Understanding the process that produces these traces is difficult. Darwin (1859) took a great step forward in recognizing one of these processes, natural selection. Natural selection takes place when inherited forms within a species duplicate themselves at different rates. Eventually, the fast duplicating forms become more numerous than the slowly duplicating forms. Variation within the species maintains the process

The human central nervous system permits another form of evolution, environmental learning. More complex species, like humans, have a large central nervous system that can receive information from the environment and alter behavior to meet the challenges of that environment. Human beings can also receive information from each other. When this happens, the learner does not have to pay the costs of the experience itself. What humans acquire culturally from each other comes at a much reduced cost than the original knowledge.

Until its evolutionary history is fully understood, we might regard religion as a type of irrational adaptation. It is irrational in the sense that it does not move individuals to solve problems rationally. However, adaptive behavior does not have to be rationally aimed at an obvious goal. Behaviors that appear irrational can evolve by increasing reproductive fitness, often called just "fitness", which is a measure of the rate at which individuals reproduce. Evolving behavior may increase the fitness of an individual or the fitness of kin, who have a high probability of carrying the behavioral gene themselves.[1] If religion is adaptive, it must be irrationally adaptive in this way. Its ubiquity and its longevity argue in favor of its being adaptive in some way at this level.

To deal with the complex gene-culture evolution of religion, E. O. Wilson (1978: 182-185) postulated three types of selection: (1) ecclesiastic (2) ecological, and (3) genetic. They are given in the order of how rapidly they respond to environmental change. Ecclesiastic selection is the quickest. It is the response that religions leaders have to changing situations. It is the most irrational: There are hard times. A message from "god" is received, and people follow a new leader, hopefully, to a better life. Now, ecological selection sets in. The lives of some are improved. The lives of others are not. Ecology may favor some religious changes and punish others. After a long while, genetic change takes place. The genes that favored the successful religious responses are propagated, and those that did not favor them are lost. Nothing in this process appears as obvious rational behavior.

Religious behavior could be selected in other ways. It could be selected sexually. In other words, it could enhance the ability to attract mates. It could also promote the successful rearing of

1. This later measure of the total fitness effect of a gene is called its *inclusive fitness* (Eberhard 1975).

offspring. Religions do meddle in sexual behavior and relations within the family. All of this needs to be kept in mind when developing an evolutionary model of religion.

A Critique of Anthropological Definitions of Religion

Many social scientists prefer a single encompassing definition of religion, an essential definition. For example, Guthrie (1993) sees anthropocentrism as the essence of religion. He sees a projection of human attributes into the perceived world as the essence of religion. Kirkpatrick (2005) sees religion as psychological attachment, a powerful emotional relationship to things. Such essentialist authors do not confine themselves to discussing the narrow range of behavior signified by their concepts, but they use the concepts as a way of organizing the information that they present, and they concentrate on those aspects of religion that support these conceptualizations. However lovely to the inquiring mind they may be, essentialist definitions such as these have not been very useful to scientific theory (Saler 1993:81). They confuse evolutionary models by lumping together traits that may have different evolutionary origins. Evolution does not create essences. It creates new genetic codes, not grand conceptions.

A definition of religion is difficult to make, because religion has many facets, many of which do not appear to be religious by themselves. For example, religion involves gathering in groups. It involves communal eating. It involves theoretical discourse about the nature of the universe, and so forth. Countless definitions have been proposed by theoreticians. The most interesting thing is that the average person can tell when others are engaging in religious behavior while many scholars and scientists have problems defining it. The concept of religion is like the concept of culture. It is easy to use in ordinary discourse, but difficult to define precisely.

Looking back at the nineteenth and early twentieth century, Bronislaw Malinowski—often regarded as the father of modern empirical anthropology—concluded that anthropological definitions of religion in mid nineteen-twenties had become frankly chaotic. Cultural anthropology was not in good shape when it came to defining religion. Malinowski wrote:

> Our historical survey of theories has left us somewhat bewildered with the chaos of opinions and the jumble of phenomena. While it was difficult not to admit into the enclosure of religion one after the other, spirits and ghosts, totems and social events, death and life, yet in the process religion seemed to become a thing more and more confused, both an all and a nothing. [Malinowski 1948:36]

The confusion of which Malinowski wrote was the result of other early anthropological theorists beginning with E. B. Tylor (1958[1871]) who defined religion as a belief in spirits. Spirits were gods, animating powers, animal-spirit companions, etc, all of which seemed to have a religious cast. He attributed the origin of these religious ideas to dreams rather than to cultural evolution. This was a back-door admission that religion had some sort of biological origin since dreams are produced in the central nervous system.[2] Tylor spent time looking at world religions and reduced their fundamentals to his concept of *animism*. He theorized that human consciousness reached out to understand the world by projecting into it beings or souls with very human-like intentions. It was a logical way of thinking, based on the experience of dreams. Tylor saw humans as always improving their intelligences through rational thought. The primitives were basically rational in their idea of souls, but they had little scientific knowledge. Animism was rational but ignorant. Tylor believed that minds would improve as they acquired more scientific knowledge. Durkheim later followed this progressive tradition by declaring that science would eventually triumph over religion as the primary human representation of reality.

Nineteenth century ideas about the origins of religion left biology behind and began to speculate about the socio-cultural evolution of religion, a process that, at that time, was clearly connected to concepts of social "progress." One of the first definitions of religion within this school of thought was proposed by Émil Durkheim (1963[1912]). He defined religion as a collective representation that made things sacred. Religion was a world view that created the sacred. The power to do this resided in the collective itself, society. So society had to create religion. Durkheim felt that religion was the foundation of society. The problem with Durkheim's definition was that some people practiced religious activities by themselves without the presence of others, without society. You could say that they were surrounded by a society in some environmental sense. But there were persons, shamans, diviners, prophets, charismatic cult leaders, and other holy folk, who communicated directly with the divine without the intervention of social convention and who created religion on their own. Durkheim lumped this sort of behavior into the category of "magic," said that it was outside of religion, and so, preserved his society-oriented viewpoint at the cost of leaving out a vast panoply of behavior that most people would call religious. Durkheim ignored the possibility that religion was coming out of the human brain, the mind, without the help of "society."

"Religion" is, in fact, a folk category in Western culture. Comparative analysis can flounder on efforts to use folk categories in scientific analysis. It is important for scientific investigation to have a clear definition of a phenomenon. Folk categories can be overgeneralized into essentialist definitions that are of little use. Today's anthropology has often been driven back to the idea that, although we can see basic human behavior in each cultural system, the cultural systems are them-

2. Tylor's book *Primitive Culture* was aimed at adducing evidence for the *great doctrine of animism*, his insightful theory about the "origins" of religion (Saler 1997). Many thinkers in the late nineteenth century were searching for these origins, a state of religion in its more pristine form out of which the more "enlightened" religions of the 19th century evolved. Thus, when so-called "primitive" beliefs such as magic or divination cropped up in 19th century European society, they were regarded as survivals carried over from earlier evolutionary forms in a manner akin to vestigial organs in animal physiology (Tylor 1958[1871]:112--159).

selves ultimately unique. This is true, but defeats the effort to discover the evolutionary history and adaptive processes in human behavior. One of the most well known proponents of an essentialist definition of religion that incorporates the idea of cultural uniqueness is Clifford Geertz, who put forward a definition that has been quite influential. Geertz (1993 [1966]: 90) defined religion as:

> (1) a system of symbols which acts to (2) establish powerful, pervasive, and long-lasting moods and motivations in men by (3) formulating conceptions of a general order of existence and (4) clothing these conceptions with such an aura of factuality that (5) the moods and motivations seem uniquely realistic.

This is a restatement of the folk category of religion in Western European culture. Talal Asad (1983) writes that the folk category that Geertz elaborates is not only Western in origin but is distinctly Christian.[3] Saler (1993: 96) disagrees with this interpretation somewhat and writes that Geertz's definition is a Western cultural idea but not necessarily a Christian one. In either case, Geertz's definition does not lead to the kind of systematic and objective observation that can support a scientific understanding of religion. Nevertheless, it feels right because it is a distillation of the Western folk category of religion. It feels right but has many problems in its application. As Geertz unpacks his definition, it appears that many of the five components of religion consist of interrelated "meanings." So the scientist is left with the task of studying and classifying meanings. Then the meanings act on people, so the observer now has to observe meanings acting on people, a very difficult task that cannot easily be carried out with objectivity.

Definitions and Theory

Definitions that are used in science are always provisional. New and better definitions of phenomena will always be accepted when they simplify and assist in the generation of better theory. The important thing to science, evolutionary and otherwise, is that the phenomena are identifiable and observable. There is no reason to start an analysis of religion with a definition that cannot be changed as later understanding develops. All that is needed is a way of identifying the phenomena being studied. For example a study of gravitational forces can start with the idea that objects fall down rather than up, to identify the phenomena to study. We don't have to specify the nature of gravitational fields. That comes later.

---

3. Asad may go too far in criticizing Geertz for leaving out power relationships. Geertz (1993 [1966]:109) does elaborate his idea of "clothing those conceptions with such an aura of factuality that ..." implying that he is thinking about the power that authorities have over believers. A difference between Geertz (1993 [1966]) and Asad (1983) is that Geertz sees religion as a phenomenon in the mind of the believer and Asad sees it as a social process.

Scientific approaches to knowledge often start with a provisional theory that points at phenomena to observe. The phenomena are observed, and the theory is developed further. The definitions are operational and not permanent. New concepts and entities with new definitions are proposed as understanding of the phenomena deepens. This paradigmatic framework has been outlined by philosophers of science such as Hempel (1966). It points in a very fruitful direction. There is proven potential in such an approach rather than the typical social science approach of making synthetic, essentialist, all-encompassing definitions that make unoperationalizable references to unmeasurable human experience. If definitions are to be made, it is better to begin with a provisional theory and then derive some categories of behavior to be observed and analyzed. The natural sciences have developed some excellent intellectual tools, and using them in the social sciences is far from foolish. Saler (2004) points the way to using natural science methods in the study of religion. He advocates abandoning a "standard social science model" that ignores human nature.[4]

Religious phenomena, religious behaviors, can better be defined by avoiding vague intuitive elements or unobservable subjective elements such as meanings. Anthony Wallace (1966:62-66), an anthropologist, in 1966 identified thirteen observable universal behavioral complexes that provide a minimal definition of religion. They are shown in Table 1. Wallace's behavioral categories show that religion can be defined by observable behavior. Every culture may not have a single concept such as *religion* to describe the totality of this collection of behavior; however these behaviors outline something that exists in most cultures. Wallace's behavioral collection puts the definition of religion on a much more operational and practical footing than definitions that refer to meanings. Myths and meanings are part of religion, but the observable behavior that goes along with them seems to be the thing that allows Westerners to perceive something that is acceptably "religious."

A Definition of Religious Behavior

Religion is a collection of behavior that is only unified in our Western conception of it. There is no reason to assume, and good reason not to assume, that this behavior evolved together at the same time in response to a single shift in the environment. For example, Atran (2002) and Boyer (2001) see religion as a great potpourri of ideas and behavior with many independent evolutionary origins outside of religion itself.

The provisional definition herein is based on three behavioral modules. I use the term *modules* to refer to these behavior complexes in keeping with the vocabulary of evolutionary psychology (Tooby and Cosmides 1992); however, the range of behavior within each one is wide, and they could

4. The standard social science model asserts that all social behavior is learned from other members of one's social group and that people have no biologically determined social behaviors. For a critique of the standard social science model see Tooby and Cosmides (1992), Wilson (1998), and Pinker (2002).

also be called *complexes* to indicate the variety of behavior to which they refer. They are modules in the sense that they are solutions to particular problems of survival and reproduction. These modules evolved at different times and actually provide three separate means for identifying religion. Behavior that is produced by any one of these modules can be considered religious.

The three modules are listed in Table 2. They are derived from modern ongoing theories about the evolution of religion. Figure 1 shows them in two dimensions. The vertical dimension represents evolutionary time. The oldest is at the bottom and the newest, at the top. The horizontal dimension represents the rapidity of the adaptive response. Genetic responses are at the left and ecclesiastical responses are at the right.

The earliest module, a *cognizer of unobservable agents*, has been proposed by Scott Atran (2002). Humans have a mental capacity to create images of unobservable agents who cause real things to happen. There are countless examples of unobservable agents, gods, ghosts, witches, angels, spirits, dead ancestors, patron saints, demons, extraterrestrials, culture heroes, etc. in religion. The idea actually goes back to Tylor's theory of souls. Souls are animating agents. The idea of agents could be a pre-human adaptation to predators. The animal who imagined a predator lurking in the bushes would have a better chance of survival than one who did not.

The second module, a *sacred category classifier*, may also not be exclusively human. We know about it among humans because it is communicated symbolically; however it may be derived from animal devotion to a herd leader or parent. This mental ability separates things into profane and sacred categories. The sacred is higher, more powerful, and must be treated with respect. Rules of behavior toward sacred and non sacred objects are established in a social context. Behavior is oriented by a moral continuum of purity, with the most sacred on one end and the most sordid and impure on the other. Roy Rappaport (1999) has discussed sacredness at length and has developed the argument that it is an evolutionary adaptation. It controls how human groups interact with their environment. Sacred signals coordinate group responses to environmental change. Many others starting with Durkheim have written about it. Sacredness was central to Durkheim's definition of religion. Mary Douglas (1966), for example, is concerned with pollution and taboo, the antithesis of the sacred that is equally religious in context. Because it is communicated syntactically it is linked to the time that humans developed language.

The third module of religious behavior, *public sacrifice*, is highly symbolic. It seems to have evolved as a defense mechanism against symbolic deception according to the theories of Irons (2001) and Sosis and Alcorta (2003). Sacrifice demonstrates a commitment to the ideology of a particular group and activates cooperation within that group. According to Irons and Sosis, hard-to-fake acts of sacrifice prove to the group that an individual can be trusted. These are costly signals akin to costly signals among other members of the animal kingdom. In the case of humans, the fitness benefits come from better cooperation. In the case of animals the benefits may come from an attraction of a better mate. Not all sacrifices are religious however. Religious sacrifices are altruistic, performed publicly, and guided by a shared philosophy of the sacred. Other non-religious sacrifices may be for the benefit of kinspersons such as offspring and are explained evolutionarily better as kin altruism. The complexity of the signaling in public sacrifice puts its development at a later date than that of sacred-symbol messaging.

Thus, religion is provisionally defined as behavior within any of these three separate modules. The systems have evolved at different times with different adaptive functions. The systems are maintained by modules in brain that are biologically reproduced in most humans. So they evoke a natural response when perceived by other people. Western culture puts all these responses into the folk category of religion, but some cultures separate them. For example, the Confucian religions of Asia seem to be focused on module 2, *acts of reverence toward sacred things*, and leave the other two modules aside.

Shamanic religions emphasize module 1, *acts and beliefs relating to unobservable agents*, and downplay modules 2, and 3, which have much more to do with maintaining cooperation in moderately-sized or large-sized social systems. Shamanism is often the only type of "religion" in hunter-gatherer cultures in which social relationships are local and organized primarily by kinship. The difference between shamans and priests is essentially the difference between religious practitioner in behavioral module 1 and practitioners in behavioral module 2. In Mesoamerica the native religious practitioners (shaman-priests) deal with both modules 1 and 2. This poses a serious conceptual problem for anthropologists who attempt essentialist definitions of shamanism (eg. Kehoe 2000; Eliade 1964[1951]). In some religions, such as folk Catholicism, the three modules are completely integrated. In Circum-Mediterranean and New-World-Iberian cultures where folk Catholicism is prevalent, Saints, supernatural agents (1), are treated with great reverence (2), and sacrifices (3) are offered to them during fiestas. This three-way synthesis is found in many peasant cultures such as those in India and China. In many cases, the agents are ancestors. There may be other behavioral modules yet delineated that are involved in "religion." The ones outlined here serve to set out a preliminary definition for the evolutionary science of religion.

These modules should be studied separately, because their evolutionary history is very different. They have responded to different selective pressures on the central nervous system as were noted by Wilson in his schema. It looks like the evolutionary science of religion will lead in different directions; however, it should eventually be able to explain how and why human beings develop and maintain a panoply of complex irrational behaviors that are very influential in their lives.

Conclusion

Religion is not a single thing. It is a body of behavior unified by our failure to find a simple rational explanation for it when seen from the perspective of the individual. However, behavioral complexes within religion do have adaptive rationality when seen in evolutionary perspective. To move ahead with the scientific understanding of religion, these complexes, modules, should be defined and studied independently. The three outlined here appear to have evolved at very different times.

To understand the module *cognizer of unobservable agents,* comparative animal-human studies of fear reactions would help to understand how images of powerful unseen beings benefit the individual and the group. The flight of a herd of animals, or more appropriately, the flight arboreal primates might be compared with reactions to images of unpredictable gods. The module *sacred category classifier* varies from culture to culture and can be related to particular cultural ecologies. It seems to have evolved at a time that cultural information was becoming an important mechanism for human adaptation. Agricultural cultures have a group of sacred categories different from those found in hunter gatherer cultures. The module *public sacrifice* is part of a complex which organizes groups around charismatic leaders.

Religion should be defined according to modular complexes that have been set up by evolution to solve adaptive problems. The three described here are probably not the only ones; however they are a starting place, a set of hypotheses that organize the search for data that will reveal why human beings engage in the behaviors that they call religious.

Ph. D. James W. Dow is Professor Emeritus
of Anthropology at the Department for
Sociology and Anthropology at Oakland
University.
Contact: dow@oakland.edu

Tables

Table 1: Wallace's Behavioral Complexes

Prayer: *Addressing the supernatural.* This includes any kind of communication between people and unseen non human entities.

Music: *Dancing, singing and playing instruments.* Although all music is not religious, there are few religions that do not include it.

Physiological exercise: *The physical manipulation of psychological state.* This includes such tools as drugs, sensory deprivation, and mortification of the flesh by pain, sleeplessness, or fatigue.

Exhortation: *Addressing another human being.* This includes preaching by a minister, shaman, or other magicoreligious practitioner.

Reciting the code: *Mythology, morality, and other aspects of the belief system.* Every religion has its myths, symbols, and sacred knowledge.

Simulation: *Imitating things.* This is a special type of symbolic manipulation found particularly in religious ritual. It is similar to Frazer's (1911) concept of sympathetic magic.

Mana: *Touching things.* This refers to the transfer of supernatural power through contact. Frazer's contagious magic is included.

Taboo: *Not touching things.* Religions usually proscribe certain things, the eating of certain foods, contact with impure things, impure thinking, etc.

Feasts: *Eating and drinking.* All celebrations are not religious, but most religions have them.

Sacrifice: *Immolation, offerings, and fees.* Sacrifice is probably the single most definitive behavior.

Congregation: *Processions, meetings, and convocations. Religions organize groups.* Their rituals identify groups and create group solidarity.

Inspiration: Wallace (1966:66) writes "all religions recognize some experiences as being the result of divine intervention in human life."

Table 2: **Provisional Determinative Behavior Modules and the Associate Behavior**

	Brain module	Behavior
1.	Cognizer of unobservable agents	Acts and beliefs relating to unobservable agents
2.	Sacred category classifier	Acts of reverence toward the sacred
3.	Motivator for religious enthusiasm and public sacrifice	Public sacrifice.

12

Figures

Figure 1: Modules in an Evolutionary Sequence

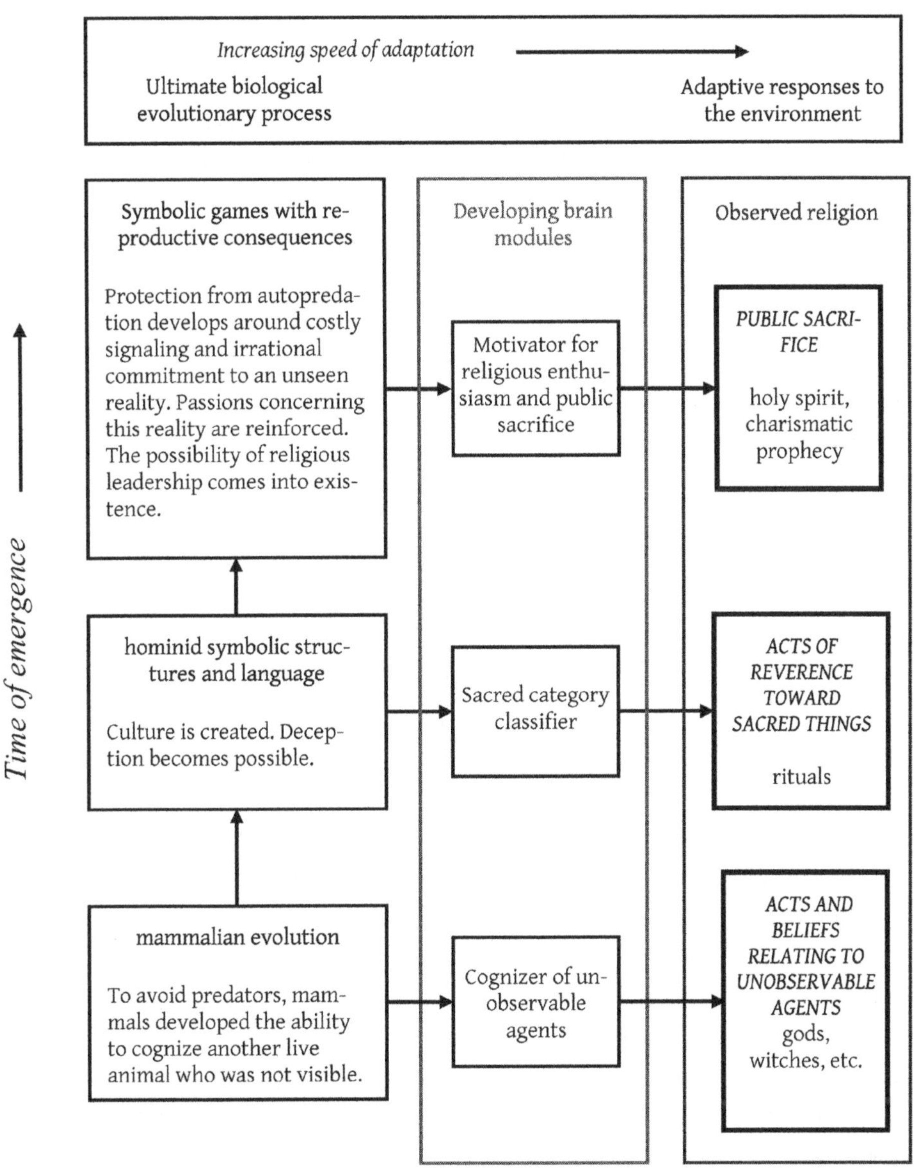

References Cited

Asad, Talal. 1983. "Anthropological Conceptions of Religion: Reflections on Geertz." Pp. 237-259 in *Man*, vol 18, nr. 2.

Atran, Scott. 2002. *In Gods We Trust: The Evolutionary Landscape of Religion.* New York: Oxford University Press.

Boyer, Pascal. 2001. *Religion Explained: The Evolutionary Origins of Religious Thought.* New York: Basic Books.

Darwin, Charles. 1859. *On the Origins of Species by Means of Natural Selection.* London: J. Murray.

Douglas, Mary. 1966. *Purity and Danger: An Analysis of Concepts of Pollution and Taboo.* Harmondsworth: Penguin.

Durkheim, Émil. 1963[1912]. *The Elementary Forms of the Religious Life.* Translated from the French by Joseph Ward Swain. New York: Collier Books.

Eberhard, Mary Jane West. 1975. "The Evolution of Social Behavior by Kin Selection". Pp. 1-33 in *The Quarterly Review of Biology*, vol. 50, nr. 1.

Eliade, Mircea. 1964[1951]. *Shamanism: Archaic Techniques of Ecstasy.* Princeton: Princeton University Press.

Frazer, James G. 1911. *The Golden Bough: A Study in Magic and Religion,* Third Edition. London: Macmillan.

Geertz, Clifford. 1993 [1966]. "Religion as a cultural system." Pp. 87-125 in Clifford Geertz, *The Interpretation of Cultures: Selected Essays.* London: Fontana Press.

Guthrie, Stewart. 1993. *Faces in the Clouds: A New Theory of Religion.* New York: Oxford University Press.

Hempel, Carl Gustav. 1966. *Philosophy of Natural Science.* Englewood Cliffs, N.J.: Prentice-Hall.

Hinde, Robert A. 1999. *Why Gods Persist: A Scientific Approach to Religion.* London and New York: Routledge.

Irons, William. 2001. "Religion as a Hard-to-fake Sign of Commitment". Pp. 292-309 in Randolph Nesse (ed.) *Evolution and the Capacity for Commitment.* New York: Russell Sage Foundation.

Kehoe, Alice Beck. 2000. *Shamans and Religion: an Anthropological Exploration in Critical Thinking.* Milwaukee: Waveland Press.

Kirkpatrick, Lee. 2005. *Attachment, Evolution, and the Psychology of Religion.* New York: The Guilford Press.

Malinowski, Bronislaw. 1948. Magic, Science, and Religion. Garden City, N. Y.: Doubleday.

Pinker, Stephen. 2002. *The Blank Slate: The Modern Denial of Human Nature.* New York: Viking Press.

Rappaport, Roy A. 1999. *Ritual and Religion in the Making of Humanity.* New York: Cambridge University Press.

Saler, Benson. 1993. *Conceptualizing Religion: Immanent Anthropologists, Transcendent Natives, and Unbounded Categories*. Leiden: E. J. Brill.

Saler, Benson. 1997. "E. B. Tylor and the Anthropology of Religion". in *Marburg Journal of Religion*. vol. 2, nr. 1.
[url= http://www.uni-marburg.de/religionswissenschaft/journal/mjr/saler.html]

Saler, Benson. 2004. "Towards a Realistic and Relevant 'Science of Religion'." Pp. 205-233 in *Method & Theory in the Study of Religion*, vol. 16. nr. 3.

Sosis, Richard and Candace Alcorta. 2003. "Signaling, Solidarity, and the Sacred: The Evolution of Religious Behavior". Pp. 264-274 in *Evolutionary Anthropology* vol. 12, nr. 6.

Stark, Rodney and Roger Finke. 2000. *Acts of Faith: Explaining the Human Side of Religion*. Berkeley: University of California Press.

Tooby, John and Leda Cosmides. 1992. "The Psychological Foundations of Culture". Pp. 19-136 in Jerome H. Barkow, Leda Cosmides, and John Tooby (eds.) *The Adapted Mind: Evolutionary Psychology and the Generation of Culture*. New York: Oxford University Press.

Tylor, Edward B. 1958[1871]. *Primitive Culture*. London: Murray.

Wallace, Anthony F. C. 1966. *Religion: An Anthropological View*. New York: Random House.

Wilson, Edward O. 1978. *On Human Nature*. New York: Bantam.

Wilson, Edward O. 1998. *Consilience: The Unity of Knowledge*. New York: Alfred A. Knopf.

[3]

ARE THERE ANY RELIGIONS? AN EVOLUTIONARY EXPLORATION[1]

Joseph Bulbulia

Common sense holds there are distinctive religions, an intuition that informs most scholarship and teaching in religious studies and the social sciences, but the intuition is somewhat misleading. In spite of apparent religious difference, recent psychological inquiry suggests that religion emerges from a single panhuman psychological design that strongly constrains variation. There is some variation in the religiosity of individuals and groups, but not the variation of "traditions". This paper uses recent research in the cognitive and evolutionary study of religion to explore some basic properties of the mental architecture that generates human religiosity, including features that enhance the illusion of religious difference.

Key words: behavioural ecology, biology, costly signalling, evolutionary psychology, game theory, healing, linguistics, prisoner's dilemma, morality, religion.

1. *Introduction*

The purpose of this paper is to examine recent advances in the evolutionary psychology of religion, and to explore how this research revises standard accounts of religion and religious variation.[2] For example, whereas ordinary language (and first year world religions courses) carve out particular doctrines and practices as belonging to distinctive religious kinds—Islam, Judaism, Christianity, Hinduism, California New Age, Satanism and so on—it seems these labels do not describe aspects of the natural world (they do not reflect the kinds of nature.) Instead, panhuman psychological architecture generates only limited, but strategically important, variation in the religiosity of individuals and groups. Religion is thus like language—for linguistic inquiry has revealed that

[1] Acknowledgements: Many thanks to Candace Alcorta and Rich Sosis for excellent comments on an earlier draft.

[2] By "religion" and its cognates I will mean, beliefs and practices relative to *supernatural* beings, places, and powers: "gods" "spirits" and "heavens" and the like. The theoretical motivation for this use of "supernatural" as a distinctive psychological kind comes from Boyer's work, discussed below (Boyer and Ramble 2001). Often I use "religiosity" to capture this intention. I am therefore interested in religion as a psychological phenomenon, a dimension of how many think and act.

variation in human languages is limited to superficial aspects, which conceal a universal biological design. I examine the similarities between language and religion in the first section below. I then describe some important aspects of the psychological design that equips us for religious thought and practice, and show how this design emerges from genotypic resources that produce religiosity within narrow parameters. Though cultural environments play a role in an agent's particular brand of religiosity, acquisition events do not explain the intricately structured understandings and motivations that emerge through religious cognition. I show how the psychological faculties that subtly prompt and guide religiosity are designed to promote biological success in the ancestral world. I conclude that in spite of striking apparent variation, core aspects of religiosity remain invariant. From the vantage point of cognitive architecture, it appears that there is only one human religion with minor but strategically important variation in its conventional expressions.

Let me first address the objection that my thesis—the idea that there is one human religion with variation at the margins—relies on an ambiguity in the similarity relation. Notice our judgment of similarity is always interest relative. Depending on our purposes, we may correctly judge that oranges and apples are both similar and different. Oranges are similar to apples when compared to black holes, but different when compared to tangerines. Black and white are distinct colours, yet both are colours, so in this respect similar. But this does not entail that there is "only one human colour with variation at the margins", or "essentially one human fruit". It seems plausible to say that religions are both similar and different depending on what we take to be the relevant comparison classes, and the practical specifications we use in judgement. Moreover it seems fruitful to discard the idea that "all religions are essentially the same" as too crude and reductive to have explanatory value. For it is unclear how we could ever gain by lumping the products of culture into gigantic categories that way. (Compare: "All games are the same;" "All marriages are the same" "All science is the same".)

I do not deny the legitimacy of using concepts of distinctive "religions" in making pragmatic sense of religious beliefs, institutions, and practices. One can truthfully (and helpfully) say, "Indonesia is largely Islamic" but not, "Indonesia is largely Hari-Krishna". The Taliban's religiosity differs from feminist Earth goddess religiosity; for example, the Taliban aren't feminists. Generalising: there are undeniable differences among agents in religious labelling and doctrinal, moral, and ritual conventions. And there are local differences in the intensity of religious commitments, and in the uses to which religious ideologies are put. Most scholars of religion labour amid these peculiarities of convention and circumstance, with illuminating results. I see nothing in the evolutionary psychology of religion that challenges this interest.

Yet there is new understanding to be gained from an evolutionary perspective that views religion as a panhuman capacity. We will see how the psychological architecture that supports religious commitment and practice is largely invariant across culture and era. And this architecture reliably produces functionally similar expressions and commitments, irrespective of labelling and convention. Though religiosity is always framed by local settings and histories, we will see that that "religion" is no more a "product of culture" than jealousy or friendship. We can talk meaningfully of the distinctive religiosity of Indonesians, the Taliban, and Earth worshipers while still appreciating that variation is strongly bound and directed by a common psychological design. The trail of this universal design runs over all religious products. The fact that this design is a common human possession that we share through biological endowment is a scientific discovery, not word play. I seek to explore aspects of this discovery below. As a way to understanding the issues, consider language.

2. *Apparent Linguistic Variation and Folk Externalism*

On the face of it, there appear to be numerous and different languages, "English," "Japanese," "Swahili," "Greek," "Hindi," and others. That they differ is made vivid when conversation partners do not share one: the tourist of China, lacking Chinese, falls on hard times. Even the same language varies over time. A speaker of Chaucer's English could not communicate with a contemporary speaker of London's English, though an unbroken chain of only twenty-odd generations of communicating English speakers separates them. 14,000 years from now no one will speak our English. Our distant successors will look on our speech as the Chinese look on the English-only tourist, incomprehensibly.

Pronunciations of words vary. Germans say, "rot" to mean "red," not intending the "rot" of "rotten wood". French speakers use "raison" to denote "correct" instead of that black dried fruit; they say "anniversaire" to mean "birthday," instead of "anniversary" and "blanc" to mean "white" not "black" [see (Harman 1998)].

Grammars also vary. English displays relatively stable word order regularity, with adjectives typically preceding nouns that they modify, and subjects usually preceding transitive verbs, and (though less frequently) intransitive verbs. Yet differences among languages in this respect seem quite extreme. Aboriginal speakers of the Australian language Warlpiri use case markers to convey grammatical relations and noun modification. In certain Native American languages, there are few noun phrases within clauses, and grammatical relations are expressed by attaching strings of agreement affixes onto verbs (Pinker and Bloom 1990). More basic differences include rules governing the designation of subject and object, either by word order, as in English and Japanese, or by case, as in German, Latin and Czechoslovakian (Pinker 1999)—a bane to those of us who study these languages as adults. Lots of apparent differences here.

We observe that children learn the language of those around them. An African born in Toronto utters English sentences ending with "eh?" instead of some African language. Language seems to be something children pick up from their surroundings.

Call *linguistic externalism* the view that explains linguistic competence as structured through local acquisition events. Many theoretical versions of linguistic externalism are possible. I'm interested in the version that most closely approximates our common sense view that language has something to do with acquiring an artefact called "language" from a community or culture. Folk linguistic externalism undertakes something like the following commitments:

1. Languages are public tools for communication shared by members of different linguistic communities.[3]
2. There are different languages.
3. These communal tools are acquired through learning. Roughly, something "external"—a language—is internalised by individuals in the course of development through cultural exposure.[4]

Each of these assumptions turns out to be false. Noticing why will help us to see where folk externalism about religion goes wrong.

[3] For example "French" is the tool used by communities in France, Corsica, Quebec, parts of Switzerland, West Africa and elsewhere (expensive restaurants, certain Cajun precincts, among poseurs, and so forth).

[4] For an example of this view, see (Dummett 1986).

3. *I-language*

Consider linguistic difference. While pronunciations vary, grammatical variation is tightly constrained. It is not the case that grammatical differences of the kind that distinguish English from French or Cherokee are rigid. English speakers can invert subject-predicate word orders, for example: "The instructor was driven to drink by the student's passive sentences". In fact, English speakers do sometimes employ case markers, for example, "'s" for possession as in: "the student's passive sentences". We can produce ergative constructions, replacing "the bottle broke" for "I broke the bottle". Moreover, there are converse orderings of English-like constructs in apparently grammatically distinct languages [For discussion see (Pinker and Bloom 1990)].

Focusing on the grammatical component more closely we find invariant principles of sentence formation children never learn. Take a descriptive generalization called the coordinate structure constraint. This constraint exposes fundamental differences in structure between sentences of the following form:

1.) Mary saw Peter with Paul.
2.) Mary saw Peter and Paul.

In English, questions can be formed by inserting a question word at the head of the sentence, followed by an auxiliary verb. Hence:

1.) Whom did Mary see Peter with?

However this transformation doesn't work when the question word is conjoined with another noun phrase, as in:

2.) Whom did Mary see Peter and?

Children never explicitly learn the coordinate structure constraint, nor do they ever say, "Whom did Mary see Peter and?" because the constraint is hard-wired.

Internal knowledge extends beyond the construction of sentences to the meanings of words (as opposed to their pronunciations). Word meanings possess extremely complicated relational properties that we never learn. Chomsky illustrates this point through the following examples. Take "house" in the sentence "Peter is near the house". Notice the implication is that Peter is outside, not standing near the inside wall. So it is with "car," "airplane," or even an impossible object like "rectangular sphere": we assume the same for "Peter is standing near the rectangular sphere" (i.e. near the *exterior* surface). Similarly, when we say "Peter painted the house red" the default assumption is that he did something to the outside, not the inside. If Peter cleans the house, however, the default assumption is that he rearranges objects on the inside. We conceive of "house" therefore as an exterior surface with internal spaces, both of which have complex properties. Chomsky notes that "home" has different implicit properties. If I have shifted my house from New York to Moscow, I have moved a massive wooden object. Notice I convey a different understanding when I say that I have shifted my "home" from New York to Moscow. "Home" has both a concrete and abstract aspect. Exploring "home" further we note its abstract properties differ from those of "book," which is also concrete and abstract but in different ways. You and I can simultaneously read the same book even if we live at opposite ends of the planet, but you and I cannot simultaneously live in the same house or home at opposite ends of the planet (Chomsky 2000: 31-37, 62-66).[5]

Not just grammar, then, but substantial semantic components of the psychological systems that produce language remain invariant. Strikingly, it appears that we all think from the same mental dictionary, with variation limited largely to conventions of pronunciation, how we say "black" "home" "Moscow" and other words.

Linguistic externalism is committed to the view that we learn these intricate meanings and grammatical rules through acquisition events, but this is implausible because competence emerges from severely limited exposure, a "poverty of stimulus". Acquiring the massive and largely tacit knowledge required for linguistic understanding and generalisation is computationally intractable within the time of childhood development. Children are simply never exposed to all that they know.

Rather than thinking of language as an acquired artefact, linguists think of language as a "mental organ", which develops along a more or less fixed schedule, in response to environmental inputs but whose intricate structure cannot be explained meaningfully in virtue of those inputs (Pinker 1994).[6] We assume as much with ordinary organs (the spleen, liver, eardrums . . .) whose organic design and development cannot be meaningfully explained in terms of "food".

[5] See also (Bigrami 1992).

[6] By "mental organs" cognitive psychologists do not mean bits of brain matter. Rather, they look at the brain as an integrated architecture of information processing modules—roughly a set of programs that supervene on the material brain, much as computer programs supervene electrical hardware. "Mental organ" describes these programs, not the underlying neurology that enables them. The classic "levels" discussion is set forth in (Marr 1980).

Clearly, it would mislead the English-speaking tourist to say: "Don't worry about China—you speak the same language". Yet we can understand the differences without appeal to artefacts called "languages". The differences adhere to the slight variation in language faculties or "language organs" of individual speakers (linguists call these "I-languages"). I say to you, "Bring your translation book". You understand these words because my I-language specifies the linguistic (phonetic, semantic, structural) properties of the sentences that you hear. The state of your I-language is similar to mine because you have been exposed to a community of speakers of a related linguistic heritage. This enables you to select an appropriate analogue with which to interpret my utterances with sufficient precision for understanding. Individuals unlike us whose I-languages develop in different environments may not be able to select an appropriate analogue if the state of their I-language differs too much (again variation in I-languages is limited pronunciations and a handful of grammatical rules).

Notice that we can account for the tourist's failure to communicate in China, or for the differences between Italian and German courses entirely on these terms, without appeal to public entities or artefact, which individuals grasp partially through "learning" aspects of them. Social exposure explains some conventional aspects of language, but the skill is structured through biological endowment.

4. *Religious Internalism*

I discuss language at length because it is a well-studied psychological phenomenon, and the results of that study cast serious doubt on the assumptions of linguistic externalism.[7] However, it is possible to accept the internalist view of language while remaining an externalist about other aspects of cognition. Take our knowledge of quantum field theory. Here complex understandings are not structured by "mental organs"

[7] It would be interesting to study the interface between the language faculty and the systems that control supernatural cognition. For at a minimum, we use language to express and understand supernatural commitments and practices. Thus, there must be legibility conditions at the interface between the systems that control each domain. I do not seek to explore these conditions in any detail here. Nor do I maintain that the faculties governing religious thought and development resemble those of the language faculty in anything but superficial ways. Such is true of ordinary organs: the liver's development and design differs in critical respects from that of teeth or sexual organs. Hence, it would be striking were the developmental parallels between language and religion to prove substantial. [For a contrasting stance see (Lawson and McCauley 1990)].

(at least not straightforwardly). Rather, knowledge hinges on scientific discovery and active transmission.

Clearly externalist projects are relevant to the explanation of many facets of human skill and intelligence. As Sterelny notes, "[c]hildren resemble their parents because of a flow of genes. But they also resemble their parents because of a flow of information" (Sterelny 2003). We are not born with blueprints for all we will know and will become: part of us is made by the culture we are born to. Cultural inheritance plays an important causal role in phenotypic expression. For this reason cultural inheritance may also play an adaptive role: for example it enables successive generations to accumulate adaptive knowledge over time. Cultural know-how—how to make a pot, how to construct a canoe, where to find water in the dessert, how to detoxify seeds—may be vital to success (Henrich and McElreath 2003). Moreover, such knowledge clearly varies between cultural lineages. Consider war technologies. The history of colonialism makes no sense without recognizing some human societies have been better equipped to savage others. Reading and writing ratchet civilization forward, because the technology enables the accurate storage and transmission of cultural know-how. Yet these abilities come too late to have been targeted by selection (Dehaene 2003). Moreover, once concocted, culturally born technologies and artefacts may alter selective landscapes. We do not merely respond to our environments, we transform them as "ecological engineers" (Odling-Smee, Laland et al. 2003). The products of culture frequently have important fitness affects on those who come after. The invention of penicillin makes the world safer for surgery, and both technologies improve the fitness of agents prone to acute appendicitis, and their descendants, altering the frequency of the relevant alleles down stream.

Reflecting on the importance of cultural endowment to all of human life, it may seem as if religion's explanation demands an approach that take us beyond genetic endowment, for religion seems to be a product of culture par excellence. There was no Buddhism before the Buddha invented it. The religion did not spring to the young Richard Gere's mind because it was prefigured in his parent's gonads.[8]

Call *religious externalism* the plausible view that religion is an aspect or product of culture. Roughly, the view asserts that:

[8] Notice, neither did Gere's English, so we must take care in externalist generalizations.

1) Religion is substantially a cultural invention or technology for living and understanding.
2) There are different religions
3) Children (and adult converts) acquire their religion from others during the course of their education or social experience.

I will show that recent naturalist inquiry casts doubt on each of these three propositions. In contrast to religious externalism, call *religious internalism*: the view that:

1) Religion is substantially structured by genetic resources (the architecture of religious cognition is not an invented technology).
2) Religious variation is minimal.
3) Children are endowed with religion, which "grows" in certain cultural environments, because it is a feature of panhuman psychological design.

Now I need to describe the sort of evidence that would count in favour of an internalist approach to religion.

Begin with following relatively uncontroversial view about nature and function of cognition.[9] Minded organisms are equipped to process information in ways that enable them to flourish. They gather, store, and access information about the external world and their various internal states, which they exploit to respond and behave to enhance biological success. ["Behave" in the widest sense: on this view, an immunological response or dying old (lifespan) is behaviour.] The functional organisation of any cognitive agent is in part specified by species-shared properties of its genotype. Thus, honeybees are equipped to communicate information about the location of nutritional resources through the subtle staging and monitoring of a figure eight dance. They furthermore possess cognitive powers to plot a course from hive to the (dance-referenced) nutritional supply, exploiting environmental cues like sun angle or (on cloudy days) light polarization and other local markers (Gould 2002). By all accounts these capacities are structured by genetic inheritance. You cannot teach these skills to drosophilae. Clearly environment plays a roll, even here. For local informational resources matter to an agent's thought. The particular angle of the sun during a particular honeybee's flight enables it to determine its course. Moreover, structural features may emerge through type-specific environmental interactions over time. Male zebra finches placed in song impoverished settings

[9] There are controversies in the details of this view, but not about its basic outlines.

never acquire the capacity to sing. These bird capacities require external stimuli. The linguistic inputs of children raised in Nepal influence their individual (token) I-languages, which do not stabilize until late adolescence. Their I-languages will differ somewhat from those of children raised in Los Angeles. In these cases, however, though structured capacities key to environmental features, structure comes from genetic resources. Zebra finches raised in L.A. will never speak like those valley kids.

The key to assessing the internalist stance on religion lies in understanding the relative importance that environment plays in scaffolding religious cognition. Language and bee-navigation (exemplars of internal design) illustrate the salience of genetic resources to cognitive structuring. We may say that such structuring is strongly *canalized.* Canalized design features are functionally organized phenotypic expressions whose end-states remain stable under a range of normal environmental conditions, ("normal" here defined in virtue of the environment of evolutionary adaptation, the EEA) (Ariew 1999). Selection targets a strongly canalized psychology because it enables agents to rapidly solve adaptive problems by narrowing their search space (Cosmides and Tooby 1992).[10]

To the extent that selection has targeted open-ended learning capacities in our species, it has done so to enable storage and transmission of abilities and knowledge relevant to understanding and navigating the world. Culture is a warehouse of good tricks, and the reality checks its stock. Yet we need not appeal to the reality of the gods to explain why sensible and rational agents become so passionately committed to them. Religious thought has not evolved to track and engage a supernatural world. Far from a warehouse, religiosity produces castles in the air. Variation in the configuration of religious doctrines, practices, and moral rules make it appear that religious understandings vary widely. Yet because these illusions are designed to solve central and invariant problems of human life, variation must be limited. Not just any religious understanding or act will clear adaptive hurdles. Religion engages fictions, but fictions with tremendous human weight and adaptive importance.[11]

[10] Two points to bear in mind. First, structure does not imply determinism: canalized structure may specify a range of behavior outputs, given one or several informational cues and triggers, none of which is exclusively determined in advance by genotype. Detecting a predator an organism may fight, flee, monitor, go about business as usual but under a higher state of alert, none of which is automatically determined by the predator cue (Sterelny 2003). Second, canalization clearly admits of degree.

[11] Nothing I say here should be taken as a wholesale rejection of the religious

We shall see that selection has canalized religious psychology to foster rapid solutions to two recurrent adaptive problems: getting along with others, and getting along with ourselves.

I suggest that the prospects for internalism in the study of religion depends on the degree of religiosity's canalization. For religiosity to be strongly canalized:

(1) Religious cognition must be structured, such that specific information is routinely acquired, stored, organised, and accessed to produce specific cognitive and behavioural outputs.
(2) This structuring must remain stable over a variety of "normal" environmental conditions, that is, conditions closely approximating the EEA (Environment of Evolutionary Adaptation).
(3) We cannot usefully explain structuring by environmental or cultural resources (for example, because there is a poverty of stimulus, or because the relevant computational problems are too complicated to solve by "learning".)

In the next section, I review research suggesting that core aspects of religiosity are highly structured, that these structures remains stable over a variety of normal environmental conditions—including a variety of cultural conditions—that structure appears to emerge from a poverty of stimulus and on a developmental schedule. Religiosity, like language, flows from a species general biological design.

5. *The Internal Architecture of Religious Cognition: Cognitive Psychology*

Numerous research programmes in cognitive psychology have converged to the view that religious concepts heavily rely on implicit folk psychological understandings [for reviews see (Barrett 2000; Atran 2002; Boyer 2003).] These folk psychological understandings govern how we think about physics and biological kinds and other minds. They are highly structured, constant, and surface from a severe poverty of stimulus. For example, before the age of one, children are aware that objects

---

dimension of human nature. At the core of most loves (and hatreds) lies the worm of bias and distortion, but one that spins the fabric of our entire existence. Some religious outlooks and practices are morally vicious, but not all (probably not most) and in my view we should only reject the vicious. We should no more discard good religion than we should good parenting or good marriages, which frequently rely on the illusion that our own children or spouses are profoundly special (they are not).

have certain stable properties (e.g. number, mass, substance) without ever explicitly learning this information. By early childhood, children are aware that human and biological kinds have properties that differ from physical objects (Spelke, Phillips et al. 1995). They know that organisms require nutrition, have life spans, and that species properties hold in spite of appearances—for example children know that a cow in horse costume is still a cow, though a table chopped up and made into a chair becomes a chair (Gelman and Markman 1987). Intuitive folk psychology, like language, is largely invariant across cultures (Petrovich forthcoming). Cognitive psychologists maintain that religious concepts are strongly determined by these folk intuitions, which guide our understandings and inferences about the supernatural (Boyer 1992; Boyer 1994; Boyer 1996; Atran 2004).

Consider a common supernatural concept: "ghost". Probing this concept, we see how heavily it relies on the tacit folk psychological understandings that organise our interpretations of "person". For example, a ghost will be interpreted to have psychological states—beliefs, desires, memories, attitudes, intentions and plans. No one ever teaches a child that a "ghost" may possess complicated intentions, beliefs, desires, and plans: we just assume this. Some aspects of folk physics and biology are also retained; it is easy to imagine that a ghost can communicate or can be seen at a distance or can move through physical space—perhaps threateningly towards you with its ghost fangs exposed. We think to these unreflective interpretations without ever noting the oddity of inference to physical properties from entities imaged as immaterial. We do not pause to consider how immaterial beings can think with no brains—though we would pause at the idea that a chunk of dirt or atmosphere could think without brains. We assume without inquiry the relevant causal connections. Nor do we do generally wonder how an immaterial spirit tooth—no matter how long and pointy—may injure (for what exactly is long and pointy?). Moreover some understandings are more salient. We think to the view that a ghost is capable of moving through walls more readily than that it may move through the centre of the sun, or a refrigerator. It knows where you are hiding is more salient that it knows you enjoy burritos. We do not learn much beyond a few encyclopaedic details of ghosts: instead folk psychological understandings do the thinking for us, generating rich inferences that allow us to represent and communicate about ghouls (Boyer 2001). Such is true of all supernatural concepts, which rely on the conceptual understandings of the natural concepts they resemble (Boyer 1998; Boyer and Ramble 2001).

Further, though supernatural concepts vary—some rely on intuitions governing persons, others places, and others, powers and forces—they always explicitly violate a few, but not many of these tacit assumptions. Ghosts may be able to move, but unlike us they can move through walls, which is startling. God may be like a man, but unlike a man he has the power to read minds, or create a universe, or punish and reward you for all you've done, also startling. Hence, any individual's religious concepts will retain much of the structure of intuitive psychology. Yet they will also contain a few violations of that implicit knowledge, and this makes religious concepts memorable (Boyer 1994). Boyer hypothesizes: it is precisely *because* religious concepts violate implicit assumptions for the relevant kinds that religious concepts are interesting and memorable, and thus easily spread in populations. Critically, the violations, though arresting, must be minimal, because religious concepts that are too complex (violate too many expectations) become too difficult to recall (Boyer 2000; Boyer 2001; Atran 2004). For each religious person grows a religiosity that is unique.

The constraints imposed by intuitive psychology on the articulation and scope of religious concepts are, in turns out, extreme, and lead to inferences that, for example, fly in the face of philosophical theologies. Barrett and Keil observe that in spite of subscribing to anti-anthropomorphic theologies, religious persons in India and the United States imagine deities as ordinary persons, different in minimal (but arresting) respects from ordinary persons. Devotees notionally understood and subscribe to theological positions that depict the gods as unlike persons, yet their interpretations belied a powerful anthropomorphising tendency (Barrett and Keil 1998). This tendency holds irrespective of culture and epoch because the intuitive psychology of persons is largely invariant (Knight, Sousa et al. forthcoming).

Thus, on the cognitivist view, religious cognition materializes from a mind already guided by expectations about the natural world it will encounter. Religious concepts violate these intuitive expectations in minimal ways, eliciting our "passion for surprise and wonder" to adopt Hume's idiom (Hume 1993).[12] The panhuman mental architecture of folk psychology heavily structures and constrains religiosity. And the methods appropriate to the study of religious cognition, are just the methods appropriate to the study of the cognitive systems that underwrite

[12] For reviews of constraints on ritual actions see: (Lawson and McCauley 1990; Whitehouse 2000; McCauley and Lawson 2002).

it [for recent reviews see (Barrett 2000; Andresen 2001; Atran 2002; Boyer 2003)].

6. *Religion as a Biological Adaptation: The Adaptationist Literature*

Most cognitive psychologists view religiosity as an after-effect of structured cognition, as cognitive noise. Here the constraints are strong enough on religious cognition to reconsider the role of cultural inheritance mechanisms or "tradition" in composing religious thought (Boyer 1990). Though we acquire a few aspects of our religion from those around us, a panhuman cognitive design configures and guides religious thought.

Adaptationists maintain that the constraints on religious cognition are even more extreme, for religious thought is the product of *dedicated* cognitive architecture. Adaptationists accept that cognitive methodologies illuminate aspects of religious cognition. But they further hold that we can deepen that understanding by approaching religiosity as functionally organised to promote biological success. For if religiosity has been targeted by selection, we can reverse engineer its design by contemplating the adaptive problems religious thought is designed to solve. Moreover, if dedicated cognitive architecture prompts religiosity, then it is likely even more internally organised and constrained than most cognitive psychologists suspect. For biological success frequently demands exquisite precision and speed, which open-ended problem solving architectures simply cannot deliver (Marr 1980; Cosmides and Tooby 1992).

Recall that cognition fuels success by tracking and responding to the world. Yet I have suggested that 1) religious cognition involves beliefs and practices relative to supernatural beings and 2) we need not suppose these supernatural beings exist. Thus on my minimal naturalistic assumptions, religiosity need not track and register features of a supernatural world, for cognitive agents may come to believe in such worlds though nature is secular. This presents a problem. If religion is an adaptation, then how is responding to things that do not exist helpful? Given religious costs, it seems selection should instead have culled religious tendencies. However adaptationists have demonstrated that religious errors and costs, if systematic, can help in at least two ways: by forging reliable social alliances and by fostering heath-inducing outlooks and practices. I consider these in turn.

7. *Religion and Reciprocity*

One of the keys to biological success in our lineage has been the ability of individuals to forge large and stable alliances among non-kin (Dunbar 1998). In anchoring individual fates to collective fates, and in co-ordinating activity to achieve common ends, individuals are able to promote their interests more effectively than by taking on the world alone. Yet such alliances are notoriously unstable, subject to invasion by agents willing to exploit these advantages without contributing. Social exchange often pays better than it costs. But it is fragile, because cheating social life often pays better still (Skyrms 1996). A society of cheats is no society but a confederation of individuals. Adaptationists urge that a core function of religiosity is to help to make social life possible. For religion fosters co-operation by enabling individuals to reliably predict and secure social exchange where there are rational incentives to defect from cooperation (Cronk 1994; Irons 1996; Sosis 2000; Irons 2001; Atran 2002; Sosis 2003; Sosis and Alcorta 2003; Bulbulia 2004). Religious cognition leads to inaccurate beliefs in gods, but religious expressions accurately lay bare motivational states in a way that enables us to forecast how others will act, and plan. Through religion we do not track gods, we track each other. And the accuracy the system affords solves co-operations problems. Thus, religion functions as a *commitment device*—a mechanism that binds us together—as the Latin root of the word "religare" ("to bind") suggests.

To understand the issues, consider an exchange among a pair of unrelated individuals that is modelled as a prisoner's dilemma. Assume that P is the fitness payoff of non-co-operative individual action (going at it alone). Call R, the reward for co-operative action (give and take). Let T = the temptation to cheat co-operation (to take but not give), and let S = the payoff of cooperating when an exchange partner cheats (to give but not take). A large class of interactions will conform to the following payoff matrix:

Natural payoff matrix: T (cheating) > R (reciprocity) > P (solitary action) > S (getting cheated)

Because T > R, it pays better to defect rather than co-operate. And because P > S it pays better to defect if the other agent defects. Thus defection strongly dominates co-operation because it pays better regardless of what the other agent does. Defection is what economists call the "Nash equilibrium" for this interaction (Nash 1951; Schelling 1960). Yet observe that co-operation is "strictly efficient". If both agents opt for fair exchange, both are better off than by going at the world alone (Skyrms 1996).

Notice that specifically structured religious understandings may solve co-operative problems by altering the expected returns of co-operative action. Whatever the actual payoff matrix, if agents believe that co-operation always pays better than defection, they will be rationally motivated to exchange. Suppose the reward for a cooperative response is eternal bliss and the punishment for defection is reincarnation as a sweat gland. Such belief is at odds with nature, for it inaccurately depicts probable outcomes. The relevant causal processes are supernatural. Yet precisely because cognition errs in just this way, it motivates cooperative exchange.

Perceived supernatural payoff matrix: R > P > S > T

Selection may equip agents with dispositions to form such moralizing illusions, though crucially, only where it is possible to reliably recognize the presence of these illusions in others. The *recognition constraint* is critical to the evolution of religious cognition. For religion to foster exchange, agents must be able to reliably discern religious commitments in their partners. Yet it is always in a defector's interest to state a religious commitment, only to defect when conditions allow. Defectors will have incentives to pollute the epistemic environment of religious co-operators with signals of authentic religious commitment, only to defect when the opportunity presents itself. Hence religionists must have a reliable safeguard from pretend co-religionists. Finding a safeguard is difficult, given the advantages to bypassing it.

Behavioural ecologists have shown that selection may outfit organisms with signalling and detection systems capable of producing and assessing co-operative commitment or worth. Frequently, such systems rely on costly expressions or "handicaps" that target the relevant property. An authentic signal will be arrayed so that only a signaller that genuinely possesses the relevant property (in this case, the presence of a motivating religious belief) will be able to afford producing the signal. Such costly signals of worth are superabundant in nature (Zahavi 1980). For example the songs of male bulbuls *(Pycnonotidae)* are carefully arranged to signal strength and reproductive fitness, and so are capable of warding off potential rivals and of attracting mates. Only fit birds can manage such ornamental song. Similarly, the whistling of skylarks when chased by predatory merlins is a costly signal of aerobic fitness, one that generally forestalls the chase. Skylarks able to sing while avoiding beak and talon are unlikely to be caught, so merlins don't generally chase them (Zahavi and Zahavi 1997). The vigorous leaping or "stotting" of Springbok gazelle authenticates their ability to outrun predatorial lions, also forestalling a pointless chase (FitzGibbon and Fanshawe 1988). The costumes and ornaments of real estate agents are configured to signal success, etc.

In our lineage, costly religious signalling has evolved as a means to authenticating the relevant religious beliefs. First, *emotional displays* of religious commitment—the thralls of devotion and piety—accurately convey a religious lifeway. It is notoriously difficult to fake emotions because they are not under conscious control. We can easily lie with words but not with our faces (Frank 1988; Ramachandran and Blakeslee 1998). Religious emotions express specific altruistic motivations and religious beliefs. That is, emotional signals communicate a *behavioural trajectory*. Weeping in prayer before LORD FOZ suggests commitment to a supernatural scheme in which LORD FOZ relates to our lives. Doing the same before LORD CHOP suggests otherwise. Moreover such displays modulate in ways that enable an audience to discern strength of commitment. Religious emotions measure a commitment's *intensity*. We can thus discern the strength of a religious commitment from the power of an emotional display. And this information can be factored into an assessment of future interactions. As defection incentives rise, the strength of an emotion sufficient to secure exchange will also rise. Knowing the bounty, Bin Laden will search the religious expressions of his cohort [For discussion see (Sosis 2003; Sosis and Alcorta 2003; Bulbulia 2004; Sosis 2004)].

Second, *religious ritual* provides a forum in which religious emotions, and other costly signals, can be produced, evaluated and inculcated. Though private rituals may function to reaffirm religious commitment, such displays are wasted signals.[13] Sosis and colleagues have shown that ritual participation is an excellent predictor of altruistic exchange with

[13] Sosis argues that private rituals "appear to be critical in keeping out potential free-riders by raising the overall costs of ritual performance. Since a skeptic will not perform private acts of devotion (because they are not observed) the overall costs of ritual behaviors imposed on a community must be significantly higher to keep out skeptics than if a community did not impose private demands on their members" (per. comm.) See also (Sosis 2003). Maybe so, but if a community has access to information about "private" ritual, then such rituals are not private.

non-kin (Sosis and Ruffle 2003; Sosis and Ruffle 2004). Religious rituals also correlates positively with communal longevity (Sosis 2000; Sosis and Bressler 2003). As with emotional display, ritual displays communicate concrete intentions and plans, enabling audiences to plot a *behavioural trajectory*. Sacrificing a lamb in the ritual of FOZ conveys a different commitment in the ritual of CHOP. Further ritual costs modulate to assess the *intensity* of commitment. As the relative expense of a ritual display rises so too does the relative degree of commitment it assesses (Sosis 2003; Bulbulia 2004). The strongly committed will be willing to spend more on ritual offerings, because they expect future supernatural returns to repay the expense. For religious agents perceive rituals as investments repaid by the gods. And because defectors perceive rituals as resource waste, they have disincentives to opt out of these costs. Moreover because they elicit powerful religious emotions, dramatic religious rituals provide experiential evidence for the supernatural understandings that bind a group together. Thus religious rituals not only afford opportunities to assess religious conviction, they instil them (Sosis 2003; Alcorta and Sosis under review).

The adaptationist approach to costly religious signalling leads to a few surprising outcomes. When facing crisis, religious agents will tend to devote more time and material resources to ritual participation than when secure. This prediction that has recently received empirical support, see (Chen 2004). Such outlays may seem puzzling. With adversity we would ordinarily predict tendencies to conserve. Yet from an adaptationist perspective the importance of group resources to individual well-being escalate during crisis. And so too do defection incentives. Hence for co-operation to be reliable, commitment costs need to be amplified. [For overviews of the adaptationist literature see (Sosis and Alcorta 2003; Bulbulia in press)].

8. *Religion and Internal Co-ordination*

"The space between skins is a no-man's land, controlled by no-one; designed for no-one; littered with the detritus of biological struggle" (Sterelny forthcoming). Religiosity has evolved to tame aspects of that external world, fostering exchange by restraining intra-group competition and promoting self-sacrifice. Yet religion may also do work within our skins. For though not contested in the same sense, an agent's internal psychological states need to be integrated and co-ordinated for it to flourish. We must make peace not just with the world, but also with ourselves. And religiosity may mend our minds as well, in its function as a *mental harmonizer*.

Making only conservative assumptions about ancestral conditions, we can appreciate how supernatural commitments foster psychological and physical well beyond reciprocity. Consider a certain idealisation, which I will call "traumatic world": agents regularly face situations where natural probabilities, if assessed accurately, paint bleak reproductive outlooks (infection and illness, childhood mortality and other family death, loss of body parts and disfigurement, dental issues, predator culling, shipwreck, famine and war, etc.) It is reasonable to assume that the ancestral world was to a substantial degree a traumatic world. Suppose further, that accurate estimations of the traumatic world have damaging side effects, which come by way of various psychological and physical stressors. For example understanding one's odds of surviving a shipwreck may support inferences to the view: "I am almost certainly going to drown". Given the rest of our psychological make-up, such an inference may present an agent with overwhelming grief, anxiety, or fear. And we know that stress is physiologically and psychologically damaging. (It is hard to maintain a state of wellbeing with a knife held to one's neck) [for and overview see (Kompier and Cooper 1999).] Averting this damage requires a refinement of response to the traumatic world. Optimising our responses, selection will restrain and dampen the effects of bad news. But it is reasonable to assume that there are few direct evolutionary pathways to dampening. If our loves drive us, then grief and anxiety at love's loss may be fairly inexorable. For a motivation easily overturned is hardly a motivation.

Yet specific varieties of religiosity may facilitate solutions, through the inaccurate conceptions of the world supernatural understandings afford. Directed supernatural distortions and biases may synchronise internal psychological states and physiological responses, providing religious agents with integrated, meaningful, and broadly optimistic understandings of the world and their prospects in it. Atran and colleagues have shown that supernatural understandings become salient to individuals confronted with stories of death. Though not an adaptationist, Atran suggests that "emotionally eruptive existential anxieties motivate supernatural beliefs" (Atran 2004). Supernaturalisms seem to help us to endure the foxholes of life.

Clearly, misconstruing the traumatic world as not hopelessly bad may reduce stress, fear, and malaise. Here salubrious religious illusions shield agents from the slings and arrows of existence by altering damaging (though accurate) assessments of the world. In this religiosity not only binds us to each other, but also to ourselves. [For a comprehensive surveys of the religion and health literature see (Pargament 1997; George, Ellison et al. 2002)]. Though the psycho-causal connections that facilitate health through religion remain somewhat unclear (Shapiro and Shapiro 1997), assuming these connections, we can compare the fitness effects of accurate judgement to inaccurate optimistic judgment, and consider the downstream biological consequences of each.

Accuracy v. supernatural illusion

Perception	Response	Outcome
Traumatic World	Stress, Dread, Fear, Malaise, Anxiety, Despair	Poor health
Enchanted World	Optimism	Good health

Given these benefits, it is easy to see how religious agents may slog up adaptive hills through very specific supernatural understandings of their prospects: belief in gods that make the world a better place.[14]

However so, religious distortions will need to be highly encapsulated. In most cases, there is nothing more lethal to an organism than misunderstanding its actual circumstances and prospects. Selection may work with these illusions, but in relative isolation from survival. Religion is like a surgeon's scalpel: effective only when precisely guided, lethal otherwise. Lost at sea, we may believe that the gods will save us—and such inspiring conviction may benefit us—but we must still swim.[15] We may believe the gods protect our children, our genetic legacy hinges on restraining our children from the jungle. We may believe that God is just and all-powerful, but to survive we must still visit harm to infidels who threaten us.

Researchers have only recently begun to examine the fitness effects of religious illusions outside of social exchange, so further adaptive functionality beyond co-operative domains remains somewhat conjectural. Yet if supernatural illusions and motivations sustain hope and meaning,

[14] Here again we need not assume supernatural reality to explain commitment to supernatural reality.

[15] This idealisation trades on the beneficial effects of optimistic belief. But clearly selection may favor other forms of misconstrual, even pessimistic outlooks, where these on average enhance life and reproduction. Though again, whatever the error vector, it will need to be encapsulated, lest the remedy turn to poison.

provide comfort and solace (Atran 2002; Atran 2004), and even facilitate healing (McClenon 2002), and is encapsulated (Bulbulia 2004) (plausible assumptions) then it seems selection will amplify and entrench supernatural dispositions.

9. *Adaptationist Accounts of Dedicated Religious Cognitive Architecture*

The story of evolutionary research on religiosity over the past decade is a story of increasing awareness of how biological constraint produces exquisite refinement in panhuman psychological design. Religiosity furnishes mental gear for grappling with the world. To facilitate success religious cognition must be delicately structured. Reverse engineering religions has produced the following crude picture of the common architectural features of the religious mind.

1) *Conviction Generator.* The cognitive architecture that employs religious concepts produces *ontological commitment* to non-existent agencies and powers. Such commitments distort expected outcomes, to motive predictable responses and behaviours. Local culture environments supply names to these agencies and powers, along with some encyclopaedic information. But the functions these commitments perform remain stable. Notice, the concepts "Dracula" and "Bugs Bunny" violate intuitive expectations governing the relevant domains (Slavic noblemen and rabbits, respectively). And as cognitive psychologists observe, these concepts produce memory effects. Observe however that few adults believe these concepts apply to reality. Yet gods and supernatural powers frequently become the objects of sincere and deep conviction, expressed as certainty. Such convictions differ from a mere memory effects [for discussion see (Atran 2002).] Our minds appear designed to respond to our cultural environments by developing the relevant supernatural commitments. We are predisposed to embrace the spirituality of our social worlds (Bering in press).

2) *Experience.* Religiosity involves inferences from a purely secular world to a robust supernatural world. To accomplish this task, there must be an active biasing and distorting of information to produce and reinforce these inferences. A psychological design that produces supernatural commitment will contain structures that modulate perception to produce supernatural belief-affirming experiences. Psychologists observe that there is the tendency to experience the world as animated by supernatural beings and powers and to interpret random events as meaningful (Heider and Simmel 1944; Guthrie 1993; Scholl and Tremoulet

2000; Guthrie 2001; Bering 2003). Such tendencies are well explained in light of a dedicated religious experience function: a mechanism that processes the world as containing supernatural elements.

3) *Love and condemnation: strategic response plasticity.* Supernatural concepts are configured to support co-operation in prisoner's dilemmas. Such concepts are thus further constrained to possess a moral aspect. The gods and godly powers take an interest in the morality and piety of mortals, to harm or reward agents relative to their behaviour. In this way, social action is related to "just" desserts through supernatural policing. I place "just" in shutter quotes to denote an inherent relativity in religious morality, for the morality of these concepts tends to be localized to the relevant exchange group (and to the gods of the relevant exchange group). For if the gods were to sanction exchange with defectors, then selection would quickly replace religious inclinations with a cold economic calculus. Religious morality is arbitrary but not random (Wilson 2002). As a corollary to intra-communal policing, god concepts may be arrayed to favour inter-group asociality or anti-sociality. With respect to those outside the community, moralizing gods may well enjoin indifference, or where these outsiders compete for resources, condemnation or violence. Notice, an optimal design would enable agents to adjust strategies to local situations. What may benefit agents in one setting may prove disastrous in another. Thinking of Lord CHOP'S absolute justice will not provide the inspiration helpful to the stowaway (unjustly, by her lights) stranded at sea. Such a Caruso needs to focus on CHOP'S loving power to assist the faithful in need. So the expectation is for response plasticity targeted to strategic settings. Notice however that the relevant information to decide how to act is often not available, or obscure. "Do we convert, or ignore or castigate these cannibals?" With respect to intra-group settings, we can predict that gods will generally enforce pro-social morality. However the epistemological waters are muddied with respect to out-groups. Where conversion clearly builds stronger coalitions the expectation is for god concepts enjoining missions to out-groups. Where out-groups are perceived to pose genuine threats, adaptationists predict religiously sanctioned caution or violence.

4) *Illusions of difference.* The similarity underlying language is not obvious largely because pronunciations vary. With respect to religious similarity, it may be strategically useful for consciousness to actively enhance the illusion of difference with moralistic overtones. Where religious competition is acute, an optimal psychological architecture will minimize and distort religious similarity to favour strong religious branding. For

holding that all religions are essentially the same stands in tension with the strategy, central to moralizing religion, that one's religion aligns to one's coalition and exchange group. An optimal design will modulate the perception of differences to map strategically to perceived exchange outcomes, with a distortion and biasing of information flow frequently obscuring religious similarity, even obvious similarity, as in the case of Northern Irish or Yugoslavian Christianities [compare (Hardin 1995)].

5) *Placebo effects.* Religious architecture may bias and distort information flow in ways that do not immediately relate to exchange concerns. For example, I have sketched an adaptive rational for a design that promotes distortions of hope, solace, meaning, and other psychological states that correlate to physical and mental well-being. It is easy to see how selection could forge channels integrating the conceptual materials of religion with life enhancing understandings of reality and forecasts, to fabricate judgements and attitudes that fuel success. For in actively distorting and misunderstanding the outlay of reality—incorrectly assessing the causal powers of a placebo, finding meaning in a random occurrence, or solace in the face of despair or horror—an organism's prospects may rise over capturing the world as it is, or is likely to be (the effect of which may trigger debilitating despair, or other damage). Religiosity appears to function a mental harmonizer with flow on effects to physical health.

6) *Informational encapsulation.* Critically, whatever the functions performed by supernatural belief, religiosity must be *informationally encapsulated* to only those functional domains it serves. Importantly, with regard to maladaptive behaviour, religiosity tends to be *practically inert.* Religious cognition rarely thwarts life. Though the religious believe as if the gods reward and punish, they do not generally alter their behaviour to reflect that commitment over the basic (non-coalitional) circumstances of life. Religiosity does not generally hinder the devout from building shelters, gathering nourishment, and battling enemies (Bulbulia 2004). Religion is thus similar to other forms of self-deception, where reproductive value comes from a systematically distorted representation of oneself and one's world [see (Trivers 2001)].

7) *Costly display.* Where religious costs do appear—the sacrifice of material and opportunity resources on behalf of supernatural agents, strong emotional displays, the undertaking of risky projects, bodily mutilation, scripture study, fasting and so forth—these costs often figure as public signals of authentic commitment to binding religious ideologies. Such costs are authenticating mechanisms that facilitate the solution of

prisoner's dilemmas. And they modulate to circumstances: cost outlays rise as religious groups face crisis (Sosis 2003; Chen 2004). Non-signalling costs are minimal, and generally relate to the education of religious emotion (private rituals)—investments that maintain the overall functionality of the relevant cognitive systems.

8) *Developmental patterns.* There appears to be a structured, internally driven developmental schedule leading to mature religious cognition. Young children readily produce supernatural understandings of the world (Barrett and Richert 2003), believing the universe around them to be inhabited by god-like creatures (Kelemen in press). Children exhibit a bias for intentional accounts of the origins of nature, a bias that strikingly holds regardless of the religiosity of their parents (Evans 2000; Evans 2001). Thus the children of atheists tend to theism, suggesting their early religiosity comes from within. Distinctively moralizing versions of invisible agents appear by the age of 2 1/2, and children seem to regulate their moral behaviour in virtue of ontological commitment to them (Bering in press). This ambient religiosity of childhood takes on strongly affective contours by late childhood and early adolescence, a time when emotional authentication of religious commitment is critical to forging reliable social exchange [for extensive discussion of the underlying neurological basis of adolescent religion, see (Alcorta and Sosis under review)]. While developmental details remain somewhat obscure, the patterns appear to hold irrespective of culture (Bering (in press); Knight, Sousa et al. forthcoming). From a severe poverty of stimulus, children grow to interpret their world as holding supernatural powers who are interested in them and what they do—in particular how they treat others. By early adolescence they display emotional responses to these beings (Francis and Kay 1995), and organise their activities in virtue of the supernatural commitments these emotions certify [for reviews of the developmental literature see (Bulbulia in press; Kelemen in press)].

In sum, the results of adaptationist inquires appear to vindicate the internalist stance on religion. For it appears that religious cognition is:

(1) is heavily structured so that supernatural conceptual information is acquired, stored, and accessed in ways linked to specific cognitive and behavioural outputs. The psychological scaffolding of religion comes mainly from biological, not cultural endowment.

(2) remains stable over a variety of "normal" environmental conditions (conditions closely approximating the EEA). This structural stability follows from the problems religiosity has been designed to solve:

only functionally specific and informationally encapsulated distortions of reality, and functionally specific hard-to-fake emotions and costly behaviours will foster the adaptive understandings and motivations religiosity provides

(3) materializes through an internally directed developmental programme: the "intuitive theism" of childhood grows through adolescence (in suitably religious contexts) to motivate reciprocity and promote physical and mental well-being. From fragmented cultural sources, religious commitment springs to a rich and intricate shape, in ways that equip religious agents to flourish.

It therefore appears that religion is strongly canalized; Religious variation operates within the narrow constraints of a psychological design optimized for biological success in the ancestral world.

10. *Conclusion: between convention and constraint*

Time to summarize the argument. There is undeniable variation in religious thought and expression. This variation flows from cultural circumstances and processes. Whether an individual worships a god named "FOZ" or "CHOP" depends on the exigencies of local history and institutional transmission. So too are the elemental features of ritual: whether to walk fires for FOZ or prostrate before icons of CHOP; what to wear to an inquisition; which melodies to sing before waging war; how to read the future from a sheep's entrails, and others. And at least some religiously motivated moral strategies appear to vary by convention. Many Aztecs worshipped gods imagined to require human blood sacrifice. California Zen Buddhists perceive different demands. Their religious brief is roughly: "sit and think nothing and do nothing". The morality of inclusiveness varies along institutional lines: Chinese Buddhists and Chinese Christians differ markedly in how they treat out-groups (Hansen and Norenzayan 2004). More basic moral differences seem to flow directly from religious understandings: some are inspired by their religion to pacifism, others to bellicosity (Atran 2003; Atran 2004). With respect to the conventional character of religion, internalist inquiry has limited predictive value (as it does in the study of language). A complete psychological understanding of panhuman religious architecture could not have *predicted* a group would call their god "Jehovah". May the scholars of local detail continue to illuminate such peculiarities.

Nevertheless current evolutionary and cognitive research suggests that peculiarity operates within the constraints of a common religious psychological design. For religiosity appears to flow from dedicated cognitive structures configured to form precise non-natural understandings that motivate intricate adaptive responses to ancestral conditions. Are there any religions? In fact there are innumerably many. For each religious person grows a religiosity that is unique. Yet religious differences among individuals and groups, though striking, are largely conventional, owing to differences in the configuration of supernatural concepts fueling the system, variation in the doctrinal and ritual proscriptions, and the perceived strategic exigencies of local happenstance. Religious development consists largely in fixing labels to pre-existing cognitive structures, and applying these understandings to local settings. Settings vary, so religious responses vary, but such is true of the most canalized capacities. (For example, settings vary, so visual responses vary, but we don't "learn" to see.)

Of course religious labeling is generally important to religious agents themselves. For it is through these labels (and associated signaling practices) that religious persons make sense of their world, understanding whom to trust and how to organize their lives. Though Hindu and Islamic gods and rituals are functionally similar, recent Indonesian history suggests that only Islamic understandings will coordinate and inspire religious agents on the ground there. As with language, religiosity is an aspect of our sociality. Slight variation in individual religiosities can have profound practical consequences in local strategic settings. Worshipping MOTHER EARTH among the Taliban may prompt a stoning. Nevertheless, although the names of gods and the specific costs and elaborations of rituals obviously vary, these relatively superficial differences obscure vast commonalities in the cognitive undergirding that supports the religious mind. Cultural inheritance explains some conventional attributes of an individual's religious outlook and practice, but biological endowment accounts for the massive cognitive structuring necessary for religion to solve adaptive problems. Without this psychological architecture, the conventions that regulate and control religiously motivated thought and exchange would not be possible: without the support of internally directed cognition, all sacred canopies fall. To comprehend this panhuman psychological architecture, we must look past a thicket of striking cultural variation to universal structures that buttress and facilitate religiosity in our species. To be sure, use of label "Jehovah" was never predictable. Yet to understand why religious perspectives employing this concept make sense to so many otherwise sensible people (instead of infallibly triggering incredulous stares and slammed doors) we need to explore the elaborate design of the Pleistocene adapted mind that produces religiosity. Through current naturalistic inquiry, this fascinating and powerful dimension of human nature is becoming less mysterious.

Religious Studies
Victoria University
P.O. Box 600
Wellington
New Zealand

References

Alcorta, C. and R. Sosis (under review). Religion, emotion, and symbolic ritual: The evolution of an adaptive complex.

Andresen, J., Ed. (2001). *Religion in Mind.* Cambridge: Cambridge University Press.

Ariew, A. (1999). *Innateness is Canalization: In Defence of a Developmental Account of Innateness. Where Biology Meets Psychology: Philosophical Essays.* V. G. Hardcastle (ed.). Cambridge: MIT Press.

Atran (2002). *In Gods We Trust: the evolutionary landscape of religion.* New York: Oxford University Press.

Atran, S. (2003). Genesis of suicide terrorism. *Science* 299: 1534-1539.

— (2004). Combating Al Qaeda's splinters: Mishandling suicide terrorism. *The Washington Quarterly* 27: 67-90.

— (2004, in press). Religion's evolutionary landscape: Counterintuition, commitment, compassion, communion. *Behavioral and Brain Sciences.*

Barrett, J. L. (2000). Exploring the natural foundations of religion. *Trends in Cognitive Sciences* 4 (1): January.

Barrett, J. L. and F. C. Keil (1998). Cognitive constraints on Hindu concepts of the divine. *Journal for the Scientific Study of Religion* 37: 608-619.

Barrett, J. L. and R. A. Richert (2003). Anthropomorphism or preparedness? Exploring children's god concepts. *Review of Religious Research* 44 (3): 300-312.

Bering, J. (2003). Towards a cognitive theory of existential meaning. *New Ideas in Psychology* 21: 101-120.

— (in press). The evolutionary history of an illusion: Religious causal beliefs in children and adults. In B. Elllis and D. F. Bjorklund (eds). *Origins of the social mind: Evolutionary psychology and child development.* New York: Guilford Press.

Bigrami, A. (1992). *Belief and meaning.* Oxford: Blackwell.

Boyer, P. (1990). *Tradition as Truth and Communication: a cognitive description of traditional discourse.* Cambridge: Cambridge University Press.

— (1992). Explaining religious ideas: Elements of a cognitive approach. *Numen* XXXIX (1): 27-57.

— (1994). *The naturalness of religious ideas: a cognitive theory of religion.* Berkeley: University of California Press.

— (1996). What makes anthropomorphism natural: intuitive ontology and cultural representations. *Journal of the Royal Anthropological Institute* (2): 1-15.

— (1998). Cognitive aspects of religious ontologies: how brain processes constrain religious concepts. *Theories and Method in the Study of Religion.* T. Alhback, Donner Institute.
— (2000). Functional origins of religious concepts: Ontological and strategic selection in evolved minds. *Journal of the Royal Anthropological Institute* 6: 195-214.
— (2001). *Religion Explained: the evolutionary origins of religious thought.* New York: Basic Books.
— (2003). Religious thought and behaviour as by-products of brain function. *Trends in Cognitive Sciences* 7 (3): 119-124.
Boyer, P. and C. Ramble (2001). Cognitive templates for religious concepts: cross-cultural evidence for recall of counter-intuitive representations. *Cognitive Science* 25: 535-564.
Bulbulia, J. (2004). Religious costs as adaptations that signal altruistic intention. *Evolution and Cognition* 10 (1).
— (in press). Area review: The cognitive and evolutionary psychology of religion. *Biology and Philosophy.*
Chen, D. (2004). Economic distress and religious intensity: Evidence from Islamic resurgence during the Indonesian financial crisis. Under review.
Chomsky, N. (2000). *New horizons in the study of language and mind.* New York: Cambridge University Press.
Cosmides, L. and J. Tooby (1992). The psychological foundations of culture. In J. H. Barkow, L. Cosmides and J. Tooby (eds).*The adapted mind: Evolutionary psychology and the generation of culture.* New York: Oxford University Press.
Cronk, L. (1994). Evolutionary theories of morality and the manipulative use of signals. *Zygon* 29 (1): 81-101.
Dehaene, S. (2003). Natural born readers. *New Scientist* 179 (2402): 30.
Dummett, M. (1986). A nice derangement of epitaphs: Some comments on Davidson and Hacking. In E. Lepore (ed.). *Truth and Interpretation.* Oxford: Blackwell: 459-476.
Dunbar, R. I. (1998). The social brain hypothesis. *Evolutionary Anthropology* 6: 178-190.
Evans, E. M. (2000). The emergence of beliefs about the origin of species in school-age children. *Merrill Palmer Quarterly* 46: 221-254.
— (2001). Cognitive and contextual factors in the emergence of diverse belief systems: Creation versus evolution. *Cognitive Psychology* 42: 217-266.
FitzGibbon, C. D. and J. H. Fanshawe (1988). Stotting in Thompson's gazelle: An honest signal of condition. *Behavioral Ecology and Sociobiology* 23: 69-74.
Francis, L. J. and W. K. Kay (1995). *Teenage Religion and Values.* Leominster: Gracewing.
Frank, R. (1988). *Passions Within Reason: The Strategic Role of The Emotions.* New York, Norton and Company.
Gelman, S. A. and E. Markman (1987). Young children's inductions from natural kinds: The role of categories and appearances. *Child Development* 58: 1532-1540.
George, L. K., C. G. Ellison, et al. (2002). Explaining the relationships between religious involvement and health. *Psychological Inquiry* 13 (3): 190-200.
Gould, J. L. (2002). Can honey bees create cognitive maps? In M. Bekoff, C. Allen and G. Burghardt (eds). *The Cognitive Animal: Empirical and Theoretical Perspectives on Animal Cognition.* Cambridge: MIT Press.
Guthrie, S. (1993). Faces in the clouds: a new theory of religion. New York: Oxford University Press.
— (2001). Why gods? A cognitive theory. In J. Andresen (ed.). *Religion in Mind.* Cambridge, Cambridge University Press.
Hansen, I. and A. Norenzayan (forthcoming). *Religious devotion, religious exclusivity, and tolerance for religious outsiders.*

Hardin, R. (1995). *One for all: The logic of group conflict.* Princeton: Princeton University Press.

Harman, G. (1998). Moral philosophy and linguistics. In K. Brinkmann (ed.). *Proceedings of the twentieth world congress: Volume I: Ethics.* Bowling Green: Philosophical Documentation Center: 107-115.

Heider, F. and M. Simmel (1944). An experimental study of apparent behavior. *American Journal of Psychology* 57: 243-249.

Henrich, J. and R. McElreath (2003). The evolution of cultural evolution. *Evolutionary Anthropology* 12: 123-135.

Hume, D. (1993). *Of miracles. David Hume: Writings on Religion.* A. Flew. La Salle: Open Court: 63-88.

Irons, W. (1996). Morality, religion, and evolution. In W. W. W.M. Richardson (ed). *Religion and science: history, method, and dialogue.* New York: Routledge: 375-399.

— (2001). Religion as hard-to-fake sign of commitment. In R. Nesse (ed). *Evolution and the Capacity for Commitment.* New York: Russell Sage Foundation.

Kelemen, D. (in press). Are children "Intuitive Theists"? Reasoning about purpose and design in nature. *Psychological Science.*

Knight, N., P. Sousa, et al. (forthcoming). Children's attributions of beliefs to humans and god: Cross-cultural evidence.

Kompier, M. and C. Cooper (1999). *Preventing Stress, Improving Productivity.* New York: Routledge.

Lawson, E. T. and R. N. McCauley (1990). *Rethinking religion: Connecting cognition and culture.* Cambridge: Cambridge University Press.

Marr, D. (1980). *Vision.* New York: Freeman and Company.

McCauley, R. N. and E. T. Lawson (2002). *Bringing Ritual to Mind.* New York: Cambridge University Press.

McClenon, J. (2002). *Wondrous Healing: Shamanism, Human Evolution, and the Origin of Religion.* DeKalb, Illinois: Northern Illinois University Press.

Nash, J. (1951). Noncooperative games. *Annals of Mathematics* 54: 289-95.

Odling-Smee, J., K. Laland, et al. (2003). *Niche Construction: The Neglected Process in Evolution.* Princeton: Princeton University Press.

Pargament, K. I. (1997). *The Psychology of Religion and Coping: Theory, Research, Practice.* New York: Guilford.

Petrovich, O. (forthcoming). The child's theory of the world.

Pinker, S. (1994). *The language Instinct.* New York: HarperPerennial.

— (1999). Words and rules: The ingredients of language. New York: Basic Books.

Pinker, S. and P. Bloom (1990). Natural language and natural selection. *Behavioral and Brain Sciences* 13: 707-784.

Ramachandran, V. S. and S. Blakeslee (1998). *Phantoms in the brain: Probing the mysteries of the human mind.* New York: Quill William Morrow.

Schelling, T. (1960). *The Strategy of Conflict.* New York: Oxford University Press.

Scholl, B. J. and P. D. Tremoulet (2000). Perceptual causality and animacy. *Trends in Cognitive Sciences* 4: 299-309.

Shapiro, A. and E. Shapiro (1997). The placebo: Is it much ado about nothing? In A. Harrington (ed.). *The Placebo Effect.* Cambridge: Harvard University Press: 12-36.

Skyrms, B. (1996). *Evolution of the social contract.* New York, Cambridge University Press.

Sosis, R. (2000). Religion and intragroup cooperation: Preliminary results of a comparative analysis of utopian communities. *Cross-Cultural Research* 34 (1): 77-88.

— (2003). Why aren't we all Hutterites? *Human Nature* 14 (2): 91-127.

— (2004). The adaptive value of religious ritual. *The American Scientist* 92: 166-172.

Sosis, R. and C. Alcorta (2003). Signalling, solidarity, and the sacred: The evolution of religious behavior. *Evolutionary Anthropology* 12: 264-274.

Sosis, R. and E. Bressler (2003). Co-operation and commune longevity: a test of the costly signalling theory of religion. *Cross-Cultural Research* 37 (2): 11-39.

Sosis, R. and B. Ruffle (2003). Religious ritual and cooperation: Testing for a relationship on Israeli religious and secular Kibbutzim. *Current Anthropology* 44(5): 713-722.

— (2004). Ideology, religion, and the evolution of cooperation: Field experiments on Israeli Kibbutzim. *Research in Economic Anthropology.*

Spelke, E., A. Phillips, et al. (1995). Infants knowledge of object motion and human action. In D. Sperber, D. Premack and A. Premack (eds). *Causal Cognition.* Oxford: Clarendon Press.

Sterelny, K. (2003). *Though in a Hostile World: The Evolution of Human Cognition.* Oxford: Blackwell.

— (forthcoming). Made by each other: organisms and their environment. In J. Odling-Smee, K. Laland and M. Feldman (eds). *Niche Construction: The Neglected Process in Evolution in Biology and Philosophy.*

Trivers, R. (2001). Self-deception in service of deceit. In R. Trivers (ed.). *Natural Selection and Social Theory.* New York: Oxford University Press.

Whitehouse, H. (2000). Arguments and icons. Oxford: Oxford University Press.

Wilson, D. S. (2002). *Darwin's Cathedral: Evolution, Religion, and the Nature of Society.* Chicago: University of Chicago Press.

Zahavi, A. (1980). Ritualization and the evolution of movement signals. *Behaviour* 72: 77-81.

Zahavi, A. and A. Zahavi (1997). *The Handicap Principle: A missing piece of Darwin's Puzzle.* New York: Oxford University Press.

[4]

Why religion is nothing special but is central

Maurice Bloch*

Department of Anthropology, London School of Economics, London WC2A 2AE, UK

It is proposed that explaining religion in evolutionary terms is a misleading enterprise because religion is an indissoluble part of a unique aspect of human social organization. Theoretical and empirical research should focus on what differentiates human sociality from that of other primates, i.e. the fact that members of society often act towards each other in terms of essentialized roles and groups. These have a phenomenological existence that is not based on everyday empirical monitoring but on imagined statuses and communities, such as clans or nations. The neurological basis for this type of social, which includes religion, will therefore depend on the development of imagination. It is suggested that such a development of imagination occurred at about the time of the Upper Palaeolithic 'revolution'.

Keywords: religion; sociality; imagination; evolution

1. INTRODUCTION

This paper reconsiders how we should approach the study of the evolution of religion. The discussion leads me, however, to a more general consideration of the way social cognition has been approached in recent literature. This reconsideration bears in mind the kind of problems that Colin Renfrew has called the 'sapient paradox'. The paper proposes a cognitively and neurologically more probable scenario for the development of religion than certain recent theories that are questioned by the problems he highlights.

The problems I am referring to are particularly thrown into focus by a series of theories that originate in Sperber's suggestion that religious-like beliefs are to be accounted for by a subtle mix of intuitive human capacities based on evolved neurological modules, and certain, very limited, representations that, because they go against the core knowledge that the modules suggest, are therefore 'counter-intuitive' and 'intriguing' (Sperber 1985). The motivation for these theories is to seek an answer to a question. How could a sensible animal like modern *Homo sapiens*, equipped by natural selection with efficient core knowledge (or modular predispositions), i.e. knowledge well suited for dealing with the world as it is, hold such ridiculous ideas as: there are ghosts that go through walls; there exist omniscients; and there are deceased people active after death? The authors who hold such a theory of religion give the following answers to this question. First, our core knowledge ensures that, however bizarre such ideas might seem at first, when they are more closely examined, they, in fact, turn out to be mainly disappointingly intuitive. Second, even though beliefs in supernatural things nevertheless do involve a *few* counter-intuitive aspects, if only by definition, these are possible owing to accidental misapplications of core knowledge to domains for which it is not designed. These limited misapplications are, however, so alluring that they make these minimally counter-intuitive beliefs spread like wildfire. They thus become key elements in religions (e.g. Boyer 1994, 2001; Pyysiainen 2001).

The problems with these theories that I shall discuss here do not necessarily imply outright rejection. They are what might be called 'upstream' objections since they occur even before we consider the main proposals. The first objection echoes a similar one long ago made by Durkheim, but it has been reformulated more recently by Barrett (2004) when he points out that it is odd to account for such a central phenomenon in the history of mankind as religion in terms of minor cognitive malfunctions. My second objection is that those who propose such theories forget the fact that anthropologists have, after countless fruitless attempts, found it impossible to usefully and convincingly cross-culturally isolate or define a distinct phenomenon that can analytically be labelled 'religion'.[1] The third problem with such theories is that they explain religion as a product of core knowledge or modular capacities, such as naive physics, number, naive biology and naive psychology, all of which, with the possible exception of the last, we share with all our anthropoid relatives. Such a proposal is therefore unconvincing simply because no other animal than humans manifests any behaviour that is remotely like what is usually called religion. This lack also seems to be the case for all hominids or hominims, apart from post-Upper Palaeolithic modern Sapiens. In other words, the explanations that I am challenging account for a highly specific and general characteristic of modern Humans, what they call religion, by general factors that have existed for millions of years before the Upper Palaeolithic revolution when the phenomenon first manifested itself.

The alternative story I propose here avoids these problems. It argues that religious-like phenomena in general are an inseparable part of a key adaptation unique to modern humans. This is the capacity to

*m.e.bloch@lse.ac.uk

One contribution of 14 to a Theme Issue 'The sapient mind: archaeology meets neuroscience'.

imagine other worlds, an adaptation that I shall argue is the very foundation of the sociality of modern human society. This neurological adaptation occurred most probably fully developed only around the time of the Upper Palaeolithic revolution.

2. THE TRANSACTIONAL AND THE TRANSCENDENTAL

For heuristic reasons, a consideration of chimpanzee society can serve as a starting point. I turn towards our nearest surviving relatives in order to stress, as is so often the case in the evolutionary literature, a major difference between them and us. Of course, we cannot assume that contemporary chimpanzee social organization is necessarily like that of early *Sapiens*. There is no way to know; especially since the social organizations of the two extant species of chimpanzees are radically different though both are equally closely, or equally remotely, related to us. In this case, it is not the similarity but the difference that is revealing and this difference provides us with something like a thought experiment that enables us to reflect on certain characteristics of human society.

Chimpanzees do not have anything which remotely resembles the many and varied phenomena that have been labelled religion in anthropology. Indeed, this was probably also true of early Sapiens. But, more importantly, there is also something else that chimpanzees, and probably early Sapiens, do not have. This is social roles or social groups, understood in one particular sense of the word social.

Of course, chimpanzee social organization is highly complex. For example, the dominant animal is not necessarily the biggest or the one who can hit the hardest. Dominance seems to be achieved as much by machiavellian politicking as it is by biting. Also, chimpanzees do create long-lasting coalitions, often of females, and these may well dominate the social organization of the group (De Waal 2000). Such roles and groupings are of a type that I call here the *transactional* social. This is because such roles and groups are the product of a process of continual manipulation, assertions and defeats. This type of social is also found in modern humans.

However, what chimps do not have is the kind of phenomenon that used to be referred to as 'social structure' in the heyday of British social anthropology (Radcliffe-Brown 1952). This I shall label here as the *transcendental* social. The transcendental social consists of essentialized roles and groups.

Essentialized roles exist separately from the individual who holds them. Rights and duties apply to the role and not to the individual. Thus, a person who is a professor should act 'as a professor' irrespective of the kind of person he/she is at any particular stage in the transactional social game. Similarly, in central Madagascar, as a younger brother, I should walk behind my older brother; as a woman, I should not raise my voice in a mixed gathering. All this applies, however powerful I have actually become, even if my prestige is greater than that of my older brother or of a man.

Essentialized groups exist in the sense that a descent group or a nation exists. These groups have phenomenal existence not because the members of the descent group or the nation are doing certain kinds of thing together at particular moments, or because they have been together doing certain kind of things at particular moments in the sufficiently recent past so that it is reasonable to assume that they retain the capacity to behave now in similar ways. One can be a member of an essentialized transcendental group, or a nation, even though one never comes in contact with the other members of the descent group or the nation. One can accept that others are members of such groups irrespective of the kind of relationship one has had with them or that one can suppose one is likely to have with them. Such groups are, to use Benedict Anderson's phrase, 'imagined communities' (Andersen 1983).

As noted above, in stressing the system of essentialized roles and groups, I am emphasizing what British social anthropologists, such as Radcliffe-Brown, were referring to when they spoke of social structure. However, my position is theoretically very different from theirs. For them, the human social was equated with the network of such roles and groups. For me, these phenomena are only a *part* of the social: they are the transcendental social.

The transcendental social is not all there is to human sociality. There is plenty of transactional social in human sociality that occurs side by side or in combination with the transcendental social. The transactional social exists irrespective of the role-like essentialized statuses and the essentialized groups of the transcendental social, though it may use the existence of the transcendental social as one of the many counters used in the transactional game. Human sociality is thus, as Durkheim stressed, double. It has its transactional elements *and* transcendental element. Chimpanzee sociality, by contrast, is single because the transcendental social does not exist among the chimpanzees.

The double character of the human social can be illustrated by the example of a Malagasy village elder I have known for a long time. By now, he is old, physically weak and a little bit senile. He has difficulty in recognizing people. He spends most of his days in a foetal position wrapped up in a blanket. Yet he is treated with continual deference, consideration, respect and even fear. Whenever there is a ritual to be performed, he has to be put in charge so that he can bless the participants. When he is treated with great respect he is being behaved to, and he accordingly behaves towards others as a transcendental elder. This does not mean, however, that he is not also within the transactional social system. While as a transcendental elder he is little different to what he was when he was in his prime several years ago, as a transactional player he has lost out completely in the machiavellian game of influence, and nobody takes much note of him anymore or of his opinions since in the continual power play of daily life he has become insignificant.

This kind of duality is impossible in chimpanzee society. There, once you are weak or have lost out in the continual wheeling and dealings of power, you lose previous status. In an instant, a dominant animal is replaced in his role (De Waal 2000). A chimpanzee's rank depends entirely on what those it interacts with believe it can do next. Chimpanzees do pay respect to each other in all sorts of ways… for instance, bowing to a dominant animal, but once this animal has lost out in the power game, this behaviour stops instantly. A social position in chimpanzee society never *transcends* the predictable achievements of the individual. This absence of transcendental roles is where the fundamental difference between chimpanzee and human sociability lies. The Malagasy in the village where this elder lives bow to him just as much now that he is weak as they ever did, even though he has become obviously without transactional influence. It is important to remember, however, that the respect shown to him does not mean that he is an elder all the time. The people, who interact with him, and probably himself, represent him in two ways. These two ways are not experienced as contradictory, but they are clearly distinguished and made visible by the behaviour of all concerned. Everybody knows that he is a weak old man whose hands shake and whose memory is going, and people sometimes behave towards him in terms of that representation, even with occasional cruelty. They also behave towards him in terms of the respect as described above. Thus, he belongs to two networks and, although the two are different, the transcendental network is taken into account in the transactional network while the transactional network affects the transcendental network only indirectly; for example, when another person is ultimately able to replace an elder in his transcendental role through revolutionary manipulation (for example in a traditional African society, convincing people that he is a witch; Middelton 1960).

In order to fully understand the role of an elder such as the one I have in mind, it is essential also to remember that, as a transcendental being, he is part of something that appears as a system, even though this systematicity may be something of an illusion. The transcendental elder implies the existence of transcendental juniors, of transcendental affines, transcendental grandchildren, etc. The transcendental network involves gender roles, thereby creating transcendental women and men. It is a system of interrelated roles and it is this complexity of interrelations at the transcendental level that most critically distinguishes the human social from the sociality of other species.

This transcendental network also includes what the structural functionalists called 'corporate groups', but which I have referred to above as essentialized groups. These are transcendental groups. By this, I mean that, for example, members of a clan are dual. At the transactional level, they differ from each other just as much or as little as they do from people of the next clan. But, in the transcendental social mode, all members of such a group are identical as transcendental members. They are, as is often said, 'one body'. As one body, they differ absolutely, and all in the same way, from those others in the other clan. The transcendental character of such groups is made all the more evident when we realize that the composition of such groups, whether they are clans or nations, may equally include the living and the dead. Thus, when in the transcendental one-body mode, members can make such bizarre statements as '*We* came to this country two hundred years ago'. The transcendental can thus negate the empirically based transactional in which people do not live for 200 years. Thus, the transactional social can as much ignore the present physical state of an elder as it can ignore death and individuality. The transcendental network can with no problem include the dead, ancestors and gods as well as living role holders and members of essentialized groups. Ancestors and gods are compatible with living elders or members of nations because all are equally mysterious invisible, in other words transcendental.

3. THE TRANSCENDENTAL SOCIAL AND RELIGION

This social indissoluble unity between the living and the dead and between what is often called the 'religious' and the 'social' has never been better explained than in a famous article by Igor Kopytoff 'Ancestors as elders in Africa' (Kopytoff 1971). Although the article is phrased as a criticism of earlier work by Fortes, it actually follows the latter author closely. Kopytoff points out how in many African languages the same word is used for living elders and for dead ancestors whom, it has often been said in the literature, Africans 'worship'. This is because in a sense, in the transcendental sense, they are the same kind of beings. Kopytoff stresses how both ancestors and elders have much the same powers of blessing and cursing. This leads him to assert that to talk of 'ancestor worship', and thereby to suggest something analogous to an Abrahamic notion of a distinction between material and spiritual beings, is an ethnocentric representation that imposes *our* categorical opposition between the natural and the supernatural, or between the 'real' and the religious, onto people for whom the contrast does not exist.

I accept much of Kopytoff's and Fortes' argument and want to expand it. What matters here is that if they are right, there is no reason why we cannot reverse his argument, something that Kopytoff himself suggests. If dead ancestors in an 'ancestor-worshipping society' are the same ontological phenomena as elders, then elders have the same ontological status as ancestors. If there is a type of phenomenon that merits the appellation ancestor worship, which suggests the kind of things that have often been called religion, then there is also elder worship or elder religion. And since elders are part of a system, there is in the traditional sense, junior religion, descent group religion, man religion, woman religion, etc.

Although to talk in this way may be fun, we have to use our words with the meanings that they have historically acquired. So it might be better to rephrase the point and say that what has been referred to above as the transcendental social and phenomena that we have ethnocentrically called religion are part and parcel of a single unity. This implies that the English word religion, inevitably carrying with it the history of Christianity, is misleading for understanding such phenomena as ancestor worship since, in such cases, there is not the same boundary between the 'supernatural' and the 'natural' as that perceived to occur in societies caught in the history of the Abrahamic religions. The boundary exists also in these cases, however, and it occurs between one type of social (the transcendental social including the phenomena that have been called religion) and the transactional social. This boundary is clear in the kind of society I am referring to and explains the two different ways of acting towards the Malagasy elder noted above.

The inseparability of the transcendental social and the religious is not only manifested in cases of so-called ancestor worship. Hinduism is a phenomenon that is often assumed to be comparable with the Abrahamic religions, but such an equation is misleading for the same reasons as apply to the African examples discussed previously. For example, Fuller begins his study of popular Hinduism by pointing out that a wife should, and indeed does, at some moments, treat her husband in the same way as she treats the gods. The same gestures and bodily positions are used in both cases in performing *puja* and the husband can thus be said to be a 'god' to his wife in the Hindu sense of god. The point is that here also the transcendental social husband and wife role is part of one single overarching transcendental hierarchical social system that includes the gods (Fuller 1992).

The societies I have discussed above clearly present a challenge for the kind of theories referred to at the beginning of this article, i.e. the theories advocated by, among others, Boyer. This is because they explain a phenomenon that can only be distinguished from a greater whole: the transcendental social, by using a contrast between the religious and the secular that is borrowed from a relatively modern system of representations that simply does not apply in their cases. Consequently, I shall argue that it is the greater whole in its totality, i.e. the transcendental, that needs to be explained. However, such a redefinition of the project presents an obvious difficulty. If Boyer is wrong to take a specific type of society, those with religion, to represent the human condition in general, is it not equally wrong to take specific other societies, those discussed in this paper so far, as representing human nature?

4. HISTORICAL EXCURSIONS

In what follows I argue that this is not so because societies with religion are the subsequent product of an inessential and superficial modification of the societies discussed above. A full demonstration of this point would require much more space than is available in this short paper. What follows is therefore nothing more than a tentative sketch of what such a proposal would look like. So, in order to explain how a certain state of affairs occurred for some, and only some, human groups, I move to a historical argument to argue that it is in certain specific historical circumstances (admittedly of great importance for the majority of mankind though not for all) that the kind of phenomena we call religious take on a separate appearance that seems to distinguish it from the more inclusive transcendental social.

The creation of an apparently separate religion is closely tied to the history of the state. It has long been noted that in early states such as Mesopotamia, Egypt, China, the early Andean states and many other places that the religious and the 'political' were inseparable. Frankfort long ago argued that in ancient Egypt, pharaoh was a visible god interacting on a compatible footing with the invisible gods. The organization of the state was part of the divine order (Frankfort 1948). The ancient Egyptian kingdom was part of an explicit cosmic ordering of space and time. The recurrence of the flooding of the Nile was represented as the consequence of the repetitive cyclic action of the gods, including the pharaoh. The world was centred on the capital with distant uncivilized, barely human, peripheral peoples far from its centre. Egypt was, to borrow the Chinese phrase, the empire of the centre. All this is the familiar attribute of what has been called divine kingship, whether it is that of the Swazi, Indic states or the Mesopotamian city-states.

The transcendental representation of such states was not all there was to political organization. There were also other available transactional representations of the state, pharaoh, time and space. In much the same way as the Malagasy elder is dual, it was also possible to see the pharaoh in more straightforward terms, and that was in spite of the prodigious efforts that were made to transform him through his palace and his tomb into an empirical manifestation of his transcendental side.

The transcendental construction of such states is also accompanied by another corollary process. The development of the Merina state in Madagascar in the eighteenth and nineteenth century shows how the *construction* of the symbolic state is accompanied by a partial *destruction* and *reformulation* of the symbolism of the subjects. Thus, certain key attributes of elders/ancestors were forcibly transferred from local descent groups to the king and his palace (Bloch 1986). Interestingly, a similar process involving the diminution of the transcendental social of subjects for the benefit and construction of the royal transcendental has been examined for early Egypt by Wengrow (2006).

Thus, the royal centralized transcendental construction depends on the partial destruction or at least transformation of the symbolical system of subjects. In Madagascar, the focus of the symbolism of the subjects migrated, thanks to violent encouragement, from the house to the tomb, as the palace became the symbolical *house* of the kingdom with the ruler as its central 'post' (in Malagasy, *Andry* the root of the word for ruler *Andriana*: lord; Bloch 1995). Similarly, and in more detail, I described how the descent group ritual of circumcision subsequently became orchestrated by the state and how certain aspects were taken away from the elders to become constitutive elements of grand-state occasions. The descent groups *lost* key elements to the representative of the state and were punished if they attempted to perform the full ritual independently (Bloch 1986).

Since in such systems the transcendental social and the religious are identical, it is not just the religious that is being reorganized in a centralized state and sucked up to a point into a centralized, organized, organic-seeming system, it is the whole transcendental social. The creation of this transcendental holistic image of the complete kingdom, including gods and men, thus requires the creation of the incompleteness and disorganization of the subjects' transcendental social, which can only be made complete in the kingdom.

After such a process a change that is different to the symbolic centralization of the state happens. States are unstable and political systems continually collapse. That causes a new problem. When the royal state collapses at the hand of its enemies, the subjects find themselves bereft because the construction of the state had previously made them transcendentally incomplete and the state, after its collapse, is not there anymore to complete them.

The same Malagasy example can again illustrate this point. The growth of the Merina kingdom in the nineteenth century had led to the circumcision ritual being partly taken out of the transcendental construction of descent groups and being placed in the realm of the symbolical construction of the kingdom. However, in 1868, when the Merina kingdom became disorganized, in part owing to the influence of Christianity, the ruler failed to perform the royal circumcision ritual. At that point, a popular movement arose which sought to *force* him to perform it.

Why did the subjects feel bereft by the royal non-performance when originally the ritual had been their privilege? Why should they seek the state that in many ways exploited them? Because, given the previous process, when the state collapsed, they were left with nothing but their incomplete transcendental social and, for reasons that I cannot explain, it seems as if the deprivation process is irreversible. Thus, when the state, having confiscated a large part of the transcendental social so as to create its own ordered pseudo totality of cosmic order, then collapsed, a totalizing transcendental representation without its political foundation remained, floating in mid air, so to speak. This begins to look like what we call religion. For example, the collapse of the political base of the transcendental social may lead to the occurrence of these ritual, sacred, pseudo-royal systems of Africa that so fascinated Frazer, where as Evans-Pritchard said, the king 'reigns but does not rule' (Evans-Pritchard 1948). It is what leads to shadow 'states' that only exist in mystical form as spirits that possess mediums. Examples of these are found among the Shona or in western Madagascar where they were caused, Feeley-Harnik argues, by the collapse of the political as a consequence of colonial rule (Feeley-Harnik 1991). This is also what explains the bizarre institutions of contemporary European monarchies. These post-state states are 'religions', i.e. phenomena apparently distinct from the rest of the transcendental social.

The Abrahamic religions offer another example of the process. The historian of Judaism and of early Christianity, J. Z. Smith, argues that Jewish monotheism must be understood as the product of a longing for the unified, centralized, holistic transcendental Mesopotamian city-states with Ziggurats at their centre. These were a kind of state that the Jews, as minor peripherals to that system, hardly ever managed to achieve for themselves, or, when they did, did so on a tiny fragile scale. Early Judaism is therefore also a transcendental incomplete residue: religion. This residue was modelled on the Mesopotamian prototype with, at its centre, the Ziggurat in a purely religious form, i.e. the temple in Jerusalem (Smith 1982).

With this sort of situation, we therefore get religions that are only apparently separate from the transcendental social state, but this separation is always uncomfortable and unfinished and it leads to the kind of flirting processes between state and religion that has characterized history in much of the Abrahamic world. At least in Europe and those great sways of Asia and Africa that are still under the ghostly spell of ancient Mesopotamia and Egypt this flirting takes various forms. One form of the process involves new states taking on, ready made, one of these politically detached religions issued from clearly different political entities. Rome was an example of the process. Imperial Rome became one of these centralized systems where political conquest led to the creation of a transcendental social representation of the state through making incomplete the transcendental social of subjects. Yet the transcendental construction never worked very well and when Rome got into even more trouble than usual, the system broke down. This led to the adoption of foreign and abandoned centre religions, therefore, 'hungry' for the recovery of their lost politico/transcendental social element, e.g. Judaism, other eastern religions and ultimately, one of the many forms of Christianity. Rome was therefore taking on the religious side of a centralized system from a collapsed tiny city-state as a late attempt at reorganization of a unified transcendental (Beard *et al.* 1998).

The process repeated itself. When, in the seventh century, the Franks began to develop centralized entities in western Europe, they picked up Christianity and, so to speak, 'put it on' with modifications to make it fit. One of the most spectacular moments was when Charlemagne in 800 invented a ritual that made him the Holy Roman Emperor with bits borrowed from the old testament, from Frankish rituals and of course, above all, from Roman rituals (Nelson 1987).

The other form of the relation between religion and the state, made necessary by their previous separation as a result of the collapse of a centralized unit, is for the religious bit to try to grow back its lost political undercarriage. Again and again, the popes tried. The Ayatollah Khomeini was more successful. Most of the movements that have been called millenarians try this sort of thing. Mormon history furnishes a particularly interesting example. Joseph Smith started the Mormon religion in the eastern USA for people who were heirs to a Christian religion that at many removes was heir to a long history of trouble between the religious-like pretensions of the state and the state-like pretensions of religion. However, the Mormons were in a place where the state was weak and, unusually, where the totalizing cosmological pretensions of post-state religion were strikingly incoherent, largely because they were meant to apply to a country not included in the cosmology of the Bible. So the Mormons put that to right by finding a Gospel that did mention the New World and its inhabitants and, in their creative enthusiasm, began to rebuild the political part of the destroyed transcendental entity. Not surprisingly, this annoyed the other state in Washington and they had to try to build it up in the desert, which, amazingly, they just about succeeded in doing. At the centre of this renewed unitary entity, where the transcendental social and religious were again to be an inseparable totality as in ancient Egypt or Mesopotamia, they built their temple: a temple that looks strikingly like a Ziggurat.

5. CONCLUSION

The point of these historical excursions is to suggest that the separation of religion from the transcendental social in general is, even in the places where it appears at first to exist, superficial and transient. In any case, this superficial phenomenon has occurred in human history only relatively recently.

It is this transcendental social in its totality that should be our focus. It is what distinguishes the human social from that of other closely related animals, such as chimpanzees. It is a unique characteristic and an essential part of human sociality, which, as often suggested, is *the* fundamental difference between humans and other anthropoids. An explanation of its occurrence cannot thus be in terms of a minor evolutionary adaptation, or misadaptation, as is suggested by Boyer-type theories.

Such a conclusion is negative, but it is possible to propose a more positive and fruitful one.

What the transcendental social requires is the ability to live very largely in imagination. We often act towards elders, kings, mothers, etc., not in terms of how they appear to the senses at any particular moment but as if they were something else: essential transcendental beings. Once we realize this omnipresence of the imaginary in the everyday, nothing special is left to explain concerning religion. What needs to be explained is the much more general question, how it is that we can act so much of the time towards visible people in terms of their invisible halo. The tool for this fundamental operation is the capacity for imagination. It is while searching for neurological evidence for the development of this capacity and of its social implications that we, in passing, will account for religious-like phenomena. Trying to understand how imagination can account for the transcendental social, and incidentally religion, is a quite different enterprise to accounting for the religious for itself in terms of modules, or core knowledge, which, in any case, we share with other primates. Unlike this, imagination does seem to distinguish us from chimpanzees and perhaps also distinguishes post-Upper Palaeolithic humans from their forebears.

A number of recent writers have suggested something along the same lines. In a book by Paul Harris about imagination, the author shows how the ability to engage spontaneously in pretend play begins very young and develops in a multitude of ways such as creating 'imaginary friends' and other forms of explicit make believe. Such imagination practice seems essential for normal human development. Nothing like that occurs in other species. Clearly, this capacity is necessary for engaging in the transcendental social as defined above, inevitably including the religious like. The selective advantage this form of sociality procures explains its evolutionary potential. It is central to human life. Harris suggests this centrality in an adventurous introduction when he notes that the first evidence for such a capacity is the cave paintings of Europe dating back to *ca* 40 000 years ago (Harris 2000). He might have gone a bit further back to what has been called the Upper Palaeolithic revolution, one feature of which was the first suggestion of transcendental roles found in grand burial.

Again, in a parallel argument, also taking empirical data on ontological development as its starting point, Hannes Rakoczy connects the imagination and the transcendental social even more explicitly (Rakoczy 2007). In that work, and that of his co-workers, this is referred to as 'status functions' but it is as yet little developed. However, the argument is strikingly similar to that proposed above and totally congruent. It does not however, like Harris, touch on the topic of religion, but according to my argument, this is inevitably subsumed under this type of discussion of the social.

To explain religion is therefore a fundamentally misguided enterprise. It is rather like trying to explain the function of headlights while ignoring what motorcars are like and for. What needs to be explained is the nature of human sociability, and then religion simply appears as an aspect of this that cannot stand alone. Unfortunately, the recent general discussion on social cognition does not succeed in doing the job that is needed to understand the transcendental social either. This is because, for the most part, it has considered the human social as an elaboration and an expansion of the type of social found in other animals, especially other primates (Dunbar 2004). This is useful but it obscures a fundamental *difference* between humans and others. Such an approach only pays attention to the transactional, or the 'Machiavellian' social, since that is what is shared by, for example, baboons and humans. It ignores the uniquely human transcendental social that represents a qualitative difference with other non-human socialities. What is essential to understand is the evolution of this specificity. Concentrating on that equally unique human capacity, imagination seems the most fruitful approach in that enterprise and, in passing, we will also account for religion since it is nothing special.

ENDNOTE

[1]Boyer insists that he is not talking about religion in the usual sense, but he does not define what he is talking about and he has no problem in entitling his books: *The naturalness of religious ideas: a cognitive theory of religion* and *Religion explained.*

REFERENCES

Andersen, B. 1983 *Imagined communities: reflections on the origin and the spread of nationalism.* London, UK: Verso.

Barrett, J. 2004 *Why would anyone believe in god?* Walnut Creek, CA: Alta Mira Press.

Beard, M., North, J. & Price, J. 1998 *Religions of Rome*, vol. 2. Cambridge, UK: Cambridge University Press.

Bloch, M. 1986 *From blessing to violence: history and ideology in the circumcision ritual of the Merina of Madagascar.* Cambridge, UK: Cambridge University Press.

Bloch, M. 1995 The symbolism of tombs and houses in Austronesian societies with reference to two Malagasy cases. *Austronesian Studies* **August 1995**, 1–26. (Taipei.)

Boyer, P. 1994 *The naturalness of religious ideas: a cognitive theory of religion.* Berkeley, CA: University of California Press.

Boyer, P. 2001 *Religion explained: the evolutionary origins of religious thought*. New York, NY: Basic Books.

De Waal, F. 2000 *Chimpanzee politics: power and sex among apes*. Baltimore, MD: Johns Hopkins University Press.

Dunbar, R. 2004 *The human story*. London, UK: Faber and Faber.

Evans-Pritchard, E. E. 1948 *The divine kingship of the Shilluk of the Nilotic Sudan*. Cambridge, UK: Cambridge University Press.

Feeley-Harnik, G. 1991 *The green estate: restoring independence in Madagascar*. Washington, DC: Smithsonian Institution Press.

Fuller, C. 1992 *The camphor flame: popular Hinduism and society in India*. Princeton, NJ: Princeton University Press.

Frankfort, H. 1948 *Kingship and the gods*. Chicago, IL: Chicago University Press.

Harris, P. 2000 *The work of the imagination: understanding children's world*. Oxford, UK: Blackwell.

Kopytoff, I. 1971 Ancestors as elders. *Africa* **41**, 129–142.

Middelton, J. 1960 *Lugbara religion*. London, UK: Oxford University Press.

Nelson, J. 1987 The Lord's anointed or the people's choice: Carolingian royal rituals. In *Rituals of royalty* (eds D. Cannadine & S. Price), pp. 137–180. Cambridge, UK: Cambridge University Press.

Pyysiainen, I. 2001 *How religion works*. Leiden, The Netherlands: Brill.

Radcliffe-Brown, A. R. 1952 *Structure and function in primitive society*. London, UK: Cohen and West.

Rakoczy, H. 2007 Play, games and the development of collective intentionality. In *Conventionality in cognitive development: how children acquire representations in language thought and action* (eds C. Kalish & M. Sabbagh), pp. 53–67. New directions in child and adolescent development, no. 115. San Francisco: Jossey-Bass.

Smith, J. Z. 1982 *Imagining religion: from Babylon to Jonestown*. Chicago, IL: University of Chicago Press.

Sperber, D. 1985 Anthropology and psychology: towards an epidemiology of representations. *Man (N.S.)* **20**, 73–89.

Wengrow, D. 2006 *The archaeology of early Egypt. Social transformations in north-east Africa, 10,000–2,650 BC*. Cambridge, UK: Cambridge University Press.

Part II
Naturalization of Religion

[5]

THE EVOLUTION OF RELIGION: THREE ANTHROPOLOGICAL APPROACHES

James W. Dow

This article examines three anthropological theories explaining how religion has evolved and continues to evolve. They are: commitment theory, which postulates that religion is a system of costly signaling that reduces deception and creates cooperation within groups; cognitive theory, which postulates that religion is the manifestation of mental modules that have evolved for other purposes; and ecological regulation theory, which postulates that religion is a master control system regulating the interaction of human groups with their environments. An assessment of the success of the theories is offered. The idea that the biological evolution of the capacity for religion is based on the group selection rather than individual selection is rejected as unnecessary. The relationship between adaptive systems and culturally transmitted sacred values is examined cross-culturally, and the three theories are integrated into an overall gene-culture view of religion that includes both the biological evolution and the cultural evolution of behavioral systems.

1. *Introduction*

In recent years, evolutionary psychology and cultural anthropology have been incorporating neo-Darwinist evolutionary models of human behavior into their thinking. This has produced some interesting new theories about the evolution of religion, which was one of the earliest concerns of cultural anthropology and which was the focus of considerable thought in the 19th century.

In the 19th century, anthropological ideas about the evolution of religion were more about the evolution of a cultural form than about behavior driven by the underlying capacities of the evolving human brain. Within these purely cultural ideas, a lower-to-higher, savagery-to-civilization sequence emerged as the primary framework for describing the evolution of religion. In this 19th century work, the religious beliefs of other cultures, which were published by anthropologists such as Frazer (1911), offended the sensibilities of many Europeans, and they acquired a low position in the hypothetical sequence of cultural evolution. An important issue within this framework was the search for the "origins" of religion, a state of religion in its more pristine form out of which the more "enlightened" religions of the 19th century evolved. Thus, when "primitive" beliefs such as magic or divination cropped up in 19th century European society, they were regarded as "survivals" carried over from earlier evolutionary forms in a manner akin to vestigial organs in animal physiology (Tylor 1958[1871]:112-159).

At the beginning to the 20th century, Durkheim moved British and American social anthropology away from evolutionary perspectives (Harris 1968:479-480). Durkheim (1961:20-21) redefined "origins" as the "ever-present causes upon which the most essential forms of religious thought and practice depend". "Ever-present" implied that the temporal sequence had become a side issue. Yet, the idea of a hierarchy remained. Durkheim felt that "origins" were more clearly revealed in "lower" cultures that had not been "complicated" by contact with civilized ones. Gradually, through ethnography and cultural contact, the religious beliefs of other cultures acquired more reason in the European mind, and the prejudicial idea of their being lower in an evolutionary sequence faded. Still, religion was thought of as a cultural thing by itself, and no one thought of it as a behavioral capacity that had previously come into existence because of some success it had for survival and reproduction.

The change of course toward evolutionary biology was very slow in coming. The big issue addressed by David Bidney (1950) in an article in the American Anthropologist was whether or not the myths of modern natives represented primitive thinking or were just imaginative thinking based on different data than the data available to modern scientists. He concluded that mythology was universal and differed only in the spread of its domain in preliterate and modern cultures. He suggested that, in the preliterate cultures, myth enjoyed a wider domain of application; however, he did not mention the neurophysiology of the brain. In 1972, Alexander Gallus (1972) proposed that religion evolved because it aided survival and reproduction. He located religion in the evolved brain. Lacking an understanding of neuroanatomy, an understanding that has developed greatly since then, Gallus peppered his theory with Jungian, Freudian, and existential speculative notions of how the brain worked and still used a lower-to-higher model in which the higher reasoning, culture-producing areas were superimposed on lower emotional areas of the brain. Because it had an emotional appeal, myth was seen as an earlier form of adaptive cognition, as an emotional sort of cognitive map that was being replaced by a more logical one, science. In the 1980s, the psychiatrist Eugene d'Aquili (1985) suggested that the brain was involved in religion and that it had evolved in such a way as to promote group solidarity through rhythmic auditory, visual, or tactile stimuli. He and others began to move anthropological theorizing about religion toward paradigms that included biology. Anthropologists and other evolutionists are presently examining the hypothesis that religion is a genetically evolved behavioral trait that is subjected to the laws of natural selection (Steadman and Palmer 1995; Boyer 2001; Atran 2002; Wilson 2002; Roes and Raymond 2003; Sosis 2003b; Sosis and Alcorta 2003; Sosis 2004).

The difference between early and recent evolutionary theories of religion is that the early theories were blank-slate theories in which the biology of the brain was not involved (Pinker 2002). They assumed that culture was progressing, at least in technological terms, from a state of savagery to one of civilization, and it was believed, or at least hoped, that science would replace religion as culture progressed; however, the facts do not support this optimistic belief. Human beings are replacing non-empirical religious beliefs with new non-empirical religious beliefs as fast as the old ones are discarded. Religion is not disappearing.

2. *The Evolutionary Process*

For the purposes of this discussion I would like to put forth a tentative set of definitions and a rough model of how the evolutionary process works in regards to religion. I define religion in an existential sense, as an object that we perceive in others. This object consists of human behavior that people in Western cultures generally put together in the category of "religion." In other cultures there are similar but not identical categories (Csordas 2004:163), but what is important to this discussion is that there is a generally perceived object that we can call religion. It can be observed. It can be summarized as a world view, a system of belief, a concept of reality, that other people communicate collectively.

The central nervous system evolved, at least in mammals, to facilitate survival and reproduction by relating input from sensory organs to an output of behavior. This is clearly seen in its anatomy. It has evolved to make this neural data processing adaptive in the sense of promoting survival and reproduction. One of the ways that it does this is to create internal models of external realities. In fact there is no absolute external reality. There are only models of it that are created by the central nervous system and used to process information coming from the senses. Somewhere before 60 thousand years ago humans added a capacity for complex symbolic communication (Mithen 1996;

Henshilwood and Marean 2003).[1] This allowed them to share through syntactic communication their internal models with each other.[2] The ability to share information increased the ability of humans to survive in groups, and a process of cultural transmission ensued (Boyd and Richerson 1985). A cultural storehouse of valuable, symbolically shared knowledge developed. One can assume that many different models with adaptive possibilities appeared. They were syntactically encoded and communicated to groups. The most popular of these shared models of an external reality seem to have become what we perceive as religion; however shared models of reality and other forms of collective consciousness do go beyond what most people would think of as religion. I am concerned with those very popular models that are clearly defined as religion.

3. *Commitment theory*

A recent evolutionary theory is commitment theory. This was sketched out by the economist Robert Frank (1988) and has been developed by the anthropologists William Irons (2001) and Richard Sosis (2004). Commitment theory starts with the paradox that religion is simultaneously rational and irrational. It is rational in that it leads people to successful cooperation within a group, but it is irrational in that it requires a belief in unverifiable superhuman entities and forces. Commitment theory proposes that religion's rationality is hidden from the individual. It postulates that people signal each other by making irrational commitments to unverifiable truths. This tends to create trust within the group. Trust is always a problem in human groups, because the syntactic symbolic communication that creates human culture also creates a potential for deception and autopredation. Irons (2001) proposes that a costly signal that cannot be faked shows that the signaler can be trusted. Thus, when an individual abandons all self-interested logic and commits himself or herself to an irrational belief, other persons are inclined to trust him or her. This can be compared to costly sig-

[1] Pinning down the point at which language appeared is difficult given the state of modern evidence and scientific technique (McBrearty and Brooks 2000:486). It is fairly certain that archaic H. sapiens developed it and that H. erectus did not have it. The FOXP2 gene research points to the area where genetic mutations can affect the development of language ability (Dunbar 2003:176; Enard et al. 2002).

[2] From a information processing point of view, syntax is an extra layer of encoding that connects a sequence of symbols with a sense of meaning (Chomsky 1957, 1965).

naling in the animal world, but the selection process is different. In the animal world it is often an advertisement of reproductive fitness, and the advantage is being chosen as a mate. Among humans the advertisement is for trustworthiness, and the advantage is better cooperation with others. The theory receives strong empirical support in observations that religiously organized groups have better internal cooperation and better success competing with non-religiously organized groups (Sosis and Bressler 2003).

Commitment theory sees religion evolving to maintain the advantages of group cooperation by means of an unselfish attachment to an empirically arbitrary system of belief. The commitment leads to behavior that optimizes group cooperation. From an evolutionary perspective, the individual is protected from autopredation that makes use of complex symbolic deception, at least in the context of a simple paleolithic society. Whether or not individual fitness is also enhanced in a modern culture with mass communication and with sophisticated methods of symbolic deception is not clear. A number of studies show that religiosity in modern industrial cultures is still associated with longevity and health (Levin 1994). Much of this association is linked to social participation (Hummer, Rogers, Nam, and Ellison 1999). More empirical data is needed to show when and where there is a real benefit to the individual while responding to costly signals in complex cultures. With the advent of mass communication, the benefits may sometimes go to the signalers who are activating a defense mechanism against autopredation that is no longer useful to the responders.

A parallel to commitment theory has been outlined by the economic sociologist Lawrence Iannaccone (1992), who is working within general sociological concepts of rational religious behavior (Stark and Finke 2000). He looks at it from the point of view of economic utility and sees a type of rationality in religious behavior (Iannaccone 1995). Since the main psychological benefits of belonging to a religious group come from being in a close-knit enveloping society, religious groups will try rationally to maintain that closeness. Small cult-like religious groups in which the closeness rewards are the greatest will often keep their members dedicated by imposing taboos that prevent their involvement with the larger society. A religious group will reject casually committed members because they dilute the intensity of the feeling of group solidarity, the fundamental store of utility that the group holds. Thus, religious groups will exact irrational acts of commitment in order to keep free riders out and to prevent the dilution of the ideological intensity that they offer their members. Iannaccone points out that the demands for irrational acts of commitment are really rational when seen as the protection of group benefits. One has to judge rationality in this case from the point of view of the believers, who value their support system, and not from the point of view of outsiders (Stark and Finke 2000). One advantage of the economic rationality theory is that it offers a prediction of religious change, whereas commitment theory does not. An individual changes his or her religion, drops out, or backslides when the utility of membership changes Calculations of utility can be affected by changing costs of membership and by market competition for the same religious services.

4. *Cognitive theory*

The cognitive theorists in anthropology, and related cognitive fields, are asking what makes religious models so popular and widely accepted? Instinctive non-rational popularity implies that there has been a selective process in the past that selected this behavior because it increased survival and reproduction. They are searching for the fitness advantages of religious behavior and have concluded that most of these advantages lie outside of religion itself. They feel that the behavior has survival and reproductive value in other contexts. Atran (2002) has noted a number of these emotionally attractive things in religion: the search for agents that cause things to happen; easily remembered stories for learning important cultural ideas; the evocation of therapeutic states in the brain; pleasant rhythms and sounds, etc. The attractive features of religion point to evolved capacities in the brain that are called modules (Barkow, Cosmides, and Tooby 1992). Whether or not these modules evolved in the context of an adaptive religious system is the issue. Most anthropologists immersed in the comparative study of religion tend to see religion as evolving biologically and culturally as a special integrated complex. However, cognitive anthropologists such as Atran (2002) and Boyer (2001), who look at cognitive structures in general, tend to see religion as composed of modules that evolved to solve different non-religious problems, for example the awareness of predators or the detection of cheaters.

One of the most concrete mental modules that Atran has pointed out is the module of agency. It is related to Guthrie's (1993) ideas of animism and anthropomorphism. Humans from an early age manufacture theories at a tremendous rate and very rarely test them with rigorous logic and careful observation. One of the most common and universal tendencies is to theorize that something happens because some agent made it happen. The agent need not be another human being. It may be any imagined natural or supernatural force or creature. Religions, not just the animistic ones, are loaded with beliefs in unseen agents, gods, spirits, and the like, who cause things to happen. Possibly this belief in agents is a pre-human module that animals developed to detect predators. A rabbit that sees a rustle in the bushes and imagines a fox has a better chance of survival than one that does not imagine a fox. Does a person who imagines a god have a better chance of survival? The success of religion in promoting individual health indicates that possibly he or she does.

This position, that religion evolved as separate modules, finds commitment theory inadequate. Atran and Norenzayan write as follows:

> Religion is not an evolutionary adaptation, but a recurring by-product of the complex evolutionary landscape that sets cognitive, emotional and material conditions for ordinary human interactions. The conceptual foundations of religion are intuitively given by task-specific panhuman cognitive domains, including folkmechanics, folkbiology, folkpsychology. . . . This enables people to imagine minimally impossible supernatural worlds that solve existential problems, including death and deception. Because religious beliefs cannot be deductively or inductively validated, validation occurs only by vritually addressing the very emotions motivating religion. (Atran and Norenzayan 2004)

Cognitive theory is an evolutionary by-product theory. By-product theories pop up in other areas of evolutionary theory. For example, killing one's sexual partner hardly is a good way to increase one's reproduction, but it occurs from time to time with great passion (Daly and Wilson 1988). It can be seen as a byproduct of mate-guarding that, by and large, increases the reproduction of aggressive genes when it occurs at a less violent level. It only appears irrational when it goes overboard. Cognitive theory propose that religion is an overboard manifestation of other behavior the adaptive rationality of which should be seen apart from religion. Religion is a culturally constructed complex of behavior that captures a number of adaptively unrelated behavioral modules.

One problem in utilizing cognitive theory is that some of the modules are pre-human, so we need data from animal as well as human studies. Most scholars of religion have avoided looking at non-human psychology. Only a few anthropologists of religion have been willing to consider animal cognition and behavior. For example, Anthony Wallace (1966), whose theories of religion were quite advanced for the time, proposed that ritual evolved among animals and that humans simply added symbolically communicated belief to make it into religion. Let us speculate for a moment. If we take the human-to-human species-specific communication out of religion, then how do we know that other mammals do not have convictions of the existence of an unseen world of beings and forces? Perhaps they have something like religious convictions developed by their central nervous system from their experience. Humans communicate this reality with their capacity for symbolic communication, but animals may have it without communicating it in symbols.

Atran's idea of religious cognition is that it formulates counterfactual counterintuitive models. What does it mean to say that a model is counterintuitive or counterfactual? Atran and Norenzayan write:

> The meanings and inferences associated with the subject (omnipotence = physical power) of a counterintuitive expression contradict those associated with the predicate (insubstantial = lack of physical substance), as in the expressions "the bachelor is married" or "the deceased is alive" (Atran and Norenzayan 2004).

But this test depends on semantic logic. The semantic structure of language varies from one culture to another, so something can be counterintuitive in one culture and intuitive in another. If a theory of religion is to be universal, then it must apply to all cultures. The counterfactual/counterintuitive characteristics are not recognized by everyone. Geertz (1966) and Boyer (1994) believe religion is inherently factual. Whether or not a religious belief is factual, counterfactual, intuitive, or counterintuitive depends on who is looking at it. Counterfactuality cannot be measured absolutely but only in the context of other beliefs. A belief is counterfactual only if we have another model of reality that is more "factual", and, of course, we have to be able to measure factuality in both domains of discourse with the same tools of measurement. Atran and Norenzayan adduce evidence from psychological tests of children that shows that the acceptance of "counterfactual" beliefs increases with age, but this could be interpreted as the result of learning culturally supplied cognitive structures.

5. *Ecological Regulation Theory*

Ecological regulation theory has been around longer than commitment or cognitive theory. It proposes that religion evolved because it sends control signals to a group telling it about the state of its interaction with the natural environment. There are several critical assumptions in this theory: (1) that religion alone has the emotional power to alter group behavior, whereas other symbolic signals lack the authority and emotional impact to get the job done, (2) that religion responds to changes in the group's relationship to the natural environment, and (3) that the group is the unit of natural selection by which religion evolves. Marvin Harris (1974) proposed a version of this theory consistent with his materialistic view of culture. He looked at various rituals that seemed to defy logic, cow protection in India for example, and found within them a reasonable logic, the need for draft animals in this case. Harris's position was that there was a material foundation to culture that was obscured from common view by ideology. However, he was not concerned with how things got that way, that is with evolution.

Rappaport added two important ideas: (1) that rituals were signals to a group, and (2) that religion was part of an ongoing adaptive mechanism embedded in culture. He saw social behavior as regulated by a system of values that were arranged in a hierarchy with instrumental values at the bottom and sacred values at the top (Rappaport 1999:425-427). Therefore, the highest level of values, the sacred ones, controlled the widest range of behavior and integrated the total adaptive system of the culture. Delineating such a system is difficult. Rappaport's (1974) masterful analysis of Maring ritual, in which pig-feasts control warfare and the exploitation of the environment stands above all others. However, many smaller connections between the material environment and religion have been found, and it has even been postulated that Christianity led to profligate environmental destruction (White 1967), a controversial idea that has now received some empirical support (Eckberg and Blocker 1989).

In Rappaport's overall scheme, religion provides a master cybernetic feedback control for the whole adaptive system of a culture. Primary sacred values provide a conceptual encoding system that people use to communicate their experience to each other. They permit some types of messages to be sent and limit others. They function as minders of cultural adaptation somewhat like individual values function as minders of individual adaptation. They create certain types of group action and limit other types.

Ecological regulation theory is a step beyond Durkheim's social theory of religion. Durkheim emphasized the power of religion to create social action, but he did not give religion the power to respond to the environment. Religion was society. Religion held society together, but an adaptive mechanism was left out. Durkheim ignored the powerful ecological symbolism expressed by the Australian Aborigines, the people he used to illustrate his social theory (Peterson 1972). It seems incredible that after writing:

> Out of more than 500 totemic names collected by Howitt among the tribes of south-eastern Australia, there are scarcely forty which are not the names of plants or animals; these are the clouds, rain, hail, frost, the moon, the sun, the wind, the autumn, the summer, the winter, certain starts, thunder, fire, smoke, water, or the sea (Durkheim 1961[1915]:124).

he rejected the idea that religion was responding to the natural environment. According to Durkheim, religion interpreted nature incorrectly. Science made the correct interpretation.

> Whatever we may do, if religion has as its principal object the expression of the forces of nature, it is impossible to see in it anything more than a system of lying fictions whose survival is incomprehensible (Durkheim 1961[1915]:100).

> However, it will be said that in whatever manner religions may be explained, it is certain that they are mistaken in regard to the real nature of things: science has proved it. The modes of action which they counsel or prescribe to men can therefore rarely have useful effects: it is not by lustrations that the sick are cured nor by sacrifices and chants that the crops are made to grow (Durkheim 1961[1915]:102).

Durkheim theorized correctly that religion organized people into social groups, but he completely missed the idea that this could have been done as part of an adaptive response to the environment.

Harris and Rappaport brought the natural environment back into the evolutionary theory of religion. Yet, they did not see the evolution of religion as occurring in the individual. They proposed instead that selection operated on groups, not on individuals. Rappaport suggested that groups in better alignment with their natural environments would out-compete other groups. The behavior would be passed on culturally, and there would be no impact on the evolution of the brain.

Rappaport's contribution to the theory of religion is tremendous. He moved the anthropological understanding of religion forward by decades. He discussed ritual, evolution, ecology, symbolism, language in one complete integrated theory. He once mentioned to me in a personal conversation that he was playing a very difficult and odd role in anthropology because on one hand he was an ecologist dealing with the material world and on the other he was a philosopher dealing with the symbolic world. The groups of anthropologist to whom he was talking were usually divided into these two camps, the materialist and the symbolist, and they seldom communicated with each other. He was clearly moving in the direction of consilience, the unification of the sciences, that E. O. Wilson (1998) has suggested as the salvation of the social sciences.

Rappaport defines adaptation as:

> the processes through which living systems of all sorts—organisms, populations, societies, possibly ecosystems or even the biosphere as a whole—maintain themselves in the face of perturbations continuously threatening them with disruption, death or extinction. (Rappaport 1999:6)

He sees adaptive dynamics as a product of evolution:

> Adaptive responses are seldom, if ever, isolated but seem, rather, to be organized into sequences possessing certain temporal and logical characteristics (Bateson 1972h, Rappaport 1971a, 1979a, Slobodkin and Rapoport 1974) commencing with quickly mobilized easily reversible changes in state (if perturbation continues), proceeding through less easily reversible state changes to, in some cases, the irreversible changes not in state but in structure that are called "evolutionary". (1999:6)

This is a way of saying that evolution produces adaptive dynamics in systems subject to selection. Adaptation is a characteristic of living systems produced by evolution, not just of evolution itself. The quick mobilizer of human adaptation is the central nervous system. Evolution has made it adaptive; yet cultural learning does much, some would say practically all of the programming of brain-mobilized behavior. Thus, symbolic communication has an effect on the organization of human adaptive systems.

Although one can see genes and culture working together to evolve adaptive systems, Rappaport (1999) saw the superorganic evolution of cultural behavior evolving as an adaptive system without biological involvement. He was concerned only with adaptation produced by cultural evolution. In order to do this he resorted to a evolutionary model based on group selection:

> First, whatever the case may be for explanations of behavior and organization of other species, and of their evolution, the extent to which concepts like "inclusive fitness" and "kin selection" can account for cultural phenomena is very limited. Secondly and related, whatever the case may be among other species, group selection (selection for the perpetuation of traits tending to contribute positively to the survival of the groups in which they occur but negatively to the survival of the particular individuals in possession of them) is not only possible among humans but of great importance in humanity's evolution. All that is needed to make group selection possible is a device that leads individuals to separate their conceptions of well-being or advantage from biological survival. Notions such as God, Heaven, Hell, heroism, honor, shame, fatherland and democracy encoded in procedures of enculturation that represent them as factual, natural,

public, or sacred (and, therefore, compelling) have dominated every culture for which we possess ethnographic or historical knowledge. (Rappaport 1999:10)

6. *The Problem of Group Selection in Ecological Regulation Theory*

I find this to be a very profound statement, but we must ask how the device that leads individuals to "separate their conceptions of well-being or advantage from their biological survival" is going to evolve. If it is a basic behavior given to humans by their species heritage then it must be in the brain and have evolved biologically. If it varies widely from culture to culture and is completely absent in some, then it could have evolved culturally. However, religion is practically universal in human groups and there are many pan-cultural features of which people have taken note (Wallace 1966; Turner 1969). Rappaport (1999) notes that this is the case too and essentially contradicts his proposition that group selection is the only mechanism for the evolution of adaptive systems. The confusion, I believe, comes from the hierarchical nature of the religious control mechanism. Religion is higher because it controls group behavior rather than individual behavior, but it is a mistake to assume that because it is a group control mechanism it must have evolved by group selection. In the animal world group cooperation, deception, and signaling have all evolved biologically. Why must humans be the only species to defy what seem to be a general pattern in biological evolution?

The idea of people working together to promote group welfare is a common theme in cultural ideology. People everywhere promote group cooperation. But what drives them to do this? Many evolutionists have suggested that group cooperation has evolved biologically through kin altruism or reciprocal altruism and that sacrifice for the benefit of the group really benefits the individual and his or her kin more than it benefits the group. However there is much human cooperation between unrelated individuals and much unselfish altruistic behavior. We do not have to go as far as postulating kin selection to explain this as a product of biological evolution.

A group, in competition with other units gains or looses reproductive advantage in relation to the other units. In the case of cultural group selection, the behavior is culturally acquired. The behavior must stick to the group so that the whole group is selected as a unit in the evolutionary process. Rappaport was probably thinking of this because he saw groups as firmly committed to an unchanging set of sacred values under which the day-to-day practical instrumental values would shift as necessary to maintain the adaptive equilibrium. If the behavior does not stick to the group and becomes acquired on an individual basis, then cultural group selection dissolves and cultural meme selection takes over. Obviously this happens often. Real human groups contain contending religious systems, and people move back and forth from one religious point of view to another. Only through commitment can religious groups be defined.

Punishing or policing behavior can evolve by individual selection and maintain group cooperation. Behavior that punishes individuals for deviating from group goals can explain the evolution of group cooperation in the absence of relatedness. Steven Frank (1995) has produced a generalized mathematical model that shows that "policing" can evolve biologically to maintain cooperation as the degree of relatedness between members of a group goes down. This kind of policing would manifest itself in humans primarily as shame, guilt, and anger at free-loaders. Making use of human group experiments, Fehr and Gächter (2002) show that altruistic punishment will actually develop in human groups in order to maintain cooperation. Boyd and Richerson also see punishment as process that can maintain cooperation (Boyd and Richerson 1992). Thus, group selection is not necessary for group cooperation of a religious type to evolve, or for a complex system of communication via sacred symbols to develop.

I do not deny the possibility of group selection, but it is really not necessary for a religiously mediated cooperative system to evolve. Sosis (2003a) points out that religious defenses against freeloaders indicate that individual selection is actively at work and can threaten cooperation if it is not policed. Policing can develop through individual selection if there are benefits to be had from group cooperation. Iannaccone (1992) points out that the suppression of free-loaders is an important feature of all religions. If religion evolved through group selection, then it is hard to explain the numerous new and quasi religious ideologies that keep cropping up in human society. Why do they keep occurring? They occur because the capacity for religious behavior has been developed in individuals not groups, although we often see its manifestation in the context of groups. Cognitive theorists would have no problem with this and neither would commitment theorists, who recognize that the impulse to commit one's self to a religious belief is an individual impulse. After commitment, people often become more intolerant of non-believers, a behavior that is consistent with the policing hypothesis.

Commitment and intolerance usually go hand in hand because of the way that religious behavior has been selected in individuals.[3]

The hypothesis that group selection is responsible for the evolution of religion is probably due to the way that people instinctively see the social world. The evolved brain leads people to think that well ordered social groups with few defectors are more successful. This is absolutely true, but individual selection is the mechanism that has made it true. What evolution has realized, and created, is not the the same thing as the mechanism by which evolution works. There is no reason to postulate benefits to a group as a whole when one can postulate benefits to individuals living in a group whose self-sacrificing members are well policed. Individual selection is a simpler, clearer, well-proven, basic Darwinian mechanism of natural selection.

7. *A Cross-Cultural Test of Ecological Regulation theory*

There is a long history of anthropological interest in the adaptive value of religion, which in earlier times was phrased in terms of functions (Wallace 1966; Malinowski 1948[1921]). Rappaport (1974) illustrated a religiously controlled adaptive system among the Tsembaga. Proving in general that religion is a critical component in culture's adaptation to natural environments requires more than simple observations of how behaviors are interrelated within particular cultures, because cognitive theorists may see religion as by-product of other adaptations and some evolutionists even see it as a parasite of cultural behavior (Dawkins 1993). If religion is adaptive in the sense that Rappaport defined it, one should find a correlation between types of adaptation to the natural environment and sacred values. Cultures with similar systems of adaptation should have developed similar sacred values. In fact, this is true and can be shown by cross-cultural analysis as follows.

Unfortunately, cross-cultural codes for religious beliefs have typically been made to measure variables suggested by pre-existing hypotheses. For example, Guy Swanson (1960), who was influenced by Durkheim's meta-theory, classified sacred values according to their social implications and not according to their relationship to the environment. Swanson wanted to test Durkheim's belief that society was the primary force

[3] One can love the unconverted and still be intolerant of their misguided beliefs. The argument that Liberal Christianity is a tolerant religious movement needs to tempered with the reality that it sometimes manifests an intolerance of intolerance that is often not recognized as a type of intolerance.

establishing religion. Swanson's work was an important effort at placing religion in the context of other features of culture, and his cross-cultural codes might have revealed a relationship between sacred values and adaptive systems. The study has been reexamined by Peregrine (1996) and clearly indicates that that a belief in a high god is more likely in a society with large communities, multiple levels of political hierarchy, and social differentiation (Peregrine 1995:59). However the relationship between religion and environmental adaptation is not examined. Swanson's cultural variables are so biased toward measuring political organization that they can't show the relationship of religion to the environment. Only one variable, Principle Source of Food, out of 39 has anything to do with the environment.

Because Swanson and others have not answered the question about a possible general linkage between adaptive systems and sacred values, I examined this possibility with cross-cultural data coded for the Standardized Cross-Cultural Sample (SCCS) (Murdock and White 1969), a selection of 186 geographically dispersed cultures. The electronic journal, *World Cultures* (Gray 2003), has currently collected 1,849 variables coded for the SCCS by various researchers.

Cultures can develop different ways of adapting to the same environment. These depend on the culture core (Steward 1963), the basic subsistence or energy-extraction pattern that a culture uses to support its populations. Therefore, in classifying adaptive systems, it is better to classify them by their culture cores rather than by their environments, because the culture cores more accurately reflect the adaptive systems than the environments alone. SCCS variables representing subsistence systems, a good measure of the culture core, are easy to find, because many cross-cultural researchers have been interested in the effect of ecological adaptation on other aspects of culture. I chose variables from a study by Whyte (1978) primarily because they were rank ordered and contained more detail than others. They are shown in Table 1.[4] It was difficult to find variables measuring sacred values, because few cross-cultural researchers have looked at the symbolic aspects of culture without tying them to a particular theory of social or material influences. I was unable to find any overall, detailed, geographically balanced, hologeistic study of sacred values connected to the Standard Cross

[4] Note that the variables have been recoded so that *1* represents the lowest level of importance and *5*, the highest. This allows positive correlations to represent positive effects. In Whyte's original codes *1* represented the highest and *5* the lowest importance.

Cultural Sample. In deciding what to take from other existing codes, I looked for variables that measured beliefs in non-empirical or supernatural objects, beings, or forces, in other words, variables that had no direct connection with the material world and consisted entirely of abstract beliefs to which people were seriously committed. I was able to find five useful sacred value variables from a cross-cultural study by Murdock, Wilson, and Frederick (1978) who were looking at beliefs in the supernatural causation of llness. They are shown in Table 2.

Table 1: Variables Representing Adaptive Systems

Source: Whyte (1978)

V727 Importance of Agriculture in Subsistence, including gardening
V728 Importance of Animal Husbandry in Subsistence
V729 Importance of Fishing, Shellfishing and Marine Hunting
V730 Importance of Hunting and Gathering in Subsistence
V731 Importance of Handicrafts, Manufacturing
V732 Importance of Trade in Subsistence

Levels and Frequencies

		Frequencies for Variable					
Code	Value	V727	V728	V729	V730	V731	V732
1	Insignificant, sporadic, or absent	22	29	31	15	6	13
2	Present, but relatively unimportant as a subsistence activity	2	14	19	24	44	44
3	Important, but not a major subsistence activity	6	32	26	28	42	33
4	Co-cominant, sharing position of principal subsistence activity with one or more other categories	20	12	13	14	1	3
5	Dominant, the principal subsistence activity	43	6	4	12	0	0

Table 2: Variables Measuring Sacred Values

Source: Murdock, Wilson and Frederick (1978)
V649 Theories of Fate
V652 Theories of Mystical Retribution
V654 Theories of Spirit Aggression
V655 Theories of Sorcery
V656 Theories of Witchcraft

Levels and Frequencies

		Frequencies for Variable				
Code	Value	V649	V652	V654	V655	V656
1	Absence of such a cause	99	26	2	16	81
2	Minor or relatively unimportant cause	27	68	18	45	24
3	An important auxiliary cause	1	32	37	45	17
4	Predominant cause recognized by the society	0	5	74	24	9

The five sacred value variables were correlated with the 6 adaptive system variables. The details are given in the Appendix. Twelve of the 30 possible correlations were significant. The number of significant correlations and the effort made by the creators of the Standardized Cross Cultural Sample to eliminate correlations due to diffusion and historical geographic links gives a strong indication that religion is playing a role in maintaining adaptive systems. The existence of intermediate variables, if any, that might explain these correlations does not invalidate the possibility of Rappaport's theory, because the intermediate variables could just be another part of the system. These results do not prove that Rappaport's postulate is correct for all cultures; yet, if the theory is correct one of the consequences would be correlations like this. More measures of sacred values need to be developed, more correlations need to be studied.

8. *Summary*

Commitment theory proposes that religion increases group cooperation by sending costly signals to members of the group. These signals overcome mistrust, lower defection, and allow the benefits of cooperation to be realized. Religion can be seen as an autopredatory defense mechanism that works in the context of groups. One does not have to deal with postulations of group selection because successful group cooperation can evolve by natural selection at the individual level if individuals consistently live in groups. Cognitive theory proposes that the underlying modules of the brain create religion. This implies that religion emerges from the unconscious without any particular function as an entity itself; however, there is empirical evidence that there are psychological and ecological adaptations that are associated with the whole entity. Cognitive theory simply states that the mental modules that make up religion have evolved because they created well being in other ways for individuals in the past. Each module has to be studied on its own. Cognitive theory tends to ignore the adaptive properties of whole religious complexes.

Ecological regulation theory proposes that religion is part of a cybernetic system that controls the exploitation of the environment in a productive way. It helps to maintain the food supply and produce well being. Religion helps to make a cultural system adaptive. Cross-cultural evidence indicates that this is probably true in a wide sample of cultures.

A critical question to be answered is how does the brain link well being to the acceptance of an internal model of reality? The common human way of thinking about survival is to apply rational logic to solving problems, but religion clearly works at other levels of consciousness. The psychological mechanisms that link stress and discomfort to religions change need to be examined more carefully.

A major question in these three theoretical orientations is whether religion is a byproduct of mental modules or whether it is evolving as a complex of its own. Actually it could be both. Religion could be evolving as a complex and still be using mental modules that evolved in other contexts. There are numerous examples in biological evolution where an organ originally developed in one adaptive context is also utilized in another. The central nervous system would not have evolved at all if there were not great efficiencies to be gained by using the same organ in a variety of different contexts.

Certainly the evolution of religion is a gene-culture co-evolutionary process. The human brain is evolving biologically, and the symbols of religion are changing culturally. Figure 1 diagrams this co-evolution. The vertical dimension expresses the time at which different religious behaviors developed. The horizontal dimension expresses the speed at which adaptation is taking place. In the left column we see the different biological evolutionary developments implied by the above theories. At

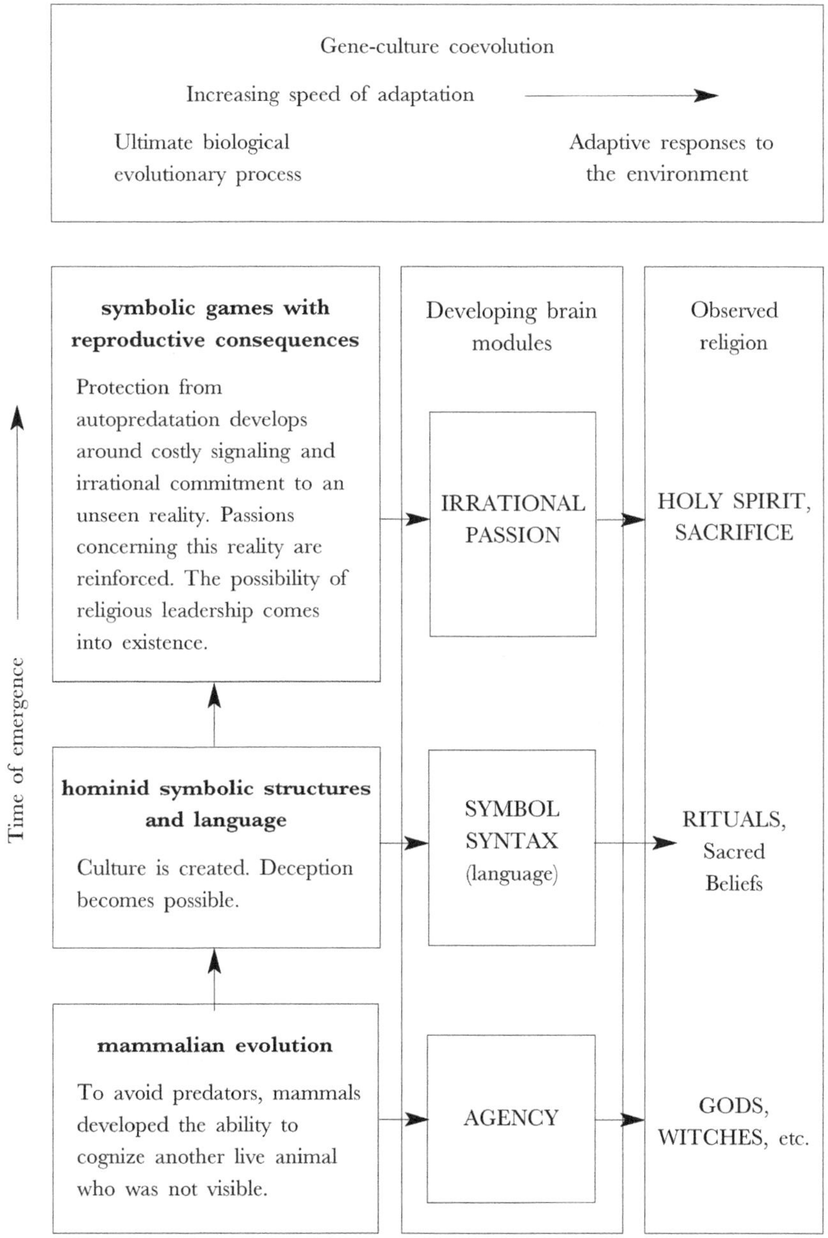

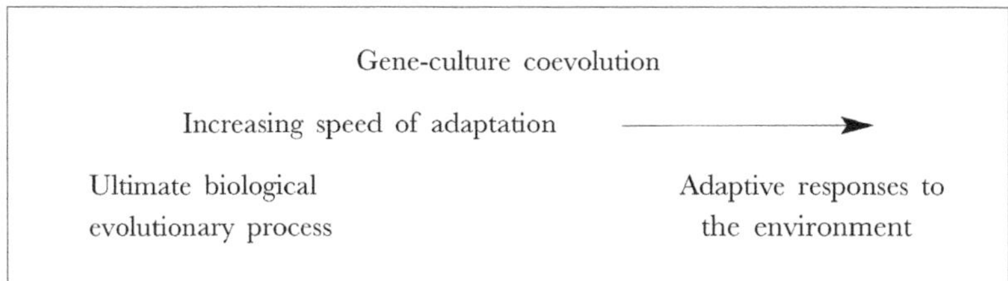

Figure 1: The Gene-Culture Coevolution of Religion

the earliest level, there is the evolution of the mammalian brain (which needs more investigation from the point of view of the study of religion). Later in time, comes the hominid development of syntactic communication, and, finally, there is the evolution of costly-signaling during evolutionary games in the context of groups held together by symbolic communication. Each of these biological developments impacts the human brain seen in the middle column. The brain enables patterns of cultural adaptation that exhibit the features that are observed as religion shown in the right column.

The idea of the sacred probably appeared at the time that syntactic communication developed. The communication of meaning in abstract belief is not possible without the development of syntax. Thus, articulated beliefs in sacred things could be a byproduct of the evolution of the capacity for syntactic communication. Rappaport refers to this as freeing signs from their significata (1999:416). The low information content but great meaningfulness of ritual symbols gives them an important function in the adaptive mechanism of a culture (Rappaport 1999:285-287). Wallace (1966) also notes that ritual is meaningful communication without information.

The evolution of syntactic communication also created a new potential for group coordination and the possibility of deception by group leaders. Costly signaling probably evolved as a defense against deception. Costly-signaling avoids the deceptive potential of syntactic communication by dealing only with unverifiable realities. These realities have no direct utilitarian value, but strengthen social bonds and trust. Religion may have utilized pre-existing behavioral modules for costly signaling developed in mating rituals in order to overcome the potential for deception created by syntactic communication. Because costly mating displays are more prevalent among males rather than among females in a polygynous species like the human one, an empirical investigation of the relative involvement of males versus females in costly religious ritual signaling might produce interesting results.

The three evolutionary theories discussed in this paper are part of an effort in the study of religion that recognizes the role of biological evolution in creating religious behavior.[5]

[5] I would like to thank Candace Alcorta for her helpful comments on an earlier version of this article.

Department of Sociology and Anthropology
Oakland University
Rochester, MI 48309
USA

Appendix: A Cross-cultural Analysis of Adaptive Systems and Sacred Values

I was able to glean five useful religious variables from a cross-cultural study by Murdock, Wilson, and Frederick (1978) who were looking at beliefs in the supernatural causation of illness. They are shown in Table 2. Two other useful religious variables, *Origin of First Creator or Ancestor* (V674), and *Mode of First-Mentioned Creation* (V675), coded by Sanday (1985), were in the SCCS data, but they had too many missing cases to give meaningful correlations with other variables.

The codes for adaptive systems were taken from Whyte (1978). They are shown in Table 1. These variables were also ordinal, so their association with sacred values could be measured by rank correlation. In the codes, *1* represents the lowest level of importance and *5*, the highest. This allows positive correlations to represent positive effects. The variables were recoded from Whyte's codes in which *1* represented the highest and *5* the lowest level of importance. In this scheme positive correlations would have represented negative effects and would have confused the discussion of the results.

All the six adaptive-system variables were correlated with the five sacred-value variables using Kendall's measure of rank correlation τ-b. Some of the correlations were highly significant and others were insignificant. Significance was determined by a two tailed test of the hypothesis that τ-b was zero. Any P value above 0.05 was considered non-significant. The statistical package used was R 1.9.1, which has a more sensitive test than SPSS. In general, the larger the value of τ-b, the more significant it is. The results are shown in Table 3.

Table 3: Kendal τ-b Significant Rank Correlations of Adaptive Systems with Sacred Values

		Sacred Value Theory				
Adaptive System		Fate V649	Retribution V652	Aggression V654	Sorcery V655	Witchcraft V656
V727	Agriculture	+.471				
V728	Animal Husb.		*-.193*	+.203	*-.167*	+.366
V729	Fishing	*-.253*			+.218	*-.356*
V730	Hunting	-.427	+.245			*-.264*
V731	Handicrafts					
V732	Trade	+.265	*-.228*			+.183

Most of the correlations are congruent with what one would expect from the ethnographic literature. At this intermediate level of analysis, data is aggregated. One has to recognize that adaptive systems in the same category may have different environmental parameters to contend with. Thus, correlations may not be strong. Agriculture in the Andes and gardening in Tikopia have different requirements although they may be classified as agricultural/horticultural systems. The purpose of this analaysis was not to ferret out the actual role that religion plays in an adaptive mechanism, but to show that there is empirical support for the conjecture that religion is involved in maintaining adaptive sysems and that there are some universal adaptive patterns arising from a common behavioral base for religion in the brain.

The two biggest, and most significant, correlations were a positive one (+.471) between *Importance of Agriculture* and *Theories of Fate* and a negative one (-.427) between *Importance of Hunting and Gathering* and *Theories of Fate*. This makes some sense because agriculture is an uncertain enterprise. Groups may recognize that their agricultural efforts are sometimes subjected to forces beyond their control and express this in religious terms. Why hunter gatherers would not have a concern over fate is an interesting question. Murdock, Wilson, and Frederick (1978), who coded the beliefs, offer some insight into this relationship. They define fate as "the ascription of illness to astrological influences, predestination, or personified ill luck" (178:453). They note that theories of fate tend to be concentrated in societies with considerable complexity. Therefore ideas about fate seem to support the more complex cultures rather than the less complex ones such as hunter-gatherer cultures. Complexity has several dimensions such as a greater social division of labor, a greater energy capture per unit area inhabited, and a more efficient energy using technology. The maintenance of such complexity in its various dimensions may be aided by religious values. Fate seems to say that if something goes wrong you just keep on doing what you are doing. This is a good sacred value to follow when survival is dependent on keeping a complex social-technological system running.

References

Atran, Scott (2002). *In Gods We Trust: The Evolutionary Landscape of Religion.* New York: Oxford University Press.

Atran, Scott & Ara Norenzayan (2004). Religion's evolutionary landscape: Counter-intuition, commitment, compassion, communion. *Behavioral and Brain Sciences* 27(6):713-730.

Barkow, Jerome, Leda Cosmides, & John Tooby (1992). *The Adapted Mind: Evolutionary Psychology and the Generation of Culture.* NY: Oxford University Press.

Bateson, Gregory (1972). The role of semantic change in evolution. In Gregory Bateson, *Steps to an Ecology of Mind.* New York: Ballantine.

Bidney, David (1950). The concept of myth and the problem of psychocultural evolution. *American Anthropologist* 52(1):16-26.

Boyd, Robert & Peter J. Richerson (1985). *Culture and the Evolutionary Process.* Chicago: University of Chicago Press.

— (1992). Punishment allows the evolution of cooperation (or anything else) in sizable groups. *Ethology and Sociobiology* 13:171-195.

Boyer, Pascal (1994). *The Naturalness of Religious Ideas: A Cognitive Theory of Religion.* Berkeley: University of California Press.

— (2001). *Religion Explained: The Evolutionary Origins of Religious Thought.* New York: Basic Books.

Chomsky, Noam (1965). *Aspects of the Theory of Syntax.* Cambridge, Mass.: M.I.T. Press.

— (2002[1957]). *Syntactic Structures.* New York: Mouton de Gruyter.

Cronk, Lee (1994). Evolutionary theories of morality and the manipulative use of signals. *Zygon* 29(1):81-101.

Csordas, Thomas J. (2004). Asymptote of the ineffable: Embodiment, alterity, and the theory of religion. *Current Anthropology* 45(2):163-185.

Daly, Martin & Margo Wilson (1988). *Homicide.* New York: Aldine de Gruyter.

Dawkins, Richard (1993). Viruses of the mind. *Free Inquiry* (Summer 1993):34-41.

D'Aquili, Eugene G. (1985). Human ceremonial ritual and the modulation of aggression. *Zygon* 20(1):21-30.

Dunbar, Robin M. (2003). The social brain: Mind, language, and society in evolutionary perspective. *Annual Review of Anthropology* 32:163-181.

Durkheim, Emile. (1961[1915]). *The Elementary Forms of the Religious Life.* Joseph Ward Swain (trans.). New York: Collier Books.

Eckberg, Douglas Lee & T. Jean Blocker (1989). Varieties of religious involvement and environmental concerns: Testing the Lynn White thesis. *Journal for the Scientific Study of Religion* 28(4):509-517.

Enard, W., M. Przeworski, S. Fisher, C. Lai, V. Wiebe, T. Kitano, A. Monaco, & S. Pääbo. (2002). Molecular evolution of FOXP2, a gene involved in speech and language. *Nature* 418:869-872.

Fehr, Ernst & Simon Gächter. (2002). Altruistic punishment in humans. *Nature* 415:137-140.

Frank, Robert H. (1988). *Passions Within Reason: The Strategic Role of the Emotions.* New York: Norton.
Frank, Steven A. (1995). Mutual policing and repression of competition in the evolution of cooperation groups. *Nature* 377:520-522.
Frazer, James G. (1911). *The Golden Bough: A Study in Magic and Religion.* 12 vols. Third Edition, revised and enlarged. London: Macmillan.
Gallus, Alexander. (1972). A biofunctional theory of religion. *Current Anthropology* 13(5):543-568.
Geertz, Clifford (1966). Religion as a cultural system. In Michael Banton (ed.), *Anthropological Approaches to the Study of Religion.* London: Tavistock Press.
Gray, J. Patrick (ed.) (2003). *World Cultures, A Journal of Comparative and Cross-Cultural Research* Vol. 14. New York: World Cultures and York College, City University of New York.
Guthrie Stewart. (1993). *Faces in the Clouds: A New Theory of Religion.* New York: Oxford University Press.
Harris, Marvin. (1968). *The Rise of Anthropological Theory.* New York: Crowell.
— (1974). *Cows, Pigs, Wars and Witches: the Riddles of Culture.* New York: Random House.
Henshilwood, Christopher S. & Curtis W. Marean (2003). The origin of modern human behaviour: Critique of the models and their test implications. *Current Anthropology* 44(5):627-651.
Hummer, Robert A., Rich G. Rogers, Charles B. Nam, & Christopher G. Ellison (1999). Religious involvement and U.S. adult mortality. *Demography* 36(2): 273-285.
Iannaccone, Laurence R. (1992). Sacrifice and stigma: Reducing free riding in cults, communes, and other collectives. *Journal of Political Economy* 100(2):271-291.
— (1995). Voodoo economics? Defending the rational approach to religion. *Journal for the Scientific Study of Religion* 34(1):76-88.
Irons W. (2001). Religion as a hard-to-fake sign of commitment. In Randolph Nesse (ed.) *Evolution and the Capacity for Commitment*, pp. 292-309. New York: Russell Sage Foundation.
Levin J. (1994). Religion and health: Is there an association, is it valid, is it causal? *Social Science and Medicine* 38(11, June):1475-1482.
Malinowski, Bronislaw (1948[1921]). *Magic, Science, and Religion.* Garden City, New York: Doubleday.
McBrearty, Sally & Alison S. Brooks (2002). The revolution that wasn't: A new interpretation of the origin of modern human behavior. *Journal of Human Evolution* 39(5):453-563.
Mithen Steven J. (1996). *The Prehistory of the Mind.* London: Thames and Hudson.
Murdock, George Peter & Douglas R. White (1969). The standard cross-cultural sample. *Ethnology* 8(4):329-369.
Murdock, George P. & Suzanne Wilson & Violetta Frederick (1978). World distribution of theories of illness. *Ethnology* 17(4):449-470.
Peregrine, Peter (1995). The birth of the gods and replications: Background to the data and codes. *World Cultures* 9(2):56-61.
— (1996). The birth of the gods revisited: A partial replication of Guy Swanson's (1960) cross-cultural study of religion. *Cross-Cultural Research* 30(2):84-112.
Peterson, Nicholas (1972). Totemism yesterday: Sentiment and local organization among the Australian Aborigines. *Man* 7(1):12-32.
Pinker, Stephen (2002). *The Blank Slate: The Modern Denial of Human Nature.* Viking Press.
Rappaport, Roy A. (1971). Nature, culture, and ecological anthropolology. In H. Shapiro (ed.), *Man, Culture, and Society.* 2nd ed. New York: Oxford University Press.

— (1974). *Pigs for the Ancestors: Ritual in the Ecology of a New Guinea People.* 2nd ed. New Haven: Yale University Press.
— (1979). On cognized models. In Roy A. Rappaport, *Ecology, Meaning, and Religion.* Richmond, CA: North Atlantic Books.
— (1999). *Ritual and Religion in the Making of Humanity.* New York: Cambridge University Press.
Roes, Frans & Michel Raymond (2003). Belief in moralizing gods. *Evolution and Human Behavior* 24:126-135.
Sanday, Peggy (1985). Female power and male dominance. Codes. *World Cultures* 1(4).
Slobodkin L. & A. Rapoport (1974). An optimal strategy of evolution. *Quarterly Review of Biology* 49:181-200.
Sosis, Richard (2003a). Review of Darwin's Cathedral: Evolution, religion, and the nature of society, by David Sloan Wilson. *Evolution and Human Behavior* 24(2):137-143.
— (2003b). Why aren't we all Hutterites? Costly signaling theory and religious behavior. *Human Nature* 14:91-127.
— (2004). The adaptive value of religious ritual. *American Scientist* 92(2):166-172.
Sosis, Richard & Candace Alcorta. (2003). Signaling, solidarity, and the sacred: The evolution of religious behavior. *Evolutionary Anthropology* 12(6):264-274.
Sosis, Richard, & Eric Bressler (2003). Cooperation and commune longevity: A test of the costly signaling theory of religion. *Cross-Cultural Research* 37(2):211-239.
Stark, Rodney and Roger Finke (2000). *Acts of Faith: Explaining the Human Side of Religion.* Berkeley: University of California Press.
Steadman, Lyle B. & Craig T. Palmer (1995). Religion as an identifiable traditional behavior subject to natural selection. *Journal of Social and Evolutionary Systems* 18(2):149-164.
Steward, Julian Haynes (1963). *Theory of Culture Change: The Methodology of Multilinear Evolution.* Urbana, IL: University of Illinois Press.
Swanson, Guy (1960). *The Birth of The Gods: The Origin of Primitive Beliefs.* Ann Arbor: University of Michigan Press.
Turner, Victor W. (1969). *The Ritual Process: Structure and Anti-Structure.* Chicago: Aldine.
Tylor, Edward Burnett (1958[1871]). *The Origins of Culture.* New York: Harper and Row.
— (1966). *Religion: An Anthropological View.* New York: Random House.
White, Lynn, Jr. (1967). The Historical Roots of Our Ecological Crisis. *Science* 55(3767):1203-1207.
Whyte, Martin K. (1978). *The Status of Women in Preindustrial Societies.* Princeton University Press. Codes published in World cultures 1(4).
Wilson, David Sloan (2002). *Darwin's Cathedral: Evolution, Religion, and the Nature of Society.* University of Chicago Press.
Wilson, Edward O. (1998). *Consilience: The Unity of Knowledge.* New York: Alfred A. Knopf.

[6]

RELIGION GENERALIZED AND NATURALIZED

by Loyal Rue

Abstract. Much of contemporary scholarly opinion rejects the attempt to construct a general theory of religion (that is, its origin, structure, and functions). This view says that particular religious traditions are unique, *sui generis,* incommensurable, and cannot therefore be generalized. Much of contemporary opinion also rejects the attempt to explain religious phenomena using the categories and concepts of the natural and social sciences. This view says that the phenomena of religion cannot be understood apart from a recognition of "the sacred," or some element of transcendence, implying that religion cannot be naturalized. This article begins to show how the phenomena of religion can be both generalized and naturalized.

Keywords: antireductionism; consilience; generalists; levels of meaning; naturalism; particularists; scientific materialism.

If religion is not about God, then what on earth *is* it about? It is about *us*. It is about manipulating our brains so that we may think, feel, and act in ways that are good for us, both individually and collectively. Religious traditions may be likened to the bow of a violin, playing upon the strings of human nature to produce harmonious relations between individuals and their social and physical environments. Religion has always been about this business of adaptation, and it will remain so.

To be sure, it is one thing to state a raw thesis of this sort and quite another to show how it all works. The purpose of this essay is to lay the groundwork for showing how the ideas, images, symbols, and rituals of religious traditions have been designed to engage and to organize human neural systems. To introduce this task, I indicate in a broad way what is at stake in this issue, and in the process hopefully come clean with respect to a few personal biases.

Loyal Rue is Professor of Religion and Philosophy at Luther College, 700 College Drive, Decorah, IA 52101. His e-mail address is rueloyal@luther.edu. This article is a version of the introductory chapter for his book, *Religion is Not About God* (forthcoming).

At stake herein is a general and naturalistic theory of religion. By a *general* theory I mean one that tells us what religion is, where it comes from, and how it functions. General theories are premised on the belief that universal properties of structure and function can be found lurking behind the varying details of religious phenomena. The goal of a general theory is to show that all religious traditions may be seen as particular variations on a set of common themes. By a *naturalistic* theory I mean one that reduces religious experiences and expressions to the status of natural events having natural causes. As such, a naturalistic theory of religion seeks to understand religious phenomena by using categories, concepts, principles, and methods continuous with the ones normally applied to nonreligious domains of human behavior. Briefly stated, the central claims are, first, that it is possible to construct a satisfying general account of religion, and second, that this can be done without invoking special (that is, supernatural) principles of explanation.

DISCLAIMERS

I begin with three important disclaimers. First, the issue is not about hostility to the idea of God. I will not be arguing either for or against the existence of God. Perhaps there are gods, perhaps not. I will not pretend to know one way or the other. The question of God's existence simply does not come into the business of understanding religious phenomena. Both the existence of God and the nonexistence of God are perfectly consistent with the claim that religion is essentially about fiddling on the strings of human nature. There is much to be said for the thesis that all theological formulations are equally and utterly dubious for the simple reason that God is inscrutable. The measure of a religious orientation is therefore not whether it gives an accurate account of divine reality but whether it effectively manages human nature. It could be argued, of course, that religion will lack the power to manage human nature unless it is believed to offer truths about God. This may be the case, but even so it is easy to see that *belief* is the thing, not the reality of the objects of belief. The religious question, then, is completely independent of the theological question. God—whatever God is—probably has no more to do with religion than religion has to do with God.

Second, the issue is not about criticizing the religious life. Indeed, I hope the opposite message comes through clearly; I regard religion generally to be a salutary thing. Religious phenomena are everywhere present in human life and will undoubtedly remain so. As far as anyone can tell—and there is plenty of evidence to the point—there has never been a coherent human culture without a religious tradition. Religion is a given, an important universal feature of human affairs, God or not. My work should therefore not be seen as an attempt to undermine religious sensibilities. If anything, I hope to kindle insights that will enable us to deepen them.

Third, the thesis I explore here is not original. The claim that religion is not about God has been advanced many times in the past, notably by the likes of Kant, Feuerbach, Marx, Durkheim, and Freud. Each of these authors believed, as I do, that regardless of what religion *says* it is about, it has to *do* fundamentally with meeting the challenges to a full life. Kant thought that religion was about achieving rational coherence in human experience; Feuerbach believed that religion was a covert way of coming to terms with self-alienation; Marx thought that religion was about coping with the dehumanizing consequences of economic exploitation; Durkheim associated religion with a veneration of the social order; and Freud described religion as the projection of deep psychological dynamics. In each of these views, the claim is that religion is about us, not about God. I simply offer a fresh iteration of the thesis on the warrant that new insights into human nature have cleared a path toward a new theory of religion.

Can Religion Be Generalized?

The feasibility of general theories of religion is open to question, and there are good reasons to discourage the pursuit of such theories from the start. Indeed, the weight of informed scholarly opinion currently favors a moratorium on general theories. In his excellent treatment of these matters, Daniel Pals claims that "the course of the most recent discussions in the theory of religion has only deepened doubts and multiplied hesitation about all general formulations" (Pals 1996, 278). The sheer diversity of religious phenomena is itself discouraging. In the course of human history, thousands of religious traditions have appeared, each with its own distinctive pattern of meaning. Some speak of millions of gods, others speak of mere dozens, still others speak of only one, and some recognize no gods at all. Some religious traditions are rich, even baroque, in symbolic and ritual convention, while others are minimalist and spare of form. Some traditions are narrowly exclusive, others broadly inclusive. Some are militantly dogmatic, others tolerant. Some religious orientations are focused on community, and others center on the solitary individual.

The complexity of religious phenomena is no less daunting. It is difficult to identify any domain of human interest and activity where religious issues are not at stake. Politics, economics, personal morality, health, education, birth, death, sexuality, art, science—all of these, in some measure, affect and are affected by religion. Religion, then, is as large and complex as life itself. Many cultures, in fact, do not possess a word for religion, apparently not needing to distinguish religious phenomena from the rest of human experience and expression. Given the extremes of diversity and complexity associated with the religious life, one might reasonably doubt the prospects for a satisfying general theory. Theories broad enough to contain such extremes are likely to sacrifice substance and insight to vagueness, while theories offering significant content will run the risk of neglecting or distorting relevant facts.

Each religious tradition is unique—unique in its cultural setting and historical development, unique in the set of challenges it has faced, and (perhaps most important) unique in the experiences and the constellations of meaning that these have generated in the lives of individual men and women. Serious regard for the unique complexity of religious orientations has provoked a scholarly reaction against general theories in favor of a "particularist" approach to the phenomena of religion.[1] If I understand the particularist view correctly, it goes something like the following. To have a theory about something is to describe and explain what the thing is about—that is, what it *means*. Thus, to have a theory of a particular religious tradition is to show what its various beliefs, values, rituals, and symbols mean to those individuals who practice the religion, within their own self-defining cultural context. Such a theory would attempt to capture the internal logic of the symbols and practices, to convey a sense of the "lived" tradition by somehow getting at what it feels like to be nurtured by it. By contrast, to have a general theory is to show what religious phenomena mean *in general*—that is, apart from their particular setting. But, say the particularists, religious phenomena are always culture-bound, which implies that their meanings will be lost when they are transposed to an alien context of meaning (such as that of the philosopher or the social scientist), where they are not self-defined. Particularists insist that each culture is *sui generis*, one of a kind, irreducible to the generalities that pretend to make the meanings of one culture commensurate with the meanings of another. In the end, there is nothing comparable to being a Christian, or a Jew, or a Buddhist. The essence of each tradition is inexorably linked to particular defining moments. These defining moments, processed by communal intercourse, emerge in self-contained patterns of meaning and a unique rationality by which the entire range of human experience may be interpreted. The sense the world makes to a Hindu is not the sense it makes to a Confucianist or an Ogallala Sioux. In a final sense, these individuals live in fundamentally different worlds of experience and meaning, worlds that cannot be unified by the artificial categories of a general theory. One cannot apprehend the meanings of a tradition from the outside, and one cannot be simultaneously inside a plurality of traditions. Broad theoretical objectivity is therefore out of the question—a religious tradition *just is* its subjective meanings. One might, perhaps, venture a coherent interpretation of a particular cultural tradition, but it is futile to generalize across the boundaries of incommensurate meanings. Thus the moratorium on general theories of religion.

Having already declared my intention to ignore this moratorium, it is only fair that I justify general theorizing in light of particularist objections.

Let it be stressed that this can be done without in any way disparaging the genuine contributions of the particularist approach. In other words, one may applaud the positive program of the particularists without accepting a negative attitude toward general theories. In fact, the particularist approach makes a substantial contribution to the general theorist by providing both heuristic and corrective insights. No one would argue that botanists and zoologists, who study the unique adaptations of particular species, contribute to a bias against general theories of evolution. On the contrary, evolutionary theorists are dependent in many ways on detailed research emphasizing the uniqueness of species. Likewise, particularist interpretations that emphasize the culture-bound nature of religious experiences and expressions are important for formulating general theories about these experiences and expressions. To the extent that we are better informed about the culture-specific meanings of religious traditions, we are better able to say something gainful concerning what religion in general is about.

The particularist bias against general theories of the nature and function of religion is tied to the claim that the essential and self-defined meanings of a tradition are lost when one assumes a perspective that transcends cultural particularity. This claim is in turn tied to certain assumptions about the nature of meaning. To escape the bias against general theories, one need only show that particularist assumptions about meaning do not constrain the general theorist. On this point it is relevant to see that meaning is open to analysis at different levels of generality.

On one level it makes sense to speak of *subjective meaning*, referring to various mental states of an individual such as beliefs, desires, hopes, fears, regrets, intentions, and the like. These are all meaning states—that is, they are *about* something or another. For example, Maggie may hope that she will win the lottery, or she may believe that her car is in the garage. The meaning of her hope is about the reality and the outcome of an event (the lottery), and the meaning of her belief is about the reality and the location of an object (her car). To *be* Maggie is in an important sense to experience her meaning states. In some measure, Maggie's meaning states will be absolutely unique—that is, it is probable that no one else is capable of thinking about her car (or her brother, or her future) in exactly the same ways as she thinks of them. The meanings involved are uniquely hers. Or consider Maggie's religious piety. It is likely that no one else has religious beliefs and experiences that are precisely identical to hers. Maggie's religious orientation—her peculiar constellation of meaning states about God, the creation, her responsibilities, her destiny—is hers alone. If Maggie is, say, a Christian, then *her* Christianity is like no one else's.

The particularist view does not deny the importance of subjective meaning states, but it wants to make a claim for an authentic level of meaning that transcends individual subjectivity. Maggie's Christianity may be unique in some measure, but not completely so. Maggie shares important overlaps of meaning with other Christians, and these overlaps define a coherent body of *conventional meanings.* Conventional meanings lie outside the realm of subjective states. They are in the public domain, making it possible to identify an objectively real thing called the Christian cultural tradition. A cultural tradition is the sum total of its conventional meanings, meanings that come to have an objective reality through a continuous historical process of social interaction and negotiation. A cultural tradition is both a repository of subjective expressions and a reservoir of symbols for nurturing subjective experiences. But its symbolic meanings are not subjective; they are independent artifacts belonging collectively and exclusively to those whose subjective meanings are drawn from them and among whom these meanings are commensurable. As objective artifacts the conventional meanings of a cultural tradition are open to the inquiry of particularists, whose purpose it is to construct coherent interpretations of symbols and their functions in the lives of those who share them.

The real question is whether theory can venture beyond these self-contained conventions. Particularists say no. It may be tempting to compare, say, Christians and Hindus on the basis of perceived similarities in their patterns of worship, but because these patterns are uniquely provoked in different cultures by distinctive meanings and experiences, they are not comparable things—coincidental, perhaps, but not commensurate. There exist no useful points of contact between the conventional meanings of Christian and Hindu cultures, which means that all attempts at generalized comparisons are off. The urge toward a general theory must therefore be resisted for the reason that religion, in general, does not exist.

If this were all there is to the business of meaning, we might readily accept the particularist bias against general theories of religion. But there is more to be said. It seems clear enough that the particularist view is committed to the doctrine that everything relevant and interesting about subjective and conventional meanings is determined at the level of cultural dynamics. Take Maggie's Christian piety, for example. It is demonstrably true that Maggie would not have the subjective meanings she does were it not for the ministrations of the particular cultural tradition in which these meanings were nurtured. Further, it is demonstrably true that Christian culture would not have the conventional meanings it does were it not for any number of Maggies who through the centuries achieved significant overlaps of subjective meanings. Thus we are left with the picture of an ongoing dialectic between individuals and the larger cultural context that both creates and is created by them. But this picture is arbitrarily narrow and circular, for it leaves out everything entailed by the recognition that subjective and conventional meanings are ultimately performed by human brains.

The various meanings that play into different aspects of human life have their origins in all those dynamics that result in brains organized to *have* meanings. A good share of these dynamics are cultural, as the particularist will be quick to point out, but the cultural matrix comes far short of telling the whole story. No less important, surely, are the dynamics of biological evolution that have assembled and organized the functions of the human brain over millions of years. Neither Maggie nor Christian culture would have the meanings they do were it not for the ministrations of natural selection. The brain sculpted in human beings by the evolutionary process has a complex modular organization. It is composed of distinct yet interactive functional units, each one adapted to perform specific operations. There are modules for perception, for memory, for emotions, for language, and much else—each with its own evolutionary subplot, each with its appointed task, but all working in a cooperative hierarchy to serve the interests of the organism. This modular brain is the biological substrate for all meanings: for all individuals, for all cultures, for all times and circumstances.

In a profound sense these various systems of the human brain participate in the construction of meanings. It seems arbitrary to deny it. They are *about* something, as surely as subjective and conventional meanings are about something. It is very difficult to specify just how the meanings embedded in neural systems become involved in the formulation of explicit beliefs, values, intentions, and the like. The aboutness of these deep meanings does not enter into conscious experience in the same way that the aboutness of subjective meanings does. Like the operating system in a computer, these subconscious meanings provide the general rules and defaults that enable and constrain the explicit meanings that eventually emerge into consciousness. I will refer to these as *adaptive meanings*, for they have been appointed by natural selection to direct on-site constructions of meaning having more specific aboutness, enabling individuals to think, feel, and act in ways that will be appropriate to local circumstances.

The obvious example is language. All normal human beings are equipped with neural modules preparing them to learn whatever particular language their culture presents to them. Thus, a Chinese infant placed in a French family will acquire the French language as surely as French children will. I believe it appropriate to say that the information states of the brain's language modules—that is, their adaptive meanings—participate in the subjective meaning of every explicit linguistic formulation.

Likewise, I believe it should be possible to specify the adaptive meanings that contribute to the formulation of specific religious beliefs and values. If so, a general theory of religion is feasible. It is true that religion in general does not exist, but the same is true of language in general, and this has not precluded the construction of insightful general theories about the nature, origins, and functions of language. A general theory of religion should focus on the adaptive meanings inherent in certain modular systems of the brain. Such a theory should be able to tell us what these meanings are about at the highest level of generality. Our best resource for such a general theory will be the discipline of evolutionary psychology. Evolutionary psychologists have been very helpful in describing the rules and strategies inherent in particular adaptations, such as perceptual systems, emotional systems, and language systems. These studies are relevant to a general theory of religion because they contribute to the larger (and perhaps more speculative) task of understanding the general purposes that are characteristic of human nature.

At the most general level of all, we can say with confidence that the ultimate goal of human beings accords with that of all other life forms—that is, to maximize reproductive fitness. Every species, however, has its own characteristic global strategies for doing this. If we can identify these broad strategies for our species, we will have in hand the general purposes of *human* nature. To know this is to know what our particular projects and their meanings (including religious ones) are really about.

What, then, are the general strategies that may be said to be both universal and exclusive to human nature? I offer the following summary. The general strategy of our species is to achieve personal wholeness and social coherence—that is, to develop healthy and robust personalities while at the same time constructing harmonious and cooperative social groups. To the extent that we succeed in these vital projects, we enhance our prospects for reproductive fitness. For other species the strategies will be slightly or vastly different, but for human beings the name of the game is personality and sociality.

How, then, do we achieve these ends? We achieve them in large measure by formulating explicit on-site meanings about how things are in the world around us and which things matter for advancing our individual and collective interests. That is, we construct and maintain shared worldviews composed of cosmological and moral elements. These shared traditions of meaning tell us who we are, where we come from, and how we should live. They give us an orientation in nature, society, and history and thus provide us with resources to negotiate our way through the many challenges to a full life. The precise contents of cosmological and moral ideas will vary somewhat with the accidents of historical experience, and this variance tells us that there are many particular ways to pursue the general goals of personality and sociality. Nevertheless, the general goals are universally and exclusively human.

Here we have the basic ingredients for a general theory of cultural and religious traditions. If the theory is correct, we may say that the neural operating systems of our species (our human nature) have prepared us to construct integrated narratives about how things are and which things matter, and that these shared meanings may be judged as more or less adaptive to the extent that they are conducive to the achievement of personal wholeness and social coherence. This leaves us in the odd position of asserting that, while subjective and conventional religious meanings may be about God, religion in general is not—it is rather about influencing neural modules for the sake of personal wholeness and social coherence. An adequate general theory of religion should be able to bring substance to these various claims, and, further, should be able to show us how conventional religious meanings have been honed to promote personal wholeness and social coherence. I intend to explore these areas more fully in future work.

Most of the tensions between particularist and generalist theories about religion arise, I suspect, from the fact that different minds find themselves attracted to different sets of facts. Consider that some facts tell us about inherited similarities and differences, and other facts tell us about acquired similarities and differences. *Inherited similarities* include all the morphological and behavioral traits that determine the uniqueness of a given species, what we might call the "nature" of the species. *Inherited differences* provide the individual variations upon which natural selection acts. Without genetic variations of this sort there would be no biotic evolution to speak of. *Acquired similarities* are determined by common patterns of learned adaptations to the environment. These patterns culminate in distinctive cultural traditions. *Acquired differences* result from unique experiences, providing the individual variations upon which cultural selection acts. Without such variations cultures would lack their most significant resource for innovative change. These four categories of facts evoke most of the interesting questions about human beings, including interesting questions about religious behavior.

People who think and write about religion are moved to do so because they become curious about facts of one sort or another. Suppose you become curious about Saint Teresa and want to make sense of her life. To do so would be to move in the direction of a theory of Saint Teresa—that is, you will try to sort out the details in a way that makes her life into an intelligible whole. This would amount, in part, to a theory of religion at the most particular level, the level of acquired differences, or biography. At a more general level, you might become curious about the larger tradition of which she was a part. Thus, you will be attracted to the facts of history and ethnography, which will lead you toward a comprehensive theory about the uniqueness of Christianity. This is the level of acquired similarities, the level of conventional meanings, where the so-called particularist theorist operates. You will find both gaps between and overlaps of ethnographic and biographic theories. A comprehensive theory of Christianity will certainly be relevant to the biographer of Saint Teresa. At the same time, however, no adequate comprehensive theory of Christianity would overlook her influence on the larger tradition. So there are overlaps. But there are irreducible gaps as well. The most erudite theory of Christian tradition could never predict the details of Saint Teresa's spiritual life. Alternatively, knowing all there is to know about Saint Teresa will not take you very far toward a theory about the larger reality of Christian tradition.

Now let us say that your curiosity heads off in a different direction. Suppose that in the course of things you become curious about the phenomenon of mysticism. You have noticed that mystics emerge in all religious traditions, and you want to understand why that happens. It will not help much to become an expert on Saint Teresa, or even on Christianity, because these theories cannot explain the independent appearance of mysticism in, say, Hinduism. Being an authority on the uniqueness of Christianity could not enable you to predict that even one Hindu mystic would exist. Nor would it help to be knowledgeable about every religious tradition save Hinduism. At best, this knowledge would embolden a guess that, since every other tradition has mystics, Hinduism can be counted on to have them too. But this would explain nothing. You cannot explain why every tradition has mystics by saying that mystics are found in every tradition. There is something real and important about the phenomenon of mysticism that clearly transcends the historical contingencies typical of ethnographic and biographic explanation. Here are two interesting facts about mysticism: (1) every tradition has its fair share of it, and (2) not all religious people are mystics. These facts suggest that mysticism is both differential and, in some decisive measure, inherited. No one inherits genes for mysticism, but people do differentially inherit genes for various psychological characteristics that predispose them to unpack their acquired religious baggage in distinctive ways. Mysticism may be characterized as one of these ways, as may legalism, fundamentalism, scholasticism, and other well-defined panreligious stereotypes. To pursue a satisfying account of these phenomena, one needs the resources of psychological theories about inherited differences in temperaments, personality types, and cognitive styles, not to mention a variety of psychopathologies.

Finally, one's curiosity about religion might be drawn to the most universal facts of all, such as why religious traditions are found in every human culture and only in human culture; why these traditions invariably develop narrative integrations of cosmology and morality; why they all address themselves to matters of personal therapy and social policy; and why they dwell on so many of the same virtues. These formal characteristics, shared by religious traditions everywhere, are not accidental. They indicate species traits, transcending the contingencies of historical existence, and yet they constitute real facts, deserving of theoretical inquiry, which cannot be explained apart from a theory of human nature grounded in the disciplines of biology, anthropology, evolutionary psychology, and the neural sciences. These domains provide the primary conceptual resources for a general theory of religion. A general theory of religion is therefore necessary if we are to understand a substantial set of undeniable facts. Yet a general theory—even the most sophisticated—could never predict the appearance of Judaism or Jainism.

I may be totally wrong about this, but I have the impression that many persons who think and write about religion find it difficult to accept that there may be interesting and important determining factors in each of several factual domains. It is natural to suppose that all the real action plays out in one's own arena of curiosity. The ethnographer might very well concede that many traits are inherited but then insist that none of these is relevant to insights about what religions are or how they work. The generalist might agree that particular traditions have unique characteristics but that these amount to insignificant details. Chauvinism like this actually exists, and wherever it does, it creates misunderstanding and defensive posturing. Generalists are then likely to endure charges of arm waving and armchair deduction, while the efforts of particularists may be trivialized by sweeping claims of reductionism. Academic umbrage of this sort is unfortunate because it tends to undermine the breadth of cooperation necessary for understanding the complexities of religion. In contrast, we might begin to envision an atmosphere in which different levels of theoretical interest are mutually supportive. The serious student of religion will be curious about all sorts of facts and will welcome resources from every direction. For example, a biographer of Saint Teresa should be well grounded in the sciences of human nature. Would it be relevant to an understanding of Saint Teresa's spiritual life to know that she suffered from epilepsy? Of course it would. It is equally important that generalist theorists of religion be well grounded in studies of particular traditions. Would it be relevant to a general theory to show that Buddhists are capable of emotional experiences having no equivalents in Christianity, and vice versa? Certainly. And would one's understanding of particular traditions be enhanced if a general theory could show what it is about human nature that makes such radical cultural differences possible? Of course.

I do not mean to obscure the difficulties. There will be much left to argue about. But disputation is, after all, the fuel of serious inquiry. In any event, I remain convinced that general theories of religion are beyond the merely possible; they are essential to the enterprise.

Can Religion Be Naturalized?

It should be evident that the quest for a general theory of religion is not without controversy. Even more controversial, however, is a second major thesis: that religion can be naturalized. This thesis asserts that the experiences and expressions constituting the religious life can be seen to result exclusively from natural causes. This view does not imply that religious phenomena can be completely explained—few events in nature can be—but only that the extent of our understanding is contingent on our efforts to reduce these phenomena to the terms of underlying natural processes.

Naturalism is a variant of metaphysical monism, the philosophical stance declaring that all meaningful distinctions pertain to observed or reasoned facts within a self-contained and continuous order of being. For the naturalist, the order of being is the order of nature: the natural is real, and the real is natural. If, therefore, we have reason to believe that some entity or event is real, then we have precisely those reasons for believing it to be natural.

Naturalism may be characterized by its rather strict application of "Ockham's razor," or the principle of parsimony. William of Ockham, the most influential Western philosopher of the fourteenth century, was famous for his intellectual crusade to eliminate pseudoexplanatory categories. Ockham's rule of elimination goes as follows: "Plurality is not to be assumed without necessity." Alternative forms are "What can be done with fewer is done in vain by more" and "Entities are not to be multiplied without necessity." The point of Ockham's razor is to achieve economy of explanation by shaving away concepts, principles, and categories that are not essential to the subject matter to be explained and are not established by rigorous methods of observation and reasoning. Thus, naturalists oppose explanations that unnecessarily assume a transcendent order of entities and events having causal influence in the order of nature. Why posit two orders of being where one is sufficient?

It is precisely on this point of sufficiency of explanation that nonnaturalists have pressed the case for supernatural accounts of religion. Naturalists, the argument goes, tend to overeconomize in their explanations to the point of distorting and ignoring facts. So argued Mircea Eliade (1957), an eloquent and prolific opponent of reductionism. Eliade believed that religion could be generalized but not naturalized—generalized because all particular forms of religion derive from human encounters with the *sacred*, but not naturalized, because the sacred does not derive from the order of nature. The sacred, a realm of absolute transcendent reality, cannot be apprehended by natural categories. Nor can the religious life be apprehended apart from the sacred. Natural processes—biological, psychological, economic—may have certain limited effects on religious phenomena, but these are peripheral. Ultimately and essentially, religion is *theogenic*—that is, its facts can be understood only by assuming the causal influence of a supernatural reality.

There are several other variations on the antireductionist thesis. Some of these maintain that there are patterns of religious meaning inherent in human history that radically transcend the dynamics of nature. Others argue that religion cannot be naturalized because it is essentially about a moral order and that moral values can be neither derived from nor explained by natural facts. In order to make sense of moral behavior, therefore, we are forced to transcend the vocabulary of naturalism. A similar argument is that religion cannot be naturalized because it necessarily involves something immaterial, such as a life force or spiritual awareness, neither of which can be sufficiently described in natural terms. Another alternative is to concede that certain *forms* of religion such as idolatry, fertility cults, and false prophets can be naturalized but that "true" religion is divinely inspired and cannot be traced to natural causes. And finally there are default arguments, insisting that the many failures of previous attempts to naturalize religion (Freud's psychogenic theory, for example, or Durkheim's sociogenic theory) warrant the conclusion that the job simply can't be done. These arguments all share in the view that religious experiences and expressions cannot be reduced to the status of natural phenomena. Even after the naturalists have taken their best theoretical shots, there remains something of decisive significance that eludes our understanding. Thus, reason ordains, in order to have an adequate theory of religion one must admit to the necessity of supernatural categories of explanation.

In the face of such arguments, the naturalist is left with a single option: to produce. The only sure way to make the case for naturalizing religion is actually to do it—that is, give an account of the origins and functions of religion that renders theogenic alternatives vapid and unnecessary. It may be observed that the history of inquiry is on the side of naturalism. Earthquakes, floods, astral displays, birth defects, diseases, and a good many other phenomena have been effectively naturalized to the undisputed satisfaction of all. And for the past century we have been slowly acquiring the theoretical resources for naturalizing human behavior. Recent progress in behavioral genetics, neuroscience, and evolutionary theory provides additional resources for extending the naturalistic program into the more sensitive areas of art, literature, morality, and religion.

It should not be assumed that all naturalists agree on a common worldview. Far from it. There have been many widely divergent variations on the naturalistic theme, including the syncretism of Confucius, the materialism of the Atomists, the substantialism of Aristotle, and the idealism of Hegel. Naturalists will agree that the natural order is ultimate, but beyond this point there is much room for dispute about the nature of nature. Everything depends upon the recognition of natural facts—that is, what entities, events, properties, and processes one accepts as real. I cast my lot with a version of naturalism I call *consilient scientific materialism.*

Materialism. A materialist worldview claims that all natural facts can be construed, in some minimal sense at least, in terms of the organization of matter. This should not be taken to suggest that all natural facts are "nothing but" physical facts, only that whatever is or happens in nature is contingent on a substrate of material reality.

The picture of nature presented to us by contemporary science reveals a cosmic evolutionary process that has unfolded in a complex hierarchy of interlocking systems and subsystems that govern the organization of matter. As one follows the arrow of time, more complex systems emerge to organize matter, bringing new entities, properties, and relations into the order of nature. At the lowest level are subatomic particles, which are organized into higher-level complex systems called atoms. Atoms have diverse properties, which account for their organization into various molecular systems. Molecules are systematically organized to form a variety of complex structures, including rocks, minerals, planets, stars, and galaxies. Molecules also may be organized into living systems, composed of cells, tissues, organs, and so on. Living organisms are systematically organized into populations, communities, and ecosystems. The most highly organized material systems are found in human beings. Here, nerve cells are organized into various functional systems, which may be integrated into coherent personalities. Persons then interact in complex patterns to form social groups and cultural systems.

Some materialists still maintain that all of this complexity in the organization of matter may be reduced to the dynamics of atomic and molecular systems. This is the strong reductionist thesis, which claims that it is in principle feasible to give a full account of higher systems in terms of physics and chemistry. This view, now very much in the minority, is a good example of taking Ockham's razor too far. A more satisfying picture recognizes that, although more complex systems are contingent on physical and chemical substrates, they nevertheless involve emergent principles of organization and manifest genuinely novel properties that cannot be fully described by the principles and properties known to the physical sciences. That is to say, not all natural facts are physical facts. There also exist biological facts, psychological facts, and cultural facts.

I assume a view of nature in which all natural facts may be resolved into four general categories, or levels of material organization: physical, biological, psychological, and cultural. The distinctions between these four levels of nature may be seen to derive from different modes of information. Physical facts describe the behavior of matter insofar as it is organized by information inherent in physical systems. Biological facts describe the behavior of matter insofar as it is organized by information preserved in the genetic code. Psychological facts describe the behavior of matter insofar as it is organized by information stored in neural systems. And cultural facts (sociosymbolic facts) describe the behavior of matter insofar as it is organized by information embodied in symbols. That's it! If something is a fact, then it is in principle reducible to these four types (or perhaps a constellation of them, in the case of complex facts, which most are).

Science. The various disciplines of science represent attempts to organize our knowledge of physical, biological, psychological, and cultural facts. If we were to construct the academic curriculum afresh, in conformity with this view of nature, we would do well to establish four major faculties. Alas, we are left to contend with the disciplines that developed haphazardly over time. Still, there is a rough correlation between the existing disciplines, subdisciplines, and interdisciplines and the four categories of natural facts.

At the level of physical facts are the corresponding disciplines of physics, chemistry, astronomy, geology, astrophysics, and cosmology. Bridging the gap between physical systems and biological systems are the interdisciplines of biophysics and biochemistry. Corresponding to the level of biological facts we find cell biology, genetics, physiology, anatomy, zoology, botany, and ecology. The neurosciences, together with evolutionary psychology, bridge the gap between biological and psychological systems. At the level of psychological facts are the various subdisciplines of cognitive psychology, developmental psychology, personality theory, and others. Social psychology connects the levels of psychological and cultural facts. Attending to the organization of cultural or symbolic facts is an unruly assemblage of disciplines, including sociology, anthropology, political science, economics, history, linguistics, mathematics, philosophy, and the various "critical" disciplines focused on literature, the arts, and religion. The major differences between these intellectual domains have to do with methods, and with lingering attachments to traditional agendas. What they share in common is their (selective) interest in the artifacts of human symbolic abilities.

Taken together, these many disciplines constitute *science*, the collaborative enterprise of systematically organizing our knowledge of the natural order. For the scientific materialist, all plausible explanations for natural phenomena will find their place among these disciplines.

Consilience. Scientific materialism claims that all natural facts involve the organization of matter and that the empirically grounded and self-correcting disciplines of science are to be considered normative in all attempts to explain natural facts. A consilient scientific materialism goes a step further to advance a thesis about the unity of science. Edward O. Wilson (1998) has rescued the concept of consilience from historical obscurity to characterize the ultimate prize of inquiry: a coherent, unified meshwork of ideas that renders intelligible the full scope of human experience. The sciences, as we have them, still tend to be fragmented into separate domains of inquiry. But such fragmentation is both artificial and unsatisfying. If nature is itself a unified meshwork of interlocking causal events, as the naturalist believes, we should expect that existing gaps between the sciences might be significantly narrowed by further inquiry. Wilson demonstrates that the unification of knowledge is already remarkably complete among the natural sciences. Consilience of theoretical explanation from physics to chemistry to biology and well into the nascent field of neuroscience has already been achieved. What remains is to explore ways of thinking that might extend the consilience program to include the social sciences and the humanistic sciences.

Thus we are returned to the original question: Can religion be naturalized? To naturalize religion is to provide explanations for religious phenomena that are consilient with our scientific explanations for other natural facts. The focus of this inquiry is on human nature. For a consilient theory of religion to succeed, it must show, first, what human nature is and how it emerges in the process of evolution, and second, how religious experiences and expressions emerge from the dynamics of human nature. To the extent that such a consilient theory succeeds, we shall have before us a satisfying naturalistic understanding of religious phenomena.

But suppose it *does* succeed. What then? What can be said about the power of religion under the conditions of understanding it? Does an understanding of religion preclude religious understanding? Does the attempt to naturalize religion also effectively neutralize it? These are delicate questions that need to be addressed. I say "delicate" because the religious life is the sanctuary of existential meaning, where attempts to naturalize may be experienced as hostile acts intended to destroy the meaning of life, to undermine faith and hope, to steal away the treasured grail. I do not attempt to deny the reality or the gravity of the experience. I have endured it myself. Nevertheless, the urge toward consilient explanation is strong. All I can promise from my own experience is that any existential losses incurred by naturalizing religious meanings may be compensated for without remainder by an acquired sense for the mystery and sanctity of nature itself.

Note

1. Pals (1996, esp. chapters 7 and 8) has an excellent discussion of the particularist position, which he associates with the influential anthropologist Clifford Geertz and his followers.

References

Eliade, Mircea. 1957. *The Sacred and the Profane.* New York: Harcourt, Brace and World.
Pals, Daniel. 1996. *Seven Theories of Religion.* New York: Oxford Univ. Press.
Wilson, Edward O. 1998. *Consilience: The Unity of Knowledge.* New York: Alfred A. Knopf.

[7]

Testing Major Evolutionary Hypotheses about Religion with a Random Sample

David Sloan Wilson
Binghamton University

Theories of religion that are supported with selected examples can be criticized for selection bias. This paper evaluates major evolutionary hypotheses about religion with a random sample of 35 religions drawn from a 16-volume encyclopedia of world religions. The results are supportive of the group-level adaptation hypothesis developed in *Darwin's Cathedral: Evolution, Religion, and the Nature of Society* (Wilson 2002). Most religions in the sample have what Durkheim called secular utility. Their otherworldly elements can be largely understood as proximate mechanisms that motivate adaptive behaviors. Jainism, the religion in the sample that initially appeared most challenging to the group-level adaptation hypothesis, is highly supportive upon close examination. The results of the survey are preliminary and should be built upon by a multidisciplinary community as part of a field of evolutionary religious studies.

KEY WORDS: Adaptation; Evolution; Evolutionary religious studies; Group Selection; Religion

Evolutionary biologists typically employ a number of major hypotheses for the study of all traits. Perhaps the most important question is whether a given trait has evolved by natural selection and adapts the organism to its environment. If so, then more specific hypotheses are needed to identify the particular selective forces. For example, a social behavior can evolve by either within-group selection (increasing the fitness of the individual relative to others in its same group) or by between-group selection (increasing the fitness of the group relative to other groups in the total population). If the trait is not a product of natural selection, then another set of specific hypotheses is needed to explain its existence. Perhaps it is an ances-

Received July 28, 2004; accepted January 11, 2005; final version received June 13, 2005.

Address all correspondence to David Sloan Wilson, Departments of Biology and Anthropology, Binghamton University, Binghamton, NY 13902-6000. Email: dwilson@binghamton.edu

tral trait that does not vary within the lineage. Perhaps it was adaptive in past environments but failed to keep pace with environmental change. Perhaps it is a costly by-product of another trait that is a product of natural selection, and so on.

These hypotheses are not mutually exclusive. Evolution is a multifactorial process, and traits usually reflect a variety of selection pressures and constraints on natural selection. Nevertheless, the different hypotheses are still needed to determine the combination of factors that operate in any particular case. To pick a paradigmatic example, morphological, behavioral, and life history traits in guppies (*Poecilia reticulata*) are influenced by a variety of selection pressures, notably predation and female mate choice (Endler 1995). Predators are both larger and more numerous in the downstream portions of rivers than the upstream portions, resulting in a corresponding gradient of traits in guppies. Downstream guppies that are transplanted into upstream tributaries that lack predators quickly evolve the suite of traits characteristic of guppies in predator-free environments. One trait that does not change is live birth, which is shared by all members of the family to which guppies belong and does not vary within the lineage. Decades of research guided by evolutionary theory has led to a comprehensive understanding of guppies, even though the story is complex and includes numerous selection pressures and constraints on natural selection.

This way of forming and testing evolutionary hypotheses, which is familiar for the study of nonhuman species such as guppies, is increasingly being used to study the human phenomenon of religion (e.g., Bulbulia 2004; Hinde 1999; Irons 2001; Sosis and Alcorta 2003; Wilson 2002). Not only can it be used to guide current research, but it can also be used to reorganize past research that was conducted without evolutionary theory in mind. Table 1 presents a classification of major hypotheses about religion, past and present, from an evolutionary perspective. It begins with the basic distinction between adaptive and nonadaptive hypotheses, with more specific hypotheses under each heading. Starting with adaptation hypotheses, one possibility is that religions are designed to function for the benefit of the religious group. This hypothesis has a long history in the social sciences, including Durkheim's *Elementary Forms of Religious Life* (originally published in 1912). In modern evolutionary terms, it needs to be understood in terms of genetic and cultural group selection. A second possibility is that religions are designed to function for the benefit of some of its members (presumably the leaders) at the expense of other members (Cronk 1994). For example, the Protestant reformation was in part a reaction to abusive practices within the Catholic Church that were clearly benefiting the elites at the expense of the laity. In modern evolutionary terms, this hypothesis needs to be understood in terms of genetic and cultural within-group selection. A third possibility is that the cultural traits associated with religion can evolve to be like parasites, infecting minds without benefiting either individuals or groups. This is suggested by the modern concept of memes (Aunger 2002; Blackmore 1999; Dawkins 1976) but can also be found in earlier theories of religion that were not explicitly framed in terms of evolution (e.g., Durkheim 1995:49).

Table 1. Major Evolutionary Hypotheses about Religion

Religion as an Adaptation	Religion as Nonadaptive
• Group-level adaptation (benefits groups, compared to other groups)	• Adaptive in small groups of related individuals but not in modern social environments.
• Individual-level adaptation (benefits individuals, compared to other individuals within the same group)	• By-product of traits that are adaptive in nonreligious contexts.
• Cultural parasite (benefits cultural traits without regard to the welfare of human individuals or groups)	

Turning to non-adaptation hypotheses, the traits associated with religion might have been adaptive in past environments, when social groups were small and composed largely of genetic relatives, but not in the large groups of unrelated individuals that characterize modern religious groups (Alexander 1987). Alternatively, the traits associated with religion might be a costly by-product of traits that are beneficial in nonreligious contexts. Two versions of the by-product hypothesis deserve special mention because they are prominent in the current study of religion. Sociologists such as Rodney Stark and William Bainbridge interpret religion as a by-product of economic thought (Stark 1999; Stark and Bainbridge 1985,1987). The basic idea is that people use cost-benefit reasoning to obtain many benefits in non-religious contexts. Some benefits cannot be obtained, such as rain during a drought or everlasting life, but that does not prevent people from wanting and trying to achieve them, so they invent supernatural agents with whom to bargain for that which they cannot have. Stated in evolutionary terms, religion is a functionless by-product of mental processes that are highly adaptive in nonreligious contexts.

More recently, evolutionary biologists such as Boyer (2001), Atran (2002), Atran and Noyenzayan (2004), and Guthrie (1995) have proposed a by-product theory of religion that differs from Stark and Bainbridge primarily in reliance upon evolutionary psychology rather than economics for the basic conception of the human mind. Instead of being general cost-benefit reasoners, humans are thought to employ numerous cognitive modules that evolved to solve specific adaptive problems in ancestral environments. These modules are adaptations, at least when they were expressed in nonreligious contexts in the past, but their expression in religious contexts, past and present, has no function. This modern evolutionary theory of religion differs from the modern economic theory in the basic conception of the human mind, but they are similar in regarding religion as a functionless by-product of traits that are functional in nonreligious contexts.

Two important insights can be derived from this classification of hypotheses about religion, even before we attempt to test them. First, all of them are plausible and might be true to some degree. Second, they make very different predictions that should be possible to test empirically. A religion designed for the good of the group must be structured differently than a religion designed as a tool for within-group advantage, which in turn must be structured differently than religion as a cultural parasite good for nothing but itself, which in turn must be structured differently than a religion for which the word "design" is inappropriate, at least within a religious context. These various conceptions of religion are so different that it would be surprising if they could not be empirically discriminated from each other. In short, evolutionary theory can be used to achieve the same comprehensive understanding of religion that we have achieved for guppies (and the rest of life), even though the emerging story will be complex and will include numerous selection pressures and constraints on selection.

Darwin's Cathedral: Evolution, Religion, and the Nature of Society (Wilson 2002) presents my own attempt to explain the subject of religion from an evolutionary perspective. My central thesis is that religions are largely (although by no means entirely) group-level adaptations. In their explicit behavioral prescriptions, theological beliefs, and social practices, most religions are impressively designed to provide a set of instructions for how to behave, to promote cooperation among group members, and to prevent passive freeloading and active exploitation within the group. The features of religion that appear most irrational and which have always made religion such a puzzle to explain from a scientific perspective can be largely understood as part of the "social physiology" (to use a term employed by social insect biologists) that enables the religious group to function adaptively.

Before continuing, it is important to explain why I stress a single hypothesis (group-level adaptation) even though I also appreciate the multifactorial nature of evolution (as also emphasized by Hinde 1999). One reason is historical. Not only was group selection rejected by many evolutionary biologists during the middle of the twentieth century, but the related tradition of functionalism was rejected by many social scientists during the same period. Serious intellectual work is required to return the basic concept of groups as adaptive units to scientific respectability (comprising chapters 1 and 2 of *Darwin's Cathedral*), even before we can apply it to the subject of religion. Another reason is based on the distinction between religion as idealized and as actually practiced. People often behave selfishly in the name of religion, as in the case of the Catholic practices that led to the Protestant reformation. However, these practices are often regarded as a "corruption" of religion rather than part of the "true" religion that is more "purely" associated with the welfare of the group. The meaning of terms such as "ideals," "corruption," "true," and "pure" requires an analysis of cultural evolution from an evolutionary perspective, which provides part of the broad theoretical background for the study of religion along with the basic concept of groups as adaptive units. Ideals are phenomena in their own right that influence actual behavior, even if they are not completely successful. Theoretically, religious ideals could reflect any of the major hypotheses outlined in Table 1. The fact that they reflect the group-level adaptation hypothesis even more than actual behavior is worth noting.

When it comes to testing the major hypotheses outlined above, it is important to recognize the importance of descriptive in addition to quantitative information. Darwin established his theory of evolution very successfully on the basis of descriptive information about plants and animals gathered by the naturalists of his day, most of whom thought they were studying God's handiwork. Traditional religious scholarship provides a comparable body of information about religious groups in relation to their environments that can be used to test evolutionary hypotheses about religion. Quantitative methods *refine* but do not *define* scientific inquiry. Thus, although I review the modern social scientific literature on religion in *Darwin's Cathedral,* I also draw heavily upon detailed descriptive accounts of particular religious systems in relation to their environments.

These accounts provide compelling evidence for the group-level benefits of religion, but they are also vulnerable to the criticism of selection bias. Couldn't someone else handpick examples that illustrate the nonadaptive nature of religion, such as the celibate Shakers or the suicidal Jonestown cult? Random sampling provides an effective solution to this problem. If the major hypotheses are evaluated for a sample of religions chosen without respect to the hypotheses, then (barring freak sampling accidents) the results for the sample will be representative of the population from which the sample was drawn.

In Chapter 4 of *Darwin's Cathedral* I initiated such a survey by selecting 25 religious systems at random from the 16-volume *Encyclopedia of Religion* (Eliade 1987). In this paper I provide a preliminary analysis of the survey, which has been expanded to include 35 religious systems. It is not the last word but rather the first step of a task that is best continued by a community of religious scholars who are qualified to evaluate in detail the "natural history" of the religions that are comprised in the sample. Even in its preliminary stage, however, it provides important insights about the nature of religion from an evolutionary perspective.

METHODS

The religions to be included in the sample were chosen by writing a computer program that selected volume numbers and page numbers within each volume at random. An entry located by this procedure was then evaluated by criteria listed below to see if it qualified for inclusion in the sample. If not, I paged forward until I encountered the first entry that met the criteria.

An entry qualified if it could be associated with a single religious system, defined as a recognizable group of people with beliefs and practices that can be distinguished from other beliefs and practices. All systems were assumed to be religious because they were included in an encyclopedia of religion. In other words, I based my definition of religion on the inclusion criteria of the encyclopedia rather than imposing my own definition. This is crucial to avoid my own selection bias, although the selection bias of the editors might well deserve scrutiny. A particular entry that met the criteria might be the name of a person who founded a new religious movement (e.g., Eisai, founder of the Rinzai school of Zen Buddhism in

Japan during the twelfth century), a god (e.g., Mithra, an Iranian deity and god of a Roman mystery religion), or the name of the movement itself (e.g., the Cao Dai cult that originated in Vietnam during the twentieth century). Minor religious movements within a larger religious tradition were included, since the larger traditions themselves started out as minor movements. Entries on general subjects such as "myth" or "polytheism" were excluded because they did not refer to a single religious system. Somewhat arbitrarily, I excluded religions associated with tribal groups that have no known starting date, even though I include them in *Darwin's Cathedral* and regard them as supportive of my main thesis.

This sampling procedure is not completely unbiased. Judgment calls were sometimes required to decide if an entry met the criteria, as described in more detail below. The procedure favors long entries over short entries. The major religious traditions might not be equally represented because some (e.g., Protestant) divide into separate movements more than others (e.g., Catholic). The entire encyclopedia might be biased in its inclusion criteria, contributors, and information available for different religions around the world and throughout history. State-level societies are probably over-represented. Nevertheless, the important point is that the religions were not chosen with the major evolutionary hypotheses in mind. The bias of choosing religions known to support a given favored hypothesis has been successfully avoided.

One potential bias deserves special mention. Religions that succeed in the sense of persisting and becoming large are more likely to be included in the encyclopedia than religions that remain small and quickly fail. This bias, if it exists, would reflect cultural evolution in action. The statement "most religions have secular utility" would not be false because it is based on a biased sample, but true because the encyclopedia reflects the winnowing process of cultural evolution. Correcting the "bias" would provide a more complete cultural "fossil record" that includes the ephemeral "losers" in addition to the persistent "winners," enabling the process of cultural evolution to be studied in even greater detail. As we shall see, the sample does include some religious "losers" in addition to "winners," which are highly instructive.

The encyclopedia was used to select the random sample and provided a small amount of information about each religion, but the main work of the survey involved gathering as much information as possible about each religion and evaluating it with respect to the major evolutionary hypotheses. This was accomplished with the help of 35 undergraduate students who enrolled in a 4-credit class entitled "Evolution and Religion." In addition to reading *Darwin's Cathedral* and discussing the general subject, each student was assigned a single religion to research over the course of the semester (which is why the size of the sample was expanded to 35), culminating in a bibliography and narrative answers to 32 questions addressing key issues (available upon request). A first draft of the answers was read in time to provide feedback for each student to correct shortcomings in the final draft. This procedure insured that the students addressed the most important evolutionary issues and facilitated comparison among the religions.

This material provided the basis for my own analysis. I did not rely exclusively on the student analyses but rather used them as a guide to my own reading of the primary literature. As I have already stressed, the use of students to gather information and the descriptive nature of my analysis are only the first steps of an enterprise that ultimately should include the scholars who are the real "natural historians" for the religions in the sample.

DESCRIPTIVE ANALYSIS

Table 2 lists the encyclopedia entries that were included in the survey and a brief description of the religions that they represent. The major traditions of Buddhism, Taoism, Judaism, Christianity, and Islam are represented, although not Hinduism or Confucianism. Jainism and Zoroastrianism are among the oldest religions that are still being practiced today, albeit among a small minority of the world's population. Also included are a cult with African roots (M'Bona) a cult based upon an ancestor (Cinggis Khan), and two modern movements that are composites of the major religious traditions and other influences (Cao Dai and the Theosophical Society). The religions span the globe and range in time from the twenty-fifth century BCE to the present.

Most of the entries refer to religious movements, large or small, that clearly meet the inclusion criteria, but a few proved to be somewhat inappropriate in retrospect. The entry "cult of saints" refers to many cults within the Catholic religion rather than to a single cult, which makes it difficult to evaluate. Saint Catherine of Siena played an important conciliatory role in the Catholic Church during the fourteenth century, helping to prevent schism rather than promoting it. Ziya Gokalp was a political rather than a religious leader who was influential in the separation of church and state for the nation of Turkey. Agudat Yisra'el is not a religious movement in its own right but a political arm of a preexisting religious movement (Orthodox Judaism). Even though these entries marginally qualify for inclusion in the survey, they are instructive in ways that will be described in more detail below. It is important to keep in mind that none of the entries could have been included in the sample without first being included in the encyclopedia of religion. Thus, they are relevant to the subject of religion writ large (as defined by the editors of the encyclopedia) even when they don't constitute a specific religious system as defined by the inclusion criteria of the survey.

The religious systems identified by the entries differed greatly in the amount of available information. Even when information was available, authors differed in the degree to which they related theology to social and ecological context. Despite these problems, a number of preliminary conclusions can be drawn that are relevant to the major evolutionary hypotheses.

Table 2. 35 Religions Chosen at Random from the 16-Volume *Encyclopedia of Religion* (Eliade 1987)

Vol	Page	Entry	Description (dates CE unless specified otherwise)
1	149	Agudat Yisra'el	Orthodox Judaism, twentieth century
1	161	Airyana Vaejah	Zoroastrianism, Persia, tenth century BCE
1	211	Allen, R.	African Methodist Episcopal Church, nineteenth century
1	492	Atisa	Ṭibetan Buddhism, tenth century
3	72	Cao Dai	Composite of traditions, Vietnam, twentieth century
3	120	Catherine of Siena	Catholic church, Italy, fourteenth century
3	230	Chen-Jen	Chinese Taoism, third century
3	328	Chinggis Kahn	Ancestor cult, Mongolia, thirteenth century
3	333	Chinul	Korean Buddhism, thirteenth century
4	172	Cult of Saints	Catholic Church, general
4	200	Dalai Laṃa	Tibetan Buddhism, general
4	236	Dan Fodio, Usuman	Nigerian Islamic revivalist movement, eighteenth century
4	326	Dge-Lugs-Pa	Tibetan Buddhism, fifteenth century
5	72	Eisai	Rinzai school of Japanese Zen Buddhism, twelfth century
5	156	Eshmun	Phoenician healer god, fifteenth century BCE
6	66	Gokalp, Z.	Turkish nationalism, twentieth century
7	119	Iman and Islam	Islam, general
7	215	Indus Valley religion	Western India, twenty-fifth century BCE
8	104	Jodoshu	Pure land sect of Japanese Buddhism, twelfth century
8	423	Lahori, Muhammad Ali	Lahori branch of the Ahmadiyah movement, Islamic, twentieth century

The Secular Utility of Religions

According to the by-product hypothesis, human psychological and social processes are clearly adaptive in nonreligious contexts but are triggered inappropriately in religious contexts. We pray to God for everlasting life, not to convey us to work in the morning. We see faces in the clouds because our minds are wired for social interactions. Going to work and engaging in social interactions have clear practical benefits, whereas praying for everlasting life and seeing faces in the clouds do not. Regardless of whether this hypothesis is framed in terms of rational choice theory or evolutionary psychology, the expectation is that religions by themselves do not produce practical benefits.

The random sample does not support this expectation, even with the limited information available (see also Reynolds and Tanner 1995). The majority of reli-

Table 2. *(continued)*

Vol	Page	Entry	Description (dates CE unless specified otherwise)
9	128	Mahavira	Jainism (India), sixth century BCE
9	188	Maranke, J.	Apostolic Church of John Maranke (Africa), twentieth century
9	287	Maurice, F. D.	Christian Socialism (England), twentieth century
9	291	Mawdudi, Sayyid Abu Al-a'la	Indian Islamic revivalist movement, twentieth century
9	303	M'Bona	African territorial cult, nineteenth century
9	579	Mithra/Mithraism	Iranian deity and god of Roman mystery religion, ca. fourth century BCE
10	290	Nagarjuna	Indian Buddhism, second century
10	297	Nahman of Bratslav	Bratslav sect of Hasidic Judaism, Ukraine, eighteenth century
10	360	Neo-orthodoxy	Protestant revivalist movement, Europe and America, twentieth century
11	226	Pelagianism	Christian doctrine opposed by Augustine, fourth century
11	324	Pietism	Protestant reformation movement, Europe, seventeenth century
12	335	Rennyo	Pure land true sect of Japanese Buddhism, fifteenth century
14	38	Spurgeon, C. H.	English Baptist Church, nineteenth century
14	464	Theosophical Society	Composite of traditions, America, nineteenth century
15	539	Young, B.	Mormonism, America, nineteenth century

gions in the sample are centered on practical concerns, especially the definition of social groups and the regulation of social interactions within and between groups. The impetus for a new religious movement is usually a situation in which a constituency is not being well served by current social organizations (religious or secular) and is better served in practical terms by the new movement. This dynamic describes the origin of Christianity and Islam and more recent religious movements within all of the major religious traditions, including the following examples from the random sample:

- Within Judaism, Agudat Yisra'el was formed in the early twentieth century to "unite under one organizational roof representatives of Orthodox communities from Germany, from Russia, Poland and Lithuania, and from Hungary" (Eliade 1987:150). Its primary goal was preserve and advance an orthodox form of Judaism, compared to more secularized forms. Agudat Yisra'el is described as the political arm of Orthodox Judaism because so many of its objectives are utilitarian, such as the economic support of distressed communities.

- Within Islam, Sayyid Abu Al-a'la Mawdudi founded an Islamic revivalist movement in the early twentieth century whose purpose was to protect Muslim interests from Hindus, secular nationalism, and Western culture. This movement was explicitly intended to define and promote the survival of a minority group threatened by competing social organizations.
- Within Christianity, Mormonism arose as one of many new movements in America during the early nineteenth century but was special in its ability to create encapsulated cooperative groups, which were persecuted for their success before undertaking their spectacularly coordinated westward migration. Mormonism continues to grow at a rate that rivals early Christianity and Islam.
- Among Eastern religions, Jainism constitutes a small fraction of the Indian population but one that has persisted for several thousand years. This impressive longevity is based on practical benefits, not some mysterious connection to traits that have functioned in a nonreligious context for such a long period, as I will describe in more detail below.

Clearly, these religious systems are about more than seeing faces in clouds and praying for unattainable goals, such as everlasting life. They are about goals that can be achieved but only through the coordinated action of groups. The practical benefits of religion might seem so obvious that they don't need to be pointed out, but then why have so many by-product theories of religion been proposed over the decades, from "animism" and "naturism" in the nineteenth century to the economic and evolutionary by-product theories of today? Somehow these theorists have managed to interpret the practical benefits of religion as "incidental," in contrast to something more "fundamental" about religion that cannot be explained functionally. This rendering can accommodate occasional practical benefits associated with religion, but not the results of this survey based on a random sample. According to my assessment, *most* of the religions in the sample are thoroughly rooted in the practical welfare of groups. In addition, the beauty of random sampling is that results for the sample apply to the entire encyclopedia from which the sample was drawn. If my assessment is correct, then the nature of religion cannot be understood without acknowledging its "secular utility," as Durkheim put it.

The practical purpose of most religious groups explains why Ziya Gokalp, a political leader who helped to separate church and state for the nation of Turkey, was included in an encyclopedia of religion. According to Heyd (1950:56):

> To Gokalp, Allah (Islamic God) was no longer the personal God. Instead to him "God was society." The sanctity of human personality is explained by its being the bearer of the "collective consciousness," the soul of society taking the place of the religious conception of the divine spirit.

Despite the use of nationalistic rather than religious imagery, it is obvious that church and state were *in the same business* of organizing the lives of a group of people. A similar process took place for the separation of church and state in American history (Cousins 1958). Framers of the constitution such as Benjamin Franklin and Thomas Jefferson realized that religions are good at organizing social life among their own members but became part of the problem with respect to the larger scale of social organization that they were trying to achieve. The separation of church and state was a remarkable piece of social engineering and the imagery of God was freely combined with the imagery of nationalism to bless the new enterprise. To summarize, religion is intimately involved with the practical commerce of life, which requires an adaptationist explanation.

The Proximate/Ultimate Distinction and the Otherworldly Aspects of Religion

If religions are so practical, then why are they also so otherworldly? Why do they flaunt the kind of practical reasoning associated with science and rational thought? Why the belief in Gods that cannot be empirically verified, costly and time-consuming rituals, and the rest? These are the elements of religion that drive theorists toward nonfunctional explanations (the right side of Table 1). However, evolutionary theory offers a robust alternative in the distinction between ultimate and proximate causation.

All adaptive traits require two complementary explanations: the environmental forces that favor the trait in terms of survival and reproduction (ultimate causation) and the mechanisms that cause the trait to exist in actual organisms (proximate causation). Most flowers bloom in spring because those that bloomed earlier were nipped by frost and those that bloomed later had insufficient time to grow their fruits (ultimate causation). The same flowers bloom in spring because they possess physiological mechanisms that are sensitive to day length (proximate causation). Both explanations are required to explain an adaptive trait fully, and one explanation can never substitute for the other.

Continuing this example, notice that day length by itself has no effect on survival and reproduction. It is merely a signal that reliably causes the flower to bloom at the best time with respect to other environmental forces. In general, a proximate explanation need bear no relationship whatsoever to the corresponding ultimate explanation, other than to reliably produce the trait that survives and reproduces better than other traits.

Returning to religion, a given belief or practice might exist because it enhances survival and reproduction—for example, by causing the group to function well relative to other groups—but this is only the ultimate explanation. A complementary proximate explanation is needed that need bear no relationship to the ultimate explanation, other than to reliably cause the trait to occur. Perhaps a religious believer helps others because she wants to help others, or perhaps because she wants to serve a perfect God who commands her to help others. As far as proximate causation is concerned, the particular psychological motivation makes no difference as long as the helping behavior is reliably produced.

The proximate/ultimate distinction has profound implications for the study of religions by providing a way to reconcile their functional and otherworldly aspects.

When trying to explain a given feature of a religion, the primary question is not "Is it rational?" or "Can it be empirically verified?" but "What does it cause people to do?" This is the *only* relevant gold standard as far as proximate mechanisms are concerned. If the feature motivates adaptive behaviors, then it is fully consistent with a functional explanation (the left side of Table 1) no matter how bizarre (to nonbelievers) in other respects. If it fails to motivate adaptive behaviors, then a nonfunctional explanation (the right side of Table 1) is warranted.

In *Darwin's Cathedral,* I attempt this kind of analysis for a few selected religions, especially Calvinism as it originated in the City of Geneva in the sixteenth century. I show that theological beliefs (such as original sin, predestination, and the nature of faith and forgiveness) and social practices (such as rules governing decision making, discipline, and excommunication) combine with explicit behavioral prescriptions to form an impressive self-reinforcing system for organizing collective behavior. The system is necessarily complex because adaptive behavior is necessarily context-sensitive. For example, adaptive forgiveness behavior cannot possibly be embodied in a rule as simple as "Turn the other cheek." Different rules of forgiveness are required for different situations and categories of people, which must somehow be specified by the religious system. These rules can appear contradictory and hypocritical (e.g., How can Christians be intolerant of various behaviors while preaching "Turn the other cheek"?) until their context-specificity is appreciated. Comparative and longitudinal studies of religion are especially helpful for revealing the adaptive nature of these proximate mechanisms. For example, early Christian communities appear to have altered their sacred stories in response to the demands of their particular social environments (Pagels 1995, 2003). In this fashion, the otherworldly side of religion can be largely explained in terms of proximate causation, rather than as forms of maladaptive behavior. I include the word "largely" because I do not claim that each and every nuance of religion is adaptive. Evolution is a messy and multifactorial process for religion in addition to the rest of life. My point is that the otherworldly side of religion does not by itself necessitate a rush to nonfunctional explanations (the right side of Table 1). The proximate/ultimate distinction provides a very robust alternative explanation, and empirical research is required to settle the issue for any particular feature of particular religions.

Readers can judge for themselves how well I have succeeded for my selected examples in *Darwin's Cathedral,* but in any case they are vulnerable to the criticism of selection bias. The random sample avoids selection bias but has other limitations, such as limited information for some of the religions and my own limited ability to evaluate the enormous amount of information for all 35 religions. Nevertheless, some preliminary observations will help set the stage for more detailed future analysis by others in addition to myself.

The otherworldly side of religion is richly represented in the random sample. Joseph Smith's encounter with heavenly messengers that marked the beginning of Mormonism is well known. Comparable examples include an encounter with the Supreme Being through a Ouidja board for the Cao Dai religion in Vietnam and an ancient secret brotherhood of adepts for the Theosophical Society. Numerous religious leaders in the sample attracted a following by their exceptional piety and indifference to worldly values. St. Catherine of Sienna had a vision of Christ at age six and took a vow of virginity against her family's wishes. Nahman of Bratslav locked himself in his parent's attic for long periods of time in an attempt to gain nearness to God. His disapproval of secular desires went so far that he didn't even want a following, which only enhanced his reputation as an enlightened spiritual leader. The ascetics of Eastern religions give up all worldly belongings and at times even fast themselves to death. At a less extreme level, numerous religious movements in the sample were envisioned as a move away from worldly secular values to more pure religious values based upon God and his commandments or the achievement of enlightenment. Finally, numerous religious movements attracted followers on the basis of miraculous claims such as bringing rain and faith healing that (based on current scientific knowledge) have no basis in fact. In short, the random sample amply confirms that religious belief includes but also goes far beyond a direct motivation to help others. The question is, do these seemingly nonutilitarian beliefs reliably cause the members of religious groups to help each other and otherwise function as adaptive units?

By my assessment, the answer to this question is primarily "yes" for the religions included in the random sample and therefore the entire encyclopedia. Saint Catherine treated love of God and love of neighbor as "inseparable commandments" (Hilkert 2001). Similarly, the encyclopedia defines the word Islam as follows:

> A noun derived from the verb aslama ("to submit or surrender [to God]"), designates the act by which an individual recognizes his or her relationship to the divine and, at the same time, the community of all of those who respond in submission. It describes, therefore, both the singular vertical relationship between the human being and God and the collective, horizontal relationship of all who join together in common faith and practice (Eliade 1987 [vol. 7]:119).

The success of Mormonism in secular terms is as famous as its otherworldly beliefs. The Cao Dai religion similarly functions as an organizer of secular life for its believers. The spread of relics associated with Saints evidently played a major role in the Christianization of the West (Eliade 1987 [vol. 4]:172). Wills (2001) provides a detailed account of how the City of Venice had its own religion based upon Saint Mark that very successfully organized secular life, frequently in opposition to the Catholic Church in Rome. Although Buddhism is often portrayed as an individualistic quest for enlightenment, most versions of Buddhism in the sample were closely involved with the organization of society through the patronage of kings and other secular rulers. In the African M'Bona cult, a shrine is constructed in a way that deteriorates over time. Members of the cult must periodically rebuild the shrine, but only after resolving their secular disputes. When offered the opportunity to build the shrine out of more durable materials, they refused (Schoffeleers 1992:75).

Pelagianism provides an excellent example of competition among alternative religious belief systems. Pelagius was a Christian monk who disagreed with St. Augustine on fundamental religious doctrines. Whereas Augustine believed that humanity was sinful by nature and must rely on God's grace for salvation by converting to Christianity, Pelagius believed that the souls of all men were created by God and that even pagans could enter heaven by their moral actions. Both doctrines motivated other-oriented behaviors, but they were not compatible with each other and Pelagianism was condemned in 431 by the Council of Ephesus.

In a contest such as this, one contender is going to win even if they are evenly matched. Alternatively, they can coexist by fissioning into separate religions that fill different socioecological "niches." Although Pelagianism ceased to exist in its original form, its elements have resurfaced throughout Christian history, for example in the Quaker doctrine of an inner light that stands in contrast to the doctrine of original sin (Ingle 1994). Religious scholar Elaine Pagels (1995, 2003) has written extensively on competition among alternative versions of Christianity, leading to the accumulation of forms that are exceptionally good at creating and maintaining strong communities. She does not frame her argument in terms of evolution, but it strongly supports the proximate/ultimate distinction as a way to reconcile the otherworldly and practical dimensions of religion.

In this section I have tried to establish two major points. First, the proximate/ultimate distinction theoretically enables the otherworldly and practical dimensions of religion to be reconciled with each other. The key question is: *What do the otherworldly elements of religion cause people to do?* Second, I have made an empirical claim based on the survey that the otherworldly and practical dimensions of religion are indeed tightly yoked to each other. If I am correct, then the major hypotheses on the right side of Table 1 are not required to explain the otherworldly side of religion. However, there is a third major point that I have not addressed: Why can't the proximate mechanisms be more straightforward? Why don't we just help our neighbor rather than believing in a perfect God who commands us to help our neighbor? This is a fundamental question but it requires a comparison of religious systems vs. nonreligious systems, where the proximate mechanisms *are* more straightforward. It is therefore beyond the scope of this survey but has been discussed by myself and others elsewhere (e.g., Wilson 2002: chap. 7; Alcorta and Sosis 2005 [this issue]).

Group-level Benefits, Individual Benefits, or Cultural Parasites?

So far I have tried to establish that most religions in the random sample are rooted in practical concerns (the left side of Table 1) and that their otherworldly aspects can be understood largely in terms of the proximate/ultimate distinction. Now it is time to discuss the three adaptationist hypotheses in more detail.

As I have already stressed, religion is inherently group- and other-oriented, as practiced and especially as idealized. The benefits produced by religion are obviously enjoyed by members of the group, and are "selfish" in that sense, but they are usually not selfish in the sense of causing some members to profit at the expense of other members of the same group. Instead, the benefits of religion tend to be public goods whose production requires time, energy, and risk on the part of individuals. When we focus on the fitness *differences* required for natural selection to act, we find the same problem for religion as for public goods in general. Producing them decreases fitness relative to those within the same group who enjoy the benefits without the costs, a negative fitness difference. The positive fitness differences that favor public good production are primarily between groups. Very simply, groups that "get their act together" outperform other groups, and this advantage outweighs the disadvantages of being a public good provider within groups. Most elements of religion are designed to favor the production of public goods and to limit the disadvantages of public good production within groups. When these mechanisms fail, the self-serving behaviors performed by religious believers tend to be regarded as a corruption of religion rather than an aspect of the "true" religion. That is why between-group selection needs to occupy a central role in the study of religion, as I argue in *Darwin's Cathedral.*

This argument is strongly supported by the random sample. Most religions in the sample are as dedicated to the production of public goods as the selected examples in *Darwin's Cathedral.* In addition to those that have already been cited, Richard Allen founded the African Methodist Episcopal Church to address the needs of African Americans that were not being met by white-dominated churches. John Maranke founded the Christian Apostolic Church in Africa for the same reason. Frederick Maurice helped to establish Christian socialism as a religious version of the socialist movement in England. These examples might seem mundane when considered individually, but they gain significance as part of a random sample by establishing the group- and other-oriented nature of religious systems in general.

No social system, religious or secular, completely solves the problems of passive freeloading and active exploitation within groups, especially by the leaders. In *Darwin's Cathedral* I discuss two major ways that religions fall apart. The first is by becoming victims of their own success. Once a religion generates wealth by collective action, its members no longer need each other and leave or try to weaken the constraints on their behavior. John Wesley, the founder of the Methodist Church, was perfectly aware of this problem when he stated, "I do not see how it is possible, in the nature of things, for any revival of religion to continue for long. For religion must necessarily produce both industry and frugality. And these cannot but produce riches. But as riches increase, so will pride, anger, and love of the world in all its branches" (1976 [vol. 9]:529).

A second way that religions fall apart is by becoming exploitative, such that some members benefit more than others. When this happens, three outcomes are possible: The exploited members can work for reform, they can be forced or deceived to remain exploited, or they can branch off to form their own church. These possibilities illustrate that religions are *not* pure products of between-group selection. They always reflect a balance between levels of selection in which the disruptive effects of within-group advantage are present and in danger of escaping social control. The religious obsession with "sinfulness," "worldliness," "attachment," and "self-will" reflects this ever-present danger.

These conflicts are amply represented in the random sample. A substantial proportion of religions in the sample are based not on a new constituency (as for the African and African-American churches) or a new social need (such as Christian Socialism) but on the need to "purify" an existing church that has become "corrupted" by worldly values. For example, Usuman Dan Fodio founded an Islamic movement in Nigeria that, according to the student reviewing the material, was "distinguished by their refutation of those who had knowledge but failed to put it into practice; those who presented an appearance of compliance with the outward religious duties, but had not eliminated such characteristics as vanity, hypocrisy, ambition, desire for political office and high rank; those who presumed that they had the exclusive right to guide the common people and yet entered into unholy alliance with the sultans, thus encouraging the sultan's oppression of the people; those who engaged in jihad but only to obtain fame and wealth; and those scholars who used false methods, such as music, to lure people into spiritual practice." With the exception of music, this list clearly focuses on behaviors that are self-serving without contributing to the welfare of the group. The new, "purified" religion was rigidly structured to avoid these problems and to turn the community of believers into an encapsulated group. Great attention was paid to matters of dress, prayer, and ritual which appear to have no functional basis when taken out of context, but which make sense in terms of the proximate/ultimate distinction discussed in the previous section.

Even the famous Taoist indifference to worldly affairs makes sense as a way to prevent political corruption. Chen-Jen is a term used in the Chuang-tzu for a person who "does not refuse all contact with human society and politics, but if he should happen to 'get involved' he will not allow himself to 'feel involved'." During this period of Chinese history, "the feudal system of the Chou dynasty was in its final agony, and interstate relationships were characterized by ruse and violence." According to chapter 21 of the Chuang-tzu, Sun Shu-ao, an exemplar of Chen-Jen, had "thrice been named prime minister without considering it glorious and thrice been dismissed without looking distressed. 'Why should I be better than anyone else? When [the nomination] came, I could not refuse it; when it left, I could not keep it. Neither getting it or losing it had anything to do with me'" (all quotes from Eliade 1987 [vol. 3]:230–231). In a world full of vested interests, who better to choose for a leader than someone who has demonstrated a lack of vested interests (Irons 2001; Sosis 2004)?

In *Darwin's Cathedral* I stress that the *product* of natural selection is adaptation but the *process* of natural selection includes many failures for each success. Religious systems reflect a degree of intentional thought but in many respects they are unplanned social experiments, only a few of which succeed. A good cultural fossil record of religions should include the failures in addition to the successes. Several religions in the random sample can be regarded as failures in the sense that they did not achieve a large following or succeed at their stated goal, such as the example of Pelagianism that I have already described. Frederick Maurice's effort to establish a form of Christian Socialism was well-meaning—no one could doubt its communitarian purpose—but never amounted to much and became a footnote to religious history. The Theosophical Society was based on a blend of science and occultism that made sense in the nineteenth century but attracts only a tiny number of followers today. Nevertheless, in its own way it provided "a new sense of purpose, mission, and service to others" (Campbell 1993:8). Thus, even the failures illustrate the fundamentally other- and group-oriented nature of religion.

Since natural selection is always based on fitness differences, group-level adaptations can evolve only by some groups contributing more to the gene-pool or culture-pool than other groups. Between-group competition can take the form of direct conflict but it can also take more benign forms, such as differences in economic efficiency. Darwin was careful to point out that natural selection at the individual level does not always take the form of nature red in tooth and claw. A drought-tolerant plant out-competes a drought-susceptible plant in the desert, even though they do not directly interact with each other. The same point needs to be made for natural selection at the group level. It is encouraging that *most* of the religions in the random sample and therefore the encyclopedia did *not* spread by violent intergroup conflict. Instead, competition among groups took place primarily through differences in recruitment, retention, and birth and death processes based on the ability of the group to function as an adaptive unit. It is undeniable that group selection sometimes takes the form of violent conflict, but the relatively small number of cases in the random sample adds a new perspective and makes it an open question whether religion per se increases or decreases the potential for violent conflict, compared with nonreligious human social organizations.

Our species is unique in its reliance on cumulative, socially transmitted information. The psychological and cultural processes responsible for the origin and spread of new traits are evolutionary in a broad sense but they differ from genetic evolution in many of their details. Even for purely genetic evolution, different traits are favored when the genes are autosomal (inherited through both parents), cytoplasmic (inherited only through the mother), or on the y-chromosome (inherited only through the father). Cultural evolution includes an even broader range of possibilities, in which a given trait can be transmitted via both parents, one parent, nonparent adults (teachers), peers, and so on. Each transmission mode is expected to favor a different set of traits, just as for purely genetic transmission modes. For these and other reasons, cultural evolution is not expected to produce exactly the same outcome as genetic evolution (Richerson and Boyd 2004).

Dawkins (1976) coined the term "meme" as a cultural analog of "gene." As he and others have developed the concept, memes can be regarded as autonomous life forms that have evolved exclusively to perpetuate themselves with no more interest in the benefit of their human hosts than a tapeworm or the AIDS virus has. Religion is sometimes cited in support of this conception (e.g., Blackmore 1999), usually with the assumption that religion is so mystifying that it cannot be explained from any other perspective and that people would be better off without it, just as if we could eradicate the common cold.

The random sample provides virtually no support for the cultural parasite hypothesis. As I have already described, most of the religions in the sample are designed to promote the welfare of their members, and their otherworldly nature can be straightforwardly explained with the proximate/ultimate distinction. In addition, the basic concept of memes as independent agents can be faulted on theoretical grounds (Richerson and Boyd 2004).

A more sensible conception of cultural evolution is provided by Richerson and Boyd (2004). Not only have the parameters of cultural evolution evolved by genetic evolution to promote biological fitness on balance, but they have evolved to increase the efficacy of between-group selection relative to within-group selection. It is thanks to cultural processes that human groups are able to function as well as they do. Potential examples discussed in the theoretical literature include social transmission rules that increase variation among groups, low-cost mechanisms for detecting and punishing norm violations (such as gossip), and so on.

This conception of cultural evolution is far more theoretically plausible and consistent with the random sample than the parasitic concept. A good religion is awesome in the degree to which it organizes behavior and replicates itself through time. The mechanisms that enable all of this nongenetic information to be encoded, expressed under the right conditions, and faithfully transmitted must be very sophisticated indeed. Theoretical models of cultural evolution have not yet grasped this degree of sophistication and can benefit as much from the study of religion as the study of religion can benefit from the theoretical models.

Jainism: A Challenge and Its Resolution

Of all the religions in the random sample, the one that initially posed the greatest challenge to the group-level adaptation hypothesis was Jainism. As old as Buddhism, Jainism is famous for its ascetic values. Jain renouncers wear masks to filter the air that they breath, carry a broom to sweep the path in front of them, and have dozens of food restrictions to avoid killing any tiny creature. They are homeless and in some sects travel completely naked. Some even accomplish the ultimate ascetic act of fasting themselves to death. How can such beliefs and practices possibly contribute to the secular utility of either individuals or groups? Remarkably, they do. The following account is based on a detailed ethnography of a modern Jain community whose title says it all: *Riches and Renunciation: Religion, Economy and Society among the Jains* (Laidlaw 1995).

The ascetic renouncers constitute a tiny fraction of the Jain religion, whose lay members include some of the wealthiest merchants of India.

As is generally the case among the Jains, these families are on the whole active and dedicated followers of the religion. This is especially so during periods when there are renouncers living among them, but even at other times the social mores of the community and the everyday lives of its members are shaped in profound ways by Jain religious values. The daily rites in local temples are well attended and local public events are almost all religious. Like the renouncers, members of Jain families engage in ascetic exercises and in periodic fasting.

But this does not mean that lay Jain communities come to resemble renouncer orders. Nothing in the latter's strict regime would prepare one for the celebration and enthusiasm which attend Jain religious ceremonies, for the colour and opulence of their collective life, for their wealth, for their frank and cheerful pride in that wealth, or for the manifold ways it is linked with asceticism. Like most Jain communities, the Khartar Gacch and Tapa Gacch Jains of Jaipur are generally affluent, and their collective religious life is presided over by members of the most successful business families—in this case, for the most part, wealthy merchants who dominate the city's emerald-trading market, which is one of the largest in the world. It would be going too far to say that it is always the richest lay Jains who have the reputation for being the most religious; but it would only be going too far. In any case it is clear that the Jain religion provides for these families a medium in which to celebrate their worldly success, and to express and affirm the continuity of both family and local community. Yet the doctrine of the religion, as expressed by local teachers and by Jain renouncers themselves, is a soteriology—a project and a set of prescriptions for how to bring one's life to an end (1995:4).

Jainism has been religion of merchants throughout its history. The parallels with Judaism, another merchant religion, are remarkable. Jains lived in diaspora communities thoughout India and surrounding regions to organize trade, developed a sophisticated system of banking and merchant capitalism, formed alliances with nobility, engaged in debt-farming, and were persecuted by resentful lower classes. This economic niche requires a large degree of cooperation and is correspondingly vulnerable to exploitation. Laidlaw describes the modern Jaipur gem trade this way:

The Jaipur emerald market is firmly oriented to international trade and connected, partly through diaspora Jain communities abroad, to markets overseas. . . . Speculative trading of stones within the market is very extensive, and there is an elaborate and active brokerage system. Emeralds are traded not only so that exporters can meet deadlines for large consignments of cut stones, but also in anticipation of the price fluctuations within the market which result from such highly time-dependent demands. The liquidity of the market depends on a system of informal banking, in which all the major gem firms participate, and which uses a version of the *hundi,* a type of promissory note used in India at least since Mughal times. Unsecured cash advances, which might have to be arranged at very short notice, are to be repaid after a fixed period of time, and this might in some cases be a matter of hours. The price at which a business can obtain money depends directly on its reputation for wealth, honesty, and prudent business practice and it depends beyond that on the public perception of its creditworthiness . . . (1995:353–354).

This reputation is based in part on status within the Jain religious community. The same beliefs that prescribe one code of conduct for the renouncers prescribes another code for the lay members that is ascetic in its own way but fully consistent with secular values. Fasting in young women demonstrates mastery over their appetites and increases their marriage prospects. Men compete for the privilege of supporting community activities. The more extreme these demonstrations of religious devotion, the more they are publicized and raise the status of the family. The connection between religion and business is so close that family shrines include account books and tools of the gem trade along with religious artifacts.

The renouncers not only set a personal example and provide guidance through sermons, but they actually enforce religious observance through their food-gathering activities. The principle of nonaction dictates that the renouncers cannot prepare their own food or cause anyone else to prepare it for them. They must drop unexpectedly into many households and take only small amounts of food that will not be missed. In addition, they must be certain that the food is sufficiently pure, which goes beyond the details of preparation to the purity of the preparer.

> The purity of food depends perhaps most of all on that of the person who cooks it. . . . A loose or impious woman puts her family in moral peril, in part through the food she feeds them. Therefore it is particularly to the moral and religious standards of the women in the household that the renouncers look. Do they fast on the auspicious days of each month? Do they attend sermons? Does the household in general, and the women in particular, follow restrictions on what they will eat, and when, that at least come close to those they follow themselves? (Laidlaw 1995:304)

These matters are so important that the renouncers must inspect the entire household before they can accept a tiny amount of the food that the family has prepared for itself. Laidlaw describes a typical visit, which demonstrates the respect and even fear commanded by the renouncers.

> If renouncers are spotted approaching the house, a family will launch into a flurry of preparation, but their manner becomes instantly formal and elaborately graceful as soon as the renouncers actually appear. They perform *vandan,* and then invite them in as they would any honoured guests, "Come Maharaj Sahab, Come." And the renouncers are as curt and perfunctory as their hosts are ingratiating. Typically, they march straight through to the kitchen without acknowledging the family's bows and greetings. . . . The women answer renouncers' sometimes sharp and repeated questions about whether a dish is acceptable for them. . . . During the whole proceedings, the renouncers keep up a constant refrain, "Enough! Finish! No, we won't take that! No more of that! Enough!" The householders counter with assurances of the purity and quality of the food. . . . On leaving the house, renouncers say the words *dharma labh* as a blessing; but on most occasions when I have been present they had already turned away from their hosts and were on their way into the street by the time they called this out behind them (1995:309–313).

It is a mark of honor for a household to be included in the daily rounds of the renouncers and a mark of shame to be avoided. Men are expected to be more lax than women in their observance of these rules, in part because of the demands of their businesses, but they have their own field of competition in the many opportunities that are provided to become financial patrons, including initiation ceremonies for new renouncers, which resemble lavish weddings. These events are so public that it would be impossible for a wealthy member of the Jain community to maintain his reputation without sharing his wealth with the community.

The high moral standards demanded for conduct within the Jain community are not always extended toward outsiders. Debt-farming "hovers between paternalism and naked exploitation" (1995:106). Questionable business practices outside the Jain community are tolerated (1995:342). Despite their obsessive efforts to avoid killing even tiny unseen creatures, Jains do not fit the Western conception of pacifists. Jain mythology includes a renouncer who converts a king to Jainism and persuades him to disinfect the arrows of his army to avoid killing invisible air-beings. No mention is made of using the arrows to kill people (1995:155). When Laidlaw asked a lay Jain about war, he received the following answer:

> No, Jain religion does not say that you should be a coward. Jains are heroes. Religion first teaches you about duty. So if it is part of your duty to go to the front in war, you should do that. It is different for renouncers, but laymen should do that duty (1995:155).

Any adaptive religion must be sufficiently flexible to prescribe different behaviors for different contexts. Jainism possesses this flexibility as well as the religions discussed in *Darwin's Cathedral.*

Even if Jainism is adaptive for the laity, isn't it clearly maladaptive for the renouncers in biological terms? It is important to think holistically when answering this question. All cultures include a fraction of individuals who do not reproduce, sometimes by choice but especially by virtue of circumstances. This is the raw material that cultural evolution has to work with for religions that include a nonreproductive caste. We can predict that the decision to become a renouncer (which is a lifelong commitment) is made primarily by people who do not have other attractive options, and that is exactly what we find. According to Laidlaw, "while non-Jain recruits are welcome as renouncers . . . access to the property, power, and prestige of lay Jain communities is not so readily extended" (1995:115). One Jain woman "was actively encouraged to seek initiation, so that her husband, as a 'widower' would be free to remarry; and the woman was happy to express her flight from 'this world of suffering' as a triumphant escape from an unhappy marriage" (1995:241). More generally,

> [In] Jainism men are much more heavily discouraged than women from entering an order. Dowry among Jains is high, and subject to a constant inflationary pull from richer sections of the community. I know of cases where young women from impoverished Jain families chose renunciation in a situation in which finding a respectable husband was proving very difficult. By contrast, the loss of a son is a financial, organizational, and emotional calamity for the typical Jain family (1995:241).

Another circumstance is age. Older men and women alike often devote more time to their religion when they can relinquish the duties of job and family to their grown children. Their commitment can rival that of the renouncers, but rather than becoming renouncers themselves they tend to adopt leadership roles in the lay religious community. To summarize, reproductive division of labor is not difficult to explain from an evolutionary perspective. It has evolved numerous times in the biological world, and cultures provide plenty of scope for individuals to contribute to the welfare of their society without themselves reproducing.

I have described Jainism in detail for a number of reasons. First, it perfectly illustrates the secular utility of religion that I have also stressed for the random sample as a whole. Second, it shows how the proximate/ultimate distinction can reconcile even the strangest religious beliefs and practices (to outsiders) with the functional side of religion. I cannot improve on Laidlaw's own wording:

> How then, is it possible to live by impossible ideals? The advantage for addressing this question to Jainism is that the problem is so very graphic there. The demands of Jain asceticism have a pretty good claim to be the most uncompromising of any enduring historical tradition: the most aggressively impractical set of injunctions which any large number of diverse families and communities has ever tried to live by. They have done so, albeit in a turbulent history of change, schism, and occasionally recriminatory "reform," for well over two millennia. This directs our attention to the fact that yawning gaps between hope and reality are not necessarily dysfunctions of social organization, or deviations from religious systems. The fact that lay Jains make up what is—in thoroughly worldly material terms—one of the most conspicuously successful communities in India, only makes more striking and visible a question which must also arise in the case of renouncers themselves (1995:7).

Third, this example shows how much progress can be made on the basis of careful descriptive studies of religious systems in relation to their environments. Jainism appears obviously dysfunctional based on a little information but becomes obviously functional based on more information. What exactly accomplished this transformation of the obvious? The most relevant facts are that the renouncers constitute a tiny fraction of the Jain religion, that lay Jains are impressively wealthy, that they occupy a particular economic niche, that the religion prescribes different (and more functional) behaviors for the laity than for the renouncers, that mechanisms of enforcement exist, and so on. Most of these facts are so basic that they are beyond dispute, once they are uncovered and put together, even if there is plenty of room for disagreement at a finer scale of analysis. This is the kind of "natural history" information that enabled Darwin to build such a strong case for his theory of evolution, and it can be used to build an equally strong case for an evolutionary theory of religion. Thanks to Laidlaw's detailed analysis of Jainism, the religion in the random sample that seemed to pose the greatest challenge to the group-level adaptation hypothesis now provides solid support.

The analogy between current religious scholarship and natural history information during Darwin's time can be taken a step further. In both cases, the information was gathered by individuals who did not have evolution in mind. Laidlaw is a cultural anthropologist who describes his own perspective this way:

> This book begins from the observation that people may hold values which are in irreducible conflict, and that logical consistency in what we casually identify as a culture, is not something which is necessarily there to be found. It takes work to create, reproduce, and maintain it, and it is always partial. In so far as people manage, in particular cultural traditions and particular local communities, to create lives which are ethically and intellectually coherent, they are not just inheriting a ready-made, complete and integrated package, but sustaining and reproducing the achievement of culture. Jainism can be made to look like the ordered execution of a single doctrinal program, and as is the case perhaps in all cultural traditions, some of its greatest minds have always wished to make it so; but looking at Jainism as an enduring form of life, one is struck by a different achievement. It seems to provide its followers with ideas, institutions, relationships, and practices—a set of ways of going on—which together make conflicting values compossible, and impossible ideals compelling. This is a considerable achievement, and one that calls for elucidation (1995:21).

Laidlaw never uses the e-word, but his metaphorical use of phrases such as "enduring life form" and his practical focus on "a set of ways of going on" converges upon the evolutionary perspective. There is every reason to use the formal theoretical and empirical tools of evolutionary biology to guide future research on religion.

Summary of the Preliminary Analysis

The initial incentive for this survey was to address the problem of selection bias in *Darwin's Cathedral.* Is the *average* religion as adaptive at the group level as the ones that I chose for detailed analysis? Random sampling potentially provides a definitive answer to this question. It might sound naïve to talk about averages for a subject like religion. Aren't they too diverse for such a simple categorization? Religions are indeed diverse, in the same sense that organisms are diverse, but both can still be evaluated in terms of the major hypotheses listed in Table 1. Despite the preliminary stage of analysis, a number of conclusions can be drawn:

- Most religions in the sample have what Durkheim called secular utility.
- The practical benefits are inherently group- and other-oriented.
- In some cases the practical side of religion is so overt that it becomes indistinguishable from politics.
- In other cases the practical side is obscured by the otherworldly side of religion, but these can be largely reconciled through the proximate/ultimate distinction.
- Evolution is a multifactorial process with many constraints on natural selection, so all of the major hypotheses have some degree of validity. However, portrayals of religion as primarily nonfunctional or individually selfish (in the sense of benefiting some members relative to others within the same group) can be rejected on the basis of the survey.

- Religions are not autonomous cultural life forms that parasitize human individuals and groups, often to their detriment.
- Instead, religions demonstrate that the parameters of cultural evolution have themselves evolved to enhance between-group selection and restrict within-group selection.
- Between-group selection can take the form of direct conflict, but it usually takes other forms.

These conclusions are tentative, based on limited information and my own limited ability to evaluate the information. Ideally, every religion in the sample would be analyzed in relation to its environment with the same thoroughness as Laidlaw's analysis of Jainism. I invite others to join this effort. A thoroughly analyzed random sample can provide a reality check for all theories of religion from any theoretical perspective in addition to my own analysis from an evolutionary perspective.

In addition to a survey based on a random sample to guard against selection bias, other surveys are needed to ask more focused questions from an evolutionary perspective. For example, it would be fascinating to compare the religions of cultures that occupy the same economic niche, such as the merchant cultures of Jains in India and Jews in Europe. Numerous merchant cultures have existed around the world and throughout history, providing the basis for a study of convergent cultural evolution (Landa 1999; Wilson 2001). As another example, people from Christian cultures often assume that belief in a glorious afterlife is a feature of all religions and even one of its main functions—to allay the fear of death. Not only is a glorious afterlife absent from many non-Christian religions, but it is even muted in Judaism, the religious tradition from which Christianity was derived. One implication of the proximate/ultimate distinction is that *any* set of beliefs and practices that motivate adaptive behavior can serve as the proximate mechanism for a human social organization. A comparative study is needed to determine why belief in a glorious afterlife is featured in some religions more than others, and why it became more prominent in Christianity than in Judaism. At a finer grain of analysis, different branches of Christianity and Judaism almost certainly vary in their reliance upon belief in a glorious afterlife, which can be measured and related to historical, social, and environmental factors.

TOWARD A FIELD OF EVOLUTIONARY RELIGIOUS STUDIES

One theme of *Darwin's Cathedral* and this article is that religions and other human social organizations can be studied with the same theoretical and empirical tools that evolutionary biologists use to study the rest of life. As I mentioned earlier for the paradigmatic case of guppies, this enterprise is complicated but manageable. It is complicated because evolution is inherently a complicated process with multiple selection pressures and constraints on selection that vary from species to species, trait to trait, and place to place—even over a scale of a few meters in the case of guppies. It is manageable because the pressures and constraints that operate in any particular case can be determined with enough hard work, and a very satisfying "big picture" can emerge for the system as a whole. This kind of enterprise requires a community of people who share the same set of theoretical and empirical tools that allow them to address a common set of issues. I will end this article by discussing how such a community can form around the subject of religion.

Ideally such a community would include people from evolutionary biology, traditional religious studies, and social scientists who are already using their own theoretical perspectives and empirical methods to study religion. The basic evolutionary principles and empirical methods are not difficult to learn. In general, the burgeoning study of human-related subjects from an evolutionary perspective is being conducted largely by people who received their formal training in other fields and picked up their evolutionary biology along the way. However, a number of major pitfalls need to be avoided.

The first is a belief that adaptationist hypotheses are hopelessly difficult to test and are destined to remain speculative "just-so stories." This belief is a pillar of skeptical arguments about evolution, as if evolution can be rejected for its difficulty rather than its falsehood. In any case, the belief is highly misleading. Functional hypotheses are as amenable to the scientific method as nonfunctional hypotheses, and in any case they cannot substitute for each other, since the proximate/ultimate distinction requires both mechanistic and functional explanations for everything that evolves by natural selection. Productive evolutionary scientists do not wring their hands about the difficulty of testing hypotheses but roll up their sleeves and get to work.

The second pitfall involves thinking about individuals and groups. Holistic conceptions of groups as being like organisms were widespread in both biology and the social sciences until the middle of the twentieth century, when individualism became the dominant intellectual tradition. Only now are evolutionary biologists achieving a middle ground that admits the possibility of adaptations at multiple levels of the biological hierarchy and provides the tools for determining the facts of the matter on a case-by-case basis. It is easy to portray a group-level adaptation as individually advantageous because groups of individuals who pull together do, after all, succeed as individuals. Identifying the appropriate level of selection requires locating the fitness *differences* that drive evolutionary change. There are no fitness differences in a win-win situation, so cooperation can increase in frequency in a large population only if groups that engage in win-win interactions out-compete groups that don't, even for a no-cost public good. To the extent that public good provision (including social control, which is a second-order public good) requires time, energy, and risk on the part of individuals, fitness differences arise within groups that weigh against cooperation, requiring even greater fitness differences among groups for cooperation to increase in frequency in the larger population. Comparing fitness differences within and among groups is as easy as riding a bicycle, once one gets the hang of it, but errors still abound in the past and present literature. A good example from the current study of religion is the idea of costly commitment, in which religions promote group cohesion by requiring members to engage in behavior that is too costly to fake (Irons 2001; Sosis 2004). This is indeed an important and adaptive feature of religion, which is well represented in the random sample, but is it adaptive at the group or individual level? Often it is portrayed as an individual-level adaptation because it is successful in general. However, breaking costly commitment into its component traits and comparing fitness differences within and among groups reveals that group-level selection is required for its evolution (Bowles and Gintis 2003). It is essential for the field of evolutionary religious studies to reach a consensus on how to identify levels of selection based on fitness differences within and among groups.

There is already a sizeable community of social scientists who study religion from an economic perspective. Economic and evolutionary theory are similar and inter-translatable in some respects but not others, creating another set of pitfalls for social scientists who wish to enter the field of evolutionary religious studies, as I discuss at length in *Darwin's Cathedral.* Once these pitfalls are avoided, a very impressive body of literature and empirical methods can be applied to the evolutionary study of religion. A number of economists who are at the forefront of the study of human genetic and cultural evolution from a multilevel perspective should be consulted by those who are currently studying religion from an economic perspective (see Hammerstein 2003 and Henrich 2004 for useful introductions).

Ironically, the fewest pitfalls might exist for scholars who conduct detailed historical and present-day studies of religion without having any particular theoretical perspective in mind. The functional nature of religion speaks for itself as soon as particular religious systems are studied in relation to their environments. That is why the selected examples in *Darwin's Cathedral* and the random sample of religions discussed in this paper are so interpretable from an evolutionary perspective. As the natural historians of religion, these scholars can provide the foundation of empirical knowledge for asking a new set of questions organized by contemporary evolutionary theory.

My hearty thanks to the students who helped me conduct the survey: D. P. Barnett, B. M. Bartholomew, R. I. Brilliant, H. Chiu, D. M. Davidson, K. E. Davies, L. L. DiAntonio, R. A. Fendrick, J. G. Flannery, S. T. Fosmire, J. J. Goldenthal, J. G. Goldshlager, M. M. Gordon, O. C. Grant, L. K. Hall, T. A. Hanke, R. E. Humphrey-Sewell, J. A. Isreal, B. G. Katz, R. M. Kindig, S. & S. Latif, K. Y. Lin, M. E. Malick, V. M. Mehta, S. O. Mohiuddin, C. R. Murolo, N. Nami, D. Oliver, S. C. Pavlides, J. M. Sherman, H. A. Vanengel, B. L. Vite, and P. S. Wirsing. I also thank A. B. Clark, M. Csikszentmihalyi, W. Greene, W. Irons, J. Neusner, S. Post, J. Schloss, E. Sober, and R. Sosis for helpful discussion. This research was supported by a grant from the Institute for Research on Unlimited Love.

David Sloan Wilson is an evolutionary biologist interested in a broad range of issues relevant to human behavior. He has published in psychology, anthropology, and philosophy journals in addition to his mainstream biological research. He is co-author with the philosopher Elliott Sober of *Unto Others: The Evolution and Psychology of Unselfish Behavior* (Harvard University Press, 1998).

REFERENCES

Alcorta, Candace, and Richard Sosis
2005 Ritual, Emotion, and Sacred Symbols: The Evolution of Religion as an Adaptive Complex. *Human Nature* 16:323–359.
Alexander, R. D.
1987 *The Biology of Moral Systems.* New York: Aldine de Gruyter.
Atran, S.
2002 *In Gods We Trust: The Evolutionary Landscape of Religion.* Oxford: Oxford University Press.
Atran, S., and A. Norenzayan
2004 Religion's Evolutionary Landscape: Counterintuition, Commitment, Compassion, Communion. *Behavioral and Brain Sciences* 27:713–730.
Aunger, R.
2002 *The Electric Meme.* New York: Free Press.
Blackmore, S.
1999 *The Meme Machine.* Oxford: Oxford University Press.
Bowles, S., and H. Gintis
2003 Origins of Human Cooperation. In *Genetic and Cultural Evolution of Cooperation,* P. Hammerstein, ed. Pp. 429–444. Cambridge: MIT Press
Boyer, P.
2001 *Religion Explained.* New York: Basic Books.
Bulbulia, J.
2004 The Cognitive and Evolutionary Psychology of Religion. *Biology and Philosophy* 19:655–686.
Campbell, B. F.
1993 *Ancient Wisdom Revised.* Princeton: Princeton University Press.
Cousins, N.
1958 *In God We Trust: The Religious Beliefs and Ideas of the American Founding Fathers.* New York: Harper.
Cronk, L.
1994 The Use of Moralistic Statements in Social Manipulation: A Reply to Roy A. Rappaport. *Zygon* 29:351–355.
Dawkins, R.
1976 *The Selfish Gene.* Oxford: Oxford University Press.
Durkheim, E.
1995 *The Elementary Forms of Religious Life.* New York: Free Press. (Originally published in 1912)
Eliade, M., ed.
1987 *The Encylopedia of Religion.* New York: Macmillan.
Endler, J. A.
1995 Multiple-Trait Coevolution and Environmental Gradients in Guppies. *Trends in Ecology and Evolution* 10:22–29.
Guthrie, S. E.
1995 *Faces in the Clouds: A New Theory of Religion.* Oxford: Oxford University Press.
Hammerstein, P., ed.
2003 *Genetic and Cultural Evolution of Cooperation.* Cambridge: MIT Press.
Henrich, J.
2004 Cultural Group Selection, Coevolutionary Processes, and Large-Scale Cooperation. *Journal of Economic Behavior and Organization* 53:3–35.
Heyd, U.
1950 *Foundations of Turkish Nationalism: The Life and Teachings of Ziya Gokalp.* London: Harvill Press.
Hilkert, M. C.
2001 *Speaking with Authority: Catherine of Siena and the Voices of Women Today.* New York: Paulist Press.

Hinde, R.
1999 *Why Gods Persist: A Scientific Approach to Religion.* New Brunswick, N.J.: Routledge.
Ingle, H. L.
1994 *First among Friends: George Fox and the Creation of Quakerism.* New York: Clarendon.
Irons, W.
2001 Religion as a Hard-to-Fake Sign of Commitment. In *Evolution and the Capacity for Commitment,* R. Nesse, ed. Pp. 292–309. New York: Russell Sage Foundation.
Laidlaw, J.
1995 *Riches and Renunciation: Religion, Economy and Society among the Jains.* Oxford: Oxford University Press.
Landa, J. T.
1999 The Law and Bioeconomics of Ethnic Cooperation and Conflict in Plural Societies of Southeast Asia: A Theory of Chinese Merchant Success. *Journal of Bioeconomics* 1:269–284.
Pagels, E.
1995 *The Origin of Satan.* Princeton: Princeton University Press.
2003 *Beyond Belief: The Secret Gospel of Thomas.* New York: Random House.
Reynolds, V., and R. E. Tanner
1995 *The Social Ecology of Religion.* Oxford: Oxford University Press.
Richerson, P. J., and R. Boyd
2004 *Not by Genes Alone: How Culture Transformed Human Evolution.* Chicago: University of Chicago Press.
Schoeffeleers, J. M.
1992 *River of Blood: The Genesis of a Martyr Cult in Southern Malawi.* Madison: University of Wisconsin Press.
Sosis, R.
2004 The Adaptive Value of Religious Ritual. *American Scientist* 92:166–172.
Sosis, Richard, and Candace Alcorta
2003 Signaling, Solidarity, and the Sacred: The Evolution of Religious Behavior. *Evolutionary Anthropology* 12:264–274.
Stark, R.
1999 Micro Foundations of Religion: A Revised Theory. *Sociological Theory* 17:264–289.
Stark, R., and W. S. Bainbridge
1985 *The Future of Religion.* Berkeley: University of California Press.
1987 *A Theory of Religion.* New Brunswick, N.J.: Rutgers University Press.
Wesley, J.
1976 *Thoughts upon Methodism: The Works of John Wesley,* vol. 9. R. E. Davies, ed. Nashville: Abington Press.
Wills, G.
2001 *Venice: Lion City.* New York: Simon and Schuster.
Wilson, D. S.
2001 Religious Groups and Homogeneous Merchant Groups as Adaptive Units: A Multilevel Evolutionary Perspective. *Journal of Bioeconomics* 2:271–273.
2002 *Darwin's Cathedral: Evolution, Religion, and the Nature of Society.* Chicago: University of Chicago Press.

[8]

Exploring Resources of Naturalism

RELIGIOPOIESIS

by Ursula Goodenough

Abstract. Religiopoiesis describes the crafting of religion, a core activity of humankind. Each religion is grounded in its myth, and each myth includes a cosmology of origins and destiny. The scientific worldview coheres as such a myth and calls for a religiopoietic response. The difficulties, opportunities, and imperatives inherent in this call are explored, particularly as they impact the working scientist.

Keywords: belief; metaphor; myth; religion; religiopoiesis; reward; science; theological reconstruction.

Since the publication of *The Sacred Depths of Nature* in 1998, I have had opportunities to present the core concepts of religious naturalism in numerous venues: bookstores, colleges and medical schools, museums, youth groups, adult-ed groups and sermons in churches and synagogues, women's forums, writers' workshops. I have also received numerous letters and e-mails from readers. Even after factoring in the obvious bias that persons who are resonant with the project are more likely to express appreciation than those who are not, it can nonetheless be said that there has been an outpouring of appreciation and gratitude, both for the scientific narrative itself ("I came to understand the nature of things for the first time") and for my personal reflections on its religious meanings ("Finally someone has written down what I have long felt").

Of particular interest has been the response of fellow scientists. Many have expressed appreciation and gratitude, but many others have expressed

Ursula Goodenough is Professor of Biology at Washington University in St. Louis, past president of the Institute on Religion in an Age of Science, and author of *The Sacred Depths of Nature* (Oxford Univ. Press, 1998). Her address is Department of Biology, Box 1129, Washington University, St. Louis, MO 63130. Her e-mail address is ursula@biosgi.wustl.edu.

incredulity: How did I have the "nerve" to write such a book? How did I "dare" to wander into the topic of religion? Wasn't I concerned, in so "exposing" myself, that I would lose respect as a professional scientist? Didn't I worry that I might not get my grants funded or my papers published?

In this essay I explore these concerns, first analyzing the religion-science landscape from the perspective of a scientist and then inviting scientists to join the conversation. I examine some of the factors that discourage persons, and scientists in particular, from making contributions to religious dialogue, and suggest ways around these difficulties.

THE THEOLOGICAL RECONSTRUCTION/ RELIGIOPOIESIS SPECTRUM

To most outsiders, and therefore to most scientists, the religion/science dialogue is perceived as a venture in what academic theologians refer to as theological reconstruction: a new insight about the nature of the universe is encountered through scientific inquiry, and adherents of traditional religious faiths then work to find ways to incorporate that understanding into the canon. This cycle of challenge and response, ongoing now for several millennia, has yielded religious traditions that are selected for their resiliency quite as much as for the potency of their myths. A conspicuous sector of the present-day dialogue continues in this vein. Scientists are to my mind correct in regarding theological reconstruction as outside their ken, since it requires a deep and nuanced knowledge of the histories and trajectories of particular faiths that most scientists (with some notable exceptions) have not begun to master.

In a second kind of venture, the scientific understanding of Nature serves as the starting point, and its religious potential is then explored. John Dewey, Teilhard de Chardin, and Julian Huxley are among those who have made important early contributions here. A key stimulus for carrying such a project into present times is the transformation that has occurred in the nature of the scientific story itself. Whereas science has, until recently, been segregated into discrete sectors of knowledge—Newton's laws, thermodynamics, Mendelian genetics—there has emerged in the past fifty years or so a coherent cosmology, fully as integrated as Genesis 1, that yields important insights into our nature, our history, and our constraints and possibilities. The second project, which can be called religiopoiesis, takes this story and works with it.

I regard *The Sacred Depths of Nature* as a contribution to present-day religiopoiesis. I stress the word *contribution*. As developed more fully below, no one person constructs a religion. But it is also the case that unless individual persons are encouraged—exhorted!—to offer contributions, there will be no "stuff" available to cohere into new religious orientations in future times.

The Perils and Opportunities of Religiopoiesis

The poiesis part of religiopoiesis comes from the Greek *poiein*, to make or craft, the same root as poetry. Religiopoiesis, then, is the crafting of religion. The term religiopoiesis has the advantage that its Greek-antiquity-ness helps disguise its meaning and hence obfuscates its baggage. The phrase "crafting religion" is in fact deeply problematic—for at least two reasons.

First, many of us have been raised to understand that religious tenets come to humankind via blinding-light revelations, either to great/divine persons in ancient times or to mentally unstable/maniacal persons in modern times, and we feel no identification with either group. "What, me, articulate a religion? You gotta be kidding!" We become embarrassed, uneasy, even talking about the idea. Indeed, to many it can seem blasphemous.

The second problem is that religion is many different things—text, response, ritual, ideology, morality—and most of these topics have not been deeply considered by persons who have devoted their intellectual lives to understanding and contributing to the scientific worldview. An honest response would be "Religion? Don't know a thing about it. Stopped going when I was eleven." A more common response is "Religion? What a lot of balderdash! I like the Gregorian chants and all that, but the rest is baloney." Neither of these responses is likely to generate enthusiasm for engaging in the project.

Counterbalancing these difficulties are the opportunities presented by a religiopoiesis project in our times.

Whereas folk wisdom holds that religious cosmologies derive from blinding-light revelation, historians of religion tell us that most are in fact the product of the interaction of cultural traditions and approaches: e.g., a story from Mesopotamia is combined with a story from Persia and modified to be coherent with Hebraic tradition. In this respect, the fashioning of our scientific cosmology has been an analogous process. We can attribute key insights to various persons—quantum theory to Bohr, evolution to Darwin, regulated gene expression to Monod—but we all know that these are incomplete attributions, that the "revelations" experienced by these men emerged from a vast cumulation of understandings.

By the same token, the crafting of religious responses to the scientific worldview can—indeed must—be a collective and dynamic project. There are huge domains of knowledge to be considered, and there are millennia of religious quests to be explored, quests that articulate what persons seek in their religious experience. Indeed, it is the collective nature of the project that can serve to deconstruct our uneasiness about engaging in it: no one person is setting himself or herself up as the guru; we're all responding from our own perspectives, offering rather than professing.

The Duality within Religiopoiesis

Granted that religions are complex, we can recognize two poles, and an intervening spectrum, in any religiopoiesis project.

The first pole can be called theology. A theologian, trained in philosophical discourse, uses this rubric to talk about ultimacy: What is the meaning of meaning? How do we know that we know? What are we talking about when we speak of purpose or evil or destiny? These intellectual questions may strike some as sterile and uninteresting, but for others they represent the core of religious life. Talmudic scholars have movingly described their studies of the Torah as deeply religious exercises in which they experience transcendence in their cognitive apprehension of God's word. The scientific cosmology certainly invites stunning opportunities for theological discourse. How do we think about ultimate reality in an evolving universe? How does our understanding of genetics inform our need to believe that we possess free will? What do computers tell us about ourselves? Is there such a thing as a metaethics? How are we to think about and decide ethical questions if no coherent metaethical framework can be found?

Scientists may argue that they lack the philosophical training to engage in such dialogue, but I disagree. Our training has honed our ability to analyze empirical data and understandings, make deductions therefrom, and integrate disparate modes of reasoning. We have much to contribute here, and if our language sounds different from theological language, this may not be such a bad thing.

The second pole can be called spirituality. It is accorded romantic adjectives: emotional, intuitive, poetic, mystical. It explores how we feel when we apprehend a cosmology—religious responses such as hope or fear or fellowship or compassion. The new cosmology invites spiritual responses as well. How does an understanding of biological evolution inform our understanding of empathy? community? gratitude? death? How do we deal with its vast nihilistic underbelly?

Theological/spiritual dualities, and their many intergradations, are inherent in all religions and are seminal to religiopoesis. It is the integration of the theology and the spirituality that forms the matrix of a viable religious orientation: the theology alone is dry as dust; the spirituality alone is self-absorbed, even autistic. Indeed, one of the important insights from contemporary neurobiology is that these distinctions are at least partially false: without an emotional or intuitive component, theological/philosophical issues may have no meaning to the thinker in the sense that he or she will not be able to assign value or importance to alternative outcomes.

There is, of course, a whole other dimension to religious life, which is how we behave and why. Religions have always been in the business of recommending or requiring modes of morality, and any new religious orientations will doubtless come to carry such directives as well. But I have come to understand that directives only work if they flow from belief. It is as we believe in the American Way and the Constitution that we obey governmental regulations, and it is as we believe in the theological/spiritual core of our religious traditions that we attempt to respond to their moral edicts. To give an example in the current context, I would propose that the most enduring form of environmentalism will emerge from a theological and spiritual apprehension of our place in the scheme of things. Scientists have important things to tell us here.

Belief and Reward

Religion is about Belief with a capital B. A religious person adopts the most compelling theology and the most satisfying spirituality on offer, frequently the constellation encountered in childhood, and allegiance to those understandings is then called Belief. There is, of course, an additional factor here, in that religious traditions have invariably included rewards for Belief: dwelling in the house of the Lord forever, the receipt of eternal grace, reincarnation into a better life, respite from plague and drought.

The scientific cosmology, authored by cosmic evolution and not by prophets or visionaries, is not inherently a proposition that calls for belief or Belief. One is not asked to believe in the Schrödinger equation or the genetic code; one is instead asked to examine the evidence for these discoveries and, if it is judged inadequate, to propose and conduct experimental tests of alternative models of reality. Where the scientific accounts evoke our belief statements, then, is in the realm of our acceptance of their findings and our capacity to walk humbly and with gratitude in their presence.

The reward component is problematic because nothing now apprehended by scientific inquiry suggests the existence of the rewards offered by the major religious traditions. One way out is to say that, because our scientific understandings remain incomplete, these rewards may still be on offer and indeed may never be perceived by our limited human faculties. A second response is to suggest that the awe and wonder generated by the understanding of scientific cosmology is itself its own inherent reward, a response that is not likely to carry much freight in our times since most persons find the scientific cosmology difficult and alienating. But it does not have to be that way. Religiopoiesis, in the end, is centrally engaged in finding ways to tell a story in ways that convey meaning and motivation.

Metaphor

Our scientific facts come to us as facts: DNA sequences, Hubble images, extinctions. But our understandings—scientific, theological, and spiritual—come to us as metaphors, either the metaphor systems we call language and mathematics or the metaphors we call the arts. The richness of our metaphors indicates the depth of our understanding.

I have explored the topic of religious and scientific metaphor in a previous essay (*Zygon* 35 [June 2000] 233–40), but it is appropriate to revisit the topic in the present context because there exists enormous confusion and misunderstanding here. I have been told, for example, that to say that the life of Christ is a metaphor for how we can best love is to commit a heresy, that one can speak reverently of Christ only by professing full Belief in the claims made for him by the authors of Christianity. I am coming to understand that this view, in fact, can itself also be considered a heresy. Christ has always been about metaphor, and Christianity has always been about the symbol systems inherent in its texts and art and ritual—and this can be said for all religions worth our attention. To be sure, billions of persons have been warned that if they fail to regard religious metaphors as inviolate they will fail to receive the rewards of faith, but those engaging in religiopoiesis can bypass these injunctions and approach the metaphors for their inherent value, for what they tell us about how and why people value what they do. Here we have in a sense come full circle, for we began by saying that theological constructionism also works with the traditional religions. In a religiopoiesis project, however, we are informed but not constrained by previous interpretations. We can ask the traditions to speak to us yet again, in whole new contexts.

Perhaps the most important act in the process of religiopoiesis, then, is to open ourselves to metaphors: those in our traditional religions, those in the poetry and art of past and present times, and those that emerge from our articulation of scientific understandings. The goal is not strict intellectual coherence, any more than the goal of a poem is to fit in seamlessly with all other poems. The goal is to come up with such a rich tapestry of meaning that we have no choice but to believe in it. This is, to my mind, the urgent project before us all.

[9]

MYTHS AS INSTRUCTIONS FROM ANCESTORS: THE EXAMPLE OF OEDIPUS

by Lyle B. Steadman and Craig T. Palmer

Abstract. The growing interest in dual-inheritance models of human evolution has focused attention on culture as a means by which ancestors transmitted acquired phenotypic characteristics to their descendants. The ability of cultural behaviors to be repeatedly transmitted from ancestors to descendants enables individuals to influence their descendant-leaving success over many more generations than are usually considered in most analyses of inclusive fitness. This essay proposes that traditional stories, or myths, can be seen as a way in which ancestors influence their descendant-leaving success by influencing the behavior of many generations of their descendants. The myth of Oedipus is used as an example of a traditional story aimed at promoting proper behavior and cooperation among kin. This interpretation of the Oedipus myth is contrasted with Freudian and structuralist interpretations.

Keywords: evolution; mythology; Oedipus; religion; tradition.

INTRODUCTION

From our ancestors we inherit both our genes and our traditions. Although the exact mechanisms involved in the "dual inheritance" of genes and traditions is quite complex and controversial (see Boyd and Richerson 1985; Campbell 1975; Cavalli-Sforza 1981; Cloak 1975; Daly 1982; Durham 1979, 1982; Flinn and Alexander 1982; Lumsden and Wilson 1981; Tooby and Cosmides 1989), we support the view that traditions are "a system of information, support, and guidance" that serve as "cultural supplements" (Hefner 1991, 123) to genetic inheritance. This view holds that traditions are passed down from ancestors to succeeding generations, because they tend to influence the behavior of descendants in ways that increase their inclusive fitness and, hence, the descendant-leaving success of their ancestors. We also suggest that myths are a crucial

Lyle B. Steadman is Assistant Professor of Anthropology at Arizona State University, Box 872402, Tempe, AZ 85287. Craig T. Palmer is an Instructor in the Department of Anthropology at the University of Colorado at Colorado Springs, CO 80903.

part of this supplemental information and guidance system, because myths "shape the most fundamental or ultimate values" (Burhoe 1979, 156; see also d'Aquili 1983, 1986) by providing each generation with "rules laying down what ought to be done or what ought not to be done." (Hefner 1991, 115). Since cooperative social behavior does not occur automatically among humans, and the ability to form and maintain cooperative social relationships, both within the nuclear family and with larger sets of distant kin and nonkin (see Alexander 1979, 1986; Campbell 1975), is crucial to human descendant-leaving success, the most important of these rules are likely to concern social behavior. The *Lugbara* people of sub-Saharan Africa state, "The rules of social behavior are 'the words of our ancestors'" (Middleton 1960, 27). For many people, the words of their ancestors are contained in myths.

Burhoe (1979, 1986) argues that religion is the key to "trans-kin" cooperation. This view is partially based on the fact that since such trans-kin cooperation involves altruism between very distantly related individuals, such altruism cannot be explained by shared genes (see Hamilton 1964). For example, Hefner states that "altruism beyond kin is transmitted culturally, not genetically, and . . . religious traditions are the chief carriers of this value" (Hefner 1991, 132). This view neglects the fact that, despite claims about the role of innate "kin recognition" in human altruism (see Taylor and McGuire 1988; Rushton 1988, 1989; Rothstein and Pierotti 1988; Seyfarth and Cheney 1988), cooperation between even close "familial" human kin is also largely a product of parental encouragement. Hence, even altruism among close human kin is "transmitted culturally" from one generation to the next. Further, there are some clear connections between the parental encouragement of cooperation among family members and the traditional encouragement of cooperation among more distant kin and nonkin. Cooperation between more distant kin is often the product of the influence of distant ancestors, transmitted through their living descendants.

Cooperative social relationships between first cousins, for example, are usually the result of parental encouragement of sibling cooperation during the previous generation. Cooperation among people not recognized as actual kin is nearly always fostered by some type of religious or political leader who engages in parentlike behavior to encourage familylike cooperation among his or her followers. The metaphorical use of close-kin terms (i. e., mother, father, brother, sister) is frequently a part of this strategy.

We argue that myths and other traditional stories can be seen as part of the human descendant-leaving strategy, because they allow even very distant descendants to experience, vicariously and safely, the consequences of the characters' actions. Specifically, the myth of Oedipus, and in particular the play *Oedipus Rex* by Sophocles, are sets of ancestral instructions prescribing proper kinship behavior for both parents and kings by emphasizing the dire consequences of improper kinship behavior.

Why Does Oedipus Suffer?

The analysis of the Oedipus myth[1] has nearly always focused on the question of why Oedipus suffers. What is intriguing about Oedipus's suffering is that it is undeserved. Even though Freud's famous interpretation implies that Oedipus suffers because of his own desires to have sexual relations with his mother and to kill his father, there is no hint in the actual myth that Oedipus sexually desires his mother or wants to kill his father (Watling 1947, 23–24, 69–70). On the contrary, he banishes himself from Corinth to avoid violating his parents and pierces his eyes the moment he discovers that he unwittingly has done so. While Freud *may* be correct in his suggestion that the reader reacts to a "secret meaning" in the myth that parallels his own unconscious sexual desires for his mother and hatred of his father, even Freud points out that according to the myth Oedipus "did all in his power to avoid the fate prophesied by the oracle, and . . . in self-punishment blinded himself when he discovered that in ignorance he had committed both these crimes" (Freud 1935, 290).

The undeserved nature of Oedipus's suffering has led to an almost endless stream of explanations (see Dodds 1968), including Levi-Strauss's structuralist interpretation, which concludes that the myth focuses on a contradiction between normal and autochthonous birth (see Levi-Strauss 1963). Although Levi-Strauss's creative analysis has profoundly influenced the study of myths in general, the most widely held literary interpretation of the Oedipus myth is that the play demonstrates man's inability to control his own destiny; no matter what he does, Oedipus cannot avoid the fate the gods have set for him. Thus, Gould writes, "The effect [of the myth and play] is to make the audience fear that perhaps the efforts of human beings to create lives for themselves may be devoid of meaning" (Gould 1970, 2; see also Bowra 1944, 175; Ehrenberg 1968, 80).

There are those, however, who challenge the idea that the aim of perhaps the greatest tragedy ever composed is to convince the audience that behavior is irrelevant—that no matter what they do, men cannot influence anything. Dodds, for example, writes:

> What fascinates us is the spectacle of a man freely choosing, from the highest motives, a series of actions which lead to his own ruin. . . . The immediate cause of Oedipus' ruin is not "Fate" or "the gods"—no oracle said that he must discover the truth—and still less does it lie in his own weakness; what causes his ruin is his own strength and courage, his loyalty to Thebes, and his loyalty to the truth. (Dodds 1968, 23)

Thus, Oedipus is more than simply the innocent dupe of the gods. The actions that lead him to disaster—to killing his own father and marrying his mother—are his own choice. Oedipus does not simply suffer *in spite of* what he does; he suffers *because of* what he does.

But there is a more fundamental problem with the "fate" explanation. If Laius and Oedipus had truly believed in the accuracy of the oracle, they would have done nothing to thwart it: if the prediction were true, it could not be thwarted, and if it were not true, there would be nothing to thwart. Thus, what the oracle predicted would be either uninfluenceable or irrelevant. The oracle's prophecy was that Laius's son would kill him and marry his wife. It did *not* say, in regard to Laius, that *unless* he killed his son, his son would kill him. And the oracle did not say that *if* Oedipus left Corinth, he would avoid killing his father and marrying his mother. Neither prediction/assertion is contingent on any action. Thus, with respect to the oracle, there is no appropriate behavior for either Oedipus or Laius; the only rational behavior would be to ignore it.

Why, then, does Sophocles have Laius and Oedipus take action in response to the oracle? One important use of oracles in myths is as a literary device to facilitate the plot. Is there a better way to make Laius appear to fear his innocent baby son sufficiently to kill him? Is there a better way to get Oedipus both to leave his Corinthian parents and, at the same time, to demonstrate his love for them? The behavior of Oedipus and Laius, as well as everyone else in the play, implies that behavior makes a difference—that it has consequences, regardless of predictions of destiny. The very efforts of Sophocles as a playwright presume this, for by writing, he himself discourages his audiences from accepting fate and doing nothing. Everything about the play encourages the opposite.

The immediate reason for Oedipus's suffering is clear. He blinds himself because he has violated his parents. He pierces his eyes the moment he realizes he has killed his father and married his mother. The key to the meaning of the myth is that his unwitting patricidal and incestuous acts leading to his self-blinding are the results of *his father's attempt to kill him.* This is seen in the following lines from the play *Oedipus Rex* by Sophocles:

MESSENGER (from Corinth): Polybus was not your father.
OEDIPUS: Not my father? . . .
MESSENGER: Long ago he had you from my hands, as a gift.
OEDIPUS: Then how could he love me so, if I was not his?
MESSENGER: He had no children, and his heart turned to you . . . [I was] your savior, son, that day.
OEDIPUS: From what did you save me?
MESSENGER: The infirmity in your ankles tells the tale.
OEDIPUS: Ah, stranger, why do you speak of that childhood pain?

MESSENGER: Your ankles were riveted, and I set you free.
OEDIPUS: It is true; I have carried the stigma from my cradle.
MESSENGER: That was why you were given the name you bear.
OEDIPUS: God! Was it my father or my mother who did it? Tell me.
(Watling 1947; Jebb [1885] 1966)

This passage is also the moment Oedipus discovers the reason for his name being Oedipus, meaning, literally, "swollen foot." Sophocles wrote his plays for an audience that already knew both the Oedipus myth and the meaning of the name (Watling 1947, 11). To them, Swollen Foot was not simply a personal name; each time it was uttered, they were reminded of the treachery of Laius, the ultimate cause of Oedipus's mutilation and tragedy.

Although Oedipus himself does not discover why he bears this name until late in the play (indeed, the entire play focuses on his gradual understanding of the events related to his name), the Greek-speaking audience, unlike most audiences of today, was aware of its significance. To the Greek audience, the name of the myth, the play, and the hero symbolized the consequences of a selfish, treacherous father. The Greek audience knew why Oedipus suffered.

The cause of his tragedy is recognized by Oedipus when he asks the question, "God! Was it my father or my mother who did it?" This question refers to the riveting of his ankles (Gould 1970, 124) and to his true, not adopted, parents. He recognizes it again, explicitly, after blinding himself, when he entreats Creon to banish him to that mountain where, as a baby, he was sent to die, saying, "My mother and my father . . . made it my destined tomb, and I'll be killed by those who wished my ruin!" (lines 1452–1455, Watling 1947; see also lines 262–75 of the play *Oedipus at Colonus*, Watling 1947).

Oedipus Rex takes place long after Laius's attempt to kill his son, and the play does not discuss why Laius was confronted by the oracle. However, from the plays *Seven against Thebes* (lines 742ff.) by Aeschylus and *The Phoenician Women* (lines 18ff.) by Euripides (Watling 1947), it is clear that the curse is a punishment for Laius for seducing and abducting Pelops's son Chrysippus. Because of this deed, Laius was told that *if* he had a son, *then* he would be killed by this son (Gould 1970, 92; Jebb [1885] 1966, x–xii). Laius, disregarding the oracle, went on to have a son, but rather than accepting his punishment, he attempted to save his own life by having his son killed. Thus Laius, in sharp contrast to Oedipus, does not suffer innocently or unselfishly; his death is the result of his own crime, and the ultimate destruction that comes to his wife and descendants is the result of his attempt to avoid his just punishment. For a father to lose his son is a tragedy; but for a father to selfishly kill his son to save his own life must rank among the greatest of evils. Nothing forced Laius to try to kill his son. He could have responded to the oracle: "As a father I shall sacrifice for my son, even at the risk of my own life."

Oedipus, in contrast, did not knowingly kill his father; he was not in any sense an evil son. He was, in fact, the epitome of a respectful son, a concerned and loving father, a good king. He valued his kin and his subjects more than himself. When he first hears of the prophecy, rather than allow the possibility that he would kill his father and marry his mother, Oedipus exiles himself from Corinth to avoid harming the parents he loves. Years later, when he learns of his (adoptive) father's death and is offered the kingship of Corinth, he declines it because of his fear that by returning he may fulfill the prophecy and marry his mother (lines 976–86, Watling 1947). As a king, Oedipus consistently shows his concern and willingness to sacrifice himself for his people. When he discovers that he himself is the cause of the plague affecting Thebes (because he had killed King Laius), he begs to be exiled (lines 1449–51, Watling 1947). Indeed, in his attempt to save Thebes, he risks his life to answer the riddle of the Sphinx. As a father, Oedipus is equally virtuous. His final line in the play emphasizes his concern for his young daughters: "Do not take them from me, ever!" Throughout the play, Oedipus sacrifices himself for others, for both his kin and his subjects, whom he calls his "children."

The tragedy of Oedipus is that despite his selfless, virtuous, courageous behavior, he is caught in the consequences of his father's selfish actions. What the audience learns from the story is that the significant consequences of a person's behavior are not limited to that person alone. Indeed, consequences of a person's behavior can influence the lives of descendants and others through endless generations. This is an important message for kings and fathers. Both children and subjects depend on parental (or parentlike) sacrifice. Laius refused to make such a sacrifice and ultimately destroyed his family and almost destroyed his kingdom. Not only was he a bad king and father but his behavior threatened the basis of society.

Discussion

The meaning of the Oedipus myth and the play *Oedipus Rex* is found in the contrasting behavior of Laius and his son, Oedipus—a selfish parent and a son who is the epitome of selflessness. But whence comes the virtue of Oedipus? In one sentence Oedipus reveals the answer: "My father . . . Polybus, to whom I owe my life" (Watling 1947, line 827). Oedipus's morality must be related directly to the influence of his foster parents. Their parental care was crucial to his survival; it was their love for Oedipus that led him to love them and others. Human behavior is influenced profoundly by the behavior of others; indeed, culture itself presumes this influence.

The behavior distinguishing Oedipus is kinship behavior—the cooperation that occurs regularly between kinsmen. It is filial love for his Corinthian parents that leads Oedipus to exile himself from Corinth, and it is a willingness to sacrifice himself for others that leads him to challenge the Sphinx. Indeed, Oedipus uses a kinship justification in his search for the murderer of Laius, a search to save his Theban "children." Although considerable attention is now being given to the possibility that humans recognize and respond to the degree of genetic similarity they detect in other individuals, kinship cooperation clearly is not merely the result of the fact that relatives share genes. That birth alone is insufficient to establish kinship cooperation is seen in the relationship between Oedipus and Laius. Because the parental behavior of foster parents toward an adopted child, like that of the king and queen of Corinth toward Oedipus, is not necessarily different than parents' behavior toward a true offspring, the resulting relationship may be identical.[2] Social relationships identical to those between kin can occur between nonkin.

People do not behave like kin automatically; they come to behave in this way because they are influenced to do so by others, especially by those who raise them. Thus, behavior normally directed toward kin creates social relationships between individuals, whether or not they are identified as kin. Thus, kinship behavior, crucial to human descendant-leaving success, is highly modifiable by the behavior of others and must be seen as the result of both a particular genetic nature and the actual behavior of other individuals. The lack of either results in the failure of individuals to behave like kin. We suggest that the ability of parentlike behavior to encourage familylike cooperation is also why kings (and other leaders) adopt parentlike behavior, including the use of kin terms, to encourage cooperation among their followers. In *Oedipus Rex*, the first words uttered by King Oedipus are "My children," as he addresses his subjects. The play encourages fatherly behavior in fathers and in kings.

Conclusion

The Oedipus myth, we propose, is neither an expression of our alleged unconscious guilt nor a contradiction between normal and autochthonous birth. And it is not aimed at convincing people that fate rules their life. Generations have chosen to encourage the repetition of the story of Oedipus because it provides an understanding of the contrasting consequences of parental selfishness and parental love. The major assumption underlying the story of Oedipus is that behavior creates social relationships and behavior can destroy them. We also suggest that the Oedipus myth is not unique in focusing on kinship cooperation and the consequences of parental behavior. Aristotle observed that in most good tragedies the "suffering" is a result of deeds done among *philoi*, or "kinsmen," "as when the murder, or whatever, is carried out or intended by brother against brother, son against father, mother against son, and son against mother" (cited in Gould 1970, 33).

Myths and other traditional stories, by teaching vicariously the consequences of social behavior, transmit social knowledge. Such knowledge is the basis of kinship cooperation and hence social cooperation. Myths, like history, use past events, whether actual or alleged, to anticipate future consequences. This tends to influence those who hear traditional myths to behave socially in ways that increased inclusive fitness in the past and to choose to repeat the myth to their own descendants. The extent to which the behavior encouraged in traditional myths will lead to descendant-leaving success in the future depends upon the similarity between future environments and those of the past. This similarity is rapidly decreasing, and the naturalistic fallacy of equating descendant-leaving success with "success" in a more general sense is painfully clear in our rapidly overpopulating world. Still, kinship cooperation between both real and fictive kin is a crucial part of most conceptions of a desirable future. Hence, myths (now including films) still appear to be able to provide future generations with useful knowledge.

NOTES

1. The following summarizes the Oedipus myth, following Watling (1947:23–24, 69–70) and Graves (1978, 9–15):

Oedipus was the son of Laius and Jocasta, the king and queen of Thebes. Laius, the son of Labdacus, was a direct descendant of Cadmus, the founder of Thebes. King Laius, because of a crime he had committed against a prince of a neighboring kingdom (Gould 1970, 72; Jebb 1966, x–xii), was told by an oracle that if he had a son, that son would kill him and marry his wife, Jocasta. Laius and Jocasta did have a son. But then Laius, in an attempt to thwart the oracle, ordered a shepherd to kill their baby (some say Laius himself did it—see Graves 1978, 9) by abandoning him on a mountainside after cruelly piercing his ankles with an iron pin to prevent him from crawling to safety.

This was done. But in a moment of compassion, the shepherd gave the injured child to a passing shepherd from Corinth to raise as his own. This Corinthian shepherd, a servant of the king and queen of Corinth, was asked by them for the child, for they were childless and wanted to raise him as their own. He gave them the child, whom they named Oedipus ("Swollen Foot") because of his injured feet. Oedipus grew to manhood, loved and honored as their true son. By chance, however, Oedipus heard a rumor that he had been adopted. Although his parents denied the rumor, Oedipus was not satisfied and traveled to Delphi to find the truth. But the oracle there did not answer his question, asserting instead that he would kill his father and marry his mother.

Like his true father before him, Oedipus attempted to thwart the oracle. In order not to violate his beloved Corinthian parents, he resolved never to see them again. While traveling on foot on the road to Thebes, he had a sharp encounter with a small group of men, one of whom was in a chariot. After some angry words and actions (Oedipus resisting their demand that he give way, saying he acknowledged no betters except the gods and his own parents), Oedipus slew several of them. Continuing along the road, Oedipus approached Thebes and found it to be in the grip of a deadly monster, the Sphinx. Risking his life, he attempted to save the city by answering the Sphinx's riddle. He was successful, thereby destroying her power, and so was received joyfully into Thebes. Because their king had been killed by someone while traveling, the grateful Thebans made Oedipus their king and gave him their queen, Jocasta, as wife. During the next fifteen years Thebes prospered, and Jocasta bore Oedipus sons and daughters. But then pestilence and famine struck again, threatening Thebes with utter extinction. And the citizens cried to their beloved king for help.

The play *Oedipus Rex* begins when Oedipus answers them, inquiring "My children, what ails thee?" During the course of the play, Oedipus, trying to discover the source of Theban affliction, gradually uncovers the hideous secret of his unwitting sins. The man he had killed in the chariot was his father, Laius; and the wife whom he had married after rescuing Thebes, and who had borne him children, was his own mother, Jocasta. In his horror at this discovery, and at the self-inflicted death of Jocasta, he chooses not to kill himself and thus continue to plague his violated parents in the afterlife. Rather, he destroys the sight of his own eyes and asks Creon, his mother's brother or brother-in-law, to banish him forever from Thebes. At the conclusion of *Oedipus Rex*, Oedipus implores Creon to protect and cherish his young, vulnerable daughters and begs, "Do not take them from me, ever!"

The Oedipus narrative is carried forward in the play *Antigone*. In this episode, Oedipus leaves Thebes forever. But then discord again rends his family; for while his daughters remain faithful to their father—Ismene remaining home, while Antigone, the younger, joins him in his wanderings—his two sons, Eteocles and Polynices, fight and kill one another in their struggle for power. Because Polynices had attacked Thebes, Creon orders that he not be buried. Antigone, in defiance, chooses to perform the burial ceremony for her brother and is sentenced to death by Creon. As a result, Creon's son, Haemon, who loves Antigone, kills himself. Finally, Haemon's mother, Creon's wife, commits suicide upon hearing of her son's death.

2. Of course, the behavior of adoptive parents is not necessarily similar to the behavior of biological parents. There is, in fact, some evidence of differences in the behavior of adoptive and biological parents, which may ultimately be explainable in evolutionary terms (see Daly and Wilson 1988).

REFERENCES

Alexander, R. A. 1979. *Darwinism and Human Affairs*. Seattle: Univ. of Washington Press.

_____. 1986. *The Biology of Moral Systems*. New York: Aldine.

Boyd, R. and P. J. Richerson. 1985. *Culture and Evolutionary Process*. Chicago: Univ. of Chicago Press.

Burhoe, Ralph Wendell. 1979. "Religion's Role in Human Evolution: The Missing Link Between Ape-Man's Selfish Genes and Civilized Altruism." *Zygon: Journal of Religion and Science* 14 (June): 135–62.

_____. 1986. "War, Peace, and Religious Biocultural Evolution." *Zygon: Journal of Religion and Science* 21 (December): 439–72).

Bowra, Maurice. 1994. *Sophoclean Tragedy*. Oxford, U. K.: Clarendon Press.

Campbell, D. T. 1975. "On the Conflicts between Biological and Social Evolution and Between Psychology and Moral Tradition." *American Psychology* 30: 1103–26.

Cavalli-Sforza, L. L., and M. W. Feldman. 1981. *Cultural Transmission and Evolution*. Princeton, N. J.: Princeton Univ. Press.

Cloak, F. T. 1975. "Is a Cultural Ethology Possible?" *Human Ecology* 3, no. 3: 161–82.

Daly, M. 1982. "Some Caveats about Cultural Transmission Models." *Human Ecology* 10: 401–8.

Daly, M., and M. Wilson. 1988. *Homicide*. New York: Aldine.

d'Aquili, Eugene G. 1983. "The Myth-Ritual Complex: A Biogenetic Structural Analysis." *Zygon: Journal of Religion and Science* 18 (September): 247–70.

_____. 1986. "Myth, Ritual, and the Archetypal Hypothesis." *Zygon: Journal of Religion and Science* 21 (June): 141–60.

Dodds, E. R. 1968. "On Misunderstanding the Oedipus Rex." In *Twentieth Century Interpretations of Oedipus Rex*, ed. M. O'Brien. Englewood Cliffs, N. J.: Prentice-Hall.

Durham, W. H. 1979. "Toward a Coevolutionary Theory of Human Biology and Culture." In *Evolutionary Biology and Human Social Behavior: An Anthropological Perspective*, ed. by N. A. Chagnon and W. G. Irons. North Scituate, Mass.: Duxbury Press.

_____. 1982. "Interactions of Genetic and Cultural Evolutionary Models and Examples." *Human Ecology* 10, no. 3: 289–322.

Ehrenberg, V. 1968. "Sophoclean Rulers." In *Twentieth Century Interpretations of Oedipus Rex*, ed. M. O'Brien. Englewood Cliffs, N. J.: Prentice-Hall.
Flinn, M. V., and R. A. Alexander. 1982. "Culture Theory: The Developing Synthesis from Biology." *Human Ecology* 10: 383–400.
Freud, Sigmund. 1935. *A General Introduction to Psychoanalysis*, trans. Joan Riviere. New York: Liverright.
Gould, Thomas. 1970. *Oedipus the King by Sophocles*. Englewood Cliffs, N. J.: Prentice-Hall.
Graves, Robert. 1978. *The Greek Myths*. Vol. 2. Harmondsworth, U.K.: Penguin Books.
Hamilton, W. D. 1964. "The Genetical Evolution of Social Behavior," parts 1 and 2. *Journal of Theoretical Biology* 7: 1–53.
Hefner, Philip. 1991. "Myth and Morality: The Love Command." *Zygon: Journal of Religion and Science* 26 (March): 115–36.
Jebb, Sir Richard C. [1885] 1966. *The Oedipus Tyrannus of Sophocles*. Cambridge: Cambridge Univ. Press.
Levi-Strauss, Claude. 1963. *Structural Anthropology*. New York: Basic Books.
Lumsden, C., and E. O. Wilson. 1981. *Genes, Mind, and Culture*. Cambridge: Harvard Univ. Press.
Middleton, John. 1960. *Lugbara Religion*. London: Oxford Univ. Press.
Rothstein, Stephen I., and R. Pierotti. 1988. "Distinctions among Reciprocal Altruism, Kin-selection, and Cooperation and a Model for the Initial Evolution of Beneficient Behavior., *Ethology and Sociobiology* 9, nos. 2–4: 189–209.
Rushton, J. P. 1988. "Genetic Similarity, Mate Choice and Fecundity in Humans." *Ethology and Sociobiology* 9, no. 6: 329–339.
_____. 1989. "Genetic Similarity in Male Friendships." *Ethology and Sociobiology* 10, no. 5: 361–63.
Seyforth, Robert M., and Dorothy L. Cheney. 1988. "Empirical Tests of Reciprocity Theory: Problems in Assessment." *Ethology and Sociobiology* 9, nos. 2–4: 181–87.
Taylor, Charles E., and Michael T. McGuire. 1988. "Reciprocal Altruism: 15 Years Later." *Ethology and Sociobiology* 9, nos. 2–4: 67–72.
Tooby, J., and L. Cosmides. 1989. "Evolutionary Psychology and the Generation of Culture. Part 1, Theoretical Considerations." *Ethology and Sociobiology* 10, nos. 1–3: 29–51.
Watling, E. F. 1947. *Sophocles: The Theban Plays*. Harmondsworth, U. K.: Penguin Books.

Part III
Cognitive Science and Religion

[10]

A Cognitive Typology of Religious Actions

Justin L. Barrett[a,*]
Brian Malley[b]
[a] Centre for Anthropology and Mind, School of Anthropology and Museum Ethnography, University of Oxford, Oxford OX2 6QS, UK
[b] Department of Psychology, University of Michigan, Ann Arbor, Ann Arbor, MI 48109–1043, USA
* Corresponding author, e-mail: justin.barrett@anthro.ox.ac.uk

Abstract

The rapid but disproportionate growth of the cognitive science of religion in some areas, coupled with the desire to meaningfully connect with more traditional, function-inspired classifications, has left the field with an incomplete and sometimes inconsistent typology of religious and related actions. We address this shortcoming by proposing a systematic typology of counterintuitive actions based on their cognitive representational structures. This typology may serve as the framework of a research program that seeks to establish (1) psychologically, whether each class of events receives different cognitive treatment within a given context and similar representation across contexts; and (2) anthropologically, whether the different classes are characterized by different performance frequencies, social functions, and kinds of interpretations, making them useful explanatory and predictive distinctions.

Keywords

Agents; cognitive science of religion; counterintuitive; magic; ritual

Biologists find it necessary to maintain two systems for the classification of organisms, one based on form, the other on function. The behaviors of a dolphin, for instance, are similar to those of a shark or any other fish near the pinnacle of the marine food chain, and the behaviors of these animals may be profitably compared. But these functional similarities are not the whole story: by virtue of its genetic composition, anatomy, physiology, and so forth, the dolphin is more similar to cows than to sharks. The two different taxonomies yield different predictive and explanatory insights.

Typologies of cultural phenomena may be similarly constructed and prove similarly productive. Cultural phenomena may be classified by their function (e.g., subsistence *vs.* artistic activities) or by their forms (e.g., transhuman pastoralism *vs.* nomadic pastoralism). And as is the case with biological kinds, cultural kinds based on function tend to produce explanations sensitive to particular contexts, times, places, or cultural ecosystems. Typologies based on the form of cultural phenomena may produce discoveries about cross-cultural recurrence and tendencies for particular form-function associations. The cognitive scientific treatment of religious rituals is a case in point (Lawson and McCauley, 1990; Barrett and Lawson, 2001; McCauley and Lawson, 2002; Malley and Barrett, 2003). Operating on the insight that cultural beliefs and practices must be represented in human minds (Sperber, 1985, 1996), scholars in the cognitive science of religion have shown that using the typical cognitive representational structures of cultural phenomena as their deeper structure (their form) produces theoretically motivated and empirically tractable typologies of cultural phenomena (Boyer, 2001; McCauley and Lawson, 2002; Barrett, 2004; Pyysiäinen, 2004) Nevertheless, the rapid but disproportionate growth of the cognitive science of religion in some areas, coupled with the desire to meaningfully connect with more traditional, function-inspired classifications, has left the field with an incomplete and sometimes inconsistent typology of religious and related actions.

Here we seek to address this shortcoming by proposing a systematic typology of counterintuitive actions (including religious and magical acts) based on their cognitive representational structures. The proposed typology may serve as the framework of a research program that seeks to establish (1) psychologically, whether each class of events receives different cognitive treatment within a given context and similar representation across contexts, thereby supporting the analytic distinctions, and (2) anthropologically, whether the different classes are characterized by different performance frequencies, social functions, and kinds of interpretations, making them useful explanatory and predictive distinctions.

We are concerned here with counterintuitive events. By counterintuitive events, we refer to events that violate observer's naïve (i.e., untutored) intuitions about causal relations (Boyer, 1994, 2001; Hirschfeld and Gelman, 1994; Sperber *et al.*, 1995). As developmental psychologists have demonstrated, children rapidly acquire (or even manifest from birth) inference systems that generate causal expectations for objects and actors. These conceptual systems include specialized inference systems that concern physical interactions ("naïve physics"), the activities of living things ("naïve biology") and intentionality ("Theory of Mind").

From infancy our naïve physics systems intuitively tell us, among other things, that an object at rest will stay at rest until it is launched by another physical object. So, if we watch as a boulder suddenly moves several feet to the left without any other object contacting it, we will represent that event as counterintuitive. It violates our intuitive expectations governing physical objects. In the domain of living things, a cat giving birth to puppies would constitute a counterintuitive event. Our intuitive inference systems generate the expectation that animals give birth to their own kind (counterintuitive actions are not necessarily counterfactual: someone might implant dog embryos in a cat so that a cat did birth to puppies, but that would not render the occurrence intuitive). The Theory of Mind systems generate intuitions concerning the activities of intentional beings or agents. So, if a tree is said to have heard a woman's gossiping, this event would qualify as counterintuitive because trees, not being intuitively classified as intentional beings, are not expected to think or communicate.

Violations are most commonly caused, we suggest, by contradictions between different intuitive causal inferences. When biological inference systems generate intuitions that contradict those generated by naïve physics systems, e.g., if a rock moves about by itself but is not a living thing, the mental representation of this event is anomalous, counterintuitive. Our first hypothesis is, thus, that observers will distinguish intuitive and counterintuitive events. We do not expect this distinction to be conscious or explicit (though it may be either in particular contexts), but rather manifest in the allocation of attention, in memorability and in inferences. We expect that counterintuitive events will be attention demanding and, perhaps as a consequence, more memorable than intuitive events. We also expect that counterintuitive events will partly confound inferences beyond those involved in the definition of the event as counterintuitive, e.g., that observers will be uncertain whether a rock that moves by itself is also of a kind with other rocks, and whether such a rock may seek sustenance of some kind.

Counterintuitive Events *versus* Counterintuitive Actions

What distinguishes counterintuitive actions (our focus here) as a special type of counterintuitive event is the attribution of intentional agency to an actor. The wind blowing a tree branch down is an event. It becomes an action (and a counterintuitive one) when someone supposes that the wind intended or in some way willed the branch to fall, or someone else used the wind to knock down the branch. Getting ill is merely an event unless getting ill is thought to be the consequence of deliberate poisoning, or the wrath of an angry spirit, or being coughed upon maliciously. Once observers represent an event as the result of deliberate action by an intentional agent, it is an action (the attribution of deliberate action is, of course, situationally sensitive, but here the contrast is between deliberate and non-deliberate actions within particular contexts). Our second hypothesis is, thus, that observers will distinguish events and actions by attributing intentional agency in the case of actions only; that is, that the identification of an event as an action will bring with it automatically the assumption of agency behind it, even if the precise causal links between agent and event are not known.

Examples of counterintuitive actions (CI actions) are abundantly supplied by religion, magic, or even modern technology. When a Yanomamo shaman curses a neighboring village because his is sick with measles, or a girl shouts at her little brother because his remote control car has rammed her leg, the causal chain involves a link or two that confounds our intuitions, but we see in the event the actions of an agent.

It is, curiously, more difficult to think of counterintuitive events that are not interpreted as actions. Perhaps if something is interesting enough to notice, we begin wondering who is behind it. Suppose someone observes a large rock seeming to move by itself. Or perhaps our theory of mind inference systems are so productive that we tend to fall back on them, even when other inference systems tell us that they don't really apply, as when we reason about nations ("What does North Korea want?") or computers ("It just hates me!). It seems we tend to represent CI events as CI actions. Though there are counterintuitive events that are not represented as actions, we suggest that they play little role in mythology, folktales, religion, or other cultural forms (*cf.*, Boyer, 2001). Figure 1 summarizes these distinctions.

Next let us consider the various kinds of CI actions. In labeling these, we use familiar categories such as magic, ritual, possession and prayer for reasons

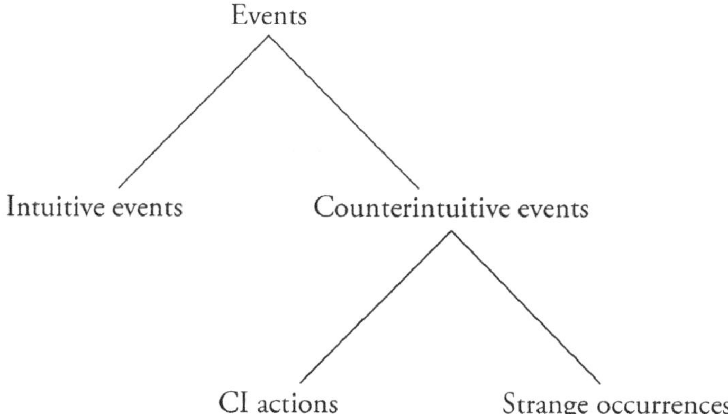

Figure 1. Major classes of events.

of precision and parsimony. If the proposed groupings prove empirically coherent, they can provide precision and clarity to these otherwise ambiguous labels. Secondly, we want to keep the typology anchored in empirical data, not merely a set of theoretically possible actions designated by some opaque code. It is easier to follow arguments about "religious rituals" than "CI actions + S-agency + observability." Yet we are not committed to these terms, nor does our proposal turn on the particular language we use.

Mundane *vs.* Supermundane Agents

Counterintuitive actions may be divided into those attributed to normal, mundane agents and those attributed to supermundane agents. Examples of actions attributed to mundane agents would be a priest sacrificing an ox to Poseidon to secure safe passage, a Shiva devotee suspending himself on skin piercing hooks so that a child will prosper, or a triumphant warrior ritually consuming his slain enemy to absorb his power. All of these cases involve someone acting in way intended to bring about counterintuitive causation, consequences that do not intuitively follow from the performed action. But many events represented by observers as being intentionally caused involve no mundane actor at all. For instance, when a child gleefully explains that God caused the snow to fall in May, the snowfall has been represented as an action, caused by a superhuman agent. Likewise, miraculous recovery from an illness may be attributed to divine intervention. These sorts of actions would constitute CI-supermundane actions.

CI-Supermundane Actions: Possessions and Miracles

CI-supermundane actions constitute a fairly limited group of phenomena. They appear in two basic forms. Possessions are those with mundane human or animal mediation. An action is carried out by a mundane agent but observers attribute it to a supermundane agent. When a superhuman is said to act directly, without mundane mediation, then the CI-supermundane action falls into the category miracles. The previously mentioned snowfall and unexpected recovery from illness exemplify this category. Nevertheless, miracles come in negative valence versions as well, as when a spirit strikes someone dead, destroys a home, or otherwise causes misery.

CI-Human Actions with and without Special Agents: Religious and Magical Acts

Inasmuch as the agents in this category are invariably human, we shall subsequently refer to CI-mundane actions as CI-human actions. The fact that the agents involved are human rather than animal or plant is a consequence of the fact that only humans have religion, and while this may be an accidental property of our taxonomy, we think it significant enough to rename the category at this level.

The most fundamental division within CI-human actions is between those that connect supermundane agents to the human action and those that do not. Let us call those that do religious actions and those that do not magical actions (in this denomination we follow an old anthropological distinction, to which testable hypotheses might lend some teeth). The distinction is simply this: some CI-human actions are thought to work by appeal, directly or indirectly, to supermundane agents; other times, certain instruments, words, people, or action-sequences are thought to possess special, counterintuitive causal power in their own right, without invoking the involvement of a supermundane agent. The former are religious actions, the latter magical.

Special Agent Actions and Special Actions

Among religious actions, we follow Lawson and McCauley (Lawson and McCauley, 1990; McCauley and Lawson, 2002) in distinguishing between those that connect supermundane agents to the human agent and those that connect the superhuman agent in some other way, between special agent actions and special procedure actions. Lawson and McCauley propose that underlying ritual performances are cognitive representations of the ritual actions, representations that include an S-agent (what we are calling a supermundane agent) somewhere in the action structure. Lawson and McCauley provide formal rules for describing the action structure and its incorporation of the S-agent, but we need not be concerned with those here. Their principle concern is to distinguish those rituals where the human actor is conceived to have been somehow – even very indirectly – the object of prior ritual transformation (e.g., the priest administering the Eucharist has himself been ritually transformed through ordination) and those where the human actor is not.

While we wish to adopt the Lawson–McCauley distinction between special agent and other rituals, we cannot do so without modifying it in two important respects. First, Lawson and McCauley limit the scope of their theory to religious rituals, which they define as religious actions in which some object or person is transformed. So, for example, their theory applies to the ritual transformation of a child into an adult, but not to prayer, which does not by itself modify the status of anything and so would be a religious action, but not a ritual. This distinction, however, has little to recommend it apart from its convenience for their predictions, and as it is inconvenient to our typology, we will ignore it. Or, to put it in hypothesis form, we suggest that the Lawson–McCauley ritual form hypothesis will apply also to religious actions generally.

Our second modification of the Lawson and McCauley's proposal is to add to their ritual structures another parameter, purpose. Lawson and McCauley's theory specifies, in considerable detail, the logical structure of ritual action. But they do not consider the fact that part of the mental representation of many rituals is their purpose. For instance, sacrifices are often performed to placate gods or ancestors. We suggest that adding the parameter purpose may help to answer our earlier finding that many religious people do not seem to know how superhuman agents are linked to the ritual's action structures (Malley and Barrett, 2003). Again, to put this in the form of a hypothesis, we suggest that when S-agents are implicated in the structure of religious actions through the actions' purpose (as cognitively represented by participants), such rituals will have the characteristics of special procedure actions.

We have taken the trouble to connect our taxonomic division with Lawson and McCauley's distinction because this allows us to incorporate the predictions of the ritual form hypothesis:

- Special agent actions are reversible, in the sense that either (1) there will exist another religious action for reversing the effects of such acts or (2) participants' will judge that such a religious action would be possible. A reversal of the effects will require another special agent action. Special procedure actions will not be reversible.
- Special agent actions have super-permanent effects, i.e., they will not require repetition for continued efficacy. Special procedure actions will have only temporary effects.
- Special agent actions carry relatively higher levels of sensory pageantry than special procedure actions within the same population.

These are interesting hypotheses, with some empirical support (Malley and Barrett, 2003), although further research is needed.

Figure 2 summarizes the typology of CI actions presented thus far.

Religious Rituals and Non-Transformative Religious Actions

Our next distinction, taken directly from the work of Lawson and McCauley, is between those religious actions in which some person, place, or thing is acted upon and those in which the actions are not performed upon anything. Lawson and McCauley restrict the term religious ritual to the former category, leaving the other category unmarked. We will adopt their

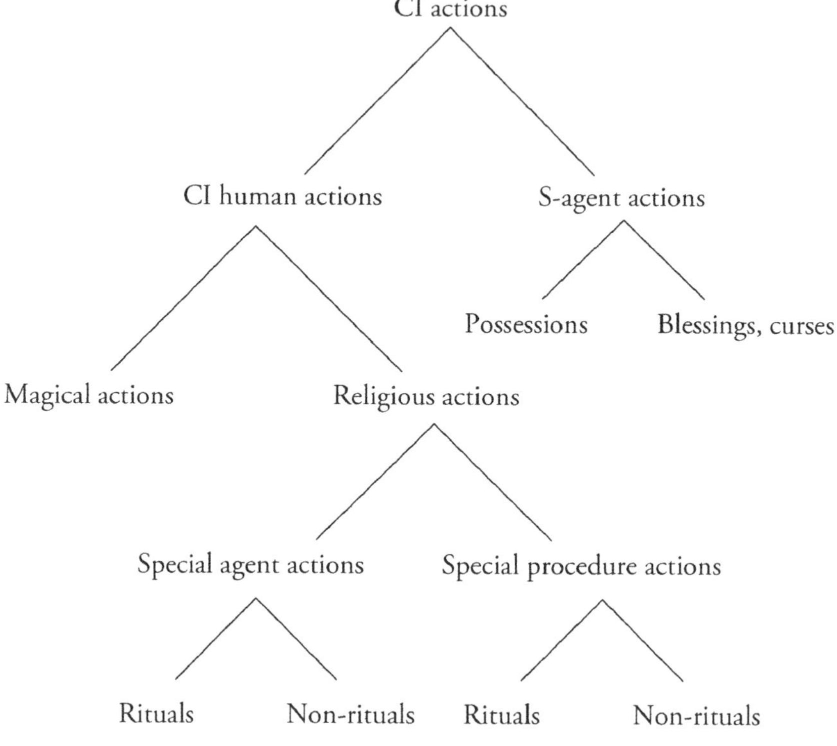

Figure 2. Classes of CI actions.

terminological convention. Lawson and McCauley suggest that the interpretation of religious rituals differ from that of religious actions more generally in that religious rituals are conceived as being transformative, that the objects acted upon will be said to be transformed in some way. As stated, this could be a trivial claim – in what does action on an object consist if not some sort of transformation of the object? – but we take it in a non-trivial sense, that the objects of ritual action will be understood as transformed in some way beyond the observable effects of the ritual action. Thus, for example, the ritual sacrifice of a cow will be understood as transforming the animal in some way beyond merely rendering it dead.

Empirical and Metaphysical Consequences within CI Actions

Pyysiäinen (2004) has proposed that religious actions are more likely to have metaphysical or unobservable consequences than are magical actions, which are often said to have practical, material effects. Although our use of "magic" hinges on a different conceptual criterion – whether appeal to an S-Agent is made – than his, a reasonable hypothesis inspired by Pyysiäinen's proposal is that though magical actions can be associated with metaphysical outcomes, empirical outcomes predominate. That is, magical actions are used more often than religious ones to try to do things like bring rain, make people sick or well, or improve crop yield; whereas religious actions tend to be used to change metaphysical states such as being forgiven of sins, becoming married, or becoming a priest. We do not know how this hypothesis may fare, but it is worth suggesting as a touch point between Pyysiäinen's work and our typology.

This concern with the class of intended outcome may also be applied as a theoretical consideration for all CI actions. We wonder, for example, whether religious rituals tend to concern metaphysical consequences more frequently than physical consequences; non-ritual religious actions like prayer tend to concern physical consequences.

Our overall typology is summarized in Fig. 3. We think that the cognitive distinctions upon which our typology is based are either well established or at least psychologically plausible. Further, we think that each gives rise to interesting hypotheses about the structure of cultural systems and their transmission. The hypotheses may be summarized as follows:

1. People will distinguish counterintuitive events – events in which causal links are contradictory or missing – from intuitive events.
   a. Counterintuitive events will be more attention demanding and memorable than intuitive events.
   b. Counterintuitive events will confound further inferences about an object, beyond those involved in the definition of the counterintuitive event.
2. People will distinguish events from actions by attributing actions alone to intentional agents.
   a. The precise causal relation between agents and the events they are said to cause may be unclear to observers.
   b. People will be far more interested in actions than in events, and actions will be more common than events in their myths and folklore.
3. When S-agents are implicated in the structure of religious actions through the actions' purpose (as cognitively represented by participants), such rituals will have the characteristics of Lawson and McCauley's special patient rituals.
4. The Lawson–McCauley ritual form hypothesis will apply also to religious acts generally.
   a. Special agent actions will be ritually reversible, in the sense that either (1) there will exist a ritual for reversing the effects of such acts or

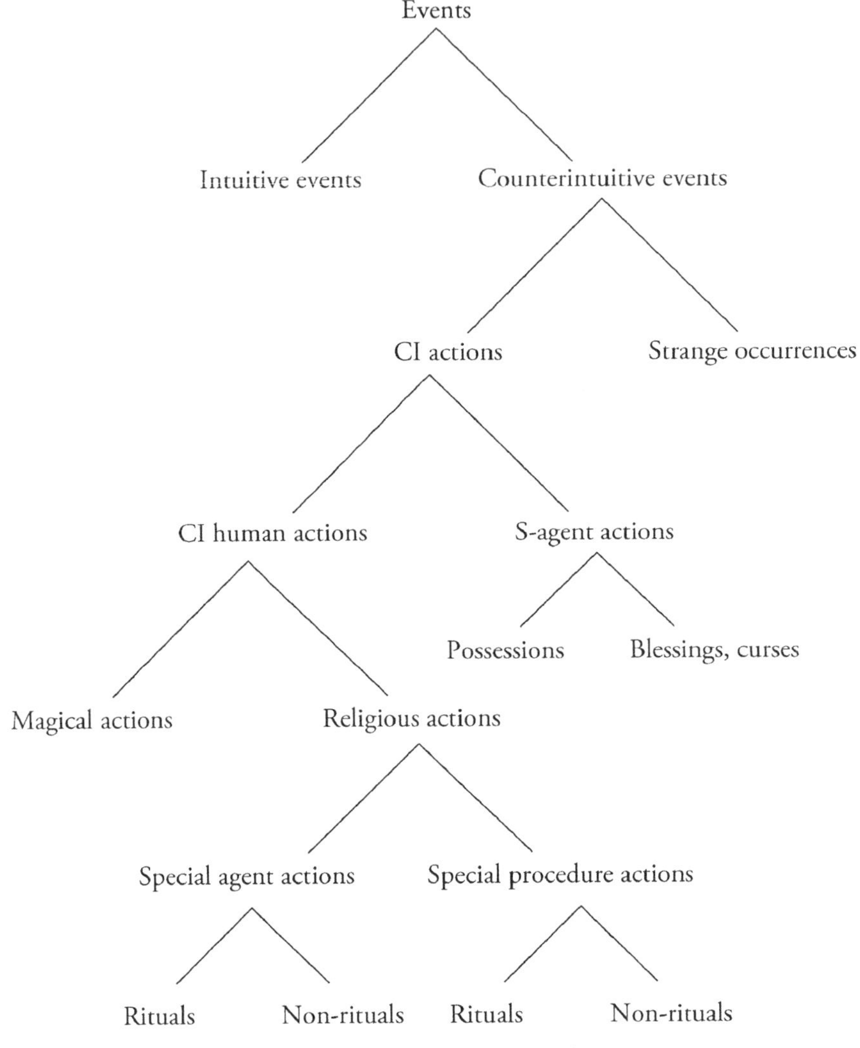

Figure 3. The complete typology.

(2) participants' will judge that such a ritual would be possible. A reversal of the effects will require another special agent action. Special patient actions will not be ritually reversible.

b. Special agent actions will have super-permanent effects, i.e., they will not require repetition for continued efficacy. Special patient rituals will have only temporary effects.

c. Special agent actions will carry relatively higher levels of sensory pageantry than special patient actions within the same population.

5. The objects of ritual action will be understood as transformed in some way beyond the observable effects of the ritual action.
6. Although magical actions can be associated with metaphysical outcomes, these actions will more often be associated with practical, material outcomes. In comparison to magical actions, religious actions will tend to be associated with unobservable, metaphysical outcomes.
7. Religious rituals will tend to concern metaphysical consequences more frequently than physical consequences. In comparison to religious rituals, non-ritual religious actions like prayer will more often concern practical, material consequences.

These hypotheses lend empirical teeth to the principled cognitive distinctions we have drawn.

Our goal has been to articulate some points of contact between the structure and function of religious actions by synthesizing distinctions and hypotheses advanced by a variety of scholars working within a cognitive approach to religious phenomena. Our hope is that this synthesis will aid in the design of future research projects and that the hypotheses advanced here, even if found false, will prove fruitfully so.

References

Barrett, J. L. (2004). *Why would anyone believe in god?* AltaMira, Walnut Creek, CA.

Barrett, J. L. and Lawson, E. T. (2001). Ritual intuitions: Cognitive contributions to judgments of ritual efficacy. *Journal of Cognition and Culture* 1, 183-201.

Boyer, P. (1994). *The naturalness of religious ideas: A cognitive theory of religion.* University of California Press, Berkeley, CA.

—— (2001). *Religion explained: The evolutionary origins of religious thought.* Basic Books, New York, NY.

Hirschfeld, L. A. and Gelman, S. A. (1994). *Mapping the mind: Domain specificity in cognition and culture.* Cambridge University Press, Cambridge.

Lawson, E. T. and McCauley, R. N. (1990). *Rethinking religion: Connecting cognition and culture.* Cambridge University Press, Cambridge.

Malley, B. and Barrett, J. L. (2003). Can ritual form be predicted from religious belief? A test of the Lawson-McCauley hypotheses. *Journal of Ritual Studies,* 17, 1-14.

McCauley, R. N. and Lawson, E. T. (2002). *Bringing ritual to mind: Psychological foundations of cultural forms.* Cambridge University Press, Cambridge.

Pyysiäinen, I. (2004). *Magic, miracles, and religion: A scientist's perspective*: AltaMira, Walnut Creek, CA.

Sperber, D. (1985). *On anthropological knowledge: Three essays.* Cambridge University Press, Cambridge.

—— (1996). *Explaining culture: A naturalistic approach.* Blackwell, Oxford.

Sperber, D., Premack, D. and Premack, A. J. (1995). *Causal cognition: A multidisciplinary debate.* Clarendon Press, Oxford.

[11]

EXPLAINING RELIGIOUS IDEAS: ELEMENTS OF A COGNITIVE APPROACH

Pascal Boyer

Summary

This paper outlines an anthropological approach to religious representations that is grounded in recent findings and hypotheses in cognitive psychology. The argument proceeds in four points. First, the main goal of this framework is to account for the recurrence of certain types of mental representations in religious systems. Recurrent features are not necessarily universal. They are the outcome of cognitive systems that make certain representations easier to acquire than others. Second, a cognitive approach must take into account the diversity of religious representations. It is argued here that religious systems bring together ontological assumptions, causal claims, episode types and social categories. These four "repertoires" may have different functional properties, and may therefore be acquired and represented in different ways. Third, universal features of tacit, intuitive systems may impose strong constraints on the variability of religious ideas. This is illustrated on the basis of ethnographic data. Finally, the type of representations one finds in religious belief-systems consists in conjectures, the cognitive salience of which is variable and should be evaluated in precise terms.

Contrary to other domains of anthropological study, theories of religious belief and action have not been much influenced so far by the remarkable development of cognitive science. True, there is in anthropology a subdiscipline known as "cognitive anthropology" (or "ethnosemantics") which focuses on the cognitive aspects of cultural representations. This approach, however, has been so far limited to representations of the everyday world: biological taxonomies, classification of daily activities, kinship terminologies, etc.[1] Religion, on the other hand, is relatively neglected in cognitive approaches. This is paradoxical, in view of the importance of religious belief and action in anthropological theory and practice. There are, obviously, some notable exceptions to this generalisation (see e.g. Dougherty 1985, Dougherty & Fernandez 1980, 1982, Lawson & Mc Cauley 1990). By and large, however, the study of religious belief and action is still conducted in the framework of anthropological theories which pay little if any attention to the findings and hypotheses of cognitive science.[2]

In this paper, I will present the elements of a possible cognitive approach to religious representations.[3] This paper is therefore largely programmatic and partly speculative. The research programme, however, is not very far removed from actual cognitive research, and the speculation is mostly consistent with both anthropological and psychological research. My aim here is to show that a cognitive approach is possible, and to examine how it can account for the representation and transmission of religious representations. The theory is about *religious ideas* rather than "religions" in the broad sense. The aim is to describe the processes whereby subjects acquire, represent and transmit certain ideas and practices. The theory may not be sufficient to account for the social dynamics of religious movements or the historical development of religious doctrines. Such "macro-phenomena" of religious transmission are not directly within the scope of a cognitive theory.

In the following pages I will put forward four main hypotheses. The first one is that the goal of a cognitive approach is to account for the *recurrence* of certain features of religious representations in many different cultures. This general objective is often misconstrued, and I will try and give a formulation which avoids certain common misunderstandings. The second point is the *cognitive diversity* of religious ideas. In a given culture, the set of representations that constitute each individual's religious ideas is distributed in several "repertoires", which have different functional properties. Here I will identify four such cognitive repertoires, which are particularly important in the description of religious representations. The third hypothesis is about the *cognitive constraints* on the content and organisation of religious ideas. Far from being pure cultural constructions, religious systems appear to be strongly constrained by universal, probably innate properties of cognitive systems, especially those properties which govern people's intuitive understanding of their everyday world. A fourth point is that we need a precise notion of *cognitive salience* in order to describe and explain the processes of acquisition and belief-fixation in the domain of cultural knowledge. For each of these four points, I will try to show that a cognitive approach provides a plausible alternative to classical anthropological theories of religion.

I

Universals, Recurrence and Explanation

It may seem the obvious first step, in the construction of a general theory of religion, to enumerate the universal features which the theory will set out to explain. Indeed, this is very much what can be found in most anthropological attempts so far.[4] Here, however, I will argue that this seemingly obvious way of proceeding is in fact mistaken and misleading, and that the search for universals is the main reason why anthropology has not produced a theoretically plausible account of religious ideas. In what follows, I will try to show how the problem of universals conceals other, important aspects of the problem.

The main starting point of the theory outlined here is that, in the variety of cultural systems of religious belief, there is a notable *recurrence* of certain precise themes or ideas, and that this recurrence ought to be explained. These themes or ideas are not universal, but they constitute a repertoire most elements of which can be found in most cultures, in one form or another. To take but a few examples, it is assumed in many (but not all) cultures that a non-physical component of persons survives after death, to become an invisible intentional being, endowed with perceptions, beliefs and intentions. In the same way, it is assumed in many (but not all) cultures that certain people are especially likely to receive direct inspiration or messages from extra-natural agencies, like gods and spirits. In many (but not all) cultures it is admitted that performing certain ritual recipes in the exact way and order prescribed can bring about changes in physical states of affairs, through causal mechanisms which are presumed but not observable. Such features are widespread in many cultures, yet they are not necessarily present in all of them. Each feature is present in most cultures, each culture has many of those features in its set of religious ideas, yet none of them should be taken *a priori* as universal.

Two obstacles to the study of recurrent features

In cultural anthropology, the recurrence of certain religious ideas is not explained in a satisfactory way, for the simple reason that is not explained at all. The few features I mentioned above are well-known to most students of religion. Their recurrence in different cultures, however, is not considered an object of scientific inquiry, for two symmetrical reasons.

A first obstacle is the pre-theoretical, instinctive form of relativism that is somehow intrinsic to anthropological investigation. Because anthropologists are professionally trained to detect and emphasize cultural differences, they naturally under-estimate the recurrence of similar ideas in different cultures. Moreover, when this recurrence is noticed, it is often treated as a deceptive appearance, which conceals underlying differences. It is widely assumed that apparently similar beliefs cannot really be similar, because they occur in different "cultural contexts". This idea, however, is vague enough to contain both a trivial truth and a profound fallacy. Take for example the widespread idea that the gods are so remote that one cannot communicate with them except through the channel of inspired mediums. Obviously, this idea can take on very different "meanings" in different cultures; more precisely, it carries rather different implications, for those who think that the gods have a direct influence on the living's well-being, and those who think that they do not. The idea that certain people are privileged "channels", however, is the same idea in both contexts. We need a theory that could account for the fact that this idea is so widespread, while others are not.

Conversely, the general study of religious ideas is often hampered by the widespread idea, that "human nature", the proper subject-matter of cultural anthropology, is manifest only in the universal features of the species. What is not universal in human cultures has therefore nothing to do with human nature. In other words, one assumes that there is a division, between cultural invariants on the one hand, which are explainable by various non-cultural factors (ecological, biological, psychological, etc.), and the rest. Cultural features which are not universal are *ipso facto* outside the influence of those various ecological, biological, psychological etc., factors. If certain traits of religious ideas are widespread but not universal, they are therefore considered outside the scope of a general theory of religion. Obviously, this argument rests on a confusion of levels, between processes and their outcome. That a universal process exists does not imply that its outcome will be the same in all possible circumstances. It is precisely the point of an explanatory theory to reduce diversity, and show in what manner diverse phenomena result from the encounter of general mechanisms, on the one hand, and a manifold of contingently diverse circumstances on the other.

Generative and selective models

This "probabilistic" approach to recurrent features in religious systems has important consequences for the type of model that is supposed to account for religious ideas. Here I must draw a broad distinction between two types of accounts that can be put forward in the explanation of the emergence of recurrent features in a population of organisms. I will label these accounts *generative* and *selective*. Given a series of recurrent features, a generative model posits an underlying mechanism such that, if it is present, it will provide a sufficient explanation for the occurrence of these features. Take for example the fact that, in all tigers, the anatomical structure of the retractile claws is exactly similar. A sufficient explanation of the recurrence is provided by models of genotype inheritance combined with models of embryological development. Selective models, on the other hand, account for recurrent features by positing (i) a set of underlying mechanisms which are necessary yet insufficient to produce the recurrence, and (ii) a set of inputs such that, given the underlying mechanisms, they will produce the recurrence. This type of explanation would be necessary if we want to explain, for example, why tigers have retractile claws rather than non-retractile ones. In such a case, an evolutionary explanation will typically invoke (i) a series of random mutations, providing the input to (ii) a fitness-maximizing mechanism (natural selection), the combination of which provides a sufficient explanation. The fitness-maximizing mechanism, on its own, provides necessary, yet insufficient conditions for the recurrence.

It should be obvious, from this example, that generative and selective models can be invoked in order to explain the same recurrence, seen from a different point of view. This does not mean, however, that choosing between these two explanatory schemes is a matter of convenience or subjective preference. The notion of a "point of view" is to be understood here in a precise sense, as implying a set of clearly defined goals for the explanation. Given such a precisely defined objective, choosing a generative account, where a selective one would have been pertinent, may generate considerable problems. Biological theory for instance would be in serious trouble if it tried to explain evolutionary trends by generative models, e.g. explaining the evolution of vertebrates by some underlying mechanisms that "pushed" fish out of the water and made them become reptiles. The evolutionary scenarios developed in biology do not require any generative models of evolutionary trends, and in fact exclude them explicitly.

Selective models and cultural transmission

Applied to cultural ideas, the notion of a selective model means that, given certian circumstances and a variety of mental representations entertained by a population of subjects, some of those representations are more likely than others to be stored in the subjects' memories and communicated to other subjects. Many anthropological models are, implicitly or explicitly, based on this evolutionary metaphor,[5] and treat the transmission of ideas as a function of their "survival" value. A set of constraints is posited, such that it will make it more likely for certain representations to "survive" (i.e. be memorized and transmitted) than for others. This is the starting point of various models of cultural evolution (Cavalli-Sforza & Feldman 1981, Cavalli-Sforza 1986), "gene-culture co-evolution" (Lumsden & Wilson 1981) or "dual inheritance" (Boyd & Richerson 1985). These various theories share the premise that the recurrence of cultural traits must be explained by selection processes rather than by generative mechanisms. They include precise hypotheses about the transmission or diffusion of cultural material. The point of departure is that, given a random input of cultural traits at generation G_i, a process can be described, such that its operation on the input will increase the relative probability of certain traits appearing at generation G_{i+1} (and of course decrease the probability of other traits). Selective models generally focus on *transmission processes* as the main cause of recurrence.

Another, qualitative approach that is based on a selective stance is Sperber's notion of an "epidemiology of ideas" (1985, 1991). The argument is based on an analogy, to the effect that, by and large, the relation of anthropology to psychology can be construed as similar to that of epidemiology to physiology. While physiology puts forward hypotheses about, e.g. the way various viruses or germs may affect the body's functioning, epidemiology on the other hand is concerned with the ways in which diseases spread. In much the same way, psychology is concerned, among other things, with the acquisition or representation of certain ideas or beliefs. Anthropology, on the other hand, specializes in observing the spread of ideas or beliefs. It may focus on either short-lived epidemics, like fashions, or more stable endemic infections like traditions. An important consequence of this notion of "epidemiology" is that psychological processes are directly pertinent to anthropological theory. In the same way as such physiological aspects have direct consequences for the spread of a given disease in given circumstances, the psychological processes of representations and transmission are bound to affect the patterns of cognitive "epidemics".

The ideas presented in this paper are, by and large, consistent with the assumptions of these selective models. I will claim that micro-processes of cognition and interaction impose strong constraints on the diffusion and transmission of religious assumptions, thereby leading to the recurrence of ideas observed in the religious domain. Before proceeding to this point, however, we must examine the types of mental representations which are the object of such selective processes of cultural transmission.

II

The Cognitive Diversity of Religious Knowledge

In anthropological descriptions, religious representations are generally presented as constituting shared, integrated, explicit, context-free general propositions, e.g. "the spirits dwell in the rivers", "the ancestors are invisible", "only shamans can negotiate with the spirits", etc. As many authors have observed, such descriptions in terms of "collective representations" are extremely ambiguous, as far as actual cognitive processes are concerned.[6] Obviously, a cognitive approach to religious ideas should provide precise answer to the questions concerning the actual representations involved, rather than work on the basis of *ad hoc* descriptions. In the following pages, I will try to put forward the elements of such an approach. My first contention is that a cognitively realistic description of religious ideas will have to take into account the *cognitive diversity* of the religious domain. That is to say, the representations we call "religious" belong to different domains and may have different cognitive properties. I will then examine the possible contributions of cognitive science to the study of the different types of representations in question.

Four repertoires of representations

Religion should not be construed as an isolated domain. This point is generally accepted by most anthropologists, who consider it a mistake to study the religious life of a given culture as though it was isolated from other domains of social life. More recently, most anthropologists have come to agree that such a division would be illegitimate for a deeper reason, namely, because the very category "religion" may well be a Western construction, of limited validity in the study of different cultures. Here I will not enter such debates; I will accept as a tentative characterisation of the domain, the fact that the cultural representations we are focusing on concern unobservable extra-human entities and processes. Such a practical definition is more or less what anthropologists have in mind when they talk about "religion", and will be largely sufficient for the purposes of the present argument.

In the general discussions concerning the coherence (or lack of coherence) of the notion of "religion", one aspect is generally left aside, that of the *cognitive unity* of the religious domain. Theories of religion, in either anthropology or sociology, are generally based on the implicit premise, that all religious ideas are acquired and represented in the same way. This assumption is itself based on a more general principle, following which most mental representations are acquired and represented in the same way. Whether based on an empiricist or rationalist epistemology, anthropological models treat all cultural representations as functionally similar. Against this view, however, it must be noted that most recent advances in experimental psychology, especially in the domain of cognitive development, insist on the *functional specificity* of certain types of representations.

In a parallel fashion, I will argue that the representations concerning extrahuman entities and processes belong to different *repertoires*, which have different functional properties. Here I will consider four such repertoires, in which the most important elements of religious belief-systems can be found. They are the *ontological*, the *causal*, the *episode* and the *social roles* repertoires, respectively. I will try to show that describing and explaining people's religious ideas consists in describing what is included, in a given culture, in each of these repertoires, and explaining how their contents are gradually made plausible. Before turning to this point, however, let me give a succint description of the typical contents of each repertoire.

The *ontological* repertoire is the set of assumptions people entertain about the existence of non-observable entities. This catalogue will include ideas about there being, e.g., a distant impersonal Creator somewhere in the skies, water-spirits near ponds and rivers, invisible ancestors lurking in the darkness of the forest, etc. This catalogue of ideas is called ontological because it consists of elementary assumptions about what sorts of things there are in the world.

The *causal* repertoire is a catalogue of ideas and assumptions about causal connections between the entities described in the ontological repertoire, on the one hand, and observable events and states of affairs on the other. Thus, a causal repertoire may include assumptions like "gods get angry if no sacrifice is performed" or "reciting this formula will guarantee good crops". It is important to note that such a repertoire may include generalisations of this kind, as well as propositions concerning singular events or states of affairs, like "So and so got disease X because he did not observe prohibition Y", or "we had no crops last year because of such and such's witchcraft".

The *episode* repertoire consists of descriptions of a certain range of event-types, which are connected to the ideas contained in the ontological and causal repertoire. In order to describe a religion, one must identify a certain set of actions and interactions which are deemed to be of particular types. Ritual performance, obviously, is the most important type of religious episodes. Again, it is important to note that the representations involved may be about singular objects, e.g. memories of ritual X being performed last week, or about generalised types, like the list of things to do in order to perform a certain rite in a proper way.

The *social roles* repertoire is a catalogue of representations concerning differences between people. In this we will include people's ideas concerning their priests, shamans, or religious specialists, but also ideas concerning other differences which are relevant in religious action. For instance it will include people's ideas about gender where relevant, about growth and maturation where an adult-child opposition is concerned, people's ideas about the effects of initiation, etc. All these ideas are used to characterize, sometimes categorize social actors, either *in abstracto* or as particular persons.

An ethnographic illustration

In order to make these explanations more intuitively clear, let me take an example of their concrete application, concerning the religious ideas of the Fang of Cameroon (see Boyer 1986, 1990 for more detail). In the following sections I will use this example to illustrate some of the general claims I want to make about religious ideas. Although the specific contents of the ''repertoires'' are, obviously, particular to Fang culture, it must be stressed that the general conclusions I will put forward are in fact based on much broader ethnographic comparisons, which cannot be included in this paper. For the sake of simplicity, I will introduce each theoretical statement as though it had been developed in order to explain the particular Fang data. It must be kept in mind that it is in fact of general relevance for the acquisition and representation of religious ideas.

Ontological ideas. Let me first take the repertoire of ontological ideas, which seems to be organized around three main points: a pair of distant personified gods, the ghosts (*bekong*) and the spirits (*minkugu*). In Fang mythology, there are two creator-gods.

Mebeghe is the name of a god understood to be the creator of all natural things while Nzame is supposed to be at the origin of most cultural techniques and social institutions. It must be stressed that the narratives concerning the origins are not the object of much attention or speculation. Nzame and Mebeghe are remote gods. Their powers are not really invoked or used in the explanation of natural or social occurrences, although there are individual variations in this domain. The role of Nzame is sometimes conceived as that of an impersonal, purposeless fate. It must be pointed out, however, that such a notion of contingency is alien to the Fang intellectual climate, as to that of most African societies. If salient events are to be explained at all, they have to be explained in terms of goals and intentions; this is precisely the type of explanations provided by the concepts of ghosts and spirits.

The term *bekong* could best be translated as "ghosts". After death a person's "shadow" is supposed to become a wandering spirit, generally malevolent until it is given appropriate funeral rites. It is then said to dwell in ghost-villages, and is supposed to protect the villagers. Correct performance of traditional rites is indispensable, lest ghosts may "throw" illnesses of various kinds at the living. Ghosts have beliefs, desires, feelings, emotions, and generally all the non-physical characteristics of humans. In Fang discourse, however, they are generally treated as a kind ("the ghosts do this", "the ghosts want that", etc.) rather than as individuals.

The mystical personnel also includes rather mysterious creatures called *minkugu*, generally glossed as "spirits". These are not clearly identified as ghosts. They are described as smaller, not related to village clans and lineages in a defined way, and rather uncanny, They too can "throw" illnesses, and some specialists say their remedies are given by the *minkugu* rather than the ghosts. Although most people insist that there is a difference in kind between *bekong* and *minkugu*, their ideas concerning the exact differences in powers, appearance, etc., are extremely vague.

Causal connections. The repertoire of causal connections includes many general connections between the entities described above and various classes of events, which however are generally less salient then representations of singular episodes. It is for instance admitted that the ghosts can trigger various kinds of misfortune if the living do not perform appropriate traditional rites. People who hold such ideas, however, rather focus on memories of singular occurences, in which a certain disease was diagnosed as "thrown" by the ghosts, and then successfully cured. In the same way, the fact that illness is commonly caused by witchcraft is stated in very vague and general terms. People, however, have precise memories of many singular cases, with the problem, the diagnosis and the outcome. The same remark applies to magical charms and recipes. While their general efficacy is stated only in the vaguest terms, people have numerous accounts of the successes or failures of precisely identified recipes.

Episodes. These representations concern categories of actions, notably ritual actions, related to the entities of the ontological repertoire. The repertoire includes names for complex scripted actions. To take but one example, *nku melan* is an initiation ritual during which neophytes are shown the skulls of their ancestors, usually concealed in special shrines. The main characteristic of such actions is that they are represented as rigidly scripted. They consist in a list of sub-actions which, from the actors' viewpoint, must be performed in the appropriate way and in the appropriate order, by the appropriate specialist. Although actual performance may display great variations, the participants are generally unaware of those changes or consider them insignificant.

Social roles. There are of course many categories designating types of people. Here I will only mention the subset that is directly pertinent to religious ideas. Some of these categories are supposedly descriptive, identifying some persons by virtue of the rituals they have learned to perform. Thus, a diviner is called *mod ngam*, "divination-man" or a story-teller specialized in myths is a *mbomm-vet*, "harp player". A crucial category is that of *ngengang* ("healer", in fact many types of ritual specialists). Most rituals connected with questions of magical connections, and relationships with the ghosts, require the intervention of a *ngengang*. Such activity-based categories, however, presuppose another classification, this one in terms of unobservable qualities. It is impossible in Fang society to be ascribed any religious role without being considered a *beyem*, that is, a person who carries an invisible organ called *evur*. Every living person either is or else is not a *beyem*, there is no intermediate point.

There is, however, no way of telling for sure whether any given person is or is not one. This is a matter of conjectures, based on the person's behaviour and ritual successes. The category *mimmie* ("simple folk") designates people who have no *evur*. Another important category is that of *ntuban nlot* ("pierced head") people, that is, people who have undergone specific initiation rites. Again, since these initiations are shrouded in secrecy, it is rather difficult to tell whether any given individual is or is not a member of that category.

Consequences of cognitive diversity

This presentation of Fang ideas differs from ordinary ethnographic presentations of religious ideas, in that the assumptions are presented as fragmented bits of information, with no indication of the "models" or "systems" they belong to.[7] This is mainly in order to avoid the "intellectualist" or "theologistic" fallacy, which takes it for granted that all cultural representations are included in coherent, systematic models. The distinction made between those repertoires is of course on an analytical nature. The actual representations which are studied by anthropologists combine elements from several or all of these repertoires. For instance, people have ideas about ancestor-ghosts which combine ontological aspects (about the existence of such beings), causal ones (about their possible role in human affairs), some identification of episodes (e.g. the rituals that may be performed to placate them) and of social categories (concerning the type of specialists who perform those rituals).

The rationale for the division between those four categories is that the elements from different repertoires are likely to behave in functionally different ways, in the acquisition and fixation of belief. This is because each of these repertoires is in fact the extension to religious matters of conceptual structures and assumptions which can be found also in other, non-religious domains. Everyday knowledge comprises, among many other things, (i) a set of ontological assumptions, about what kinds of objects there are in the world and how different they may be; (ii) a set of principles on which causal connections, in various domains of experience, are evaluated; (iii) some ways of dividing the continuous flow of action into discrete episodes; (iv) some general representations about the possible differences between persons. My main contention here is twofold: (i) that some implicit assumptions and principles from non-religious knowledge are carried over in religious representations; (ii) that these principles and assumptions play a crucial role in the acquisition and transmission of religious representations.

Furthermore, this ''fragmented'' description is more realistic from the point of view of acquisition. In most societies people do acquire their religious ideas in a fragmented, disjointed and often inconsistent form. They are faced with a mass of utterances and actions, as well as comments about those singular occurences. As I said above, they are seldom if ever faced with systematic presentations of the local stock of religious presentations. Utterances are made because they have a precise point, actions are performed with a definite purpose. Now producing a consistent description of the religious world is very rarely the point of people's utterances or the purpose of their actions. Only tangentially do they serve such a purpose. An obvious consequence is that we must have a precise description of the processes whereby people build more general models on the basis of such fragmented material.

All this leads us to the main question posed at the beginning of this paper. Some features of religious ideas appear to be particularly recurrent. A cognitive approach is based on the assumption that this recurrence may be at least partly explained by cognitive constraints on the acquisition and transmission of cultural representations. In other words, the theory would assume that certain representations, or combinations of representations, are optimally learnable or memorable. It is therefore no surprise that such features become recurrent properties of religious systems. This condition of ''optimal learnability'' is the fundamental point of a selective model of religious representations. In the next section I will try and present some possible directions for a cognitive approach to this problem.

III

Intuitive Knowledge and Religious Ideas

Most anthropological theories are based on the assumption that cultural knowledge is only weakly constrained by cognitive processes, and that all its important aspects are culturally transmitted. Cultural anthropology, however, does not seem to have a very precise account of how this cultural transmission is taking place. The discipline generally adopts what Bloch (1985: *passim*) calls the "anthropological theory of cognition", that is, the idea that people brought up in a culture are given a ready-made conceptual scheme, which is absorbed, as it were, in a mysterious way that is never described. As Bloch points out, this account of acquisition is obviously insufficient in the case of simple everyday concepts, like names for natural kinds and artefacts; one might think, *a fortiori*, that it is rather implausible as an account of the transmission of complex cultural ideas. This "theory of cognition" includes two particularly implausible assumptions. One is that cultural transmission is, by and large, a passive process. Minds are conceived as containers of ideas, which are more or less empty at the onset of cultural acquisition, and are gradually filled with whatever ready-made products are given by "the culture". The other assumption is that this filling process is simple. Both assumptions, however, fly in the face of all the experimental evidence available in the domain of concept acquisition and belief-fixation (see e.g. Markman 1989 for a general survey).

Against this widespread notion of cultural transmission, I will contend that universal properties of human minds are likely to impose strong *cognitive constraints* on the range and organisation of cultural representations that can be transmitted from generation to generation. To be more precise, my hypothesis is that in each of the four repertoires described above, the religious assumptions are constrained, in non-trivial ways, by the assumptions that govern the non-religious or everyday representations of the repertoire. In other words, people's representations about (religious) ontologies are constrained by their ordinary ontology, their ideas about (religious) causation are constrained by ordinary causal representations, and so on. In the following pages I will first survey the type of psychological data and theories on which this assumption is based. This will be done mainly by considering some aspects of children's conceptual development; developmental studies provide particularly clear illustrations of the ways in which those assumptions are represented and constrain the range of beliefs subjects can entertain about a given object. I will then apply this kind of method to one of the four repertoires mentioned above, that of religious social roles.

Intuitive ontologies

A number of recent studies in conceptual development tend to shed light on processes which are directly relevant to the questions concerning religious ideas. In order to understand this, it may be of help to describe the theoretical background of those studies and hypotheses. In a "classical", Piagetian understanding of conceptual development, the child is assumed to apply general learning heuristics to a variety of experienced phenomena. The developmental stages identified in Piagetian frameworks are mainly described in formal, non-domain specific terms. At each stage the child is described as applying certain structural procedures, which do not depend on the type of phenomena concerned. These formal procedures are gradually modified on the basis of experience, leading to the next stage, at which point other formal procedures can be applied, again across conceptual domains. For instance the child's causal thinking develops from an "animistic" stage, in which all causal connections are attributed to intentions, to a later differentiation of intentional and mechanistic causation (see e.g. Piaget 1954, 1974, Laurendau & Pinard 1962, and a critique in Leslie 1979, Carey 1985, Leslie & Keeble 1987).

There are some reasons, however, to doubt the relevance of such structural cross-domain principles. First, cognitive studies tend to show that there are important functional differences between cognitive domains. To take but a few examples, the way subjects represent and remember faces is very different from the treatment of other types of visual stimuli; the way they make inferences about living kinds differs from the processes focused on artefacts or other non-living objects. Notions of causal connections may be very different, depending on the animate or inanimate nature of the objects concerned, from a very early age. Instead of describing human minds as "general processors" endowed with some all-purpose cognitive structures fed by perceptual inputs, psychological studies uncover *domain-specific* structures, which apply only to a limited domain of external phenomena. Furthermore, most recent research in developmental processes tends to show that such specialised structures appear very early in conceptual development. The way children reason, at any stage, depends on the conceptual domain to which the intellectual operations are applied. This type of research indicates that children's conceptual development proceeds on the basis of strong *theoretical presumptions*, concerning the type of properties and generalities to be expected in different ontological domains (Keil 1979, 1987, Atran 1989). The presumptions concern large ontological domains such as physical objects, artefacts, living kinds, persons. They constitute "naïve" theories of the domains in question (R. Gelman 1990).

To take a simple example, even 4 year olds seem to make instance-based inductive generalisations in a different way, whether the objects concerned are artefacts or exemplars of living kinds (Gelman & Markman 1986: 203-205, Gelman & Markman 1987, *passim*, Keil 1986, Gelman & Coley 1991). Even small children assume that members of a living kind share undefined properties, that make them similar in spite of superficial differences. This assumption has a variety of consequences. Children for instance seem to make spontaneous inductive generalisations over natural kinds on the basis of single exemplars, even when their biological knowledge is minimal. Moreover, they seem to "project" certain properties rather than others. While the property of having certain organs is spontaneously projected from an exemplar to the kind in general, certain surface properties, such as weight, are not projected (Gelman & Markman 1986, 1987, Gelman 1988). Also, children find unnatural the suggestion that members of a living kind might be "transformed" into members of another kind, whereas such transformations are considered possible in artefacts (Keil 1986 *passim*, 1989: 183-215).

The child also appears to develop, from the early stages of cognitive development, strong expectations concerning the behaviour of physical objects (Spelke 1988, 1990). For instance, the principles of continuity (objects move in continuous paths) and solidity (objects do not coincide in space) seem to be present in children as young as 4 months (Baillargeon 1987, Baillargeon & Hanko-Summers 1990).

Another domain which is the object of early theoretical development (roughly at the "pre-operational" stage in Piagetian terms) is that of mental processes. The child gradually develops a set of theoretical principles to do with the non-physical nature of mental entities such as thoughts, desires and dreams (Wellman & Estes 1986, Wellmann & Gelman 1988). Also, the child's understanding of mental processes constitutes a naïve implicit theory, with causal assumptions concerning the relations between perceptions, thoughts and intentions (Astington, Harris & Orson 1988, Wellmann 1990, Perner 1991, Whiten 1991).

Such theories can be said to be "naïve" in that they are only partly congruent with adult notions, and constrain further conceptual development. The following are a series of tentative conclusions that can be drawn from this experimental work on such presumptions:

(i) they are *domain-specific*, and trigger functionally different cognitive processes, depending on the domain. In other words, acquiring knowledge does not imply applying an all-purpose, "theory-making" cognitive device to a variety of available stimuli. On the contrary, it implies applying significantly different cognitive heuristics to different domains;

(ii) they seem to develop *spontaneously*, independently of tuition or objective changes in the available information: for instance, children seem to shift from a "desire-based" naïve psychology to a "belief + desire" type of psychological explanation when they are about 4, without any changes in the kind of explanations or implicit explanations offered in their social environment. To put it very crudely, people learn much more than they have been taught;

(iii) such naïve conceptual presumptions are *connected in a complex way to later theoretical development*. In some cases, such as commonsense or "folk"-psychology, adult conceptions seem to flesh out the conceptual skeleton provided by naïve conceptions. They provide more material, more explanatory schemes, but never go against the spontaneous assumptions. In other cases, such as the acquisition of mathematics or physics, it is necessary for subjects to acquire counter-intuitive principles (such as e.g. the difference between force and motion, which even in schooled adults is often a domain of uncertainty).

(iv) The spontaneous assumptions seem to constitute *cross-cultural universals*; that is to say, as far as there is evidence, that evidence bears out the hypothesis that such naïve conceptual constructions are universal (see for instance Avis & Harris 1991 on Pygmy children's theories of mind, Jeyifous 1985 on natural kinds and artefacts in Nigeria). This of course seems a direct consequence of point (iii). If such assumptions do not depend on cultural tuition, how could they differ from culture to culture?

To sum up, studies of concept acquisition seem to show that, from a very early age, a number of principles orient the subjects' attention to certain aspects of experience, and constitute the basis of quasi-theoretical intuitive principles, on which empirical knowledge can be built. Most of these specific principles remain tacit, even in adults. Such data are extremely important for a study of cultural representations, and especially of religious ideas. They make it possible to put forward a more precise cognitive account of religious transmission.

My hypothesis is that the intuitive knowledge principles described here impose constraints on the content and organisation of religious ideas. Obviously, most religious assumptions focus on objects which seemingly violate commonsense assumptions. It can be shown, however, that in order to acquire these notions, subjects have to rely, implicitly, on the intuitive principles described above. In this framework, a religious idea would be described as *cognitively optimal* if (i) it contains an explicit violation of commonsense thinking and (ii) it makes implicit use of the intuitive principles of commonsense knowledge. The hypothesis is that religious representations which are cognitively optimal will be the most recurrent ones. Being easier to learn and memorise, they would have a greater "survival value", in terms of cultural transmission, than other ideas. In the next section I will illustrate these hypotheses by describing the combination of intuitive and non-intuitive principles in the Fang case.

Social roles and intuitive principles

Let me now return to the Fang repertoire of social categories. Categories like *ngengang* (specialized healer) are connected to the umbrella concept *beyem* (persons having a magical capacity). Now the problem is that, as I said above, there is very little explicit discourse or implicit suggestions as to exactly what constitutes the difference between *beyem* and other people. The identification of a person as a *beyem* is based on two types of general features. There is, on the one hand, a series of observable traits, like the fact that a person is said to have undergone a certain initiation, that he or she performs certain rituals, etc. These elements, however, are neither necessary nor sufficient conditions for judging that someone is a *beyem*. People who do not have these traits are sometimes said to be *beyem*, and some persons who perform the rituals are not said to be *beyem*. The idea that a person is a *beyem* is based on the assumption that he or she has "something more" than the superficial features, something that all members of those categories have. No one can represent what it consists of, but it has to be there, otherwise, whatever one's activities, one is not an exemplar of the category. External typical criteria are just indirect (and insufficient) evidence of the fact that people really belong to the category. This "essentialist" interpretation of the group of *beyem* makes it possible to understand both the vagueness of people's statements about what makes *beyem beyem*, and the idea that any particular person either is or else is not one.

These features of the category "beyem" led me to put forward the hypothesis that such categories are represented in the same way as certain natural kind terms (Boyer 1990: 101-105, forthcoming (b)). The representation and use of a natural kind term always involves two types of general assumptions: (i) some assumptions about the *typical features* of the exemplars of the kind. Although most of these features are present in most exemplars, they are neither necessary nor sufficient conditions; (ii) the presumption of an *underlying trait* that is common to all exemplars of the kind (Schwarz 1979 *passim*).

It is important to stress that the implicit assumption is about the *existence* of an underlying trait, not about what it consists of. Few people except biologists bother to represent what makes giraffes giraffes, although everyone does suppose that there is some such underlying trait. This implicit hypothesis is a necessary aspect of the everyday use of natural kind terms. The idea of an undefined common essence is a powerful cognitive mechanism, universally available to human minds; together with basic principles of taxonomic ordering, it organizes biological knowledge in all cultures (see Atran 1987 *passim*). There is good experimental evidence that the assumption of an underlying essence, together with the idea that typical traits are, precisely, only typical, is involved in the representation of living kinds at the earliest stages of cognitive development (Medin & Ortony 1989).

As I observed above, the features mentioned as typical of *beyem* do not constitute necessary and sufficient conditions for membership. At the same time, however, people suppose that there is some underlying feature that is common to all persons in the position. In other words, there is a presumption that persons occupying a certain position share an essence, although what the essence consists of is left undefined. People who are *beyem* are thus considered *naturally* different from others. In other words, the representation of the position seems to be based on the extension to social differences of spontaneous assumptions which prove extremely successful in dealing with the natural world.

To sum up, my contention here is that the way people acquire and represent the categories denoting certain types of position is very similar to the way they acquire and represent certain categories about the natural world. To be more specific, certain crucial assumptions spontaneously applied to natural discontinuities, concerning e.g. the presence of an unobservable essence or the plausibility of instance-based generalisations, are directly applied to those social domains.

In the domain of religious positions, we find a set of ideas which have no natural referent; nor are they the object of any explicit tuition. My first claim is that, in such cases, the transmission process is in part assured by the fact that subjects spontaneously apply to the material at hand some assumptions which are highly salient in natural domains. My second claim is that the representation of certain religious positions as the consequence of inherent, unobservable natural properties has a very high "survival potential", as it were, precisely because it makes use of assumptions carried over from core knowledge.

Many social categories related to religious performance seem to be construed in a way that is similar to the Fang *beyem*, that is, with an "essentialist" hypothesis. People know that the persons included in the category (shamans, priests, diviners, etc.) are actually different from the others, although they tend to take external criteria, like the performance of certain rituals, as only symptomatic of the underlying difference. This does not mean, obviously, that all social categories to do with religious performance are implicitly conceived as quasi-biological. The biological analogy is only a particularly salient way of expressing a deeper, "essentialist" principle, following which the difference between members of the category and other people must be postulated, cannot be observed, and usually is not reducible to external criteria. Also, this essentialist hypothesis makes it possible to understand the fact that in many groups, the idea of a transformation is absurd, as far as such religious social categories are concerned. For instance, take the case of a Fang diviner, who for some reason does not manage to convince people of his genuineness, and is subsequently declared to be a non-*beyem*. No-one in such a case would ever think that the person in question first was a *beyem* and then lost that quality. They would assume that the person had been a non-*beyem* all along, and wrongly identified as a *beyem*.

Surprisingly, there is very little reliable data on the ways in which social categories are actually represented, for instance on the way people interpret those revisions and re-identifications which inevitably happen in any social group. Because of the scarcity of data, it is difficult to evaluate the general relevance of the hypothesis of "social essentialism" put forward here. It is particularly difficult to understand in what way essential hypotheses can be combined with formal criteria for membership, for instance in churches, wher there exist formal criteria for priesthood. The question, whether such institutional criteria as ordination actually represented by members of a congregation as sufficient conditions for membership of the social category PRIEST, is a moot one. Most studies of doctrinal religions take explicit theologies as their main object, and neglect to examine to what extent those theologies are congruent with people's actual representations. In the absence of such data, however, the hypothesis of social essentialism provides at least an example of the type of result a cognitive study of religious roles may lead to, even if in this case the relevance of the hypothesis may be limited to traditional, non-scriptural religious systems.

IV

Cognitive Salience

Let me now turn to the organizing principles of the conceptual structures developed. As I mentioned above, anthropological descriptions often take it for granted that religious ideas invariably come in a *theoretical* format. It appears, however, that the sets of representations concerned cannot really be called "theoretical", unless one has a rather metaphorical understanding of that term. To return to our example, the Fang ideas about ritual specialists (*beyem* for instance) are not really integrated in a theoretical format, for all the reasons listed above. For one thing, they are not consistent and they come in different formats (some ideas are about singular occurences and others about general principles, etc.). Differences between "theories" and the type of conceptual structures we are focusing on, can be apprehended in terms either epistemic or cognitive. From an epistemic viewpoint, some anthropologists have pointed out that the "theoretical" format is singularly inappropriate to a description of religious beliefs. Sperber for instance (1982 *passim*) contends that describing such ideas in terms of propositional attitudes is misleading, in that the objects of belief are not amenable to a propositional description. They can be called "semi-propositional" representations, in that they provide only fragmentary elements of propositional identification. Here, however, I will leave aside these epistemic questions and focus on the cognitive aspects of religious conceptual structures. In many respects, their most salient characteristics make the "theoretical" description rather inadequate. Here I will insist on an aspect that is generally ignored in anthropological models, namely the *conjectural* nature of these assumptions.

Induction and uncertainty

The conjectural nature of religious ideas is often a direct consequence of the way they are produced. The identification of social roles, for instance, inasmuch as it is based on assumptions of shared essence, naturally leads to the fact that every singular identification is a conjecture. Categories are thought to denote intrinsic differences between kinds or sorts of people. That such underlying differences exist is taken as true. That they apply to particular people *hic et nunc* is necessarily a matter of conjectures and non-demonstrative inferences. That the person really occupies the position in question (i.e., really has the natural properties posited) is only the most plausible conclusion, given the observable properties at hand, but corrections and re-identifications are always possible, and do in fact happen. The Fang who re-identify someone as not being a *ngengang* would say that although the person did have the relevant surface features, he/she was not really a *ngengang* from the beginning. The main point here is that there is no possibility to tell the difference on the basis of external, observable differences. And since there are no such differences, deciding one way or another is a matter of corrigible guesses and inferences.

That religious ideas command variable commitment is a commonplace of religious anthropology. Every anthropologist knows from experience that even the most basic tenets of religious systems may be represented as extremely convincing conjectures rather than intuitively obvious facts. Another aspect of this question, however, is less often commented on, namely the fact that, in any cultural environment, there is a constant competition between alternative religious interpretations of any event or state of affairs. Even in simple cases where only one religious framework is available, that framework inevitably leaves considerable latitude in the interpretation of any single event. A consequence is that, in most circumstances, any subject is provided with several interpretative schemes, the difference between which is a matter of plausibility or "naturalness". For instance, the societies where illness is interpreted as the result of malevolent human witchcraft invariably have alternative etiological schemes, linking illness e.g. to purely contingent somatic disorders, to some powerful god's intervention, etc. The only way of explaining why a certain causal scheme is (conjecturally) taken as relevant in a particular case is to explain what makes it intuitively more natural. A description of the acquisition and fixation of religious assumptions must take into account the fact that there are intuitively obvious differences in certainty between thoughts; moreover, those differences are not immutable, they may and do evolve. Thoughts may become more (or less) intuitively plausible after certain processing episodes. These aspects are crucial for the acquisition and representation of religious ideas.

Cognitive salience

It seems difficult to account for the transmission of the latter type of conceptual structures without some precise notion of *cognitive salience*. The point of such a notion would be (i) to give a more precise formulation of the intuitions concerning partial credal states, i.e. states in which a belief is held only partially, (ii) to provide an account of the strengthening (and weakening) of commitment, and (iii) to describe the type of conceptual structures that are based on assumptions of variable salience. Here, obviously, I cannot give a full-blown formulation of such a framework. Rather, I will try to give a more detailed description of the "brief" which a theory of cognitive salience would have to fulfill.

Some notion of degree of commitment has always been a necessary ingredient in theories of subjective probability (cf. for instance Ramsey 1931, Carnap 1950). In such theories, having a partial commitment to a proposition "p" means believing that the probability of "p" being true is 1/x, where x<1. Such models, however, do not necessarily provide a good model for degrees of belief, or even for actual inductive reasoning.[8] The salience of an assumption should be approached in terms that link subjective intuitions of credal states to functional properties of the assumption. Such models are available, for instance in the "framework for induction" put forward by Holland *et al.* (1986). The main consequence of such models is to define the cognitive salience of an assumption as the objective probability of its activation, given a certain set of conditions, notably given that a certain set of other representations are already activated.

We cannot explain the actual transmission and fixation of cultural representations without a framework for cognitive salience. This should be so designed as to provide a precise account of the interaction of two types of mechanisms in the construction of conceptual structures. First, some assumptions are directly imported from intuitive knowledge, and are often domain-specific assumptions. Second, the material provided by any cultural environment provides the input for mechanisms of strengthening-weakening which are not domain-specific. They concern the general effects of confirmation, refutation and revision on the salience of assumptions.

V

Final Remarks

To sum up, I have put forward in this paper a series of hypotheses that constitute only tentative elements of a cognitive framework for the explanation of religious ideas. The first one is that the main point of a cognitive approach is to account for the recurrence of non-trivial properties of religious representations. Although this goal has often been misconstrued in anthropological research, it is now becoming increasingly clear that we must take into account important findings and hypotheses of psychology and cognitive science, in order to go beyond the present state of anthropological descriptions. Another assumption is that we cannot understand the transmission of religious ideas if we do not take into account the functional specificities of different types of mental representations. A third hypothesis is that intuitive knowledge structures impose constraints on religious representations. More precisely, if religious representations combine explicit violations of some intuitive principles, and implicit confirmation of other intuitive principles, then they are cognitively optimal. This means that they are more likely to be acquired, stored and transmitted than representations which do not include this particular combination of violated and confirmed assumptions. A final hypothesis is that the processes whereby religious ideas are made intuitively plausible to human minds are inductive processes of belief-fixation, which cannot be studied unless we have precise models of the strengthening of non-demonstrative inferences.

The main point of this paper was to show that recent advances in experimental psychology, notably in the field of conceptual development, can throw light on the processes whereby subjects develop intuitive understandings of religious notions, even in contexts where cultural transmission is fragmentary. Obviously, it is not possible to decide on the validity of such hypotheses, on the sole basis of the type of data produced in ordinary anthropological fieldwork. The various forms of "participant observation", intuitive hypothesis-testing and informal interview techniques favoured by anthropologists constitute an indispensable grounding for the study of religious representations. However, they are not designed to provide the kind of fine grain description of mental representation that is the necessary prerequisite of a cognitive study. This will probably require a different mode of data-gathering, in which traditional fieldwork methods are completed with more constrained experimental studies. Such experimental studies may not solve all the traditional problems of the anthropology of religion. They will, however, provide a precise answer to a long-neglected question, that of the processes whereby certain types of ideas are made "natural" and intuitively obvious to human subjects in different cultural settings.

King's College
University of Cambridge
GB - Cambridge CB2 1ST

PASCAL BOYER

[1] As Keesing points out, the cultural models described in modern cognitive anthropology "comprise the domain of (culturally constructed) common sense. They serve pragmatic purposes. They explain the tangible, the experiential [...] the probable" (1987: 374).

[2] See Boyer 1990: ch. 1, Lawson & McCauley 1990: 32-44, Boyer forthcoming (a) for detailed treatments of these questions.

[3] The hypotheses presented in this paper constitute a summary of the main points of Boyer forthcoming (c).

[4] See Morris 1987 for a detailed survey of anthropological theories of religion. Lawson & McCauley (1990: 122-123) give a succint description of the three types

of universals (substantive, formal and functional) that can be posited in a cognitive theory.

[5] This approach, of course, is not really new in anthropological theory, and the intellectual prestige of Darwinian theory led to many models of cultural evolution based on some notion of selection, before and after Tylor's famous statement that "to the ethnographer, the bow and arrow is a species" (Tylor 1871[I]: 7). See Ghiselin (1973 *passim*) and Ingold (198: 33-47) for an analysis of these theories and the multiple misunderstandings they often produced.

[6] See for instance Harris & Heelas 1979 for a general survey of the ambiguities in the psychological implications of descriptions couched in terms of "collective representations", and Boyer 1987 for a discussion of the problems generated by such descriptions in the study of cultural transmission.

[7] The anthropological literature on religious ideas is of course more diverse, although the point generally holds (see Boyer 1987 *passim* for a detailed argument). For examples of monographs that take into account the "fragmentedness" or incompleteness of the input, see e.g. Keesing 1982, Toren 1987, 1988.

[8] There is no space here to review the vast literature on inductive belief-fixation and the various problems generated by the application of "inductive logics" to actual reasoning. See Nisbett & Ross (1980) for a description of the "vividness" phenomena which make it necessary to have some notion of cognitive salience. Kahnemann, Slovic and Tversky eds. (1982) is a classical source on non-logical heuristics in the evaluation of subjective probabilities, and Osherson *et al.* (1986) on the relevance of such problems for argumentation theory. Holland *et al.* (1986) give a survey of the literature and put forward a detailed framework, from which most of my remarks are inspired.

REFERENCES

Astington, J.W., Harris, P., & Olson, D.R. (eds.)
1988 *Developing theories of mind*, Cambridge: Cambridge University Press.

Atran, S.
1987 Ordinary constraints on the semantics of living kinds. A commonsense alternative to recent treatments of natural-object terms. *Mind and Language* 2: 27-63.
1989 Basic conceptual domains, *Mind and Language* 4: 5-16.

Avis, J. & Harris, P.L.
1991 Belief-desire reasoning among Baka children: evidence for a universal conception of mind, *Child Development* 62: 460-467.

Baillargeon, R.
1987 Young infants' reasoning about the physical and spatial characteristics of a hidden object, *Cognitive Development* 2: 179-200.

Baillargeon, R. & Hanko-Summers, S.
1990 Is the object adequately supported by the bottom object? Young infants' understanding of support relations, *Cognitive Development* 5: 29-54.

Bloch, M.
1985 From Cognition to Ideology, in R. Fardon (ed.), *Power and Knowledge. Anthropological and Sociological Approaches*, Edinburgh: Scottish Academic Press.

Boyd, R. & Richerson, P.J.
1985 *Culture and the evolutionary process*, Chicago: University of Chicago Press.

Boyer, P.
1986 The 'Empty' Concepts of Traditional Thinking. A Semantic and Pragmatic Description, *Man (n.s.)* 21: 50-64.
1987 The Stuff 'Traditions' Are Made Of. On the Implicit Ontology of an Ethnographic Category, *Philosophy of the Social Sciences* 17: 49-65.
1990 *Tradition as Truth and Communication. A Cognitive description of traditional discourse*, Cambridge: Cambridge University Press.
[forthcoming (a)]. Introduction: Cognitive Aspects of Religious Symbolism, in Boyer, P. (ed.), *Cognitive Aspects of Religious Symbolism*, Cambridge: Cambridge University Press.
[forthcoming (b)]. Pseudo-natural kinds, in Boyer, P. (ed.), *Cognitive Aspects of Religious Symbolism*, Cambridge: Cambridge University Press.
[forthcoming (c)]. *The Naturalness of Religious Ideas: Outline of a Cognitive Theory of Religion*, Berkeley/Los Angeles: University of California Press.

Carey, S.
1985 *Conceptual Change in Childhood*, Cambridge, Mass.: The M.I.T. Press.

Carnap, R.
1950 *Logical foundations of probability*, London: Routledge and Kegan Paul.

Cavalli-Sforza, L.L.
1986 Cultural Evolution, *American Zoologist* 26: 845-55.

Cavalli-Sforza, L.L. & Feldman, M.W.
1981 *Cultural transmission and evolution: a quantitative approach*, Princeton, N.J.: Princeton University Press.

Dougherty, J.D.W. (ed.)
1985 *Directions in Cognitive Anthropology*, Urbana/ Chicago: University of Illinois Press.

Dougherty, J.D.W. & J.W. Fernandez (eds.)
1980 *Cognition and Symbolism*, [special issue of] *American Ethnologist*, 7.
1982 *Cognition and Symbolism II*, [special issue of] *American Ethnologist*, 9.

Gelman, R.
1990 First principles organizing attention to and learning about relevant data: number and the animate-inanimate distinction as examples, *Cognitive Science* 14: 79-106.

Gelman, S.
1988 The development of induction within natural kind and artefact categories, *Cognitive Psychology* 20: 65-95.

Gelman, S. & J. Coley
1991 Language and categorization: the acquisition of natural kind terms, in Gelman, S. & Byrnes, J.P. (eds.) *Perspectives on Language and Thought: Interrelations in Development*, Cambridge: Cambridge University Press.

Gelman, S. & E. Markman
1986 Categories and Induction in Young Children, *Cognition* 23: 183-209.
1987 Young children's inductions from natural kinds: the role of categories and appearances, *Child Development* 58: 32-41.

Ghiselin, M. T.
1973 Darwin and evolutionary psychology, *Science* 179: 964-968.

Harris, P. & Heelas, P.
1979 Cognitive processes and collective representations, *European Journal of Sociology* 20: 211-41.
Holland, J.H., K.J. Holyoak, R.E. Nisbett & P.R. Thagard
1986 *Induction. Processes of Inference, Learning and Discovery*, Cambridge, Mass.: The M.I.T. Press.
Ingold, T.
1986 *Evolution and Social Life*, Cambridge: Cambridge University Press.
Jeyifous, S.
1985 *Atimodemo: Semantic Conceptual Development among the Yoruba*, PhD Thesis, Cornell University.
Kahnemann, D., Slovic, P. & Tversky, A., (eds.)
1982 *Judgment under uncertainty: heuristics and biases*, Cambridge: Cambridge University Press.
Keesing, R.
1982 *Kwaio Religion. The Living and the Dead in a Solomon Island Society*, New York: Columbia University Press.
1987 Models, "folk" and "cultural", in D. Holland & N. Quinn (eds.), *Cultural Models in Language and Thought*, Cambridge: Cambridge University Press.
Keil, F.C.
1979 *Semantic and Conceptual Development*, Cambridge, Mass.: Harvard University Press.
1986 The acquisition of natural kind and artefact terms, in A. Marrar & W. Demopoulos (eds.), *Conceptual Change*, Norwood, N.J.: Ablex.
1987 Conceptual development and category structure, in U. Neisser (ed.)., *Concepts and Conceptual Development: Ecological and Intellectual Factors in Categorization*, Cambridge: Cambridge University Press.
1989 *Concepts, Kinds and Conceptual Development*, Cambridge Mass.: The M.I.T. Press.
Laurendau, M. & A. Pinard
1962 *Causal Thinking in the Child*, New York: International Universities Press.
Lawson, E.T. & R. McCauley
1990 *Rethinking Religion. Connecting Culture and Cognition*, Cambridge: Cambridge University Press.
Leslie, A.
1979 *The Representation of Perceived Causal Connection*, D.Phil. thesis, University of Oxford.
Leslie, A. & Keeble, S.
1987 Do six-months old infants perceive causality?, *Cognition* 25: 265-288.
Lumsden, C.J. & Wilson, E.O.
1981 *Genes, Minds and Culture*, Cambridge Mass.: Harvard University Press.
Markman, E.M.
1989 *Categorization and Naming in Children: Problems of Induction*, Cambridge, Mass.: The M.I.T. Press.
Medin, D.L., & Ortony, A.
1989 Psychological essentialism, in Vosniadou, S. & Ortony, A. (eds.), *Similarity and Analogical Reasoning*, Cambridge: Cambridge University Press

Morris, B.
1987 *Anthropological theories of religion*, Cambridge: Cambridge University Press.

Nisbett, R. & Ross, L.
1980 *Human Inference. Strategies and shortcomings of social judgement*, Englewood Cliffs, N.J.: Prentice-Hall.

Osherson, D.N., Smith, E.E. & Shafir, E.B.
1986 Some origins of belief, *Cognition* 24: 197-224.

Perner, J.
1991 *Understanding the Representational Mind*, Cambridge, Mass.: The M.I.T. Press.

Piaget, J.
1954 *The child's construction of reality*, New York: Basic Books.
1974 *Understanding causality*, New York: Norton.

Ramsey, F.P.
1931 *The foundations of mathematics*, London: Kegan Paul.

Schwarz, S.P. (ed.)
1979 Natural Kind Terms, *Cognition* 7: 301-315.

Spelke, E.S.
1988 The origins of physical knowledge, in L. Weizkrantz (ed.), *Thought without Language*, Oxford: Oxford University Press.
1990 Principles of object perception, *Cognitive Science* 14: 29-56.
[forthcoming] Physical Knowledge in Infancy: Reflections on Piaget's theory, in Carey, S., & Gelman, S. (eds.), *Biology and Cognition*.

Sperber, D.
1982 Apparently Irrational Beliefs, in M. Hollis & S. Lukes (eds.), *Rationality and Relativism*, Oxford: Basil Blackwell.
1985 Anthropology and Psychology: towards an epidemiology of representations, *Man* 20: 73-89.
1991 The epidemiology of beliefs, in C. Fraser (ed.), *Psychological Studies of Widespread Beliefs*, Oxford: Oxford University Press.

Toren, C.
1987 Children's perceptions of gender and hierarchy, in G. Jahoda & I.M. Lewis (eds.), *Acquiring Culture*, London: Routledge.
1988 The continuity and mutability of tradition as process, *Man* n.s. 23: 696-717.

Tylor, E.B.
1871 *Primitive Culture*, London: John Murray.

Wellmann, H.M.
1990 *The Child's Theory of Mind*, Cambridge, Mass.: The M.I.T. Press.

Wellmann, H., & Estes, D.
1986 Early understanding of mental entities: a re-examination of childhood realism, *Child Development* 57: 910-923.

Wellmann, H.M. & Gelman, S.
1988 Children's understanding of the non-obvious, in Sternberg, R.J. (ed.), *Advances in the Psychology of Human Intelligence* (vol. 4), Hillsdale N.J.: Lawrence Erlbaum Associates.

Whiten, A., (ed.)
1991 *Natural Theories of Mind: the Evolution, Development and Simulation of Everyday Mindreading*, Oxford: Blackwell.

[12]

Are Children "Intuitive Theists"?

Reasoning About Purpose and Design in Nature

Deborah Kelemen

Boston University

ABSTRACT—*Separate bodies of research suggest that young children have a broad tendency to reason about natural phenomena in terms of purpose and an orientation toward intention-based accounts of the origins of natural entities. This article explores these results further by drawing together recent findings from various areas of cognitive developmental research to address the following question: Rather than being "artificialists" in Piagetian terms, are children "intuitive theists"—disposed to view natural phenomena as resulting from nonhuman design? A review of research on children's concepts of agency, imaginary companions, and understanding of artifacts suggests that by the time children are around 5 years of age, this description of them may have explanatory value and practical relevance.*

Piaget's (1929) claim that children are "artificialists" who draw on their subjective intentional experience to conclude that all things are made by people for a purpose has encountered substantial skepticism in the past few decades of cognitive developmental research. This is because, at core, Piaget's proposal embodied not just the suggestion that children misunderstand the limits of human creative power, but a stronger claim about the profound incommensurability of children's and adults' conceptual systems. Specifically, Piaget believed that young children indiscriminately generate artificialist explanations because they are psychologically incapable of conceiving of physical causes, a shortcoming that he argued rendered them insensitive to the fundamental distinction between natural kinds and artifacts.

Research since Piaget has challenged these assumptions. Not only can children reason in physical-causal terms from infancy (e.g., Baillargeon, 1993), but they also recognize that people make artifacts, not natural entities (e.g., Gelman & Kremer, 1991). But although these results may put some aspects of Piaget's interpretation to rest, recent research has raised the specter of Piaget's findings once more. Consistent with Piaget's results, contemporary studies have found that, although children are not entirely indiscriminate, they do indeed evidence a general bias to treat objects and behaviors as existing for a purpose (Kelemen, 1999b, 1999c, 2003; but see Keil, 1992) and are also broadly inclined to view natural phenomena as intentionally created, albeit by a nonhuman agent (Evans, 2000b, 2001; Gelman & Kremer, 1991). This article explores these findings further by drawing them together with other recent cognitive developmental research to address the following question: Even if children are not artificialists, as Piaget conceived of the term, are they perhaps "intuitive theists"—predisposed to construe natural objects as though they are nonhuman artifacts, the products of nonhuman design?

Address correspondence to Deborah Kelemen, Department of Psychology, Boston University, 64 Cummington St., Boston, MA 02215; e-mail: dkelemen@bu.edu.

PROMISCUOUS TELEOLOGY AND "CREATIONISM" IN CHILDREN

Contemporary research on teleological reasoning—the tendency to reason about entities and events in terms of purpose—was initiated in the context of the debate on the origins of biological understanding. Consistent with the view that children's reasoning about living things is constrained by teleological assumptions from a very early age, studies have found that young children attend to shared functional adaptation rather than shared overall appearance (or category membership) when generalizing behaviors to novel animals (Kelemen, Widdowson, Posner, Brown, & Casler, 2003), judge whether biological properties are heritable on the basis of their functional consequences rather than their origin (Springer & Keil, 1989), and explain body properties by reference to their self-serving functions and not their physical-mechanical cause (Keil, 1992; Kelemen, 2003).

Results like these lend support to the idea that a purpose-based teleological stance might, therefore, be humans' innate adaptation for biological reasoning (Atran, 1995; Keil, 1992). This conclusion has been complicated, however, by findings that children see not only the biological but also the nonbiological natural world in teleological terms. For example, when asked to identify unanswerable questions, American 4- and 5-year-olds differ from adults by finding the question "what's this for?" appropriate not only to artifacts and body parts, but also to whole living things like lions ("to go in the zoo") and nonliving natural kinds like clouds ("for raining"). Additionally, when asked whether they agree that, for example, raining is really just what a cloud "does" rather than what it is "made for," preschoolers demur, endorsing the view that natural entities are "made for something" and that is why they are here (Kelemen, 1999b).

These kinds of promiscuous teleological intuitions persist into elementary school, particularly in relation to object properties. For instance, when asked to conduct a "science" task and decide whether prehistoric rocks were pointy because of a physical process (e.g., "bits of stuff piled up for a long period of time") or because they performed a function, American 7- and 8-year-olds, unlike adults, preferred teleological explanations whether they invoked "self-survival" functions (e.g., "so that animals wouldn't sit on them and smash them") or "artifact" functions (e.g., "so that animals could scratch on them when they got itchy"; Kelemen, 1999c; but see Keil, 1992). This bias in favor of teleological explanation for properties of both living and nonliving natural objects occurs even when children are told that adults apply physical kinds of explanation to nonliving natural entities (Kelemen, 2003). In American children, the bias begins to moderate around 9 to 10 years of age, and this pattern now has been found also with British children for both object properties and, slightly less markedly, natural object wholes. These British findings are relevant because they weigh against interpretations that promiscuous teleological intuitions are a simple reflection of the relatively pronounced cultural religiosity, or religious exceptionalism (in postindustrial, international context), of the United States (see Kelemen, 2003, for discussion of religiosity differences).

So, if ambient cultural religiosity is not the obvious explanation, what does cause this promiscuous teleology? A study of responses young children receive when asking questions about nature indicates parents generally favor causal rather than teleological explanation, so current evidence suggests the answer does not lie there, at least, not in any straightforward sense (Kelemen, Callanan, Casler, & Pérez-Granados, 2002). Another hypothesis being explored in my lab is, therefore, as follows (e.g., Kelemen, 1999b, 1999c). Perhaps children's generalized attributions of purpose are, essentially, side effects of a socially intelligent mind that is naturally inclined to privilege intentional explanation and is, therefore, oriented toward explanations characterizing nature as an intentionally designed artifact—an orientation given further support by the artifact-saturated context of human cultures. Specifically, the proposal is that the human tendency to attribute purpose to objects develops from infants' core, and precociously developing, ability to attribute goals to agents (as discussed later): Initially, on the basis of observing agents' object-directed behavior, children understand objects as means to agents' goals, then as embodiments of agents' goals (thus "for" specific purposes in a teleological sense), and, subsequently—as a result of a growing understanding of artifacts and the creative abilities of agents—as intentionally caused by agents' goals. A bias to explain, plus a human predilection for intentional explanation, may then be what leads children, in the absence of knowledge, to a generalized, default view of entities as intentionally caused by someone for a purpose.

Details aside, the basic idea that children are disposed to view entities in terms of intentional design, or as "quasi-artifacts," is similar to one independently developed by Evans in her work on origins beliefs (Evans, 2000a, 2000b, 2001). Evans has found that regardless of the religiosity of their home background, children show a bias to endorse intentional accounts of how species originate. Thus, when asked questions like "how do you think the very first sun bear got here on earth?" 8- to 10-year-olds from both fundamentalist and nonfundamentalist American homes favored "creationist" accounts whether generating their own answers or rating agreement with the following responses: (a) God made it, (b) a person made it, (c) it changed from a different kind of animal that used to live on earth, or (d) it appeared (Evans, 2001). This preference was also found in 5- to 7-year-old children's agreement ratings for animate and inanimate entities. Indeed, it was only among 11- to 13-year-old nonfundamentalist children that divergence from the theist position emerged. Evans's results do not stand in isolation. Gelman and Kremer (1991) found that although American preschoolers recognize that artifacts rather than natural entities are human made, they favor God as the explanation of the origin of remote natural items (e.g., oceans). Petrovich (1997) found similar results with British preschoolers (although see Mead, 1932, on Manus children's disinclination to use supernatural explanation).[1]

Considered together, current data on children's promiscuous teleology and explanations of origins might therefore suggest an obvious affirmative answer to the question of whether children are intuitive theists: Children view natural phenomena as intentionally designed by a god. Not coincidentally, they therefore view natural objects as existing for a purpose. But before embracing, or even entertaining, this conclusion, we must look first at whether it is actually defensible. What evidence is there that children possess any of the conceptual prerequisites that intuitive theism might entail? What evidence is there that their intuitions display any coherence at all?

CONCEPTUAL PREREQUISITES TO INTUITIVE THEISM

Piaget (1929) found that when asked how natural objects originated, children frequently identified "God" as the cause. Piaget argued that these statements were simply further cases of artificialism: Unable to entertain an abstraction such as God, and egocentrically focused, children used "God" to refer to a person who was fundamentally similar to the dominant authority in children's own lives—their parent.

Once again, however, Piaget's assumptions about the concreteness of children's concepts have been challenged. Research now suggests that rather than being anthropomorphic, children's earliest concept of agency is abstract, and is invoked by a range of nonhuman entities from the time when overt signs of children's sensitivity to mental states are becoming increasingly robust. Thus, 12-month-old infants will follow the "gaze" of faceless blobs as long as they have engaged in contingent interaction with them (S.C. Johnson, Booth, & O'Hearn, 2001) and will attribute goal directedness to computer-generated shapes (e.g., Csibra & Gergely, 1998). By 15 months, infants will complete the incomplete actions of a nonhuman agent by inferring its goals (S.C. Johnson et al., 2001). From infancy, we are, then, excellent "agency detectors" (Barrett, 2000; Guthrie, 2002).

But, although relevant, these indications that children attribute mental states to perceivable nonhuman agents while watching them are still nonevidential with respect to young children's ability to reason about the creative intentions of intangible, nonnatural agents like gods. Presumably several capacities are minimally prerequisite in order to reason about such special causal agents: first, the capacity to maintain a mental representation of such an agent despite its intangibility; second, the ability to attribute to that special agent mental

[1] Mead explored attributions of consciousness to inanimate entities by children from a small-scale animist society. However, the nature of Mead's data (e.g., drawings, queries about inanimate malintentions) makes children's nonreference to supernatural agency difficult to interpret. Furthermore, although her data suggest the children were not animists, they do not rule out possible intuitive theism.

states distinguishing it from more commonplace agents; and third—and particularly pertinent to the question of nonnatural artifice—the basic ability to attribute design intentions to agents and understand an object's purpose as deriving from such intentions.

CONCEPTIONS OF INTANGIBLE AGENTS

Several lines of research are suggestive of young children's abilities regarding the first two prerequisites. First, Taylor's (1999) research on children's propensity to maintain social relationships with imaginary companions suggests that by age 3 to 4 years, children are already conceptually equipped to vividly mentally represent the wants, opinions, actions, and personalities of intangible agents on a sustained basis. Like supernatural agents, such companions are found cross-culturally and are often distinguished from more commonplace agents by special biological, psychological, and physical traits beyond invisibility. Examples are animals that talk and individuals who understand gibberish, hear wishes, or live on stars (Taylor, 1999). Interestingly, ideas about imaginary companions, like ideas about gods, can be culturally transmitted, at least, within families.[2]

Imaginary companions, then, provide some indications of young children's ability to symbolically represent and reason about immaterial individuals. But research explicitly focused on children's understanding of God has also found that by 5 years of age, children can make quite sophisticated predictions as to how a more widely recognized nonnatural agent's mental states are distinguished from those of more earthly individuals. Specifically, Barrett, Richert, and Driesenga (2001) cleverly capitalized on the well-documented shift in 3- to 5-year-olds' ability to pass false-belief tasks—tests that putatively measure children's theory-of-mind understanding that beliefs are mental representations and, as such, can mismatch with physical reality. In their study, Barrett et al. used a standard form of the task: Children were shown a cracker box, asked what they believed it contained, allowed to peek inside and see the actual contents (pebbles), and then asked the test question, What would someone (who had not been shown) believe was inside the container? As is typical in such studies, Barrett et al. found that 3-year-olds failed the test, giving an answer that, in some sense, assumes that people are all-knowing; that is, 3-year-olds answered, "pebbles." In contrast, an increasing percentage of 4- and 5-year-olds passed, saying "crackers"—an answer recognizing the fallibility of beliefs. Interestingly, however, a different pattern emerged when these Protestant-raised children were asked what God would believe. At all ages tested, children treated God as all-knowing, even when they clearly understood that earthly agents would have a false belief. This developmental pattern led Barrett et al. to provocatively suggest that children may be innately attuned to "godlike" nonhuman agency but need to acquire an understanding of the limitations of human minds. Similar results have now also been obtained with Yukatec Mayan children, who discriminated not only the Christian God but also other supernatural agents as less susceptible to false belief than people (Knight, Sousa, Barrett, & Atran, 2003; also Atran, 2002, for a description).

In sum, then, these findings suggest that around 5 years of age, children possess the prerequisites to make advanced, distinctive, attributions of mental states to nonnatural agents. But are children truly conceptually distinguishing these agents from people or just representing these agents as humans augmented with culturally prescribed, superhuman properties inferred from adults' religious talk? The answer to this question is unclear. Certainly children's supernatural concepts, like those of adults, are likely to be influenced by culturally prescribed, systematically counterintuitive properties (Atran, 2002; Boyer, 2001) and may also be anthropomorphic in many ways. But, even if children's concepts of nonnatural agency do have human features, this does not undermine the claim that children conceive of such agents as distinct: We do not question adults' capacity to conceive of supernatural agents, and yet research indicates that even when adults explicitly attribute gods with properties like omnipresence, they assume, in their implicit reasoning, that gods act in accordance with human temporal, psychological, and physical constraints (Barrett, 2000).

Even so, perhaps applying the phrase "intuitive theists" to children—given all that the term "theism" implies to adults—might seem misplaced, if not irreverent. After all, although young children might conceive of nonnatural agents and hypothesize about their mental states, presumably they do not contemplate the metaphysical "truth" of which such agents can be part, or experience emotions concomitant with endorsing a particular metaphysical-religious system. Intuitively, these assumptions seem correct although, again, there are reasons to equivocate—not only because research suggests adult religious belief systems are often not particularly coherent or contemplated (e.g., Boyer, 2001), but also because the question of when children begin to develop metaphysical understanding in the adult self-reflective sense is debated (e.g., Evans & Mull, 2002; Harris, 2000; C.N. Johnson, 2000). Specifically, although children might not explicitly demarcate their musings as special, it has been found that even from very young ages, children pose questions about the nature of things that echo adult metaphysical themes (Harris, 2000; Piaget, 1929). Furthermore, we actually know little about young children's emotions concerning self-generated or culturally derived concepts of nonnatural agency, outside of their emotional relationships with imaginary companions. Gaps in our knowledge therefore preclude general conclusions as to children's capacity to entertain adultlike religious feeling.

However, for the present purpose, such issues are, to a large extent, irrelevant because in the current context the term intuitive theist embodies no claims regarding children's emotional or metaphysical commitments. All that is under question is whether children make sense of the world in a manner superficially approximating adult theism, by forming a working hypothesis that natural phenomena derive from a nonhuman "somebody" who designed them for a purpose—an intuition that may be elaborated by a particular religious culture but derives primarily from cognitive predispositions and artifact knowledge.[3] This point circles us back to the third conceptual

[2]I do not intend to suggest that children's relationships with imaginary companions are akin to adults' relationships with gods. An important difference is that the latter are experienced as real (Boyer, 2001), whereas evidence suggests that (American) children's imaginary companions are experienced as fictions (Taylor, 1999).

[3]Some form of folk religion appears to exist in all human cultures, but not all religions are theist (e.g., animism), raising the interesting possibility that children's intuitions may sometimes mismap with the dominant adult culture's religious ideas. However, because all known folk religions involve nonnatural agents and intentional causation—the substrate of intuitive theism—such mismappings need not represent an ongoing conceptual conflict, but instead leave children's intuitions open to coexist with and be influenced by cultural religious ideas.

prerequisite for intuitive theism—children's ability to understand that an object's purpose derives from the designer's goals.

CHILDREN'S UNDERSTANDING OF ARTIFACTS AND DESIGN

Adult reasoning about artifacts is anchored by intuitions about the designer's intended function (e.g., Keil, 1989; Rips, 1989), but although behavioral measures suggest that from around 3 years of age children will teleologically treat artifacts as "for" a single privileged function (Casler & Kelemen, 2003a; Markson, 2001), the question of when children adopt an adultlike teleological construal based on reasoning about the creator's intent (the "design stance") is debated (Kelemen & Carey, in press).

One reason for the lack of consensus is studies suggesting that, until they are quite old, children apply category labels to artifacts on the basis of shared shape, not shared function (e.g., Gentner, 1978; Graham, Williams, & Huber, 1999; Landau, Smith, & Jones, 1998). Such studies have found that until around 6 years of age, children will judge that if an object looks similar to an artifact called "a wug," it is also "a wug" even though it does not do the same thing. Children's apparent indifference to what artifacts did in these categorization studies seemed to render it unlikely that the deeper principle of intended function could play much of a role in their concepts of artifacts.

However, recent findings suggest that the stimuli in earlier studies may have significantly contributed to children's categorization failures in that experimenters unnaturally dissociated artifact form from artifact function, an approach leading to uncompelling "functions" equivalent to general object properties (e.g., capacities to rattle, roll, absorb). In current research using artifacts that look designed in that their structural properties clearly relate to their functional affordances, children from around the age of 2 years have generalized labels on the basis of function rather than shape similarity (e.g., Kemler Nelson, Frankenfield, Morris, & Blair, 2000; Kemler Nelson, Russell, Duke, & Jones, 2000). Furthermore, evidence also suggests that even when children categorize artifacts by shape, rather than being a superficial perceptual strategy, this approach reflects the valid conceptual assumption that shape predicts the creator's intent. Thus, Diesendruck, Markson, and Bloom (2003) found that if 3-year-olds have the shape similarity between two artifacts pointed out to them but then hear that the objects have different intended functions, they eschew classifying them as the same kind of artifact, instead forming categories based on shared function and perceptual dissimilarity. This shift from a shape to a function strategy happens only if children hear about *intended* functions—information about possible function is not sufficient.

These findings provide suggestive evidence that young children have a sensitivity to intended function from around the age of 3 years. They are particularly interesting when considered alongside research explicitly focused on when children weigh overt information about intended design. In studies in my own lab, this tendency is increasingly evident between ages 4 and 5 years. For example, in one study, 4- and 5-year-old children were told stories about depicted novel artifacts that were intentionally designed for one purpose (e.g., squeezing lemons), given away, and then accidentally or intentionally used for another activity (e.g., picking up snails). When asked what each object was "for," the children, like adults, favored the intended function, even in experimental conditions in which the alternative use occurred frequently rather than just once (Kelemen, 1999b). A subsequent study replicated this effect using manipulable, novel artifacts. In contrast to 3-year-olds, groups of 4- and 5-year-olds not only judged the objects as "for" their designed function rather than their everyday intentional use, but also favored intended function when judging where items belonged in a house (Kelemen, 2001).

Research by Matan and Carey (2001) also reveals some early sensitivity to intended function. In their study, children were told about artifacts that were made for one purpose (e.g., to water flowers) but used for something else (to make tea in). When asked which familiar artifact category the object belonged to (e.g., watering can or teapot), 4- and 6-year-olds, like adults, had a preference for the design category. However, 4-year-olds' tendency to be influenced by the order of forced-choice response options on some trials led Matan and Carey to conclude that an understanding of designer's intent does not organize children's artifact concepts until around 6 years of age.[4]

According to German and Johnson (2002), however, even the design bias that Matan and Carey's (2001) results did reveal offers no real indication of children's understanding of the designer's role in designating function. Instead, German and Johnson argued, naming results such as these reveal little more than children's more shallow knowledge that the designer has the right to designate an object's category name and membership ("baptism rights").

Although it is not clear that this explanation accounts for Matan and Carey's (2001) results,[5] German and Johnson's (2002) results were consistent with the notion that this is the limit of children's understanding. Using function-judgment methods similar to those used in my lab, they found that although 5-year-olds weigh designer's intent over another agent's intentional action when determining what a novel artifact's category name is, they do not reliably use designer's intent when judging what a novel object is "really for"—a lack of design-based construal that is also reflected, German and Defeyter (2000) argued, in 5-year-olds' relative success at function-based insight problem solving: Specifically, employing methods classically used to explore functional fixedness, German and Defeyter found that although 6- and 7-year-olds find it difficult to disregard an artifact's design function when asked to solve a problem creatively with it, 5-year-olds do not have this difficulty, more readily seeing how an artifact can be used unconventionally to achieve a goal (seeing a box as a platform and not a container; also Defeyter & German, 2003). Such a lack of the design stance in 5-year-olds is, in fact, no surprise, suggested German and Johnson, when the computations involved in reasoning about design intentions are actually considered; that is, design attributions require recursive reasoning about second-order mental states—"maker intends (that user intends) that X will perform Y"—something acknowledged as difficult for children.

However, this explanation of 5-year-olds' lack of design sensitivity in German and Johnson's (2002) tasks is challengeable: Design in-

[4]Matan and Carey's children made fewer design-based judgments when the design category name was presented second rather than first—an effect perhaps caused by the use of familiar artifacts as stimuli and pretrial procedures for familiarizing children with these stimuli that may have subsequently prompted prepotent responding to the first function information heard, reducing design-based reasoning overall.

[5]Half of Matan and Carey's stimuli had names encoding intended function, rendering it unlikely that participants processed only intended category membership.

tentions may not require second-order computation (they may reduce to "maker intends that user does X with Y" or "maker intends that X does Y"), and reasoning about mental-state content of a more complex form than the goal states of design intentions has been documented among 3- and 4-year-olds (e.g., Chandler, Fritz, & Hala, 1989; Siegal & Beattie, 1991). Furthermore, although in combination German and his colleagues' findings might suggest that a design-based grasp of artifact function is not present until age 6 or 7 years, some patterns across their various studies raise questions: For example, in German and Johnson's function-judgment task, even adults' tendency to judge that the novel artifacts were "really for" the designed function rather than an intended use was weak—more than half the adult subjects made design-based judgments 50% or less of the time. Perhaps, then, unintended qualities of the stimuli had a particular impact on children's judgments across all of German and Johnson's studies. Additionally, studies directly exploring whether there is a relationship between 3- to 5-year-olds' susceptibility to functional fixedness and their tendency to construe artifacts in terms of original design have found no correlation between the two abilities, suggesting that other factors (e.g., age- or education-related changes in conventionality) might account for 5-year-olds' advantage in German and Defeyter's (2000) insight tasks (Kelemen, 2001).

These disparities aside, an underlying developmental pattern does emerge across all of these studies. With some reliability, the findings suggest that beginning some time around the kindergarten period, children adopt a design-based teleological view of objects with increasing consistency. In light of this work, and the earlier-described research on children's reasoning about nonnatural agents' mental states, the proposal that children might be intuitive theists becomes increasingly viable.

However, an issue still remains: Just because children can consider objects as products of design does not mean this ability has any actual connection to children's attributions of purpose to nature. It is possible, after all, that, like some adults, children view supernatural agents as originators of nature but consider the functionality of many natural phenomena as deriving from an entirely different, nonintentional cause (e.g., evolution). Thus, although children may invoke God in their explanations of origins (e.g., Evans, 2001) and view natural phenomena as existing for a purpose (e.g., Kelemen, 1999b), the two sets of intuitions may have no systematic relation.

A recent study addressing this question suggests that this is not the case. Six- to 10-year-old British children were first asked to generate ideas about why various animals, natural objects, and events exist, and then consider other people's explanations, indicating their preference between teleological and physical explanations for each item. Subsequently, the children were also asked questions probing their ideas about intentional origins and whether they thought the earlier items originated because they "just happened" or because they were "made by someone/something." The design of the study precluded children from tracking their answers and aligning their answers to earlier and later questions in the absence of intuitions of their own. Nevertheless, the results revealed correlations between children's teleological ideas about nature and their endorsements of intentional design. Furthermore, no artificialism was found: Children identified people as the designing agents of artifacts (control items), distinguishing God as the designing agent of nature (Kelemen & DiYanni, in press).

SUMMARY

This article began by posing a question: Given findings regarding children's beliefs about purpose and their ideas about the intentional origins of nature, is it possible that children are intuitive theists insofar as they are predisposed to develop a view of nature as an artifact of nonhuman design?

A review of recent cognitive developmental research reveals that by around 5 years of age, children understand natural objects as not humanly caused, can reason about nonnatural agents' mental states, and demonstrate the capacity to view objects in terms of design. Finally, evidence from 6- to 10-year-olds suggests that children's assignments of purpose to nature relate to their ideas concerning intentional nonhuman causation. Together, these research findings tentatively suggest that children's explanatory approach may be accurately characterized as intuitive theism—a characterization that has broad relevance not only to cognitivists or the growing interdisciplinary community studying the underpinnings of religion (Barrett, 2000), but also, at an applied level, to science educators because the implication is that children's science failures may, in part, result from inherent conflicts between intuitive ideas and the basic tenets of contemporary scientific thought.

Further research is required, of course, to clarify how well the description really holds across individuals and cultures (reliable, empirical cross-cultural research is limited), how robust the orientation to purpose and design is, and how it interacts with education over time. A significant theoretical goal is to empirically discriminate the present hypothesis that children are inherently predisposed to invoke intention-based teleological explanations of nature and find them satisfying (see Bering, 2002, for a related stance) from the milder hypothesis that children's teleological orientation arises primarily from their possession of the kind of cognitive machinery (e.g., agency detection) that renders them susceptible to the religious representations of their adult culture—a position that predicts children would not independently generate explanations in terms of designing nonnatural agency without adult cultural influence.

A proper discussion of the pros and cons of each position, along with how to empirically distinguish them, is beyond the scope of this short article. However, it is worth emphasizing that the kind of research program proposed here is one that involves focusing on adults as much as children because although the question "are children intuitive theists?" implies a dichotomy between child and adult thought, the current proposal tacitly assumes that the idea of such a fundamental dichotomy is false: If, as suggested here, the tendency to think in teleological quasi-artifact terms is a side effect of human mental design (and pan-cultural experience with artifacts) rather than socialization, it is likely to remain as a default explanatory strategy throughout life, even as other explanations are elaborated. This idea contrasts with the notion that through conceptual change (e.g., Carey, 1985), such an explanatory approach is revised and replaced by a physical-reductionist view of nature in cultures endorsing such ideas.

Several factors provide support for this suggestion of developmental continuity. First, reasoning about all aspects of nature in nonteleological physical-reductionist terms is a relatively recent development in the history of human thought (see Kelemen, 1999a, for a brief history of the "design argument"), and contemporary adults are still surprisingly bad at it. For example, evolution is generally misconstrued as a quasi-intentional needs-responsive designing force, indicating that even when adults elaborate alternative scientific explanations, signs of intention-based reasoning about nature are still in evidence (see Evans, 2000a, for review). Second, recent research with American college undergraduates has found that although such populations endorse teleological explanation in a selective, scientifically appropriate way in the evaluative context of a forced-choice "scientific" experiment, in a less evaluative environment they will more promiscuously generate teleological explanations of why animals and inanimate natural objects exist. These results suggest that even in a post-Darwinian culture, continuity rather than conceptual change may be at play in educated individuals' preference for teleological explanation (Kelemen, 2003). Finally, and significant to the conjecture that scientific educations suppress rather than replace teleological explanatory tendencies, research with scientifically uneducated Romanian Gypsy adults has found that they have promiscuous teleological intuitions much like scientifically naive British and American elementary-school children (Casler & Kelemen, 2003b). In conclusion, the question of whether children and adults are intuitive theists provides fertile ground for future research.

Acknowledgments—This work was supported by National Institutes of Health Grant HD37903-01. I thank Justin Barrett, Krista Casler, Cara DiYanni, Liz Donovan, and an anonymous reviewer for very helpful comments on earlier drafts. Special thanks go to Kim Saudino.

REFERENCES

Atran, S. (1995). Causal constraints on categories. In D. Sperber, D. Premack, & A.J. Premack (Eds.), *Causal cognition: A multi-disciplinary debate* (pp. 263–265). Oxford, England: Clarendon Press.

Atran, S. (2002). *In gods we trust: The evolutionary landscape of religion*. New York: Oxford University Press.

Baillargeon, R. (1993). The object concept revisited: New directions in the investigation of infants' physical knowledge. In C.E. Granrud (Ed.), *Visual perception and cognition in infancy* (Carnegie Mellon Symposia on Cognition Vol. 23, pp. 265–315). Hillsdale, NJ: Erlbaum.

Barrett, J.L. (2000). Exploring the natural foundations of religion. *Trends in Cognitive Sciences, 4*, 29–34.

Barrett, J.L., Richert, R., & Driesenga, A. (2001). God's beliefs versus mother's: The development of non-human agent concepts. *Child Development, 72*, 50–65.

Bering, J. (2002). Intuitive conceptions of dead agents' minds: The natural foundations of afterlife beliefs as a phenomenological boundary. *Journal of Cognition and Culture, 2*, 263–308.

Boyer, P. (2001). *Religion explained: The evolutionary origins of religious thought*. New York: Basic Books.

Carey, S. (1985). *Conceptual change in childhood*. Cambridge, MA: MIT Press.

Casler, K., & Kelemen, D. (2003a). *Teleological explanations of nature among Romanian Roma (Gypsy) adults*. Unpublished manuscript, Boston University, Boston.

Casler, K., & Kelemen, D. (2003b). *Tool use and children's understanding of artifact function*. Unpublished manuscript, Boston University, Boston.

Chandler, M., Fritz, A.S., & Hala, S. (1989). Small-scale deceit: Deception as a marker of two-, three-, and four-year-olds' early theories of mind. *Child Development, 60*, 1263–1277.

Csibra, G., & Gergely, G. (1998). The teleological origins of mentalistic action explanations: A developmental hypothesis. *Developmental Science, 1*, 255–259.

Defeyter, M., & German, T. (in press). Acquiring an understanding of design: Evidence from children's insight problem-solving. *Cognition*.

Diesendruck, G., Markson, L.M., & Bloom, P. (2003). Children's reliance on creator's intent in extending names for artifacts. *Psychological Science, 14*, 164–168.

Evans, E.M. (2000a). Beyond Scopes: Why Creationism is here to stay. In K.S. Rosengren, C.N. Johnson, & P.L. Harris (Eds.), *Imagining the impossible: The development of magical, scientific and religious thinking in contemporary society* (pp. 305–333). Cambridge, England: Cambridge University Press.

Evans, E.M. (2000b). The emergence of beliefs about the origin of species in school-age children. *Merrill Palmer Quarterly, 46*, 221–254.

Evans, E.M. (2001). Cognitive and contextual factors in the emergence of diverse belief systems: Creation versus evolution. *Cognitive Psychology, 42*, 217–266.

Evans, E.M., & Mull, M. (2002). *Magic can happen in that world (but not this one): Constructing a naïve metaphysics*. Manuscript submitted for publication.

Gelman, S.A., & Kremer, K.E. (1991). Understanding natural cause: Children's explanations of how objects and their properties originate. *Child Development, 62*, 396–414.

Gentner, D. (1978). What looks like a jiggy but acts like a zimbo? A study of early word meaning using artificial objects. *Papers and Reports on Child Language Development, 15*, 1–6.

German, T., & Defeyter, M. (2000). Immunity to functional fixedness in young children. *Psychonomic Bulletin & Review, 7*, 707–712.

German, T., & Johnson, S.A. (2002). Function and the origins of the design stance. *Journal of Cognition and Development, 3*, 279–300.

Graham, S.A., Williams, L.D., & Huber, J.F. (1999). Preschoolers' and adults' reliance on object shape and object function for lexical extension. *Journal of Experimental Child Psychology, 74*, 128–151.

Guthrie, S. (2002). Animal animism: Evolutionary roots of religious cognition. In I. Pyysiainen & V. Anttonen (Eds.), *Current approaches in the cognitive science of religion* (pp. 38–67). London: Continuum.

Harris, P. (2000). On not falling down to earth: Children's metaphysical questions. In K.S. Rosengren, C.N. Johnson, & P.L. Harris (Eds.), *Imagining the impossible: The development of magical, scientific and religious thinking in contemporary society* (pp. 157–178). Cambridge, England: Cambridge University Press.

Johnson, C.N. (2000). Putting different things together: The development of metaphysical thinking. In K.S. Rosengren, C.N. Johnson, & P.L. Harris (Eds.), *Imagining the impossible: The development of magical, scientific and religious thinking in contemporary society* (pp. 179–211). Cambridge, England: Cambridge University Press.

Johnson, S.C., Booth, A., & O'Hearn, K. (2001). Inferring the goals of a non-human agent. *Cognitive Development, 16*, 637–656.

Keil, F.C. (1989). *Concepts, kinds, and cognitive development*. Cambridge, MA: MIT Press.

Keil, F.C. (1992). The origins of an autonomous biology. In M.R. Gunnar & M. Maratsos (Eds.), *Minnesota Symposia on Child Psychology: Vol. 25. Modularity and constraints in language and cognition* (pp. 103–137). Hillsdale, NJ: Erlbaum.

Kelemen, D. (1999a). Beliefs about purpose: On the origins of teleological thought. In M. Corballis & S. Lea (Eds.), *The descent of mind: Psychological perspectives on hominid evolution* (pp. 278–294). Oxford, England: Oxford University Press.

Kelemen, D. (1999b). The scope of teleological thinking in preschool children. *Cognition, 70*, 241–272.

Kelemen, D. (1999c). Why are rocks pointy? Children's preference for teleological explanations of the natural world. *Developmental Psychology, 35*, 1440–1453.

Kelemen, D. (2001, April). *Intention in children's understanding of artifact function*. Paper presented at the biennial meeting of the Society for Research in Child Development, Minneapolis, MN.

Kelemen, D. (2003). British and American children's preferences for teleo-functional explanations of the natural world. *Cognition, 88*, 201–221.

Kelemen, D., Callanan, M., Casler, K., & Pérez-Granados, D. (2002). *"Why things happen": Teleological explanation in parent-child conversations*. Manuscript submitted for publication.

Kelemen, D., & Carey, S. (in press). The essence of artifacts: Developing the design stance. In S. Laurence & E. Margolis (Eds.), *Creations of the mind:*

Theories of artifacts and their representation. Oxford, England: Oxford University Press.

Kelemen, D., & DiYanni, C. (in press). Intuitions about origins: Purpose and intelligent design in children's reasoning about nature. *Journal of Cognition and Development*.

Kelemen, D., Widdowson, D., Posner, T., Brown, A., & Casler, K. (2003). Teleo-functional constraints on preschool children's reasoning about living things. *Developmental Science*, *6*, 329–345.

Kemler Nelson, D.G., Frankenfield, A., Morris, C., & Blair, E. (2000). Young children's use of functional information to categorize artifacts: Three factors that matter. *Cognition*, *77*, 133–168.

Kemler Nelson, D.G., Russell, R., Duke, N., & Jones, K. (2000). Two-year-olds will name artifacts by their functions. *Child Development*, *71*, 1271–1288.

Knight, N., Sousa, P., Barrett, J.L., & Atran, S. (in press). Children's attributions of beliefs to humans and God: Cross cultural evidence. *Cognitive Science*.

Landau, B., Smith, L.B., & Jones, S.S. (1998). Object shape, object function, and object name. *Journal of Memory and Language*, *38*, 1–27.

Markson, L.M. (2001, April). *Developing understanding of artifact function*. Paper presented at the biennial meeting of the Society for Research in Child Development, Minneapolis, MN.

Matan, A., & Carey, S. (2001). Developmental changes within the core of artifact concepts. *Cognition*, *78*, 1–26.

Mead, M. (1932). An investigation of the thought of primitive children with special reference to animism. *Journal of the Royal Anthropological Institute of Great Britain and Ireland*, *62*, 173–190.

Petrovich, O. (1997). Understanding of non-natural causality in children and adults: A case against artificialism. *Psyche and Geloof*, *8*, 151–165.

Piaget, J. (1929). *The child's conception of the world*. London: Routledge & Kegan Paul.

Rips, L.J. (1989). Similarity, typicality and categorization. In S. Vosniadou & A. Ortony (Eds.), *Similarity and analogical reasoning* (pp. 21–59). Cambridge, England: Cambridge University Press.

Siegal, M., & Beattie, K. (1991). Where to look first for children's knowledge of false beliefs. *Cognition*, *38*, 1–12.

Springer, K., & Keil, F.C. (1989). On the development of biologically specific beliefs: The case of inheritance. *Child Development*, *60*, 637–648.

Taylor, M. (1999). *Imaginary companions and the children who create them*. New York: Oxford University Press.

(Received 2/21/03; Revision accepted 5/1/03)

[13]

Whence Collective Rituals? A Cultural Selection Model of Ritualized Behavior

PIERRE LIÉNARD
PASCAL BOYER

ABSTRACT *Ritualized behavior* is a specific way of organizing the flow of action, characterized by stereotypy, rigidity in performance, a feeling of compulsion, and specific themes, in particular the potential danger from contamination, predation, and social hazard. We proposed elsewhere a neurocognitive model of ritualized behavior in human development and pathology, as based on the activation of a specific *hazard-precaution system* specialized in the detection of and response to potential threats. We show how certain features of collective rituals—by conveying information about potential danger and presenting appropriate reaction as a sequence of rigidly described precautionary measures—probably activate this neurocognitive system. This makes some collective ritual sequences highly attention-demanding and intuitively compelling and contributes to their transmission from place to place or generation to generation. The recurrence of ritualized behavior as a central feature of collective ceremonies may be explained as a consequence of this bias in selective transmission. [Keywords: ritual, cognition, evolution, epidemiology, cultural transmission]

WHY DO PEOPLE, the world over, seem compelled to engage in ritual practices? Why invest time and resources in such behaviors? We suggest here that we may have the rudiments of an answer. Rituals are compelling because specific aspects of human cognitive architecture make these behavioral sequences attention-grabbing, intuitively appropriate, and compelling. Specifically, we consider that particular sequences of collective rituals activate a cognitive-emotional system focused on the detection of and reaction to potential danger. This *hazard-precaution system* responds to a specific set of cues in people's environments and makes certain types of precautionary action seem intuitively appropriate. The system is manifest not just in reactions to potential danger but also in individual ritualization, either normal or pathological (Boyer and Liénard in press).

We propose that instructions and actions typically found in collective rituals share core features with the information that normally activate the hazard-precaution system. This makes the ritual procedures attention-grabbing and compelling, which in turns explains their good cultural transmission.

NO "THEORY OF RITUAL": A MODEL OF RITUALIZED BEHAVIOR

According to the late Roy Rappaport, a proper account of ritual should address the question "why do human beings engage in rituals at all?," which remains unanswered in anthropological or psychological theories (see Rappaport 1999). There are specific reasons for this failure but also a general problem with the very notion of a "theory of ritual."

The problem lies with the concept of ritual itself. There is no clear criterion by which cultural anthropologists or other scholars of religion or classics determine that a particular type of behavior is or is not an instance of a ritual. True, there seem to be many "definitions" of ritual in anthropology (see, e.g., Gluckman 1975, among many others). But these so-called definitions are, in general, summaries of causal theories ("ritual expresses symbolism," "ritual is the manifestation of social status," etc.) rather than behaviorally precise criteria. "Ritual," like "marriage" or "religion," is not a proper analytical category; instead it is more of what Rodney Needham (1975) described as a "polythetic" category, in which, typically, ritual types A and B may share features [*m*, *n*, *p*], types Y and Z may share [*p*, *q*. *r*], Z and

W share [q, r, s]. And although A and W apparently do not share any major feature, they are both called "rituals." That is why it is certainly futile to collect many instances of what are commonly called "rituals" and to tabulate their common features. This too often results in very vague formulations that would potentially apply to any social institution.

In this project we focus on ritualized behavior, a concept that we construe, in a manner directly inspired by Rappaport, as a specific way of organizing the flow of behavior, characterized by compulsion (one must perform the particular sequence), rigidity (it must be performed the right way), redundancy (the same actions are often repeated inside the ritual), and goal demotion (the actions are divorced from their usual goals; see Bloch 1974; Humphrey and Laidlaw 1993; Rappaport 1999). Note that although ritualized behavior in this precise sense is typically the hallmark of ceremonies we call "rituals," it certainly is not found in all of those. Conversely, there may be many contexts outside "rituals" that include ritualized behavior.

Why should we abandon "rituals" and focus on ritualized behavior? Because the latter is characterized in terms that make the identification of particular instances empirically tractable. There may be ambiguous cases at some point, but this is nothing to worry about, as long as our characterization lends itself to empirical investigation.

So we are effectively asking the question "what are the effects of ritualized behavior, such that individuals find collective rituals attention-grabbing and participation in such ceremonies compelling?" Obviously, there are many specific reasons (including coercion, commitment, habit, or belief) justifying why a particular person should find a particular ceremony of interest and participate in it. These factors vary enough between cultures, periods, and individuals that they cannot explain the general recurrence of ritualized behavior. Some general features or effects of this kind of scripted, rigid, and so forth behavior should explain why, all else being equal, it appears with such frequency in human cultures.

We address this question in a cultural selection framework. The main points of such a framework have been explained by others (Boyd and Richerson 1985; Durham 1991; Sperber 1985), so we only mention two important points that impinge on our argument. First, recurrent features of human cultures are the winners in a constant process of generation and selection of new variants. What we observe as cultural representations and practices are the variants found in roughly similar forms in a particular place (Boyd and Richerson 1985). Those particular sets of representations have resisted better than other changes and distortions through innumerable processes of acquisition, storage, inference, and communication (Boyer 1998; Sperber 1985). This may be because they constitute cognitive "attractors"—that is, optimal activation of particular mental resources (Sperber 1996). Second, cultural transmission does not consist in the downloading of information from cultural elders, but in the inferential construction of conceptual structures from available (and generally fragmentary) information (Sperber 1996). So the acquisition, storage, and communication of those representations we call "cultural" are crucially affected by general features of human minds that should and can be independently established. Inferential processes are generally not accessible to conscious inspection, which is why experimental methods are required. Applied to the problem at hand, this suggests the following:

- There are collective rituals in human groups because certain sets of actions are selected through cultural transmission as more compelling or "natural" than other possible sets of actions. We need not assume a specific human need or capacity to perform collective rituals. All we have to assume is that, in given circumstances, these sets of actions seem more appropriate than others—certain ritual sequences are found more attention grabbing or memorable than others.

and

- The selection can be explained in terms of specific features of human psychological architecture. Rituals are not performed simply because "that's the rule" and because people absorb the conceptual schemata of their cultures. Ritual performances produce specific effects in participants that result in subsequent performance. So we must document those cognitive systems most likely to be activated by ritual performances and gather independent evidence on the effect of such activation.

THE "OBVIOUS" FEATURES OF RITUALIZED BEHAVIOR

An Example

Let us start with an example of the kind of behavior we consider here. This is taken from the ethnographic fieldwork of one of us (Liénard) among the Turkana of Kenya. As this vignette serves as illustration only, we deliberately omitted all the rich cultural background that is associated with this particular sequence of actions (see Liénard in press for the relevant information).

The ritual sequence partly described here is called *ariwo*. For this specific instance, it entails the sacrifice of an ox. Its coat should be of a specific color and shine. The animal should ideally be sacrificed by a left-handed twin. In the sequence preceding the sacrifice, ritual participants circumambulate the ritual scene three times and then gather in a semicircle, facing East. The animal is made to go around the dancers three times counterclockwise. At some point in the ritual, the members of the clan offering the ox approach one at a time the sacrificial ox and carefully rub their body from forehead to loin on the animal's forehead, in a gentle upward thrust, an operation made difficult by the animal's attempts to get loose and to shake its head violently.

The ritual officer cuts the animal lengthwise at the level of the diaphragm and upper abdomen. The body is then spread at the center of the ritual scene. During the next phase of the ritual, clans regroup and people line up to cross finally the ritual field from west to east walking right through the ox's split body, being careful to tread on a puddle of chyme—taken from the animal's stomach—in which has been placed the axe used to give the ultimate death blow. Among the crowd of each clan, elders and men are first to pass, followed by adolescent girls and girls of marriageable age, then come the mothers with children, and finally the young unmarried men. The sacrificer and his assistants make sure that everyone passing through the carcass steps on the plat or the handle of the axe placed in the chyme before proceeding.

Now ethnographers observing this kind of behavior would proceed to do two things. They would describe the particular reasons why people engaged in these specific acts on that occasion and they would detail the many representations that Turkana participants are likely to associate with these different actions. These two kinds of descriptions are essential to Turkana ethnography and are indeed provided elsewhere (Liénard in press). Neither description, however, will be sufficient to address the fundamental question raised here: Why engage in ritualized behavior in this context? To do that, we must first outline the general features of ritualized behavior that this sequence illustrates.

The Obvious Features

When Rappaport enjoined anthropologists to explain ritual, he listed what he called the "obvious" (i.e., obvious to all anthropologists) aspects of ritual—those frequent features that a decent model should explain (Rappaport 1979):

- *No obvious empirical goals: "meaningless" acts.* There is no clear point in walking the ox round the dancers—or, for that matter, the dancers around the scene. More generally, in rituals one typically washes instruments that are already clean, one enters rooms to exit them straightaway, one talks to interlocutors that are manifestly absent, and so forth. Frequent repetition bolsters this intuition that actions are disconnected from their ordinary goals. People bow or kneel repeatedly; they walk around an animal seven times, which clearly signals that no effect is achieved by any specific iteration of the action. Also, many rituals include actions for which there could not possibly be any clear empirical goal, such as rubbing an animal's forehead with one's body, passing a chicken from hand to hand in a circle, or going round a temple several times. True, a given ritual generally has a specific purpose (e.g., healing a particular person) but the set of sequences that compose the ritual are not connected to this goal in the same way as subactions connect to subgoals in ordinary behavior (Boyer 1994). In other words, the standard connections between means and ends seem broken in ritual. Practitioners themselves often concur that the actions are "meaningless," although efficacious (Humphrey and Laidlaw 1993). Obviously, in many religious traditions some scholarly specialists and theologians can produce justifications for each particular action but these rationalizations are absent in most nonliterate societies (Barth 1987) and remain inaccessible or uninteresting to most participants in all traditions (Boyer 2001). Moreover, far from providing a straightforward rationale for ritual actions, the specialists' exegesis often creates mysteries that require further symbolic exegesis (Sperber 1975).
- *Compulsion.* In the Turkana example, a diviner has established that people of a particular clan should offer, on behalf of everyone, the sacrifice of an ox as a remedy against misfortune but also as a protection against potential risks like enemy raids, intrusions, and pandemics. Even though the actual risk incurred is loosely specified employing commonplace threats, the decision to act is impervious to scrutiny. More generally, given certain circumstances, people just feel that they must perform a specific ritual, that it would be dangerous, unsafe, or improper not to do it. It is important to distinguish these feelings, this compulsive character of ritualized action, from the explanations people may have about the reasons for performing the ritual (Boyer 2001).
- *Literalism and rigidity.* Turkana people will be careful to walk the ox three times counterclockwise round the dancers, to have specific participants rub in a particular way their body on the animal's forehead, and to have everybody properly crossing the ritual field, stepping at one point on the axe in the chyme. The intricacy of the details and the prescriptions is the hallmark of ritualized behavior. People feel that they should perform the ritual in the exact way prescribed and generally in the way it was performed on previous occasions. This obviously does not mean that ceremonies are actually performed in the same way (Goody 1972, 1986). What is important is that people strive to achieve a performance that matches their representation of past performances, and that they attach great emotional weight to any deviation from that remembered pattern (Boyer 1990).
- *Repetition, reiteration, redundancy.* Repeated enactments of the same action or gesture—as well as reiterations of the same utterances—are typical of many collective rituals. A given sequence is executed three or five or ten times. What matters is the exact number. What matters too is that the action should seem identical in all these iterations. This makes many ritual sequences clearly distinct from everyday action, in which there is either no repetition of identical sequences (e.g., in assembling a musical instrument, one performs a series of unique actions), or each repeated sequence has a specific outcome (in weaving, the warp is changed at each step), or repetition is cumulative (the egg whites rise only after a long period of stirring). In ritual action, repetition itself is not motivated but strongly prescribed and perceived as intrinsically efficacious.
- *Order and boundaries.* This was visible in our Turkana vignette. There is a special space of the sacrifice, where normal behavior is suspended and particular acts must be performed. The dancers must be aligned in a particular way relative to the animal. They must walk through the body of the sacrificed animal along a specific path and

direction. In many rituals, people create an orderly environment that is quite different from the one of everyday interaction. People line up instead of walking, they dance instead of moving, they wear special clothes or makeup, they build alignments of rocks or logs, they create elaborate color and shape combinations, and so forth. There is a lot of ordering in rituals that is quite distinct from the comparatively unpredictable patterns of nonritual environments. Related to this is the recurrent concern with delimiting a particular space (a sacred circle, a taboo territory) often made visually distinct from the other, unmarked space. People also often emphasize the boundary between this space and the rest, for instance by special prohibitions (only men enter the sacred circle, only women sit on the left side, etc.) or by restrictions on communication between marked and unmarked spaces.

- *Specific concerns.* Pollution and cleansing, protection against invisible dangers, and the creation of a special space and time are common themes associated with ritualized behavior. It is clearly the case for our Turkana ritual. The whole endeavor is understood and justified as a way of countering and chasing away a menace. The danger is understood as a threatening intrusion, an amalgam of overexposure to potential enemy onslaught, sickness, and other likely threats. Whipping the dancers is done "to chase the disease away." Profusely splashing people at regular intervals is intended to "disconnect" people from potential ills by "cooling" them. By stepping on the axe buried in the chyme, participants should "sever" themselves from the disease on the spot. In many cultural contexts, the ritual space or instruments are often described as "pure" (or on the contrary as the locus of concentrated "pollution"), or the point of the ritual is to "purify" people or objects or to "cleanse" minds or bodies. This is not just a matter of metaphors. In many rituals, blood, semen, or excrement are a primary concern, the miasma or smells of decaying corpses are important, and the use of water or fire as possible ways of getting rid of pollution and contaminants is also recurrent. There are also innumerable examples of allusions to purity and pollution in ritual requirements. People must wash before prayers, they immerse themselves in water to rid themselves of pollution, they must wear spotless garments, the sacrificial animal must be absolutely clean, menstruating women (supposedly polluted) are barred from rituals spaces, and so on. This concern with pollution and cleansing is so prevalent that it has been considered a foundation of religious ritual (Douglas 1982).

BACKGROUND TO RITUALIZATION

Rappaport rightly pointed out that most cultural anthropological accounts of ritual simply ignore the "obvious" properties listed above (Rappaport 1979). Many ethnographic accounts focus on the specific reasons for participation in ritual in a given cultural context (e.g., a specific ritual legitimizes claims to high status or marks territorial boundaries) rather than the general features of this kind of behavior (Gluckman 1975). Also anthropological theories often emphasize processes (the transmission of norms or of shared cultural symbolism, or a demonstration of social commitment, etc.) that could and do occur also outside ritualized behavior; these, therefore, do not explain the specific features of rigidity, scriptedness, and so forth (Rappaport 1999; Staal 1990). This is particularly true of functionalist accounts. Some theories claim that rituals provide ethnic affiliation or reinforce religious belief but do not provide any account of why or how these functions require rigid sequences or repeated episodes or any other of the recurrent features of collective rituals. From a different perspective, rituals are often said to be "symbolic" (Basso and Selby 1976), which has led anthropologists to see rituals as conveying culturally coded meanings. Again, this interpretation begs the question of why such meanings would require ritualized actions—why they would require those actions to be expressed or sustained. Besides, most rituals do not actually convey coded meanings except in the vaguest sense (Sperber 1975). Typical features of rituals—such as the use of standardized formulaic speech, repetitions, and redundancy as well as the great number of obviously pointless actions—would seem to reduce the information potential of rituals (Bloch 1974; Staal 1990).

Beyond these very general points, only a few theoretical frameworks actually address the central question of the kinds of action typical of ritualized behavior. We briefly discuss them before proceeding to our cognitive framework.

Cognitive Background: Ritual and Ordinary Action

Ritual action should be studied in the context of human dispositions for organized action in general. Understanding the cognitive processes engaged in the ritualization of behavior surely requires at least some minimal understanding of the cognitive underpinnings of action representation. Although it may seem self-evident, the requirement is ignored in most accounts of ritual.

A notable exception is E. Thomas Lawson and Robert N. McCauley's cognitive account of religious ritual (1990), which inaugurated the field of cognitive studies of religion. Although most of the argument is about the specific features of religious ritual, and therefore only partly relevant to our problem, it makes sense to take this framework as a starting point. A central assumption of the model is that ritual actions are only a subset of actions, and therefore a large number of principles that govern action-representation are inherited in the specific domain of ritual. These principles specify, for instance, that the structure of action-representation is *partonomic* (small units with specific goals are included in larger units with larger goals but do not straddle the boundaries of two larger units; see Newtson 1973; Zacks et al. 2001) and that there are only limited possible mappings between specific ontological categories (person, object, animal) and "slots" in action-description (agent, patient, instrument; see Lawson and McCauley 1990). Religious believers' intuitions about the format of their rituals, as well as about the format of possible rituals they never witnessed, are governed by these general rules of action-representation (Barrett and Lawson 2001). What is specific to religious ritual, then, is not a way of representing action but the insertion of supernatural agents in the agent and patient positions of otherwise ordinary rules (McCauley and Lawson 2002).

Phylogeny of Ritual

What is the connection between collective rituals practiced by human beings, on the one hand, and animal displays and routines, on the other hand? Some authors have chosen to emphasize the common features, which are particularly important as regards the sequencing and repetition of action (Gluckman 1975; Staal 1990). However, it is quite clear that human rituals include specific preoccupations (e.g., with ritual cleansing, with unseen dangers) that do not seem to be present in animal displays or other rituals (Fiske and Haslam 1997). Debating whether animal and typically human rituals are continuous or not may not be the most fruitful strategy, unless we understand what these parallels could explain and how.

The main point of continuity between collective and animal rituals may lie in the simplification of communication, in the ostension of intentions that is accomplished through rituals. Rituals in many animals reduce the cost of communication by transmitting honest signals of fitness or dispositions (Bradbury and Vehrencamp 2000; Gintis et al. 2001). For instance, ritualized fights save males the fitness cost of actual fights (Watanabe and Smuts 1999). More generally, displays and rituals make the transmission of simpler messages more efficient (Payne 1998). Collective rituals too set up a special form of communication with greatly impoverished propositional content (Bloch 1974, 1998). The relationship between expressive displays, on the one hand, and ritualized behavior, on the other hand, is not as straightforward as one might imagine. Expressive behavior finds its raison d'être in the communication of signals relevant for interaction. Although collective rituals do fulfill specific social functions and goals (among which the transmission of specific signals should be counted), we believe that the characteristic features and contents of human ritualized behavior cannot be accounted only in those terms.

Furthermore, phylogenetic considerations do not by themselves provide a full evolutionary account of any behavior (Tooby and Cosmides 1989)—that is, even if we consider that ritualization in several species is a single phenomenon, it remains to explain whether rituals in humans are only an atavism or constitute a specific adaptation. What would be the possible adaptive value of this propensity for rigid scripting of joint action?

Adaptationist Models: Group Cohesion and Signaling

Some evolutionary anthropologists have suggested that rituals create coordination and therefore group cohesion (Sosis 2000). Collective rituals typically require group-scale coordination between agents. However, although it is true that rituals require coordination, it is also true that many domains of everyday social interaction in humans require it too but are not accompanied by ritualized behavior. Indeed, part of our human-evolved cognitive equipment consists of cognitive capacities that allow large-scale coalitional alliances, which imply some measure of coordinated action (Harcourt and de Waal 1992; Kurzban et al. 2001). This would suggest that there is no need for humans to use rituals as a way of creating social alliances: They already have the tools for that. It is clear, however, that rituals may seem intuitively appropriate for demonstrations of commitment (Gintis et al. 2001; Sosis 2003). To perform the ceremonies, people must not only agree on a particular script of what actions must be performed when, but also they must collaborate in the performance and they display this cooperation. In this way, rituals may reinforce not just cohesion but public commitment to cohesion, which in itself constitutes a powerful incentive for other agents' display of commitment (Kuran 1998).

This is related to the fact that many collective rituals are exceedingly expensive, either in resources or in direct fitness, as they put some participants in great danger. Male initiation, for instance, is said to turn boys into full-blown men, but its often painful ordeals typically transmit none of the skills or knowledge associated with manhood in the groups concerned (Bloch 1992; Houseman 2002). What would motivate people to engage in such rituals? One possible answer is that these rituals constitute "commitment" devices (Frank 1988; Schelling 1960). Expensive or fitness-reducing rituals may constitute an elaborate form of signaling fitness, not despite their costs but because of them (Zahavi and Zahavi 1997). Evolutionary anthropologists have noted that expensive religious rituals in particular constitute hard-to-fake signals of commitment to statements, the only guarantee for which is other people's commitment (Bulbulia 2004; Irons 2001; Sosis 2003).

RITUALIZATION PROPER, INDIVIDUAL AND COLLECTIVE

We consider all these findings and hypotheses of great interest—indeed, they provide the foundations of the model presented below. However, it is also clear that they still leave a crucial piece of the explanation missing. There is simply no description of the capacities involved in ritualized behavior. There is no general account of how and why human minds find these sequences attention demanding and compelling. In other words, Rappaport's question remains largely unanswered. Cognitive models focus on continuities between ordinary and ritual action, and it remains to explain the differences. Phylogenetic models provide little explanation for the possible adaptive value of a human disposition to perform rituals. Adaptationist models are still tentative, as they do not include a precise description of the cognitive machinery underpinning ritual action. So they fail so far to provide the proximate cognitive mechanisms required for testable evolutionary psychological models (Ketelaar and Ellis 2000). This is precisely what we aim to provide here.

Diverse Domains of Ritualization

In our view, features of collective rituals can be explained only if we leave aside cultural institutions for a while and turn to ritualized behavior that can be observed in diverse circumstances and with different consequences. This is why we must review the occurrences of such behavior in contexts that are neither collective nor "ritual" in the usual anthropological sense—behaviors whose psychological underpinnings may illuminate collective ritualization.

- *Children's rituals.* Most children engage in ritualistic behaviors at a particular stage of development, starting at age two, peaking at age five, and subsiding around age seven. These behaviors are usually considered part of normal development, distinct from the severe symptoms of OCD (see below; Leonard et al. 1990). The age of onset is similar in different cultures, as are the themes of ritualistic behavior: perfectionism, attachment to favorite objects, concerns about dirt and cleanliness, preoccupation with just-right ordering of objects, preferred household routines (Zohar and Felz 2001). Specific action sequences must be performed—there is a compulsion to engage in the activity. The ritual must also be performed in a precise way, as deviations are intuitively felt to be dangerous (Evans et al. 1999).
- *Obsessive-compulsive disorder.* In some people, intrusive thoughts and compulsions can evolve into full-blown obsessive-compulsive disorder. The main feature of the pathology is a strong compulsion to engage in stereotyped and repetitive activities with no rational justification (American Psychiatric Association 1994). Some patients engage in bouts of washing or cleaning tools or utensils (Hodgson and Rachman 1972). Others verify that they locked their door, rolled up the car window, or turned off the gas knobs over and over again (Hodgson and Rachman 1977). Still others engage in constant counting activities or need to group objects in sets of particular numbers (Radomsky et al. 2001). In most cases the ritual seems to be an intuitive response to obsessive thoughts about potential danger, notably contamination and contagion (e.g., fear to catch other people's germs, to ingest contaminated substances, to pass on diseases to one's children or others), possible harm to others or to oneself (e.g., handling kitchen utensils and wounding people), as well as social ostracism following shameful or aggressive acts (thoughts about assaulting others, shouting obscenities, exhibitionism, etc.).
- *Life-stage-relevant intrusive thoughts.* Thoughts about potential danger and appropriate precautions are not confined to the clinical population. On the contrary, systematic studies of the themes of OCD have shown that most normal people experience the same kind of intrusive thoughts as patients and to some degree generate the same ritualized action-plans to avoid such dangers (Rachman and de Silva 1978). Intrusive thoughts often become more focused and more bothersome at particular phases in the lifespan, notably the final stages of pregnancy and motherhood and fatherhood. A review of these phenomena suggests that senseless, intrusive, unacceptable ideas, thoughts, urges, and images about infants are common among healthy parents of newborns. The content of these intrusions resembles that found in clinical obsessions (Abramowitz et al. 2003). A common underlying theme is uncertainty and doubt concerning whether one may be responsible for harm to the infant (Abramowitz et al. 2003). In the same way as in OCD, some ritualized behaviors develop, repetitive checking in particular (Leckman et al. 1999).

The Fiske Hypothesis

Despite the common features, most cultural anthropologists do not consider individual and collective rituals in the same explanatory framework. A notable exception is Alan Fiske (Dulaney and Fiske 1994; Fiske and Haslam 1997), who reopened an issue famously framed by Sigmund Freud (1928, 1948) a long time ago. Freud had commented on the obvious similarities, in terms of repetition and compulsion, between individual obsessive compulsion and the prescriptions of religious ceremonies. Although Freud's general theory of neuroses would certainly provide an interpretation for some features of rituals, in particular the obsessive concern with purity and pollution, this remains implicit in his treatment of ritual, which concludes with the tantalizing observation that obsessive neurosis should be seen as a private cult and religion as a collective form of neurotic obsession (Freud 1948).[1] It does not, however, solve the question of why this type of behavior appears in either context.

Many features of collective rituals, beyond those emphasized by Freud, make the comparison tempting. First, collective rituals are often centered on themes related to potential danger. Second, a constant theme in collective rituals is that nonperformance is highly dangerous, although in many cases no coherent account is available. Third, the actions are highly scripted and there is a definite intuition that deviations from the script are dangerous. Fourth, the actions are highly repetitive and redundant. Fifth, the specific actions performed are divorced from their usual goals. So it would seem perverse to try to explain collective ritual without considering this and the other domains of ritualization as (clearly different) manifestations of a common set of cognitive processes.

Comparing hundreds of ritual sequences with clinical descriptions of OCD cases, Fiske and colleagues showed that the same themes recur over and over again in both domains (Dulaney and Fiske 1994; Fiske and Haslam 1997). OCD-typical features that also enter into rituals include specific (lucky or unlucky) numbers, use of special colors, repetition of actions, measures to prevent harm, ordering and symmetry, stylized verbal expressions, washing, concern with contagion, and so forth (Fiske and Haslam 1997). Those thoughts and practices are "egodystonic" in personal rituals, perceived as unwanted, unpleasant, shameful, or irrational. But the very same thoughts and practices are socially approved in collective rituals.

To Fiske, the similarities and differences between individual and socially acquired rituals suggest that a specific human capacity to perform rituals, usually channeled toward socially approved contexts, becomes hyperactive in pathology (Fiske and Haslam 1997). In this view, humans develop a capacity to engage in coordinated social action and also develop a subset of this as a capacity for collective rituals. The capacity helps channel personal fears or doubt into culturally transmitted conceptual schemes, thereby making them shared and probably less anxiogenic. This is because rituals, at least during performance, seem to make environments simpler, more predictable, and more meaningful than ordinary action (Fiske and Haslam 1997).

In our view, the Fiske model provides some important elements of a general account of collective rituals—not least of which is the crucial comparison with the cognitive processes engaged in individual rituals. However, we also think the hypothesis may be less than parsimonious in terms of explaining ritualization in general.

Specifically, it seems difficult to postulate a general capacity for collective ritualized behavior. As far as cognitive processes are concerned, most aspects of collective ritual can be explained in terms of capacities documented in other, nonritual contexts. Collective rituals make use of scripted actions but so do recipes and other routinized behaviors (Abelson 1981; Zacks et al. 2001). They focus on security-related issues but so do many intrusive thoughts outside rituals (Rachman and de Silva 1978). Collective rituals engage low-level parsing of action, but that is also true of other circumstances (Lassiter et al. 2002). What is specific to ritual is the combination of all these elements in a process that remains to be explained.

A NEUROCOGNITIVE MODEL OF INDIVIDUAL RITUALIZED BEHAVIOR

On the basis of anthropological and neuropsychological evidence, we proposed elsewhere (Boyer and Liénard in press) a synthetic model of individual ritualized behavior that we can only summarize here. This comprises the following: (1) a description of the neurocognitive systems involved in individual ritualized behavior, derived from specific neuropsychological models (Saxena et al. 1998; Szechtman and Woody 2004); (2) a speculative description of the type of action-representation engaged by ritual prescriptions; and (3) a description of the short- and long-term cognitive effects of ritualized behavior.

Potential Danger and Relevant Cognitive Systems

A variety of findings converge to suggest a specific system for dealing with potential danger in the brains of humans and many other mammal species (Szechtman and Woody 2004). We call this the "hazard-precaution system." Its neural correlates are distinct from those of fear-systems that respond to actual danger (LeDoux 2003). For humans as well as many other species, survival and reproductive success require not just avoidance of present danger but also detection of indirect clues for fitness threats. So it is not surprising to find that the human hazard-precaution system seems to be specifically focused on such recurrent threats as predation, intrusion by strangers, contamination, contagion, social offence, and harm to offspring (Szechtman and Woody 2004). The system does not seem to respond in the same way to more recent and actually far more dangerous stimuli such as tobacco or cars (Mathews et al. 2004). We call this system "hazard precaution" because it comprises both (1) some specific reactions to potential danger clues and (2) rudimentary descriptions of appropriate precautions, including avoidance (of other people), contact avoidance and disgust (against contamination), and attention to traces and indirect signals (against intrusion and predation).

Individual intrusive thoughts—pathological or not—are often focused on a small range of items and concepts connected to these evolutionary threats. Individual rituals—pathological or not—often include the range of appropriate reactions to these specific potential dangers (Mataix-Cols et al. 2005):

> Thoughts about contamination and contagion trigger compulsions centered on washing and cleansing, as well as precautionary measures, such as protecting oneself from intrusive material by moving to a different place, avoiding contact, avoiding breathing, and so forth.
>
> Thoughts about possible harm to one's own offspring trigger fears of handling tools and utensils in a dangerous way, smothering or dropping the infant, and forgetting about the baby or losing it (particularly in stores and other public places) with predictable precautionary measures: constant monitoring or repeated checking.
>
> Thoughts about possible acts that would offend or upset other people, resulting in social exclusion trigger a hyperactive monitoring of one's own actions, in particular the minutiae of one's own behavior, well beyond the "normal" limits. Another common feature is that people choose to avoid social contact lest they insult or assault others, which again is intuitively appropriate as a precautionary device.

Obviously, the workings of this system—and of other neurocognitive systems of this kind—are not accessible to conscious inspection. The systems' hidden computations result in a specific level of anxiety (or absence thereof) and a heightened attention focused on particular objects (features of a landscape that make predation possible, features of a body that make contagion likely, etc.), as well as the compulsion to engage in particular courses of action (washing, avoiding contact, checking one's surroundings repeatedly, etc.). Neurocognitive models explain OCD pathology as a form of hyperactivation of this system (Szechtman and Woody 2004). In our view, nonpathological activation would explain compelling precautionary behaviors in many situations; it may also explain specific intrusive precaution-related thoughts at particular stages of the life cycle; children's ritualized behavior may correspond to a phase of calibration of this security-focused system (Boyer and Liénard in press).

Complex Ritual Rules and Working Memory

Another part of the model is a speculative explanation for the presence of complicated prescriptions. As we noted above, a hallmark of ritualized behavior is the activation of complex rules: Do x but not y, walk in a particular gait, handle an object in a particular way, and so on. The variations seem potentially infinite, but there is a recurrent feature in all these rules. They generally turn usually automatic or routinized behavior (walking, washing, getting dressed, etc.) into highly controlled action that requires sustained attention. An example is having to tie one's shoelaces three times on the right foot and four times on the left. The difference with simply tying one's shoelaces is that the latter can be done without attending to the action, whereas the former requires sustained focus.

Neuropsychologists have reported that some patients' compulsive rituals result in "swamping" of working memory, so that the person cannot attend to stimuli and situations outside the ritualized action (Ursu et al. 2003; Zalla et al. 2004). Also, many patients state that performing the ritual is one way of inhibiting or repressing unwanted thoughts (Salkovskis 1985). In our view, these facts legitimate the conjecture that complicated prescriptions may constitute a spontaneous and moderately efficient form of thought suppression, with some similarities to the suppression processes studied experimentally by psychologists (Wegner 1994; Wegner et al. 1987). In other words, patients with complicated compulsions spontaneously design a kind of activity so demanding in cognitive control that intrusive thoughts can be, at least for a while, pushed away from consciousness (Boyer and Liénard in press).

Ironic Outcomes

Finally, the model specifies that ritualized behavior has particular effects on the salience of the thoughts that elicited this form of behavior. Here we take inspiration from the study of voluntary thought suppression. Studying normal subjects instructed not to think about a particular item, Wegner showed that thought suppression typically results in a "rebound," a higher salience of the unwanted thoughts (Wegner and Schneider 2003). In Wegner's model, this is caused by the combination of two distinct processes engaged in thought suppression. Although an explicit process directs and monitors the suppression, implicit processes are engaged that detect material associated with the target item (Wegner and Erskine 2003). Clinical models of OCD also concur in the conclusion that rituals often seem to exacerbate the obsession they temporarily appease (Rachman and de Silva 1978). The patients who perform more rituals are typically more anxious and also more bothered by their intrusive thoughts. The thoughts also seem to become more frequent with higher ritualization. In other words, the long-term effects of ritual performance may be the opposite of its short-term results.

A MODEL OF COLLECTIVE RITUALS

Our interpretation of ritualized behavior in collective ceremonies is not that these constitute individual rituals writ large, or even that they are generated in the same way as the individual compulsions described so far. Our model is based on cultural selection assumptions. We start from the premise that particular collective rituals are culturally successful (i.e., remembered by people and found compelling enough to perform again) to the extent that they activate specific individual neurocognitive systems more or more specifically than other possible variants. Specifically, we want to suggest that many details of collective rituals, when observed by normal human minds, activate the information-processing and motivation systems described above, which makes these behaviors attention-grabbing and compelling to participants, more so than alternates eliciting weaker operation of the hazard-precaution system.

Cognitive Capture of the Hazard-Precaution System

The hazard-precaution system only responds to information that is couched in a specific input format. This, in fact, is a general point that applies to many other functionally specific neural systems. Linguistic parsing systems, for instance, require words of a known language as an input; stereoscopic vision systems require slightly different retinal projections. To say that the hazard-precaution system requires a particular input format is simply a more specific way of saying that efficient information processing requires at least some filtering out of information, so that the system does not "fire" in each and every situation the organism faces, as this would be grossly maladaptive. The hazard-precaution system is activated by information about potential danger (in a narrow potential hazard repertoire) for which indirect clues can be found in the environment.

Another assumption is that any functional system with a specific input format is vulnerable to cognitive capture—that is, activation by signals that are not part of its intrinsic functional repertoire. There is a difference between the proper (evolutionary) domain and the (usually broader) actual domains of a system (Sperber 1996). Mimicry and camouflage use this noncongruence between the functional domain and the actual domain of inputs that activate a system. Nonpoisonous butterflies evolve the same bright colors

822 American Anthropologist • Vol. 108, No. 4 • December 2006

as poisonous ones to avoid predation by birds. The proper (evolved) domain of the birds' bright-colored bug avoidance system is the set of poisonous insects, the actual domain is that of all insects that look like them (Sperber 1996).

Our contention is that many aspects of collective rituals activate the hazard-precaution system by including typical clues for relevant potential dangers. In other words witnessing or performing the prescribed actions should result in cognitive capture of the hazard-precaution system.

Information Available and Likely Inferences

To understand the cognitive effects of collective rituals, we must describe the kinds of information available to various people who participate in them. At first sight, it would seem that most people who participate in most rituals do not have much information at all. In most societies, they certainly do not seem to have an explicit, coherent justification for most aspects of the ritual—for the actions included or their sequencing. People do not generally hold a "theory" of their own rituals; this is what makes ethnography indispensable and difficult.

However, this is not to say that people participate in a ritual on the basis of mere imitation, peering at their cultural elders and simply performing similar gestures. This would be implausible, given that very little human communication actually involves such mindless imitation (Sperber 2000). Remember that conveying information about one's behavior does not necessarily imply explaining the behavior, explicitly commenting on it, or even having the deliberate intention of conveying it. All that is required is that some manifest behavior (including but not necessarily verbal behavior) triggers nonrandom inferences about the behavior. We propose that the information made manifest in connection with performance of a particular ritual consists of the following elements:

- *Available information about the background situation.* People are told that a ritual should be performed and led to infer that nonperformance is a dangerous option. For instance, one is told that because of a particular event (someone's illness, a death or a birth, the change of seasons, a war with another group, possible damnation), it is necessary to go through a particular ritual sequence. People also receive information and produce inferences about the kind of danger against which the ritual is supposed to protect the group ("pollution" by invisible substances, attacks by invisible predators like witches or spirits, threat of disease, possible famine, etc.) These themes substantially overlap with the potential hazard repertoire.
- *Available information about required course of action.* People are instructed to participate in the ritual in particular ways—that is, people are generally not allowed to just add to their ritual whatever action they think fit. They are enjoined, more or less explicitly, to follow a particular script. Information about the script has the following properties. First, action descriptions include themes that mimic some of the typical outputs of the hazard-precaution system: actions such as cleansing, washing, and checking. Second, there is great emphasis on the details of each action, inducing what we called "low-level parsing" of the action flow above. This is made even more salient by insistence on repetition, redundancy, apparently pointless acts, and so forth. Third, descriptions of prior conditions for the ceremony (such as particular taboos, substances to avoid, etc.) reinforce activation of hazard-precaution system.

In the next pages we present some evidence for these various claims and for the psychological and cultural effects of the processes.

Recurrent Features of Collective Rituals and Cognitive Capture

Given the input format described above, various common features of collective rituals should trigger activation of the hazard-precaution system. This would be the case in particular of information concerning the occasion for the ritual, the danger of nonperformance, and the details of performance.

- *Occasion for the ritual.* The occasions for ritual often provide clues of possible danger that overlap with the potential hazard repertoire. This is quite clear in most apotropaic rituals, the explicit point of which is to prevent disasters to fitness such as famine or illness. But the same can be said of most therapeutic rituals, in the sense that the ultimate cause of illness is generally described in terms of nonobservable processes (from germs or miasma to witchcraft). Rites of passage also include such themes as removing pollution from newborn infants (as in baptism) or protecting people from the danger of indirect contact with corpses (Bloch and Parry 1982; Metcalf and Huntington 1991).
- *Danger of nonperformance.* This is a very general feature of collective rituals. In cases in which the goal of the ritual is very clearly specified (healing rituals for instance), the danger is also clearly specified. But this is not always the case for rituals—that is, most people develop the intuition that it would be really wrong and certainly dangerous not to perform a ritual without specific thoughts about what the risk is.
- *Detailed prescriptions.* Details of prescribed performance are of course the major source of security-related motifs. As we stated in our introduction, most rituals include such operations as washing and cleaning, checking and rechecking that a particular state of affairs really obtains, or creating a symmetrical or otherwise orderly environment. As we said above, Fiske and colleagues have documented these features extensively (Dulaney and Fiske 1994; Fiske and Haslam 1997), so we will not comment any further.

Our contention is that all these items of communicated information result in a weak activation of the hazard-precaution system. What we mean by that is that the activation is probably (in most participants, in most rituals, most of the time) less intense and direct than in situations where people actually encounter clues for potential danger. The activation is bound to be weak because it is mostly indirect—from other people—and mostly through verbal communication. For instance, people are told that witches may be lurking near the village, but this is not the same as having the direct experience of seeing evidence of a predator's recent intrusion.

Consequences: Action Representation in Collective Rituals

Activation of hazard-precaution systems may explain further characteristics of collective rituals, in particular the way participants represent the actions they perform and their connection to possible goals.

- *Forcing goal demotion.* Ritual prescriptions seem to resort to particular "tricks" to make actions attention grabbing. Among these features are repetition, which creates action chunks but without the goal ascription that is usually associated with natural breakpoints in action flow. Another such device, obviously, is to borrow a sequence from ordinary scripts and perform it in a context that makes goal ascription impossible (wash objects without using water, pretend to trace an imaginary line, etc.). What results from these "tricks" is what we called "goal demotion" above. Attention is focused on the low-level, fine-grained description of action, so that sequences of actions are represented without attaching a goal to each behavioral unit, as would be the case in nonritual contexts. This may be why the phenomenology of collective ritual is sometimes described as analogous to a behavioral "tunnel" in which the only action considered is the one that will follow the present one, but one does not and cannot focus on the motives of each action and especially not on possible alternatives (Bloch 1974).
- *Swamping of working memory.* Many ritual prescriptions resemble the tasks designed by cognitive psychologists in the study of working memory. They require focused attention on a set of different stimuli and their arrangement. For instance, a requirement to turn round a ritual pole three times clockwise without ever looking down imposes executive control of two tasks at the same time. Generally, the frequent combination of a positive prescription ("do x") and a negative one ("while avoiding doing y") would seem to engage working memory and executive control in a way that is not usually present in everyday action flow. The combinations of positive and negative prescriptions generally make it difficult to perform actions by engaging automatic routines. For instance, in everyday contexts one can tie one's shoelaces in a fairly automatic way; if, however, one adds the requirement that the laces should at no point touch the front of the shoe or that the fingers should never touch the sole, this imposes a high degree of attention and disengagement of routinized action patterns. Although rituals can become routinized, especially for ritual specialists who perform them repeatedly, the focus on details of action precludes automatized performance.

COLLECTIVE RITUALS IN CULTURAL TRANSMISSION

So far we discussed the psychological underpinnings of collective ritual in terms of individual responses to socially transmitted rituals. If accepted as (provisionally and roughly) valid, the model would seem to raise the questions "why are collective rituals so organized?" and "where and what is the engineering process that made them so effective in terms of creating conceptual associations and motivation?"

Transmission: Relevance and Selection

Collective rituals are generally not "engineered" in the sense of a deliberate process. In our model, scripts for collective ceremonies enjoy a transmission advantage, to the extent that they include ritualized behavior as described here—that is, they include enough hazard-precaution cues to activate the relevant systems by cognitive capture. This obviously does not imply that anyone is deliberately including such themes in collective rituals; it only suggests that variants of the collective rituals that do include them should, all else being equal, be more attention grabbing and compelling that variants that do not, and therefore they should be potentially better transmitted. This raises the questions "how much activation of the system is necessary?" and "is it really the case that all participants share this orientation toward the ritual ceremony?"

The model only predicts a potential activation of the hazard-precaution system—that is to say, we are not stating that participation in collective rituals invariably triggers full activation of the relevant circuitry. What we are saying is that collective rituals with the kind of prescribed actions described here are, all else being equal, more likely to result in some activation of these systems than ceremonies that do not include these specific kinds of prescriptions (turning automatic action into controlled action) and these specific themes (potential danger, pollution, etc.). We are furthermore saying that this potential difference would ensure the cultural prevalence of the kind of collective ceremonies we have described here.

This does not mean that ritual's success rests on such activation at every occurrence of the ritual in all individuals. Our argument is probabilistic and not determinist. What we are saying is that normal minds that process the cues contained in ritual recipes are unlikely to treat them as straightforward false alarms. An efficient hazard-precaution system should probably be slightly oversensitive, and therefore prey to the kind of cognitive capture described here. In collective ritual, people's insistence on the potential danger of not following the rules—expressed as moral reprimand (moral threat), as possible exposure to gossip or ridicule (threat of social exclusion), or as worry about misfortune—is very likely to activate the hazard-precaution system.

Turning to individual differences, a model that derives cultural transmission from individual cognitive effects must address how these effects are spread among the participants (McCauley and Lawson 2002). Nothing in our probabilistic model of cultural transmission requires that the hazard-precaution system be engaged in all participants. It is enough that such conceptual associations occur in some participants on some occasions. Indeed, it would be unreasonable to expect identical reactions to the available information. First, many participants do not seem to participate very much, as it were, and in fact do not accomplish much in the way of actions described here as ritualized behavior. Some may be attentive to the ritual prescriptions (e.g., making sure they pass the sacrificial objects clockwise, only using their left hand) whereas others just play the thing by ear, as it were, until corrected by others. Second, in most collective rituals different roles are assigned to different categories of participants (men vs. women, old vs. young, agnatic vs. maternal kin, etc.). Indeed, in many cases participants from these different categories may have substantially different views of what is going on in the ritual. These differences in perspective may be essential to the attention-grabbing potential as well as the dynamics and transmission of some collective rituals (Houseman 2002; Houseman and Severi 1998). Here we must leave aside this important question, as it is essentially orthogonal to our transmission claim.

In cultural transmission, as in other descent-and-modification models, very small effects aggregated over many cycles of transmission are sufficient to create massive trends. That is, even in the (probably rare) limiting case in which few people produce inferences about hazard precaution, the fact that they do and that no other equally salient template for understanding the collective ritual is available should result in a slightly better transmission of hazard-precaution themes over other ones.

Implications: Ritualization as the Opposite of Routinization

In this model, ritualized behavior is described as quite different from routinized action; indeed, it is conceived as its opposite. The point should be emphasized as the two are usually confused in the theoretical literature on ritual. As a result, theories of ritual handle these two very different kinds of behavior as if they originated in the same capacities or processes, which may contribute to the vagueness of many pronouncements on "ritual" in general. In most human ceremonies, one finds an alternation or combination of what we define as *ritualized behavior* (high control, attentional focus, explicit emphasis on proper performance) and *routinized action* (possible automaticity, low attentional demands, lesser emphasis on proper performance). Differences in attentional focus in particular show that these are diametrical ways of regulating behavior. Note, again, that sometimes the same episode that is ritualized behavior for a participant (e.g., a novice about to be inducted into a secret society) may eventually become somewhat routinized for another (e.g., the long-time member who has witnessed dozens of such ceremonies, or the ritual officer who has conducted them).

Previous cognitive models of ritual have emphasized differences that are relevant to this issue of routinization versus ritualization. The action-representation model of religious ritual (Lawson and McCauley 1990) predicts a fundamental difference between high- and low-intensity rituals. In the model, this is connected with participants' assumptions about the ways in which superhuman agents are involved in the ritual, as agents or patients respectively. Harvey Whitehouse put forward a similar distinction between what he called distinct "modes of religiosity": an imagistic mode that requires rare and salient events versus a doctrinal or routinized mode that emphasizes repetition and conceptual consistency (Whitehouse 2000, 2004).

In our view, a fundamental difference that may be of greater impact on cultural transmission is between the two behavioral modes described here, ritualization and routinization respectively, defined in modes of attentional focus and the division of the action flow. As we emphasized, the two modes can be found in the same series of ceremonies or indeed in the same episode as performed by different participants. So the distinct kinds of perspectives on the prescribed behavior, with their consequences on conceptual associations, can be found inside each ritual or religious tradition. Again, we must leave for further empirical investigation inquiry as to whether this view explains cultural transmission more precisely than the distinctions between kinds of traditions or kinds of rituals put forward in other models.

COLLECTIVE RITUALS IN HUMAN EVOLUTION

Adaptation or Recruitment of Other Adaptations?

Is there a general human capacity for collective rituals, as a product of evolution by natural selection? Some features of ritual seem to cry out for such an interpretation. Some form of ritual is found in all human groups. Also, we can infer distinct features of ritualization from the archaeological record, especially in burial procedures, for the earliest modern human groups and for Neanderthals (Mithen 1996; Trinkaus et al. 1993). Moreover, human ritual is sufficiently different from its phylogenetic cousins in animal behavior to suggest a specific human adaptation. Seeing collective ritual in the light of evolution makes even more sense once we consider its connection to security motivation, the restricted range of concerns that intrusive and anxious thoughts focus on, as well as the restricted range of compulsive responses.

However, we consider the hypothesis of a specific capacity for rituals fraught with difficulties. To be evolved through natural selection, a capacity for rituals should be beneficial to individual reproductive potential. Or, more specifically, one would need to consider that a higher or more focused involvement in ritual performance would result, on average, in better reproductive potential. One possible explanation would be that ritual is a good "trick" that associates individual, unmanageable anxieties with coordinated action with others and thereby makes them more tolerable or meaningful. But, again, it seems to us that we have very little direct evidence for the reproductive benefit of such a tool. Even though excessive anxiety may be maladaptive, anxiety in general is not. Indeed, evolutionary anthropologists have noted that most anxiety states seem functional rather than dysfunctional in terms of individual fitness (Nesse 1998, 1999).

Atavistic Procedures or Evolved Precautions?

In our model the rituals themselves are not the direct expression of an atavistic behavioral repertoire. Washing incessantly or spending all one's time checking the environment for intruders probably are not evolved behaviors. Rather, the rituals combine, repeat, and accumulate single behaviors (avoiding contact with certain items, checking the environment for possible traces of intrusion, washing, etc.) that are appropriate given evolutionary hazards.

The ritualization process consists in this special accumulation and distortion of originally appropriate actions. This is manifest in animal models of the condition. Henry Szechtman and colleagues observed a high rate of repetitive and pointless "ritualistic" behavior in rats treated with quinpirole (a dopamine agonist, potentially acting on some of the neurotransmitter circuits involved in OCD pathology) in comparison to controls (Szechtman et al. 1998). The animals checked and rechecked particular locations and kept inspecting the same objects. The effect of test treatment was the higher recurrence, repetition, and redundancy of actions that were all performed (but with greater flexibility and no redundancy) by controls. In other words, rats that are induced to "ritualize" combine in a new way typical components of the rat's species-specific precautionary repertoire.

Capacity or Disposition?

We proposed that collective ritual is not necessarily the outcome of an adaptive capacity but may well be the predictable by-product of adaptive capacities. In our model a collective ritual typically activates the hazard-precaution system. Given this system and its input format, a pattern of interaction that activates them may well become attention demanding and intuitively compelling. This would make ritual a by-product of evolved cognitive architecture. We are confident that this is a valid description of many of the behavioral sequences ordinarily described as "rituals" in the anthropological literature. However, as we noted above, particular rituals may or may not comprise ritualized behavior; characterized as the evocation of evolutionary danger cues, the construction of appropriate action sequences, and the association of goal demotion and precaution cues.

In this view, rituals can be considered highly successful cultural "gadgets" whose recurrence in cultural evolution is a function of (1) how easily they are comprehended by witnesses and (2) how deeply they trigger activation of motivation systems and cognitive processes that are present in humans for other evolutionary reasons. To say that a cultural creation is a "gadget" does not entail that it is unimportant. Many of the most important human cultural creations—such as literature, complex technology, music, or the visual arts—can also be considered by-products. For us, ritualized behaviors are another by-product of a species-specific human cognitive architecture and therefore an indirect consequence of its evolution by natural selection.

Pierre Liénard Department of Psychology, Washington University in St. Louis, St. Louis, MO 63130

Pascal Boyer Departments of Psychology and Anthropology, Washington University in St. Louis, St. Louis, MO 63130

NOTE

Acknowledgments. Pierre Liénard's research was supported by a fellowship grant from the Belgian American Educational Foundation. Pascal Boyer's research was supported by a grant from the Henry Luce Foundation. We are grateful to Leda Cosmides, Dan Fessler, Thomas Lawson, Robert McCauley, Pascale Michelon, Mayumi Okada, Tom Oltmanns, Ilkka Pyysiäinen, Howard Waldow, Dan Wegner, Harvey Whitehouse, Jeff Zacks, and four anonymous *AA* referees for detailed comments on a previous draft of this article.

1. Here is Freud's conclusion: "Nach diesen Analogien [...] könnte man sich getrauen, die Zwangsneurose als pathologisches Gegenstück zur Religionsbildung aufzufassen, die Neurose als eine individuelle Religiosität, die Religion als eine universelle Zwangsneurose zu bezeichnen." ["Given these analogies [...] it is possible to consider the obsessive-neurosis as the pathological counterpart to religious culture, and to see the neurosis as individual religiosity and religion as collective obsessive-neurosis"; Freud 1948:138.]

REFERENCES CITED

Abelson, Robert P.
1981 Psychological Status of the Script Concept. American Psychologist 36(7):715–729.

Abramowitz, Jonathan S., Stefanie A. Schwartz, Katherine M. Moore, and Kristi R. Luenzmann
2003 Obsessive-Compulsive Symptoms in Pregnancy and the Puerperium: A Review of the Literature. Journal of Anxiety Disorders 17(4):461–478.

American Psychiatric Association
1994 Diagnostic and Statistical Manual of Mental Disorders. 4th edition. Washington, DC: American Psychiatric Association.

Barrett, Justin L., and E. Thomas Lawson
2001 Ritual Intuitions: Cognitive Contributions to Judgments of Ritual Efficacy. Journal of Cognition and Culture 1(2):183–201.

Barth, Frederik
1987 Cosmologies in the Making: A Generative Approach to Cultural Variation in Inner New Guinea. Cambridge: Cambridge University Press.

Basso, Keith H., and Henry A. Selby
1976 Meaning in Anthropology. 1st edition. Albuquerque: University of New Mexico Press.

Bloch, Maurice
1974 Symbols, Song, Dance, and Features of Articulation: Is Religion an Extreme Form of Traditional Authority? European Journal of Sociology 15:55–81.
1992 Prey into Hunter: The Politics of Religious Experience. Cambridge: Cambridge University Press.
1998 How We Think They Think: Anthropological Approaches to Cognition, Memory and Literacy. Boulder, CO: Westview Press.

Bloch, Maurice, and Jonathan Parry, eds.
1982 Death and the Regeneration of Life. Cambridge: Cambridge University Press.
Boyd, Robert, and Peter J. Richerson
1985 Culture and the Evolutionary Process. Chicago: University of Chicago Press.
Boyer, Pascal
1990 Tradition as Truth and Communication: A Cognitive Description of Traditional Discourse. Cambridge: Cambridge University Press.
1994 The Naturalness of Religious Ideas: A Cognitive Theory of Religion. Berkeley: University of California Press.
1998 Cognitive Tracks of Cultural Inheritance: How Evolved Intuitive Ontology Governs Cultural Transmission. American Anthropologist 100(4):876–889.
2001 Religion Explained: Evolutionary Origins of Religious Thought. New York: Basic Books.
Boyer, Pascal, and Pierre Liénard
In press Why Ritualized Behavior in Humans? Precaution Systems and Action-Parsing in Developmental, Pathological and Cultural Rituals. Behavioral and Brain Sciences.
Bradbury, Jack W., and Sandra L. Vehrencamp
2000 Economic Models of Animal Communication. Animal Behaviour 59(2):259–268.
Bulbulia, Joseph
2004 Religious Costs as Adaptations That Signal Altruistic Intention. Evolution and Cognition 10(1):19–42.
Douglas, Mary
1982 Natural Symbols: Explorations in Cosmology. New York: Pantheon Books.
Dulaney, Siri, and Alan P. Fiske
1994 Cultural Rituals and Obsessive-Compulsive Disorder: Is There a Common Psychological Mechanism? Ethos 22(3):243–283.
Durham, William H.
1991 Coevolution: Genes, Cultures and Human Diversity. Stanford: Stanford University Press.
Evans, David W., F. Lee Gray, and James F. Leckman
1999 The Rituals, Fears and Phobias of Young Children: Insights from Development, Psychopathology and Neurobiology. Child Psychiatry and Human Development 29(4):261–276.
Fiske, Alan P., and Nick Haslam
1997 Is Obsessive-Compulsive Disorder a Pathology of the Human Disposition to Perform Socially Meaningful Rituals? Evidence of Similar Content. Journal of Nervous and Mental Disease 185(4):211–222.
Frank, Robert
1988 Passions within Reason: The Strategic Role of the Emotions. New York: Norton.
Freud, Sigmund
1948[1906] Zwangsbehandlungen und religionsübungen (Obsessive behavior and religious practices). *In* Gesammelte Werke von Sigmund Freud, chronologisch geordnet [Collected works of Sigmund Freud in chronological order], vol. 7. Anna Freud et al., eds. Pp. 129–139. London: Imago Publishing.
1928 Die Zukunft einer Illusion [The future of an illusion]. 2nd edition. Leipzig: Internationaler Psychoanalytischer Verlag.
Gintis, Herbert, Eric A. Smith, and Samuel Bowles
2001 Costly Signaling and Cooperation. Journal of Theoretical Biology 213(1):103–119.
Gluckman, Max
1975 Specificity of Social-Anthropological Studies of Ritual. Mental Health and Society 2(1–2):1–17.
Goody, Jack
1972 The Myth of the Bagre. Oxford: Clarendon Press.
1986 The Logic of Writing and the Organization of Society. Cambridge: Cambridge University Press.
Harcourt, Alexander H., and Frans. B. de Waal, eds.
1992 Coalitions and Alliances in Humans and Other Animals. Oxford: Oxford University Press.
Hodgson, Ray J., and Stanley Rachman
1972 The Effects of Contamination and Washing in Obsessional Patients. Behaviour Research and Therapy 10(2):111–117.
1977 Obsessional-Compulsive Complaints. Behaviour Research and Therapy 15(5):389–395.
Houseman, Michael
2002 Dissimulation and Simulation as Forms of Religious Reflexivity. Social Anthropology 10(1):77–89.
Houseman, Michael, and Carlo Severi
1998 Naven and Ritual. Leiden, the Netherlands: Brill.
Humphrey, Caroline, and James Laidlaw
1993 Archetypal Actions: A Theory of Ritual as a Mode of Action and the Case of the Jain Puja. Oxford: Clarendon Press.
Irons, William
2001 Religion as a Hard-to-Fake Sign of Commitment. *In* Evolution and the Capacity for Commitment. R. Nesse, ed. Pp. 292–309. New York: Russell Sage Foundation.
Ketelaar, Timothy, and Bruce J. Ellis
2000 Are Evolutionary Explanations Unfalsifiable? Evolutionary Psychology and the Lakatosian Philosophy of Science. Psychological Inquiry 11(1):1–21.
Kuran, Timur
1998 Ethnic Norms and Their Transformation through Reputational Cascades. Journal of Legal Studies 27(2):623–659.
Kurzban, Robert, John Tooby, and Leda Cosmides
2001 Can Race Be Erased? Coalitional Computation and Social Categorization. Proceedings of the National Academy of Sciences of the United States of America 98(26):15387–15392.
Lassiter, G. Daniel, Andrew L. Geers, and Kevin J. Apple
2002 Communication Set and the Perception of Ongoing Behavior. Personality and Social Psychology Bulletin 28(2):158–171.
Lawson, E. Thomas, and Robert N. McCauley
1990 Rethinking Religion: Connecting Cognition and Culture. Cambridge: Cambridge University Press.
Leckman, James F., Linda C. Mayes, Ruth Feldman, David W. Evans, Robert A. King, and Donald J. Cohen
1999 Early Parental Preoccupations and Behaviors and Their Possible Relationship to the Symptoms of Obsessive-Compulsive Disorder. Acta Psychiatrica Scandinavia Supplementa 100(Suppl. 396):1–26.
LeDoux, Joseph
2003 The Emotional Brain, Fear, and the Amygdala. Cellular and Molecular Neurobiology 23(4–5):727–738.
Leonard, Henrietta L., Erica L. Goldberger, Judith L. Rapoport, Deborah L. Cheslow, and Susan E. Swedo
1990 Childhood Rituals: Normal Development or Obsessive-Compulsive Symptoms? Journal of the American Academy of Child and Adolescent Psychiatry 29(1):17–23.
Liénard, Pierre
In press The Making of Peculiar Artifact: Living Kind, Artifact and Social Order in the Turkana Sacrifice. Journal of Cognition and Culture 6(3–4).
Mataix-Cols, David, Maria Conceiçao do Rosario-Campos, and James F. Leckman
2005 A Multidimensional Model of Obsessive-Compulsive Disorder. The American Journal of Psychiatry 162(2):228–238.
Mathews, Carol A., Kerry L. Jang, Shadha Hami, and Murray B. Stein
2004 The Structure of Obsessionality among Young Adults. Depression and Anxiety 20(2):77–85.
McCauley, Robert N., and E. Thomas Lawson
2002 Bringing Ritual to Mind: Psychological Foundations of Cultural Forms. Cambridge: Cambridge University Press.
Metcalf, Peter, and Richard Huntington
1991 Celebrations of Death: The Anthropology of Mortuary Ritual. 2nd edition. Cambridge: Cambridge University Press.
Mithen, Steven J.
1996 The Prehistory of the Mind. London: Thames and Hudson.
Needham, Rodney
1975 Polythetic Classification: Convergence and Consequences. Man (n.s.) 10(3):349–369.
Nesse, Randolph M.
1998 Emotional Disorders in Evolutionary Perspective. British Journal of Medical Psychology 71(4):397–415.
1999 Proximate and Evolutionary Studies of Anxiety, Stress and Depression: Synergy at the Interface. Neuroscience and Biobehavioral Reviews 23(7):895–903.
Newtson, Darren
1973 Attribution and the Unit of Perception of Ongoing Behavior. Journal of Personality and Social Psychology 28(1):28–38.

Payne, Robert J. H.
1998 Gradually Escalating Fights and Displays: The Cumulative Assessment Model. Animal Behaviour 56(3):651–662.
Rachman, Stanley, and Padmal de Silva
1978 Abnormal and Normal Obsessions. Behavior Research and Therapy 16(4):233–248.
Radomsky, Adam S., Stanley Rachman, and David Hammond
2001 Memory Bias, Confidence and Responsibility in Compulsive Checking. Behavior Research and Therapy 39(7):813–822.
Rappaport, Roy A.
1979 Ecology, Meaning and Religion. Berkeley, CA: North Atlantic Books.
1999 Ritual and Religion in the Making of Humanity. Cambridge: Cambridge University Press.
Salkovskis, Paul M.
1985 Obsessional-Compulsive Problems: A Cognitive-Behavioural Analysis. Behavior Research and Therapy 23(5):571–583.
Saxena, Sanjaya, Arthur L. Brody, Jeffrey M. Schwartz, and Lewis R. Baxter
1998 Neuroimaging and Frontal-Subcortical Circuitry in Obsessive-Compulsive Disorder. British Journal of Psychiatry 173(Suppl. 35):26–37.
Schelling, Thomas
1960 The Strategy of Conflict. Cambridge, MA: Harvard University Press.
Sosis, Richard
2000 Religion and Intragroup Cooperation: Preliminary Results of a Comparative Analysis of Utopian Communities. Cross-Cultural Research: The Journal of Comparative Social Science 34(1):70–87.
2003 Why Aren't We All Hutterites? Costly Signaling Theory and Religious Behavior. Human Nature 14(2):91–127.
Sperber, Dan
1975 Rethinking Symbolism. Cambridge: Cambridge University Press.
1985 Anthropology and Psychology: Towards an Epidemiology of Representations. Man (n.s.) 20(1):73–89.
1996 Explaining Culture: A Naturalistic Approach. Oxford: Blackwell.
2000 An Objection against Memetics. *In* Darwinizing Culture: The Status of Memetics as a Science. R. Aunger, ed. Pp. 163–173. London: Oxford University Press.
Staal, Frits
1990 Rules without Meaning: Ritual, Mantras, and the Human Sciences. New York: P. Lang.
Szechtman, Henry, William Sulis, and David Eilam
1998 Quinpirole Induces Compulsive Checking Behavior in Rats: A Potential Animal Model of Obsessive-Compulsive Disorder (OCD). Behavioral Neuroscience 112(6):1475–1485.
Szechtman, Henry, and Erik Woody
2004 Obsessive-Compulsive Disorder as a Disturbance of Security Motivation. Psychological Review 111(1):111–127.
Tooby, John, and Leda Cosmides
1989 Adaptation versus Phylogeny: The Role of Animal Psychology in the Study of Human Behavior. International Journal of Comparative Psychology 2(3):175–188.
Trinkaus, Erik, Pat Shipman, and History E-Book Project
1993 The Neandertals Changing the Image of Mankind. Electronic document, http://name.umdl.umich.edu/HEB02320, accessed June 1, 2006.
Ursu, Stefan, V. Andrew Stenger, M. Katherine Shear, Mark R. Jones, and Cameron S. Carter
2003 Overactive Action Monitoring in Obsessive-Compulsive Disorder: Evidence from Functional Magnetic Resonance Imaging. Psychological Science 14(4):347–353.
Watanabe, John M., and Barbara B. Smuts
1999 Explaining Religion without Explaining It Away: Trust, Truth, and the Evolution of Cooperation in Roy A. Rappaport's "The Obvious Aspects of Ritual." American Anthropologist 101(1):98–112.
Wegner, Daniel M.
1994 Ironic Processes of Mental Control. Psychological Review 101(1):34–52.
Wegner, Daniel M., and J. A. K. Erskine
2003 Voluntary Involuntariness: Thought Suppression and the Regulation of the Experience of Will. Consciousness and Cognition: An International Journal 12(4):684–694.
Wegner, Daniel M., and D. J. Schneider
2003 The White Bear Story. Psychological Inquiry 14(3–4):326–329.
Wegner, Daniel M., David J. Schneider, Samuel R. Carter, and Teri L. White
1987 Paradoxical Effects of Thought Suppression. Journal of Personality and Social Psychology 53(1):5–13.
Whitehouse, Harvey
2000 Arguments and Icons: Divergent Modes of Religiosity. Oxford: Oxford University Press.
2004 Modes of Religiosity. Walnut Creek, CA: AltaMira Press.
Zacks, Jeffrey M., Barbara Tversky, and Gowri Iyer
2001 Perceiving, Remembering, and Communicating Structure in Events. Journal of Experimental Psychology: General 130(1):29–58.
Zahavi, Amotz, and Avishag Zahavi
1997 The Handicap Principle: A Missing Piece of Darwin's Puzzle. New York: Oxford University Press.
Zalla, Tiziana, Isabelle Verlut, Nicholas Franck, Didier Puzenat, and Angela Sirigu
2004 Perception of Dynamic Action in Patients with Schizophrenia. Psychiatry Research 128(1):39–51.
Zohar, Ada H., and Levia Felz
2001 Ritualistic Behavior in Young Children. Journal of Abnormal Child Psychology 29(2):121–128.

[14]

GOD, GENES, AND COGNIZING AGENTS

by Gregory R. Peterson

Abstract. Much ink has been spilled on the claim that morality and religion have evolutionary roots. While some attempt to reduce morality and religion to biological considerations, others reject any link whatsoever. Any full account, however, must acknowledge the biological roots of human behavior while at the same time recognizing that our relatively unique capacity as cognitive agents requires orienting concepts of cosmic and human nature. While other organisms display quasi-moral and proto-moral behavior that is indeed relevant, fully moral behavior is only possible for organisms that attain a higher level of cognitive ability. This, in turn, implies a significant role for religion, which has traditionally provided an orientation within which moral conduct is understood.

Keywords: altruism; cognition; evolution; morality; proto-moral; quasi-moral.

In recent years, there has been continued interest and growth in evolutionary accounts of morality and religion (e.g., Pinker 1997; Donald 1991; Burkert 1996). Most of these accounts draw on work done in genetics and sociobiology, examining either aspects of altruism in biological organisms or the influence of genes on behavior, or both. In debates on the issue, often the goal is either to reduce moral impulses to genetic imperatives or, more broadly, to attack "religion" in favor of an evolutionary ethic that, paradoxically, transcends our genetic imperatives (e.g., Dawkins 1989).

My goal here is twofold. First, I suggest that most evolutionary accounts of moral behavior do not fully take the emergence of cognition and culture into consideration, with the result that many evolutionary accounts

Gregory R. Peterson is Assistant Professor of Religion at Thiel College, Greenville, PA 16125. His e-mail address is gpeterso@thiel.edu. This paper was originally presented to the American Academy of Religion Science and Religion Section, November 1999.

of morality are improperly skewed. Second, I propose that we can understand human morality in a biological framework, recognizing both the continuities and discontinuities with other organisms in the animal world. I suggest that religious accounts of human goals and behaviors have an appropriate and necessary place in such a framework. There may also be a sense in which religious accounts compete with secular accounts in a way that has a tangible effect on human conduct.

Ethical Prolegomena

It is worth noting at the outset that most accounts of morality from the scientific side, and sometimes even accounts from within the religion-and-science dialogue, give little attention to defining what counts as moral behavior and what exactly is the relation of moral behavior to religion. In philosophy and religious studies, respectively, both of these remain highly contentious issues. Yet, they are vital for any claim to give a "genealogy of morals" within an evolutionary context. Without going into great detail, I will take the following to be broadly true of most ethical theories.

First, ethics is concerned with prescribing and proscribing certain behaviors, most of which deal with social relationships and social behavior. Second, ethics is concerned with promoting and discouraging values and dispositions, particularly those with an interpersonal or social impact. In the current Western philosophical tradition, approaches that emphasize rules (characteristic of deontological and utilitarian approaches) and those that emphasize virtues (characteristic of Aristotelian approaches) are mutually exclusive. In practice, however, both kinds of approaches are utilized to varying degrees and, it seems to me, both have their place in the moral life. Likewise, while ethics is primarily concerned with our obligations toward others, there has often been room for ideas of self-cultivation that stand relatively independent of social relationships. Indeed, religious traditions often focus on the role of self-cultivation, sometimes to the exclusion and expense of direct responsibility toward others.

Beyond this, I also would argue that most ethical systems are teleological in character, that is, there is an inherent end to be achieved, and that ethical codes and encouragements are designed to help us reach that end. This is obviously the case with utilitarian and Aristotelian approaches. While deontological approaches may deny the teleological character of ethics, they often still retain an implicitly teleological approach. Thus, Immanuel Kant's formulation of the categorical imperative in terms of universalizing one's actions implicitly appeals to a calculation of what the results would be if, for instance, everyone stole or told lies. In most ethical systems, this telos is concerned with the happiness or fulfillment of oneself and of other moral agents. What counts as a state of happiness is, again, a matter of some dispute. Among utilitarians, this has ranged from mere calculation of pleasure and pain to a sorting of "higher" and "lower" pleasures, which must be properly weighted. Among religious systems, this can range from Dante's beatific vision of God to Buddhism's desireless state of Nirvana. In practice, there can be considerable overlap. Murder, lying, and stealing are almost always wrong.

What counts as "other moral agents" can also vary. It is tragically a human tendency to exclude from moral consideration those who are different from one's own group. In the modern period, the development of animal rights advocates and environmental ethics has pushed the pendulum the other way. In theological systems, God is also an "other" that needs to be taken into account.

This teleological character of ethics, in turn, requires meta-ethical foundations in terms of human nature and cosmic nature. While "is" and "ought" may be distinct, they are nevertheless related. Conceptions of human nature dictate those goods that we desire and evils that we avoid. Libel and slander are considered injurious because we are capable of taking injury at such things. Dogs, however, are immune to assaults on their character and, at best, are only able to register our displeasure with them. Concepts of human nature may also affect issues of culpability. People who suffer from mental illness are often not held fully responsible for their actions, as reprehensible as they might be. The use of brain chemistry and genetics in court cases indicates the influence of scientific concepts on our ideas of human nature. This use of science also indicates how powerful concepts of human nature are, both for good and ill. Claims that men are by nature rational and that women are by nature less so have been used to confine both men and women to specific roles and conceptions of happiness and fulfillment that have almost always been detrimental to women.

Alongside claims regarding human nature, claims about cosmic nature also play an important role. By cosmic nature I mean general conceptions of the universe, its origins and fate—in short, metaphysics. Conceptions of cosmic nature situate conceptions of human nature. Conceptions of the cosmos inform us of the kind of moral agents that exist, whether God, animals, or aliens from Alpha Centauri. Conceptions of cosmic nature also inform us of the kinds of goods that are achievable and those that are not. Augustine's distinction between the "city of God" and the "city of man" [sic] portrays a cosmos in which all human kingdoms and goods are fleeting and corruptible. Communist and utopian ideologies, by contrast, conceive of a world where human perfectibility is possible, given the right social and technological tools. For religious systems, cosmic nature is not limited to the world as we experience it (that is, what we generally call "nature") but often includes a conception of super-nature or a super-natural realm as well. Traditionally, conceptions of a super-nature and a life after death have played a hugely significant role in religious conceptions of morality, for they allow a redefinition of human nature in such a way that new moral possibilities exist.

It is my hope that none of this seems tremendously earth-shattering. I develop these ideas, however, for two reasons. First, it is rare that an explicit understanding of ethics is given when sociobiologists speak of altruism and ethics, and the same is sometimes true for the religion-and-science dialogue as well. Second, these issues, particularly those regarding human nature and cosmic nature, play a crucial role in understanding human morality and its distinctiveness and relation to our biological and cognitive heritage. It is crucial, therefore, that we make some effort to understand them from the outset.

Quasi-Moral Systems

I am a moral agent. Bacteria are not. Chimpanzees might be. It is often claimed that one thing that distinguishes us from all other creatures is our capacity to act morally. Only human beings consciously weigh their actions in terms of abstract principles and theories of virtue. A bee may sacrifice itself for the hive, but no one thanks it for doing so.

Despite this, we are biological, embodied beings in a biological, embodied world. As such, every assertion of uniqueness must be accompanied by a recognition of basic similarity that provides the context from which uniqueness can emerge. While we may claim that we are the only organisms on Earth that conceive of moral codes, this does not mean that the kinds of situations that require moral thinking are absent in the rest of nature or that the types of solutions that moral systems provide lack analogues among other organisms. Indeed, quite the opposite is the case. Animals share, cheat, steal, deceive, and sacrifice. They make alliances, dethrone leaders, and form friendships as well as rivalries. Of course, when animals do these things, they lack many of the higher-order motivations that prompt human beings to perform the same types of deeds. As such, researchers often put scare quotes around such terms, denoting that when an animal "steals" or "shares," the animal is not stealing or sharing in the human, moral sense of those words. While such an awareness can be laudable, it also can obscure the real resemblance present, both in the behaviors themselves and in the situations that give rise to the opportunity or necessity for such actions.

It seems appropriate, then, to call these situations and resultant behaviors "quasi-moral." Quasi-moral situations arise when two or more organisms are able to influence the well-being of one another, with well-being defined either in terms of reproductive fitness or in terms of pleasure and pain. Quasi-moral behavior, as a consequence, is behavior that provides a solution or strategy to resolve such interactions when they arise. Quasi-moral behavior need not be intentional or conscious in character. It may occur at the level of "genetic programming" or rigid instinctual drives. One may even abstract further from the organisms themselves to their genes. Indeed, since genes are widely perceived to be the unit of selection, it is at the level of genetics that evolutionary biologists concentrate much of their attention. Thus, not only can organisms be selfish or altruistic, so too can their genes, which "program" such behavior to begin with.

Such quasi-moral behavior has been the central focus of sociobiology for more than two decades. For most sociobiologists, amorality is the natural state of things. Organisms inhabit a Hobbesian world of all against all. Nature is red in tooth and claw, genes are selfish (or, rather, "selfish"), and the genetic duty of every organism is to maximize its fitness at the expense of others. Altruistic behavior that benefits the reproductive fitness of others at the expense of oneself should be weeded out. Thus, the existence of altruistic behavior in the animal kingdom has provided, as Edward O. Wilson (1975) remarked, the central problem for sociobiology.

Interestingly, this is a problem that most sociobiologists take to be solved, at least in general principle. Altruism, they claim, is nothing but selfishness disguised. The primary model used for explaining altruistic behavior is kin selection. On this model, an organism that helps out another related to itself does so because such help enhances the survival of the genes that they both share. Social insects represent the prime example of kin selection theory. Insect societies work together because they share the same set of genes. It is not the survival of the individual organism but the survival of the genes that is important. In this framework, it makes perfect sense for a male worker drone to sacrifice its reproductive capacity for the sake of the queen, because at the genetic level drones and queen are not truly individuals at all—they are the same. From the sociobiologists' perspective, altruism at the level of the organism amounts to selfishness at the level of the genes.

Kin selection explanations have been tremendously successful in explaining a wide range of biological phenomena, from mole rats to sibling aid in parenting. In some situations, however, altruism occurs between organisms that are unrelated. These cases, in turn, are often explained in terms of reciprocal altruism. First proposed by Robert Trivers (1971), reciprocal altruism can be summed up in the phrase, "If you scratch my back, I'll scratch yours." A famous instance of this phenomenon occurs in vampire bats, who may share some of their feast of blood with other bats not related to themselves. According to studies, cooperation is based upon previous generosity. Bats who have been generous previously are more likely to receive a helping of blood than those who are persistently stingy. Similarly, Barbara Smuts (1999) has demonstrated that male baboons who aid female baboons with care for their young and other duties are more likely to be favored as sexual partners in the future.

The success of such apparently altruistic strategies was effectively modeled in a series of computer tournaments by Robert Axelrod (1984). In the tournament series, selfish and altruistic tendencies were modeled in terms of discrete payoffs and penalties. If two individuals cooperated, they would both get a modestly high payoff (3 points); if they both refused to cooperate (acted selfishly), they would receive only 1 point. But if one chose to cooperate (act altruistically) while the other chose not to ("defected" or acted selfishly), the selfish individual would receive 5 points, and the altruist would receive 0 (the "sucker's payoff"). This game would be played several times, until losers were eliminated and a clear winner emerged. Player strategies ranged from complete altruism to complete selfishness, with many variations in between. The most successful strategy was one called "tit for tat," in which individuals cooperated when others cooperated and defected when others defected. This was seen as a vindication of the survival value of reciprocal altruism by its advocates, for it implied that reciprocal altruism could result in a genetic payoff across generations. It also implied that, once reciprocal altruism established itself in a population, it became an evolutionary stable strategy that would be difficult to invade. Reciprocal altruists would always help one another while at the same time shunning those who "cheated" and attempted to live off the generosity of others.

Of course, the question remains, Is this really moral behavior? To that, the answer can only be no. That is, biologists are not supposing that animals or their genes have a concept of "the good" that they are aiming at, are not concerned with happiness in the moral sense, and are certainly not concerned with abstract concepts of justice and fairness. But it does qualify as quasi-moral. That is, whatever vampire bats may think about helping one another, the issue that they face (the sharing of resources with others) is precisely the same situation for which, when applied to the human context, we require moral reasoning. Likewise, the possible range of solutions is largely the same. One can choose to be selfish, to share unconditionally, or to find some intermediate response. We also find ourselves giving preferential aid to kin, even though the reasons we give for doing so may be completely different from the biological motivations plainly present in other species.

Many have sought to distance human morality from the work of sociobiologists because of the sometimes unsavory conclusions that stem from this research. The presumption of sociobiology is that selfishness is the primary behavior and altruism secondary and, in the end, only apparent. Thus, Michael Ghiselin remarks, "Scratch an 'altruist' and watch a 'hypocrite' bleed" (1974, 207). More than this, some sociobiologists have been eager to apply such analyses to the human context as well, so that humans are seen as being primarily selfish creatures who cooperate out of necessity and the need to enhance personal fitness. Thus, philandering by males is a "natural" effort to spread one's genes as much as possible, and free markets are a natural extension of biological competition.

For these sorts of claims, sociobiologists have been rightly castigated (Midgley 1995; Rolston 1999). Even when sociobiologists make explicit claims that when they are speaking of, for instance, "selfish" genes, they are not implying anything about human morality, they often end up being sloppy in their language and freely cross the line back and forth between human and animal behavior (Dawkins 1989). Likewise, granting the state of nature to selfishness and requiring altruism to be the second order phenomenon in need of explanation and reduction is hardly innocent of ideological presuppositions, and the relevance of Charles Darwin's own Victorian background is now well noted. Despite this, the baby should not be thrown out with the bath water. Whether or not we are innately selfish and whether or not we are biologically driven to help kin over strangers, we still find ourselves in situations where it is possible to be selfish and possible to help kin over strangers. Our biological context may not be able to provide us with moral solutions, but it does provide the context in which moral solutions are needed. To deny this, it seems to me, is to deny our embodiedness altogether.

PROTO-MORAL SYSTEMS

At some point, quasi-morality gives way to what I shall call "proto-morality." Proto-moral systems occur when animals begin to be able to rationally deliberate actions and their consequences. Such deliberation can occur only with the acquisition of a variety of cognitive skills, including enhanced memory and planning abilities, the ability to map social relations and social hierarchies, some awareness of how one's actions affect others, and the ability to form goals and roughly weigh pains and pleasures. It may also include the ability to deceive and to develop what cognitive ethologists refer to as a "theory of mind." Obviously, most of these behaviors are limited to social mammals, and the richest proto-moral behavior is largely limited to primates. While many may be skeptical that nonhuman animals are capable of such a range of behaviors, cognitive ethology has been accumulating data to suggest that a number of species are able to perform these activities in a flexible and goal-directed manner. Vervet monkeys track social and kin hierarchies within a group. High status for a matriarch means higher status for her kin as well (Cheney and Seyfarth 1990). Without a doubt, the great apes, our closest living relatives, provide the best examples of this sort of proto-moral behavior. Chimpanzees, for instance, live in societies that require both cooperation and vigilance. Chimpanzees form alliances, subvert hierarchies when the opportunity arises, and engage in reconciliation and peacemaking (de Waal 1996).

It may be asked how this differs from the quasi-moral systems already described. I maintain that the difference lies in the element of rational deliberation present in proto-moral systems but absent in quasi-moral systems. Whereas the system of gene transmission may be described as quasi-moral, genes cannot be called proto-moral. They do not deliberate, plan, or remember past transgressions. Proto-morality can occur only when an organism can begin to deliberate ends and means, experience the consequences of its actions, and modify its behavior accordingly. Proto-morality is still distinct from a genuine moral system. There are no abstract principles, no awareness of virtues to be cultivated. It is not even clear that awareness or consciousness is needed in all situations. But the resemblance is much greater than in the case of quasi-moral systems. Not only are the situations the same (e.g., the distribution of resources), but the means of resolving them are much closer. Actions can and do result in pleasure and pain. Individuals receive group censure and group approval. Ends and means are calculated. The actions of others are taken into account.

It may be noted that, although we consider ourselves to be moral agents, we often conduct foreign policy in precisely this proto-moral fashion. While the United States is strongly influenced by a Wilsonian approach to foreign policy as moral crusade (often to spread democracy, human rights, and the free market), foreign policy is frequently conducted in terms of national interest and balance of power, typified in the United States by Henry Kissinger and dominant in Europe during much of the nineteenth century and in much of the world today. This approach is even justified by its proponents, who argue that nation states cannot be treated as moral agents and that we must be pragmatic when dealing with countries such as China and North Korea. In this approach, foreign policy is conducted in terms of self-interest and mutual benefit. Risks are evaluated, and those who go beyond the pale are punished. Chimpanzee politics may lack the subtlety of international politics, but the same principles often apply.

Genuine Morality

Genuine morality, composed of the type of considerations set forth at the beginning of this paper, requires the kind of capacities beyond those needed for proto-morality. Genuine morality is practiced only by persons, fully integrated individuals that, for the most part, we only get a hint of in the nonhuman animal world. Persons are capable of abstraction and symbolic expression. The human facility for language instantly separates us from all other animals (language-trained apes being the exception that proves the rule). Our capabilities for abstraction of moral principles potentially set our deliberations above mere calculations of self-interest. Our rich emotional life provides a repertoire of moral instincts that likely do not have equivalents in the animal world. I doubt that gazelles feel guilt or that salamanders feel shame. While often ignored, these rich emotional states are subtly linked to our decision-making process (see, for example, Damasio 1994).

The claim for uniqueness can be overstated (for a corrective, see Peterson 1999), but at the very least, the human experience in its full form represents a new, emergent level of cognition. This does not deny our strong links with the biological community and our evolutionary heritage, but it does put them in proper perspective. It also is the wedge for understanding the weakness of traditional sociobiological accounts of human morality.

Sociobiologists, in their effort to sweep away what they consider to be the disorganized clutter of the social sciences, seek to explain both moral behavior and religion in terms of biological principles and genetic influence. Because human beings are biological organisms, we too are subject to natural selection, and it is only natural to suspect that our drives and values are dictated by our genes. Michael Ruse presents a fairly typical attitude of sociobiologists when he writes, "Morality, more strictly, our belief in morality, is merely an adaptation put in place to further our reproductive ends" (Ruse and Wilson 1985, 51–52). Morality is a biological adaptation, like opposable thumbs and bipedal locomotion. Notions of good and evil are illusions fobbed on us by our genes. In the end, moral law must bend to genetic law, and genetic law reduces to the principle of survival of the fittest. For sociobiologists, then, human altruism is best explained in terms of kin selection theory and reciprocal altruism. Religious and philosophical ideas of "the good" and of "moral law" can, as a consequence, be disposed of.

While this sort of argument has many faults, which have been amply pointed out, I believe that a crucial point is often missed. While human morality shares basic elements with quasi-moral and proto-moral systems found elsewhere in nature, our nature as cognitive, cultural agents requires a different sort of analysis. Human beings have a behavioral flexibility not found anywhere else in nature. We are capable of inhabiting virtually every biome on the planet. Because of our abilities to process information, to weigh alternatives, to remember, and to abstract, the range of our biological drives is insufficient to determine our behavior. It is notable that evolutionary biologists, in their search for the evolutionary roots of cognition, emphasize the universal quality of human nature and human drives, requiring them to eschew differences among human cultures as being insignificant. Thus, to give but one example, studies have been done indicating a universal preference of males for "young, nubile women," while women universally prefer older, wealthier men (Buss 1992). While this may turn out to be true, it stops well short of explaining what many people actually do. Many marry within their own age range, matters other than wealth and beauty often come into play, and some choose to remain celibate, often for religious reasons.

In short, any explanation of human behavior is incomplete without culture. More important, any explanation of human moral behavior is incomplete without the kind of worldview that culture provides. Often, we call this aspect of culture "religion."

This, then, is what sets off genuine morality from quasi-moral and proto-moral systems. Quasi-moral and proto-moral systems do not require a global framework that guides decision making. They are always proximate and pragmatic. In these systems, there is no long-term goal or ideal state to be achieved. Yet, genuine morality is virtually inconceivable without such conceptions. As already explored, conceptions of human nature and cosmic nature are part and parcel of any moral system. This would suggest that religion is a necessary part of any embodied moral system, for a primary task of religion is to provide an orientation to cosmic and human nature. By specifying what ends are desirable and attainable, religions function to provide a framework in which moral action takes place. At the same time, the means for attaining those ends arise in the kind of situations already specified in quasi- and proto-mortal systems. Thus, despite the differences, there are interesting links as well.

It seems to me that a number of interesting consequences follow from such a view. First, it is worth remarking that, in this analysis, any sociobiological account of human morality is incomplete. Sociobiology, game theory, and behavioral genetics each has a contribution to make to our understanding of human nature. Analyses of quasi- and proto-morality also enrich our understanding of the contexts and conditions in which human moral action can take place. Genuine morality, however, cannot be reduced to sociobiology, because sociobiology operates out of a framework that moral systems either presume or compete with. When Michael Ruse, for instance, argues for an evolutionary ethic, evolution has ceased to be simply a scientific hypothesis; it has become a religious one as well.

The claim that religion is profoundly entwined with moral action also has consequences. The truth of this claim is variously taken to be trivial or controversial. It seems trivially true that religious belief affects how people behave. Religious justification undoubtedly has played an important role in the persistence of the caste system in India as well as in the high incidence of vegetarianism among Buddhists. It is trivially true that Christianity as a worldview has supported patriarchy as well as humanitarian efforts on behalf of the poor or the sick. It would be more controversial, however, if one claimed that Christianity entailed patriarchy or that holding a Christian worldview entailed a higher incidence of helping the poor. It is, after all, an academic truism that Christians (or Buddhists, or Muslims) are no different from anyone else. For every good Christian there is a good atheist, and for every bad agnostic there's a religious believer somewhere of equal notoriety. Or is there?

Either we must admit that the ideas of cosmic and human nature do make a difference in moral action or we must admit that culture adds little to moral action and arises, rather, out of evolutionary considerations, as sociobiologists claim. If we admit that such basically religious ideas do make a difference, then we must also admit that different religious systems impact moral behavior differently. We might even be able to evaluate the different moral impact such religious systems have, both for good and for ill. We might even be able to measure such differences.

These last two statements will no doubt make many religion scholars nervous, for they bring up a host of bad memories of interdenominational and interreligious polemics. No one wants to return to the horrid nineteenth-century attempts to rank religions according to their respective worth and truth value. Fortunately, we do not need to, for unlike many nineteenth-century scholars we recognize that we cannot simply speak of "the Christian worldview" or "the Hindu worldview" or "the Muslim worldview." We now recognize that religious traditions are highly symbolic in character, diverse, and malleable over time. There can be no comparisons of religious traditions *in toto*, only comparisons of living incarnations of those religious traditions, and usually any comparison that takes place will not be between religious traditions but within different versions of religious traditions.

Of course, in one sense, this is not anything new. Feminist scholarship has catalogued in detail how the ideas concerning the nature of women have impacted the real lives of women over the ages. While fundamentalists fulminate over the culture wars, liberals such as Bishop John Spong claim that Christianity must change or die. Indeed, the reason most of us became scholars in the first place is because of our conviction that ideas do, in fact, change lives. What is needed, however, is for such discussions to move into the next, more sophisticated stage, a stage that includes an understanding of the roles that biology and cognition play.

REFERENCES

Axelrod, Robert. 1984. *The Evolution of Cooperation.* New York: Basic Books.

Burkert, Walter. 1996. *Creation of the Sacred: Tracks of Biology in Early Religions.* Cambridge: Harvard Univ. Press.

Buss, David M. 1992. "Mate Preference Mechanisms: Consequences for Partner Choice and Intrasexual Competition." In *The Adapted Mind: Evolutionary Psychology and the Generation of Culture,* ed. Jerome H. Barkow, Leda Cosmides, and John Tooby. New York: Oxford Univ. Press.

Cheney, Dorothy L., and Robert M. Seyfarth. 1990. *How Monkeys See the World: Inside the Mind of Another Species.* Chicago: Univ. of Chicago Press.

Damasio, Antonio. 1994. *Descartes' Error: Emotion, Reason, and the Human Brain.* New York: Avon Books.

Dawkins, Richard. 1989. *The Selfish Gene,* 2d ed. New York: Oxford Univ. Press.

De Waal, Frans. 1996. *Good Natured: The Origins of Right and Wrong in Humans and Other Animals.* Cambridge: Harvard Univ. Press.

Donald, Merlin. 1991. *Origins of the Modern Mind: Three Stages in the Evolution of Culture and Cognition.* Cambridge: Harvard Univ. Press.

Ghiselin, Michael T. 1974. *The Economy of Nature and the Evolution of Sex.* Berkeley: Univ. of California Press.

Midgley, Mary. 1995. *Beast and Man: The Roots of Human Nature.* New York: Routledge.

Peterson, Gregory R. 1999. "The Evolution of Consciousness and the Theology of Nature." *Zygon: Journal of Religion and Science* 34 (June): 283–306.

Pinker, Steven. 1997. *How the Mind Works.* New York: Norton.

Rolston, Holmes, III. 1999. *Genes, Genesis, and God: Values and Their Origins in Natural and Human History.* Cambridge: Cambridge Univ. Press.

Ruse, Michael, and Edward O. Wilson. 1985. "The Evolution of Ethics." *New Scientist* 108:50–52.

Smuts, Barbara. 1999. *Sex and Friendship in Baboons.* Cambridge: Harvard Univ. Press.

Trivers, R. L. 1971. "The Evolution of Reciprocal Altruism." *Quarterly Review of Biology* 46:35–57.

Wilson, Edward O. 1975. *Sociobiology: The New Synthesis.* Cambridge: Harvard Univ. Press.

[15]

"O Lord . . . You Perceive my Thoughts from Afar": Recursiveness and the Evolution of Supernatural Agency*

Jesse M. Bering** and Dominic D.P. Johnson***

ABSTRACT

Across religious belief systems, some supernatural agents are nearly always granted *privileged epistemic access* into the self's thoughts. In addition, the ethnographic literature supports the claim that, across cultures, supernatural agents are envisioned as (1) incapable of being deceived through overt behaviors; (2) preoccupied with behavior in the moral domain; (3) punitive agents who cause general misfortune to those who transgress and; (4) committed to an implicit social contract with believers that is dependent on the rules of reciprocal altruism. The present article examines the possibility that these factors comprise a developmentally based, adaptive information-processing system that increased the net genetic fitness of ancestral human beings living within complex social groups. In particular, the authors argue that fear of supernatural punishment, whether in this life or in the hereafter, encouraged the inhibition of selfish actions that were associated with "real" punishment (and thus real selective impairments) by actual group members.

KEYWORDS

Evolutionary theory, theory of mind, cognitive development, morality, intentionality, cooperation, reciprocal altruism.

*Acknowledgements: We wish to thank Harvey Whitehouse, Robert McCauley, Joseph Bulbulia, Todd Shackelford, and David Schroeder for helpful comments on an earlier draft.

**Department of Psychology University of Arkansas. Correspondence to: Jesse M. Bering, Department of Psychology, University of Arkansas, Fayetteville, AR 72701; Email: jbering@uark.edu; Phone (479) 575-3489; Fax: (479) 575-3219.

***Center for International Security and Cooperation Stanford University.

When you close your doors, and make darkness within, remember never to say that you are alone, for you are not alone; nay, God is within, and your genius is within. And what need have they of light to see what you are doing?

Epictetus

O Lord, you have searched me and you know me. You know when I sit and when I rise; you perceive my thoughts from afar, you discern my going out and my lying down, you are familiar with all my ways.

Psalm 139

I will govern my life and thoughts as if the whole world were to see the one and read the other, for what does it signify to make anything a secret to my neighbor, when to God, who is the searcher of our hearts, all our privacies are open?

Seneca

If only God would give me some clear sign! Like making a large deposit in my name in a Swiss bank.

Woody Allen

Mental states are highly labile and volatile abstract entities, comprised of all the self's continually changing intentions, desires, beliefs and emotions that are often responsible for its observable behaviors. Although inferences regarding others' mental states are frequently plagued with errors, the human mind is expert at taking into account the information that affects others' subjective views of the world (Baron-Cohen 1995; Dennett 1987; Lillard 1998). Because having causal knowledge of others' behaviors may be adaptive in that it enabled ancestral individuals to control events that had important consequences for one's own genetic fitness (e.g., through implanting false beliefs, repairing false beliefs, manipulating emotions), such a mental representational system might have been subjected to intense selective pressures in evolutionary history (Cosmides & Tooby 2000; Tomasello 1999).

However, it is hard to fathom *any* evolutionary conditions that would have given rise to a completely infallible mental representational system, one in which agents could literally read the minds of social others (indeed, the selection pressures on mechanisms to deceive others were likely as strong as those for detecting them). Rather, we bear witness to the overt actions and consequences of what we presume are other people's mental states, but we cannot perceive others' mental states directly. As the philosopher Bertrand Russell (1948) pointed out in his exposition of solipsism, we cannot be certain that other minds exist at all – all we can do is logically infer that others experience as we do. Philosophical arguments aside, an observer can attempt to explain the self's overt behaviors and may be fairly successful in predicting its future actions, but only the self enjoys the conscious derivatives of its neural systems and gains *privileged epistemic access* to its own mental states (e.g., Damasio 1999; Humphrey 1992).

Privileged Epistemic Access and Supernatural Agents

Perhaps, however, there are specialized culturally-postulated agents that also enjoy such access. Enter supernatural agents. Although there may currently be insufficient ethnographic data to formally test the hypothesis, we submit that a central component of religious systems are concepts of supernatural agents that have privileged epistemic access to the self's mental states. The idea is a prominent one in Judeo-Christian thought – both in formal theology ("Before a word is on my tongue you know it completely, O LORD." Psalm 139) and in folk conceptions of God's omniscience. Although in any culture there may be a host of deities, spirits, and demons populating a common religion, religious scholars would be hard-pressed to find a religious system that does not have within its ranks some supernatural agent that, among other impressive feats, is envisioned as *knowing*, rather than merely inferring from observable behaviors, the self's true intentions. Consider Pettazzoni's (1955, p. 20) analysis of this topic:

> Divine omniscience has another field of activity; besides the deeds and besides the words of mankind, it examines even their inmost thoughts and secret intents. In the prophecies of Jeremiah we are told that the Lord tries "the reins and the heart" (Jer. Xi, 20). The same thought is found among many other peoples, savage and civilized. Karai Kasang, the Kachin Supreme Being, "sees" even what men think. The Haida say that everything we think is known to Sins sganagwa. The Great Manitu of the Ankara knows everything, including the most secret thoughts. Tezcatlipoca knows men's heart; Temaukel, the Supreme Being of the Ona-Selknam, knows even our thoughts and most private intentions. In Babylonia, the god Enlil knows the hearts of gods and men, and Shamash sees to the bottom of the human heart. Zeus likewise knows every man's thought and soul.

Such cross-cultural and cross-temporal evidence raises important questions about the relations between cognitive development and religious concept acquisition and also about the evolutionary history of supernatural causal beliefs. If implicitly granting some supernatural agents privileged epistemic access is in fact a culturally recurrent phenomenon, then perhaps it has played a more important role in human social cognition than psychologists have thus far bothered to consider (Bering, in press). It is striking that epistemic access seems both universal among cultures and unique within them (supernatural agents are the *only* ones to be attributed such powers).

The psychological mechanisms that underlie representations of supernatural omniscience are seemingly dependent on social-cognitive and motivational factors that emerge during development. Young children may not automatically perceive supernatural agents as having a direct link to all their hidden thoughts, at least not in the same manner older children and adults do. This suggestion may strike some readers as implausible because young children frequently are portrayed as operating on the basis of an egocentrism that should lead them to view supernatural agents as sharing their own private perspectives (*cf.* Bovet 1928). Recent findings by Barrett, Richert, and Driesenga (2001), for example, show that 3-year-olds who have not yet developed a comprehensive belief-desire psychology reason that God does not harbor false beliefs and does not suffer from the mundane epistemological rule that seeing leads to knowing. According to the authors, the fact that children this age also reason about natural agents in the same way does not compromise their controversial position that "children may be better prepared to conceptualize the properties of God than for understanding humans" (Barrett et al. 2001, p. 60). This is because although children's conceptions of human beings change with the emergence of a representational theory of mind, allowing them to view others as being susceptible to false beliefs, children's "theologically correct" conception of God's infallible knowledge is in place from the start.

However, although these findings convincingly demonstrate that young children overextend their knowledge to other agents, both natural and supernatural, the data may not display children's cognitive readiness for religious concepts as the authors wish to argue. This is because notions of supernatural omniscience are inextricably tied up with moral sanctions; they seldom, if ever, are associated with amoral questions such as the ones that are at issue in the study by Barrett et al. (2001) (e.g., "If God wanted to show you some crackers, what would God show you the inside of?" "What does God see inside the darkened box?"; Atran 2002; Boyer 2001; Hinde 1999; Reynolds & Tanner 1995). Rather, even a modest glance at the comparative religion literature shows that people are much more concerned with what supernatural agents know about their moral behaviors than they are about anything else. Following Boyer (2001), we may say in theory that people implicitly assume that God knows what is inside of their refrigerators or that the cat is hiding in the attic, but unless the refrigerator contains the severed head of an enemy or the cat is hiding in the attic because it is afraid of its abusive owner, it is unlikely that people would ever stop to consider whether God knows these trivial facts.

Strategic Social Information and the Moral Domain

When dealing with the attribution of epistemic access to the self's private mental states, what is really at stake is the strategic information that can be barred from social others through deceptive behaviors but that is envisioned to be transparent to special agents such as ancestral spirits and gods. From an evolutionary perspective, strategic information is that which must be fully retained from public exposure or only selectively shared with specific social others because of its capacity to interfere with or facilitate genetic fitness (Bering & Shackelford, in press; Dunbar 1993). Because social adaptations are ontogenetically "fixed" to fit the demands of particular socioecological conditions, what is considered to be strategic information is somewhat variable across cultures. Nonetheless, developmental psychologists have shown that children are intuitively equipped to differentiate *moral imperatives* (e.g., thou shalt not hit or steal), which protect the mutually agreed upon rights of social others, from *social conventions* and customs (e.g., one should remove one's hat upon entering a church; it is impolite to eat spaghetti with one's hands), which deal with expectations for cultural behaviors (for a recent review, see Turiel 2002).

Generally speaking, the former are acquired earlier and with much less effort than the latter, which because they are more culturally versatile come only with enculturation and training. In addition, breaching moral rules is accompanied by feelings of guilt and shame on the part of the child and harsher punishment on the part of the parents, whereas violations of social conventions, while they may cause embarrassment, are handled with more tolerance by both the child herself and by her parents (see also Gilbert 1998; Tangney 2001). They are far less consequential for both the perpetrator and "victim," and only loosely related to fitness.

Thus, it is typically information dealing with moral breaches that is the stuff of secrets and social anxieties. Nevertheless, it oftentimes requires more effort for both children and adults to obey moral imperatives than to participate in social conventions because impinging on the rights of others often means immediate gains in resources and enhanced opportunities for the self. Yet because the benefits of living in social groups significantly outweigh the disadvantages, desires to engage in selfish behaviors (e.g., stealing from others, aggressing against enemies, sexual coercion, and so on) that can reap immediate rewards must be kept in check by adaptations for group living (Alexander 1987; Axelrod & Hamilton 1981; Hamilton 1964; Humphrey 1976; Trivers 1981; Williams 1992).

The Problem of Other Human Minds

Unlike other species, the intentionality system, which made the human organism aware of the existence of other minds and therefore highly sensitive to what others "know" or "do not know" about the self, rendered humans' reproductive success contingent upon the opinions of others (for a general comparative review of the intentionality system, see Povinelli, Bering & Giambrone 2000). As such, human behaviors have come to bear little resemblance to actual human desires. For members of other social species, such as most nonhuman primates, whether such short-term strategies are deployed appears solely a function of who is in the immediate environment. When faced with dominant social others or those who might recruit dominant others to the scene through various alarm calls, nonhuman primates tend to inhibit selfish actions that can lead to direct punishment (e.g., see Cheney & Seyfarth 1990). However, when such conditions are not present, many primate species capitalize on their surroundings and rape and plunder with equal ease. This makes sense in light of the absence of an intentionality system in other species – a system that functions in large part to allow individual group members to collect, retain, and share strategic social information. Without such a system in place, strategic decisions should resemble those a thief makes when refraining from his thieving because there is a Doberman Pinscher in the room, but paying no mind to the cat on the sofa.

In the short run, stealing from a neighbor may be an adaptive decision – we may use the stolen money, for instance, to buy ourselves expensive jewelry and thus capture the attention of potential mates by falsely advertising our available resources. In the long run, however, risking the possibility of detection and the consequent social exclusion and tarnishing of our reputation by engaging in this short-term reproductive strategy makes this a very poor decision indeed (e.g., see Schelling 1960; Johnson, Stopka & Knights 2003; Ketelaar & Au 2003; Frank 1988; Wedekind & Malinski 2000). This is not to say that we do not *want* things that are not ours, and it certainly does not mean that we are not tempted to cheat. Rather, such impulsive desires are merely tempered – usually – by human intelligence, which weighs in on the costs and benefits of risky moves. Any psychological mechanism capable of aiding our ancestors in escaping ancient social adaptations that evolved prior to the emergence of the human intentionality system would have therefore been adaptive because it was capable of striking a careful balance between the old and the new.

Supernatural Agents Helped Individuals Cope with the Problem of Other Human Minds

At first blush, it might not appear that concepts of supernatural agents endowed with privileged epistemic access would provide any aid. For example, of what evolutionary significance is it that the gods and spirits are privy to all the perfidious and felicitous motives that happen to pop into our heads, so long as our overt behaviors are maladaptive or adaptive in the eyes of the group? Our social success, after all, is defined by our treatment by other group members, who have real epistemic limitations, not in any veridical sense by what culturally postulated supernatural agents "know" about us. Therefore, the fact that supernatural agents are seen as having privileged epistemic access, in and of itself, does not pose much of a threat because these agents have no direct means by which to communicate potentially damaging social information about the self with other ingroup members.

However, the fact that they cannot gossip or, for that matter, the fact that they may not even exist does not make supernatural agents perceived to be any less dangerous. On the contrary, they are the source of tremendous worry and fear because they are considered to have the unique power to cause general misfortune for social transgressors, and this was likely the case in the ancestral past just as it is in contemporary times. Following W.I. Thomas' Dictum, people's behavior is explained not so much by what is real, but rather by what they believe is real. Johnson and Krüger (2004) have argued that the threat of supernatural punishment, whether in this life or in the hereafter, induces cooperation because religious beliefs often serve as literal truths, and this deterrent effect was likely especially strong for societies for whom many natural phenomena remained inexplicable. Supportive evidence for this line of reasoning comes from a recent study by Roes and Raymond (2003), who demonstrated that group size is positively correlated with the presence of "moralizing gods," supernatural agents interested in human moral affairs and who adjudicate upon such matters through vengeance toward the disobedient. As evolutionary psychologists, however, we must explain not only the theoretical biology underlying adaptive processes, but we must also understand the cognitive hardware – the information-processing systems – that are designed to engage organisms in adaptive behaviors. Moralizing gods can only find their way into large social groups insofar as individuals are capable of envisioning these gods as enforcing their morals through the occurrence of positive and negative events. A moralizing god who fails to "communicate" with its followers would not be a very effective one.

This is not an entirely new theoretical concept. In his book *Primitive Law*, the early anthropologist Hartland (1924) wrote that "the general belief in the certainty of supernatural punishment and the alienation of the sympathy of one's fellows generate an atmosphere of terror which is quite sufficient to prevent a breach of tribal law (p. 214)." Social ostracism may have devastating fitness consequences within socially dependent hunter-gatherer cultures.[1] What is novel about our current thinking on this topic

[1]There is reason to believe that religious commitment on the part of individual group members serves as an important signal of commitment to the group more generally. For

is that it attempts to reconcile such recurrent causal beliefs with the fundamentals of evolutionary biology and natural selection. If such current thinking is able to explain group processes by focusing on the adaptations of individual group members, or more specifically by concentrating on the level of the gene, it may lead to the most successful account of supernatural causal beliefs to date.

Supernatural Agents and Moral Regulation: Ethnographic Literature

Although it is impossible to cover all such beliefs in the present article, following is a very small sampling of punitive religious causal beliefs as reported in the Human Relations Area Files (eHRAF). For example, among the Ndyuka of South America, "all misfortunes afflicting the community, ranging from illness and death to scarcity of game and poor harvests, are due to the withholding of divine favor occasioned by sin" (van Velzen & van Wetering 1988, p. 197). For the Chuuk of Oceania, death and illness are almost always attributed to supernatural causes. "True, the bite of a fish may be the actual cause of your death, but then [one] would not have been bitten, says the native, if women had not been along on the fishing trip" (Fischer 1950, p. 52). Among the Lao Hmong, much human illness and injury was attributed to "the wrathful punishment of an ancestral spirit for social impropriety" (Scott 1986 [1990], p. 99). For Ugandans around the turn of the last century, "cases cited of behavior which was liable to anger a spirit were failure to pay a debt or to make some gift, particularly in connection with marriage ceremonies" (Mair 1934, p. 229). The Bemba of Zambia "are deeply convinced that relatives who die with a sense of injury have the power to return and afflict the living with misfortune, illness, and even a lingering death from a wasting disease ... there is no doubt that the fear of supernatural punishment is a very strong sanction enforcing the sharing of food and the provision for

example, Atran (2002) argues that religious behaviors act as a "green beard" advertising to others that the individual is unlikely to defect from the group or become prone to jeopardizing the genetic fitness of other members (see also Bulbulia 2003; Sosis 2003). Thus, for someone to reject the notion of supernatural punishment should raise a red flag that such a person is more likely than a religious adherent to threaten the genetic fitness of other ingroup members by engaging in morally proscribed activities.

dependents" (Richards 1939, p. 199). Associating moral violations with the subsequent occurrence of physical illness is a pattern of causal belief that appears among many traditional peoples. For example, "The notion that a human being may be struck by enchantment or sorcery is quite common in America. The conception that the disease is caused by transgression of a taboo is found among the Eskimo, Athabascan, Ge, and Tupi peoples, and within the high cultures, among others. The disease is often abolished after the patient's 'confession' of the taboo offense to the medicine man" (Hultkrantz 1967, p. 88).

In many societies, not only does supernatural punishment fall upon the heads of the wicked, but is also sanguineous in nature and ostensibly disastrous for fitness because supernatural agents are often seen as unforgiving and merciless, inflicting lasting and far-reaching punishments. That is, perhaps the worst punishment of all would be to have biological relatives, including one's own children, cursed for the self's misdeeds. Thus, in many cultures, supernatural punishments extend down the generations. For the Lepcha of Asia, "anti-social acts are graded by the number of people they may affect; only for acts of minor importance is there personal and individual punishment which falls on the evil-doer" (Gorer 1938, p. 183). The threat of calamity affecting one's own offspring for criminal behavior is a particularly recurrent theme and is illustrated very clearly in the following brief passage on the Pagai from a Dutch missionary publication. "A missionary once acted emphatically against various prohibitions in order to demonstrate their inefficacy. Actually this made a totally wrong impression on the people because they said: 'The man knows perfectly well that he himself won't be punished but that the punishment will fall on his children'" (Anonymous 1939, p. 9). The Okinawans similarly believed that "the group exists in time as well as space; current living generations are centrally placed on a continuum extending from the earliest ancestors through generations as yet unborn. Accountability, in the final analysis, encompasses the entire range of the collectivity through time; thus, a child may suffer punishment for the action of his parents or ancestors" (Lebra 1966, p. 42). Because human behaviors are unconsciously motivated by genetic interests, individuals should have evolved to be motivated to refrain from any behaviors that are believed to threaten inclusive fitness (i.e., the representativeness of one's genes in current and future generations) (Alexander 1987; Dawkins 1976; Trivers 1981).

In addition, if threats of supernatural punishment indeed serve to socially regulate group norms, foster cooperative behaviors, and discourage antisocial behaviors, then it should be salient in (and thus promoted by) parenting practices as well. In that parents and other biological kin are heavily invested in the reproductive success of offspring, they should be highly concerned about children's ability to abstain from engaging in socially disapproved actions, particularly behaviors deemed immoral by the group. Failure to indoctrinate a child in this regard may ultimately contribute to the offspring's antisocial behaviors and, as a result, various forms of social exclusion, both for the child and possibly for biological relatives as well. As in the foregoing analysis, there are negative consequences for inclusive fitness, here as a result of the offspring's misdeeds rather than those of the individual him- or herself.

To find evidence of parental threats of supernatural punishment, we need only take a cursory look at the ethnographic data once again. The Delaware Indians of North America, for example, informed their children that "supernatural powers punished disobedient children by causing them to become weak and sickly" (Newcomb 1956, p. 34). A formerly delinquent Hopi recalled the corporal punishment he received as a boy. "His parents beat him, held him over a smoky fire, threatened to leave him in the dark to be carried off by a white, or by a Navaho, or a coyote. His grandfather whipped him twice. His father's brothers stood ready to punish him at the request of either parent" (Aberle 1951, p. 33). And still, "it was the threats of supernatural punishment that were considered most frightening" (p. 33).

Intuitive Moral Contracts with Supernatural Agents

Deontic expressions of how one *ought to behave* and what one *should do* are deeply embedded within the world's social frameworks and are often believed to stem directly from the expectations of the gods. Precisely *why* the gods should want us to engage in particular behaviors has never been the subject of much scrutiny by practitioners of religion. Rather, the gods simply have their expectations for human behaviors – it is their policies that we follow, and to transgress is to directly challenge their authority. In that the relationship is an entirely social one, the gods must therefore seek retaliation when confronted by moral failings. The relationship appears to function via the basic principles of reciprocal altruism, when people live up to their end of the bargain they expect to be rewarded with a good life. In other words, prosocial actions, or actions that foster cooperation between in-group members and generally help grease the wheels of social harmony, should lead actors to have expectations for positive life events. Many religions are founded (and recruit members) on the basis of the carrots of reward, as well as the sticks of punishment.

The belief in this just worldview is so strong, in fact, that among many groups personal calamities and hardships are taken as evidence that the individual must have done something horribly wrong. Often the only suitable remedy for these hardships is spiritual extirpation by way of public confession. For the Igbo of Nigeria, "adultery by a wife is regarded as bringing supernatural punishment upon herself and her husband ... thus if a woman experiences difficult labor, it is assumed that she has committed adultery and she is asked to give the name of her lover in order that the child be born. If a man falls sick, his wife may be questioned as to whether she has committed adultery" (Ottenberg 1958 [1980], p. 124). The Kogi of Amazonia force individuals to confess their most unseemly and personal thoughts to the máma, or local priest. "The máma, by means of confession, can ascertain all of the crimes which [people] commit, committed, or intend to commit in his town" (Reichel-Dolmatoff 1950, p. 142), and the person is told that he or she will be felled with misfortune should they be dishonest.

An obvious selective advantage of accruing information about social others in this manner is the increased likelihood of evading threats to genetic fitness before they happen (see Bering & Shackelford, in press; Shackelford & Buss 1996). Having knowledge of a potential mate's history of alleged physical abuse against his ex-wife can help a woman to make an informed (i.e., ancestrally-adaptive) decision when this man decides to propose to her. Indeed, much like Pascal's wager, even if such information is potentially unreliable, the risks associated with ignoring these rumors should be much greater than the risks of allowing them to influence one's decision-making. In addition, in the ancestral past, possessing such information about others could have provided a considerable degree of social leverage in the context of status-striving and resource acquisition, affording power over others who feared their social exposure (e.g., blackmail). Therefore, being under the impression that others' misfortunes are diagnostic of a supernatural agent's displeasure may have contributed to the self's genetic fitness *if* this belief encouraged the person to confess his or her hidden misdeeds and, in doing so, provided the self with strategic information it otherwise would not have been afforded. Also, fitness benefits may have accrued *if* such misfortunes or actual confessions removed someone from the favored social pool – thereby leaving more potential mates and effectively raising one's own *relative* standing simply by highlighting someone else as a black sheep.

Implicit Existential Beliefs: Punishment and Expectations for Justice

Although religious systems that make such processes of supernatural punishment explicit tend to illustrate these patterns of causal reasoning most clearly, they appear just as fundamental to human behavior in the abstract (see Bering 2002). Lerner has found extensive evidence for "just world beliefs" in both religious and nonreligious samples, in which people often tacitly assume that people get what they deserve (for a review, see Lerner & Montada 1998). When these just world beliefs are apparently violated, such as when an innocent person becomes the victim of a crime, individuals seem to go to great effort to reassert their just worldviews. For example, some individuals may begin to derogate the victim and perceive the victim as instigating the crime. Children seem to display similar just world expectations – perhaps even stronger than those of adults (Stein 1973). And Piaget ([1932] 1965) argued that young children evidence a belief in "immanent justice" in which "the child must affirm the existence of automatic punishments which emanate from things themselves" (p. 251). Thus, in his classic experiment, Piaget (1932 [1965]) presented children aged 6-12 years with the story of a child who steals or disobeys and then, upon traversing a bridge, falls into the water when the bridge collapses. Nearly all (86%) of the youngest children in the study reasoned that the accident would never have happened were it not for the character's earlier misdeeds.

In addition to seeing others' misfortunes as related to their immoral behaviors, people who have violated some moral rule themselves often appear to "expect" punishment in the form of negative life events, and those guilty parties who find themselves more or less untarnished by their wrongdoing may feel as though their current happiness is undeserved. This is a common theme in literature, exemplified by the works of Victor Hugo (e.g., *Les Miserables*) and Fyodor Dostoyevsky (e.g., *The Brothers Karamazov, Crime and Punishment*). In an interesting essay, Landman (2001) describes the real life case of Katherine Power, a Vietnam War era fugitive radical who drove the getaway car in a 1970 bank heist that left a security guard – a husband and father of nine – dead. Twenty three years later, in September of 1993, Power, chronically depressed and "obsessed with a desire to be punished, to seek expiation" (Franks 1994, p. 54), turned herself in to the authorities without any provocation on the part of the FBI, who admitted having had no idea where she was all those years. Asked why she ultimately confessed, she told her lawyer that "my strongest weapon against suicide is my contract with God and my family" (Franks 1994, p. 42).

The Primacy of Intentions in the Moral Domain: Deception Fails with Supernatural Agents

Taken together, there is good reason to suppose that these various forms of causal reasoning involve endowing supernatural agents, or their fuzzier "just world" counterparts, with privileged epistemic access to the self's psychological states. This is because social psychologists have long recognized that overt actions are by their nature morally ambiguous – what distinguishes prosocial from antisocial behaviors are the intentions of the actor (Eisenberg & Fabes 1998; Loeber 1985). For instance, the act of a man holding the door open for a woman may be prosocial if the man wishes to reduce uncertainty about who is to proceed first through the door, but if it is done to reinforce the gender stereotype that women are physically inferior to men, the act would likely be considered antisocial (Hart, Burock, London & Atkins 2003). In other words, an observation of the act alone would not allow for classification of the behavior as pro- or antisocial. Rather, one would have to have some insight into the actor's motivations for holding the door open.

For certain supernatural agents, however, the actor's underlying intentions are immediately known. Unlike other people who through various forms of deception can be inveigled into thinking that the self is a more selfless character than it may in fact be, even the most skillful legerdemain will be wholly ineffective when dealing with an all-knowing god. Therefore, any supernatural punishment that arises as a consequence of antisocial actions must have its origins in the supernatural agent's knowledge of the self's truly bad intentions. A man, for instance, may be lavishing much-needed attention on a lonely widow, bringing her the happiness that she so desperately seeks; many may be blinded by these "good deeds" and unwise to his true intentions – which are to get at this lonely widow's sizeable fortune. Although this man may escape punishment from his peers, the attitude in many traditional societies is that this is inconsequential because this man will ultimately "get his" in the form of supernatural punishment. For example, in Borneo, "the Iban believe that anyone who successfully cheats another, or escapes punishment for his crimes, even though he might appear to profit temporarily, ultimately suffers supernatural retribution" (Sandin & Sather 1980, p. xxviii).

In most societies, of course, moral franchises have probably always been constructed entirely on the basis of divine will. What makes causal attributions of divine will cognitively complex is the fact that supernatural agents' desires often stand in stark contrast to those of the self; they center on the belief that the supernatural agent is aware of the self's internal psychological states. In order to see a sibling's illness as a form of punishment or a season's good harvest as a reward, one must implicitly assume that the supernatural agent knows that he or she *does not want* the sibling to become sick or *will enjoy* the prosperity of a good harvest. But then one can certainly imagine idiosyncratic situations in which these events would not be seen as necessarily bad or desirable. Perhaps one's older brother stands to inherit the family goods, and the self is next in line for this inheritance; thus the brother's unexpected illness might be less a form of punishment and more a serendipitous reward. In this case, punishment for such private filial scheming might be the sibling's astoundingly good health. Or perhaps one wishes that the harvest would be poor this year because it would give one an excuse to leave the natal group and pursue more fertile fields, which the individual has secretly wanted to do for some time. If the crops yield a good harvest, then this might be viewed as supernatural punishment for some other moral transgression.

The point is that these examples involve more than simple godly omniscience that resembles the preschoolers' egocentric bias, as Barrett et al. (2001) attempt to argue for children's reasoning about the infallible knowledge of God. If this were the case, then individuals might grow up to believe that God was complicit in their malingering – he would be their personal partner in crime. This is because, for the cognitively egocentric individual, any immoral thoughts and desires would be God's as well. If one wished to covet thy neighbor's wife, then God would personally draw up the bed sheets. But this is not the way it works. Instead, the individual has to first *possess knowledge* that its desires/actions are socially unacceptable (e.g., "I know that it's wrong to feel this way/do this deed") rather than simply *having* socially unacceptable desires/engaging in transgressions; otherwise, naiveté obtunds the expectation of supernatural punishment. Moreover, the individual must attribute such knowledge not only to him- or herself, but also to supernatural agents (e.g., "I know that it's wrong to feel this way/do this deed, and God knows that I know this too. What's more, He doesn't like it."). The heart of supernatural causal beliefs is the notion that the gods are emotionally invested in their own apparently arbitrary rules for human conduct, emotions, and desires that are often unshared by the self. And the gods, of course, are seen as being wise to this opposition of mental states.

Second-Order Mental State Representation and Supernatural Causal Beliefs

Consider the mental representational abilities that would be required to entertain such causal beliefs. To begin with, the individual must possess the capacity to attribute psychological states to a supernatural agent. Because supernatural agents are agents nonetheless, this is a relatively easy process; they are simply understood to have minds and to experience psychological states. Developmentally, this capacity to engage in first-order mental state attribution (A mentally represents B's *belief* [or *knowledge*] [or *intentions*] [or *desires*] [or *emotions*] about X ...) is present even in very young children. Starting in infancy, humans evidence an understanding that other agents' behaviors are driven by hidden goals and intentions (Meltzoff 1995; Gergely, Nadasdy, Csibra & Biro 1995). Just a few short years later, preschool-aged children build substantially on these early intentionality mechanisms, and evidence a belief-desire psychology that provides them with an understanding that others see the world differently than the self. Thus, by age 4 or 5, children pass a variety of "false-belief" and "appearance-reality" tests in which they are asked to reason how another agent will see things from a naïve point of view (see Wellman, Cross & Watson 2001; Flavell 1999). Prior to this age, children assume that because they themselves know something (for example that a sponge that looks like a rock is in fact a sponge, or that a chocolate bar was secretly moved to another location), then so too must others, despite the fact that these others are not privy to this special information. Only around preschool age do children reason that others' subjective views of the world will not conform to reality if they lack veridical information. This, of course, enables children to begin developing their deceptive aptitude – purposely implanting false beliefs in others' minds or depriving others of specific sources of information for personal benefit – and also to construct more meaningful social relationships that involve reparations of others' confused mental states by offering them useful information (Gopnik & Meltzoff 1997; Wellman 1992).

But the development of higher-order cognition does not end here. There is at least a small developmental gap between first-order theory of mind (A [self] thinks that B [other] thinks that *X*) and second-order theory of mind (A [self] thinks that B [other] thinks that C [other] thinks that *X*), which does not appear to come online until around 6 or 7 years of age (Núñez 1993; Perner 1994; Perner & Howes 1992; Sullivan, Zaitchik & Tager-Flusberg 1994). And, we argue, it is only with the appearance of a second-order theory of mind that the child begins to see the natural events occurring in his life as *meaningful* or, more specifically, as symbolic and declarative of an abstract intentional agent's desire to share social information with him. Thus, events take on the same quality of "aboutness" that is also characteristic of behaviors – here, events come to be *about* the supernatural agent's moral judgment in reference to the self's actions which are in themselves *about* the self's psychological states. Suddenly the random occurrences of the natural world become signs of a cognitively intrusive and authoritarian eye in the sky.

There is recent laboratory evidence to support the hypothesis that second-order representation is required to view natural (i.e., random) events as communicative messages from a supernatural agent. Bering (2003) had 3- to 7-year-old children play a guessing game in which they were to place their hand on top of one of two boxes that they believed contained a hidden reward. Prior to these "guessing" trials, however, children were informed of an invisible princess ("Princess Alice") present in the room with them who would "tell them, somehow, when they chose the wrong box." On several of the trials, the experimenters triggered an unexpected event (a picture falling, a light flashing on and off) precisely at the moment the child's hand made contact with one of the boxes, and children's responses to these events were recorded.

Only the oldest children – the 7-year-olds – reliably (82%) moved their hand to the opposite box after experiencing these unexpected events. Moreover, when asked why these events occurred, only the oldest children stated that Princess Alice caused the event because she was attempting to tell them where the ball was in fact hidden. Although neither the 3- or 5-year-olds in the study reliably (16% and 31%, respectively) moved their hands on the behavioral portion of the task after experiencing the unexpected events, many of the 5-year-olds verbally stated that Princess Alice caused the events to happen. However, very few children from this age group reported that she did so in order to tell them that they had their hand on the wrong box. In other words, it was as if children in the middle age group merely saw Princess Alice as an invisible woman running around in the laboratory and "making things happen" – pulling the cord to the table lamp, knocking the picture off the wall – for no other purpose than to convey her presence in the room. The oldest children, in contrast, exploited her "antics" as a source of information; to them, the light turning on and off was analogous to her pointing to the correct box, and the picture falling off the door was treated as if it were Princess Alice saying, "no, not that one, the other one." For the youngest children, the picture falling was merely "the picture falling," and the light turning on and off was merely "the light turning on and off."

Although additional data must be gathered before a strong case can be made, these findings indirectly support the idea that the ability to see natural events as symbolic and declarative of a supernatural agent's mental states is dependent on a second-order theory of mind. In order to see the events in the study as Princess Alice's way of sharing strategic information about the hidden ball, children must view this supernatural agent as viewing their own behavior (i.e., placing their hand on the "wrong" box) as being caused by a state of ignorance. In light of previous findings showing that the ability to engage in such recursive thinking about psychological states develops at about age seven (e.g., Perner & Howes 1992; Sullivan et al. 1994) the second-order hypothesis fits these data well.

Concluding Remarks

Because divinely imposed rules are designed to disarm the (more selfish) psychological mechanisms that were adaptive for human ancestors long before the intentionality system came along, they are not always so very easy to follow. As stated earlier, social success requires that those ancient, but still present, heuristics designed to maximize genetic success for the short-term must be effectively thwarted by more novel, long-term inhibitory strategies designed to preserve and enhance social reputation. Therefore, if belief in supernatural agents with whom the self holds an implicit social contract increases behavioral inhibition under conditions whereby the self underestimates the likelihood of detection by actual ingroup members, then the gods themselves are slaves to genes. In that deities and ancestral spirits are so often the purported authors of human morality, H.L. Mencken (1949) preceded us in evolutionary theory when he stated that "conscience is the inner voice that warns us somebody may be looking." These arguments are somewhat difficult to see, perhaps, from an egocentric perspective, but the Western notion of God is no exception to the rule. Consider the biblical passage that "whosoever looketh on a woman to lust after her hath committed adultery with her already in his heart" (Matt. 528) in relation to the foregoing arguments. Any moralizing supernatural agency is but one more expression of an ancestrally adaptive psychological mechanism that was explicitly designed to cope with the sudden awareness that other minds in the community are keeping careful tabs on the self's actions in the moral domain.

In conclusion, if such beliefs are reflective of a true psychological adaptation, then barring any atypical developmental experiences that alter the expression of the human cognitive phenotype, it should be next to impossible to entirely lose the feeling that the self has a private audience that is intimately associated with its most secret thoughts. We have inherited the general template for religiosity because those early humans who abandoned the prospect of supernatural agents, or who lacked the capacity to represent their involvement in moral affairs, were likely met with an early death at the hands of their own group members, or at least reduced reproductive success. Those who readily acquiesced to the possibility of moralizing gods, and who lived their lives in fear of such agencies, survived to become our ancestors.

REFERENCES

ABERLE, D.F.

1951 *The psychosocial analysis of a Hopi life-history*. Berkeley, CA: University of California Press.

ALEXANDER, R.

1987 *The biology of moral systems*. New York: Aldine de Gruyter, 110.

ANONYMOUS

1939 *Wolanda Hindia. Zendingstijdschrift voor jongeren*, Vol. 13. Amsterdam: Liebenswürdige Wilden.

ATRAN, S.

2002 *In gods we trust: The evolutionary landscape of religion*. Oxford: Oxford University Press.

AXELROD, R. & W.D. HAMILTON

1981 The evolution of cooperation. *Science* 211, 1290-1296.

BARON-COHEN, S.

1995 *Mindblindness: An essay on autism and theory of mind*. Cambridge, MA: MIT Press.

BARRETT, J.L., R. RICHER & A. DRIESENGA

2001 God's beliefs versus mother's: The development of nonhuman agent concepts. *Child Development* 72, 50-65.

BERING, J.M.

2002 The existential theory of mind. *Review of General Psychology* 6, 3-24.

August, 2003 *On reading symbolic random events. Children's causal reasoning about unexpected occurrences*. Paper presented at the Psychological and Cognitive Foundations of Religiosity Conference, Atlanta, GA.

(in press) The evolutionary history of an illusion: Religious causal beliefs in children and adults. In B. Ellis & D. Bjorklund (Eds.), *Origins of the social mind: Evolutionary psychology and child development*. New York: Guilford Press.

BERING, J.M. & T. SHACKELFORD

(in press) The causal role of consciousness: A conceptual addendum to human evolutionary psychology. *Review of General Psychology*.

BOVET, P.

1925 *Le sentiment religieux*. Paris: E. Delachaux et Niestle.

BOYER, P.

2001 *Religion explained: The evolutionary origins of religious thought*. New York: Basic Books.

BULBULIA, J.

August, 2003 *Religious costs as adaptations that signal altruistic intention.* Paper presented at the New England Institute 2003 Conference on Religion, Cognitive Science, and Evolutionary Psychology. Portland, ME.

CHENEY, D. & R. SEYFARTH

1990 *How monkeys see the world.* Chicago: University of Chicago Press.

COSMIDES, L. & J. TOOBY

2000 Consider the source: The evolution of adaptations for decoupling and metarepresentations. In D. Sperber (Ed.), *Metarepresentations: A midtidisciplinary perspective* (pp. 53-115). New York: Oxford University Press.

DAMASIO, A.R.

1999 *The feeling of what happens: Body and emotion in the making of consciousness.* New York: Harcourt Brace & Company.

DAWKINS, R.

1976 *The selfish gene.* Oxford, UK: Oxford University Press.

DENNETT, D.

1987 *The intentional stance.* Cambridge, MA: Bradford Books/MIT Press.

DUNBAR, R.I.M.

1993 Coevolution of neocortical size, group size and language in humans. *Behavioral and Brain Sciences* 16, 681-735.

EISENBERG, N. & R. FABES

1998 Prosocial development. In N. Eisenberg (Ed.), *Handbook of child psychology* (5th ed.). Vol. 3. *Social, emotional, and personality development* (pp. 779-840). New York: Wiley.

FISCHER, A.M.

1950 *The role of Trukese mother and its effect on child training.* Washington, DC: Pacific Science Board.

FLAVELL, J.H.

1999 Cognitive development: Children's knowledge about the mind. *Annual Review of Psychology* 50, 21-45.

FRANK, R.H.

1988 *Passions with reason: The strategic role of the emotions.* New York: W.W. Norton.

FRANKS, L.

1994 June 13. The return of the fugitive. *The New Yorker,* 40-59.

GERGELY, G., Z. NÁDASDY, G. CSIBRA & S. BIRÓ

1995 Taking the intentional stance at 12 months of age. *Cognition* 56, 165-193.

GILBERT, P.

1998 What is shame? Some core issues and controversies. In P. Gilbert and B. Andrews (Eds.), *Shame: Interpersonal behavior, psychopathology, and culture. Series in affective science* (pp. 3-38). New York: Oxford University Press.

GOPNIK, A. & A.N. MELTZOFF

1997 *Words, thoughts, and theories.* Cambridge, MA: MIT Press.

GORER, G.

1938 *Himalayan village: an account of the Lepchas of Sikkim.* London: Michael Joseph, Ltd.

HAMILTON, W.D.

1964 The genetical theory of social behavior. *Journal of Theoretical Biology* 7, 1-52.

HART, D., D. BUROCK, B. LONDON & R. ATKINS

2003 Prosocial tendencies, antisocial behavior, and moral development. In A. Slater and G. Bremner (Eds.), *An introduction to developmental psychology* (pp. 334-356). Maiden, MA: Blackwell.

HARTLAND, S.E.

1924 *Primitive law*. London: Methuen.

HINDE, R.A.

1999 *Why gods persist: A scientific approach to religion*. London: Routledge.

HULTKRANTZ, A.

1967 *The Religion of the American Indians*. Berkeley: University of California Press.

HUMPHREY, N.K.

1976 The social function of intellect. In P.P.G. Bateson & R.A. Hinde (Eds.), *Growing points in ethology* (pp. 303-317). Cambridge: Cambridge University Press.

1992 *A history of the mind: Evolution and the birth of consciousness*. New York: Simon & Schuster.

JOHNSON, D.D.P., P. STOPKA & S. KNIGHTS

2003 The puzzle of human cooperation. *Nature* 421, 911-912.

JOHNSON, D.D.P. & O. KRÜGER

2004 The Good of Wrath: Supernatural Punishment and the Evolution of Cooperation. *Political Theology* 5, 159-176.

KETELAAR, T. & W.T. AU

2003 The effects of feelings of guilt on the behaviour of uncooperative individuals in repeated social bargaining games: An affect-as-information interpretation of the role of emotion in social interaction. *Cognition and Emotion* 17, 429-453.

LANDMAN, J.

2001 The crime, punishment, and ethical transformation of two radicals: Or how Katherine Power improves on Dostoyevsky. In D.P. McAdams, R. Josselson & A. Lieblich (Eds.), *Turns in the road: Narrative studies of lives in transition* (pp. 35-66). Washington, DC: American Psychological Association.

LEBRA, W.P.

1966 *Okinawan religion: belief, ritual, and social structure*. Honolulu: University of Hawaii Press.

LILLARD, A.S.

1998 Ethnopsychologies: Cultural variations in theory of mind. *Psychological Bulletin* 123, 3-32.

LOEBER, R.

1985 Patterns and development of antisocial behavior. *Annals of Child Development* 53, 1431-1446.

MAIR, L.P.

1934 *An African people in the twentieth century*. London: Routledge & Sons.

MELTZOFF, A.N.

1995 Understanding the intentions of others: Re-enactment of intended acts by 18-month-old children. *Developmental Psychology* 31, 838-850.

MENCKEN, H.L.

1949 *A Mencken chrestomathy*. New York: Knopf.

MONTADA, L. & M.J. LERNER

1998 *Responses to victimizations and belief in a just world*. New York: Plenum Press.

NEWCOMB, W.W.

1956 *The culture and acculturation of the Delaware Indians*. Ann Arbor, MI: University of Michigan Press.

NÚÑEZ, M.

1993 *Teoría de la mente: Metarrepresentación, creencias falsas y engaño en el desarrollo de una psicología natural* (Theory of mind: Metarepresentation, false beliefs and deception in the development of a natural psychology). Unpublished doctoral thesis. Universidad Autónoma de Madrid, Spain.

OTTENBERG, P.

1958 [1980] *Marriage relationships in the double descent system of the Aflkpo Igbo of southeastern Nigeria*. Ann Arbor, MI: University Microfilms.

PERNER, J.

1994 The necessity and impossibility of simulation. *Proceedings of the British Academy* 83, 145-154. (In C. Peacocke (Ed.). *Representation, simulation and consciousness: Current issues in the Philosophy of Mind*. Oxford, UK: Oxford University Press.)

PERNER, J. & D. HOWES

1992 "He thinks he knows": and more developmental evidence against the simulation (role-taking) theory. *Mind & Language* 7, 72-86.

PETTAZZONI, R.

1955 On the attributes of God. *Numen* 2, 1-27.

PIAGET

1932 [1948] *The moral judgment of the child*. Glencoe, EL: Free Press.

POVINELLI, D.J., J.M. BERING & S. GIAMBRONE

2000 Toward a science of other minds: Escaping the argument by analogy. *Cognitive Science* 24, 509-541.

REICHEL-DOLMATOFF, G.

1950 *The Kogi: a tribe of the Sierra Nevada de Santa Marta, Colombia*. Vol. 1. Bogota: El Institute.

REYNOLDS, V. & R. TANNER

1995 *The social ecology of religion*. New York: Oxford University Press.

RICHARDS, A.I.

1939 *Land, labour and diet in Northern Rhodesia: an economic study of the Bemba tribe*. London, UK: Oxford University Press.

ROES, F.L. & M. RAYMOND

2003 Belief in moralizing gods. *Evolution & Human Behavior* 24, 126-135.

RUSSELL, B.

1948 *Human knowledge: Its scope and limits.* London: Unwin Hyman.

SANDIN, B. & C. SATHER

1980 *Iban adat and augury.* Penang: Penerbit Universiti Sains Malaysia for School of Comparative Social Sciences.

SCHELLING, T.C.

1960 *Strategy of conflict.* Cambridge, MA: Harvard University Press.

SCOTT, G.M., JR.

1986 [1990] *Migrants without mountains: the politics of sociocultural adjustment among the Lao Hmong refugees in San Diego.* Ann Arbor, MI: University Microfilms.

SHACKELFORD, T.K. & D.M. BUSS

1996 Betrayal in mateships, friendships, and coalitions. *Personality and Social Psychology Bulletin* 22, 1151-1164.

SOSIS, R.

Why aren't we all Hutterites? Costly signaling theory and religious behavior. *Human Nature* 14, 91-127.

STEIN, G.M.

1973 Children's reactions to innocent victims. *Child Development* 44, 805-810.

SULLIVAN, K., D. ZAITCHIK & H. TAGER-FLUSBERG

1994 Preschoolers can attribute second-order beliefs. *Developmental Psychology* 30, 395-402.

TANGNEY, J.P.

2001 Self-conscious emotions: The self as a moral guide. In A. Tesser, D.A. Stapel, and J.V. Wood (Eds.), *Self and motivation: Emerging psychological perspectives* (pp. 97-117). Washington, D.C.: American Psychological Association.

TOMASELLO, M.

1999 *The cultural origins of human cognition.* Cambridge, MA: Harvard University Press.

TRIVERS, R.

1981 Sociobiology and politics. In E. White (Ed.), *Sociobiology and human politics* (pp. 1-43). Lexington, MA: D.C. Health.

TURIEL, E.

2002 *The culture of morality: Social development, context and conflict.* Cambridge, UK: Cambridge University Press.

VAN VELZEN, H.U.E. & W. VAN WETERING

1991 *The Great Father and the Danger: religious cults, material forces, and collective fantaises in the world of the Surinamese Maroons.* Leiden, The Netherlands: KITLV Press.

WEDEKIND, C. & M. MALINSKI

2000 Cooperation through image scoring in humans. *Science* 288, 850-852.

WELLMAN, H.M.

1992 *The child's theory of mind.* Cambridge, MA: MIT Press.

WELLMAN, H.M., D. CROSS & J. WATSON

2001 Meta-analysis of theory-of-mind development: The truth about false belief. *Child Development* 72, 655-684.

WILLIAMS, G.C.
1992 *Natural selection.* New York: Oxford University Press.

Part IV
Sociocultural Accounts

[16]

God Is Watching You

Priming God Concepts Increases Prosocial Behavior in an Anonymous Economic Game

Azim F. Shariff and Ara Norenzayan

University of British Columbia, Vancouver, British Columbia, Canada

ABSTRACT—*We present two studies aimed at resolving experimentally whether religion increases prosocial behavior in the anonymous dictator game. Subjects allocated more money to anonymous strangers when God concepts were implicitly activated than when neutral or no concepts were activated. This effect was at least as large as that obtained when concepts associated with secular moral institutions were primed. A trait measure of self-reported religiosity did not seem to be associated with prosocial behavior. We discuss different possible mechanisms that may underlie this effect, focusing on the hypotheses that the religious prime had an ideomotor effect on generosity or that it activated a felt presence of supernatural watchers. We then discuss implications for theories positing religion as a facilitator of the emergence of early large-scale societies of cooperators.*

Many theorists have suggested that the cognitive availability of omniscient and omnipresent supernatural agents has had a dramatic impact on the development of large-scale human societies. The imagined presence of such agents, along with emotional ritual and costly commitment to the social group they govern, may have been the major development that allowed genetically unrelated individuals to interact in cooperative ways (e.g., Atran & Norenzayan, 2004; Irons, 1991; Sosis & Ruffle, 2004). The research reported in this article experimentally investigated this link between two broad classes of culturally widespread phenomena of interest to social science—religious beliefs and cooperative behavior among unrelated strangers.

Although anecdotes documenting religion's prosocial and antisocial effects abound, the empirical literature has produced mixed results regarding religion's role in prosocial behavior. Sosis and Ruffle (2004) examined levels of generosity in an experimental cooperative pool game in religious and secular kibbutzim in Israel and found higher levels of cooperation in the religious ones, and the highest levels among religious men who engaged in daily communal prayer. Batson and his colleagues (Batson et al., 1989; Batson, Schoenrade, & Ventis, 1993) have shown that although religious people report more explicit willingness to care for others than do nonreligious people, controlled laboratory measures of altruistic behavior often fail to corroborate this difference. Furthermore, when studies demonstrate that helpfulness is higher among more devoted people, this finding is typically better explained by egoistic motives such as seeking praise or avoiding guilt, rather than by higher levels of compassion or by a stronger motivation to benefit other people.

However insightful these findings are, research on religion and prosocial behavior has been limited by its overwhelming reliance on correlational designs. If religiosity and prosocial behavior are found to be correlated, it is just as likely that having a prosocial disposition causes one to be religious, or that some third variable such as guilt proneness or dispositional empathy causes both cooperative behavior and religiosity, as that religious beliefs somehow cause prosocial behavior. Only rarely have studies induced supernatural beliefs to examine them as a causal factor. Bering (2003, 2006) inhibited 3-year-old children's tendencies to cheat (i.e., open a "forbidden box") by telling them that an invisible agent ("Princess Alice") was in the room with them. In a different study, college students who were casually told that the ghost of a dead graduate student had been spotted in their private testing room were less willing to cheat on a computerized spatial-reasoning task than were those told nothing (Bering, McLeod, & Shackelford, 2005). These studies suggest that explicit thoughts of supernatural agents curb cheating behavior.

In the research reported here, we examined the effect of God concepts specifically on selfish and prosocial behavior. Our research design was novel in two ways. First, we introduced an

Address correspondence to Azim F. Shariff or to Ara Norenzayan, Department of Psychology, University of British Columbia, 2136 West Mall, Vancouver, BC, Canada V6T 1Z4, e-mail: azim@psych.ubc.ca or ara@psych.ubc.ca.

experimental procedure to activate God concepts implicitly, without having subjects consciously reflect on these concepts. Second, in lieu of relying on self-report measures, we used a paradigm of cooperative behavior that is well researched in psychology and economics: the dictator game. So that we could obtain an honest indicator of prosocial tendencies, rather than artifacts of impression management, the game was conducted in a strictly controlled anonymous setting with real monetary consequences.

The purpose of the first study was to implicitly prime God concepts among student subjects and examine how this affected their generosity. The second study was intended to replicate our main finding from the first study in a more heterogeneous community sample and to compare the strength of the religious prime with that of a secular prime of social institutions enforcing morality.

STUDY 1

Method

Subjects

Fifty subjects (mean age = 21 years; 34 females and 16 males) were recruited through posters displayed at the University of British Columbia, Canada, and randomly assigned to either the religious-prime or the no-prime condition. Twenty-six indicated identification with a religion, and 24 did not. Of the religious subjects, 19 identified themselves as Christians, 4 as Buddhists, 2 as Jews, and 1 as a Muslim. Of the remaining 24 subjects, 19 were categorized as atheists and 5 as theists without an organized religion. Subjects were defined as atheists if they both indicated "none" for religion and scored below the midpoint of the scale on a question assessing belief in God. Subjects who did not indicate a religious identification but nonetheless scored higher than the midpoint on the belief-in-God question were categorized as theists, along with those who did state specific religious identifications.

Procedure and Materials

All subjects were seated in private rooms behind closed doors for the duration of the experiment. Half of the subjects were implicitly primed with God concepts using the scrambled-sentence paradigm of Srull and Wyer (1979). The other half received no prime. Following this task, each subject played a one-shot, anonymous version of the dictator game (Hoffman, McCabe, Shachat, & Smith, 1994) against a confederate posing as another subject. All actual subjects were given the following instructions:

> You have been chosen as the **giver** in this economic decision-making task. You will find 10 one-dollar coins. Your role is to take **and keep** as many of these coins as you would like, knowing that however many you leave, if any, will be given to the receiver subject to keep.

To free subjects from reputational concerns, we assured them that only the other subject would know what they decided and that their identity would be hidden from that subject. Once they had made their decision, they completed a number of measures assessing religious belief and requesting demographic information. Each subject was then debriefed (both in writing and verbally) regarding the deception and the true aims of the experiment, compensated for participating, thanked, and dismissed.

For the priming manipulation (Srull & Wyer, 1979), subjects were required to unscramble 10 five-word sentences, dropping an extraneous word from each to create a grammatical four-word sentence. For example, "felt she eradicate spirit the" would become "she felt the spirit," and "dessert divine was fork the" would become "the dessert was divine." Five of the scrambled sentences contained the target words *spirit, divine, God, sacred,* and *prophet*, and the other 5 contained only neutral words unrelated to religion, and forming no other coherent concept.

Results and Discussion

Previous research has demonstrated that the majority of givers act selfishly in this anonymous game, leaving little or no money for the receiver, although some prosocial behavior is observed even in anonymous one-shot games (Haley & Fessler, 2005; Hoffman et al., 1994). This selfish tendency was confirmed in our control condition. Subjects who received no prime left, on average, $1.84 for the other subject, with 52% leaving $1 or less, only 12% leaving $5, and none leaving more than $5. Those who were primed with God concepts left, on average, $4.22, with 64% leaving $5 or more. The average amount of money left was $2.38 more in the religious-prime condition, a considerable difference, $t(48) = 3.69$, $p < .001$, $p_{rep} = .99$, $d = 1.07$. A comparison of subjects who left either nothing or $5 showed that a higher proportion of subjects behaved selfishly (offering nothing) in the control condition (36%) than in the religious-prime condition (16%), whereas a higher proportion behaved fairly (offering exactly $5) in the religious-prime condition (52%) than in the control condition (12%), $\chi^2(1, N = 29) = 7.5$, $p = .006$, $p_{rep} = .96$, shifting the modal response from selfishness to fairness (see Fig. 1).

This effect was present for both theists (prime-control difference of $1.88), $t(29) = 2.25$, $p = .032$, $p_{rep} = .91$, $d = 0.84$, and atheists (prime-control difference of $2.95), $t(17) = 2.70$, $p = .015$, $p_{rep} = .94$, $d = 1.31$. Although unprimed atheists left slightly less than did unprimed theists ($0.97), this trend was weak and was not statistically significant, $t(23) = 1.34$, $p = .19$, $p_{rep} = .73$. Self-reported belief in God, as a continuous measure, was not a good predictor of how much subjects left in the control condition, $r(24) = .23$, $p = .29$, $p_{rep} = .65$. In summary, implicit priming of God concepts did increase prosocial behavior (i.e., increased how much subjects left for an anonymous stranger), and this effect was observed for both theists and atheists. The

Volume 18—Number 9

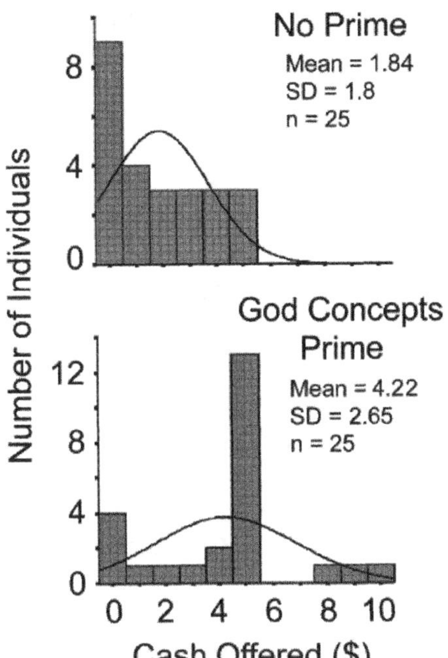

Fig. 1. Frequency distribution of money offered in the no-prime and religious-prime conditions of Study 1.

implicit religious prime proved to be much more effective at curtailing selfish behavior than was explicit religious belief.

Although these findings are compelling, their generalizability is limited by our reliance on a student sample. The behavior of such samples in economic games can be unrepresentative of larger, more heterogeneous populations in the world (Henrich et al., 2005). Moreover, the results of this study are open to the criticism that the control group did not receive a neutral prime. It is conceivable, although implausible, that merely being primed with words, rather than with religious concepts specifically, led to the difference between the control and religious-prime conditions. Moreover, we did not specifically establish that the implicit religious prime indeed affected behavior without reflective awareness of the subjects. All of these concerns were addressed in the second study.

STUDY 2

Overview

In the second study, we sought to replicate and expand the findings of the first. There were four main changes. First, instead of relying on a college sample, we recruited subjects from the larger community in Vancouver, Canada. Second, we replaced the no-prime control condition with a neutral-prime condition. Third, and most important, we introduced an additional priming condition to examine the strength of the religious prime relative to a prime of secular institutions of morality. Political philosophers since at least Voltaire (1727/1977) and Rousseau (1762/1968) have suggested that any moral benefits provided by religion could be gleaned just as easily, if not more easily, from nonreligious sources. The ideas of a justice system and, more generally, a social contract could be considered the strongest modern examples of such secular sources of moral influence. Thus, in Study 2, we added a secular-prime condition so we could examine the relative effects of religious and secular primes. Finally, we examined whether subjects reported any awareness that the primes activated religious thoughts.

Method

Subjects

Seventy-eight subjects were recruited via a combination of posters placed around Vancouver and newspaper ads. Of these subjects, 3 were dropped from analysis because of their suspicions about the study's hypothesis, leaving 75 participants equally distributed among the three conditions. Only 22% were students, and the sample was quite diverse. Ages (M = 44 years) ranged from 17 to 82. Yearly household incomes (M = $35,000) ranged from under $10,000 to over $80,000. Categorization of subjects in this study was based in part on responses to a new question, not used in Study 1, asking them specifically to mark whether they considered themselves religious, spiritual, agnostic, or atheistic. Of the subjects who indicated their religion, 25 identified themselves as Christians, and 3 as Jews. Of the remaining 47 subjects, 21 reported being "spiritual," 22 reported being agnostics or atheists, and 4 declined to answer. Subjects were categorized as atheists only if they both identified themselves as atheist or agnostic and scored below the midpoint of the scale on a question assessing belief in God—a more stringent criterion than in the first study. All other participants were classified as theists.

Procedure

Subjects followed the same procedure as in the first study, with a few notable exceptions. They were led to believe that subjects were alternately assigned to be givers and receivers and that they randomly happened to be givers, so that whatever decision they made would affect the following subject. Subjects in the control condition received a neutral prime; they completed the same scrambled-sentence task, but the scrambled sentences did not contain any target words that deliberately evoked a specific concept. Subjects in the secular-prime condition unscrambled sentences that contained the target words *civic, jury, court, police*, and *contract*. Subjects in the religious-prime condition unscrambled the same sentences as in Study 1.

At the end of the study, subjects completed demographic measures, including questions asking about their religiosity and belief in God. At the very end of the questionnaire, we asked the following two questions: (a) "Please briefly speculate on what you think this study was about so far," and (b) "Has there been anything that you do not understand or find odd about this study so far?" In addition, subjects were interviewed orally, and any suspicions expressed about the scrambled-sentences task were recorded.

Results and Discussion

Effect of Primes

The main effect from the first study was replicated. Subjects in the religious-prime condition offered an average of $4.56, whereas those in the control condition offered $2.56, a $2.00 difference, $t(48) = 2.47, p < .02, p_{rep} = .93, d = 0.71$. Perhaps because of the more heterogeneous sample, there was much greater variance in the amount of money offered than there was in the first study (see Fig. 2). Note that as in Study 1, the religious prime shifted the modal response from selfishness to fairness. A higher proportion of subjects behaved selfishly (offering nothing) in the control condition (40%) than in the religious-prime condition (12%), whereas a higher proportion behaved fairly (offering exactly $5) in the religious-prime condition (44%) than in the control condition (28%), $\chi^2(1, N = 31) = 4.40, p = .036, p_{rep} = .90$.

Unlike in the first study, there was a weak religiosity-by-prime interaction in this sample, $F(1, 46) = 2.22, p = .14, p_{rep} = .78$, indicating that the effect of the religious prime appeared to be stronger among theists than among atheists; the effect for atheists only was in fact nonsignificant, $t < 1$. We consider this inconsistency between the studies more fully in the General Discussion. Again, in the control condition, atheists did not differ from theists ($t < 1$). Self-reported belief in God, as a continuous measure, was not a good predictor of how much money subjects offered in the neutral-prime, control condition, $r(25) = -.12, p = .58, p_{rep} = .50$.

The secular prime had nearly as large an effect as the religious one. Subjects who received the secular prime left, on average, $4.44, or $1.88 more than those in the control condition, $t(48) = 2.29, p < .03, p_{rep} = .92, d = 0.67$.

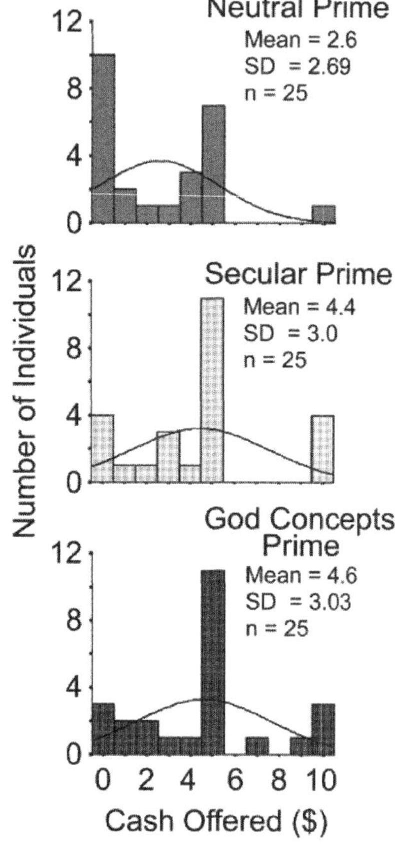

Fig. 2. Frequency distribution of money offered in the neutral-prime, secular-prime, and religious-prime conditions of Study 2.

Suspicion Probe

The key question was whether subjects reported any awareness that words in the unscrambled-sentences task reminded them of religious concepts or that this reminder was somehow related to the economic decision-making task. As in past research using this and related priming procedures (for reviews, see Bargh & Chartrand, 1999; Fazio & Olson, 2003), the vast majority of subjects did not report any awareness of this connection. In response to the probe questions, only 3 subjects (2 in the control condition and 1 in the secular-prime condition) mentioned anything related to the experimental question of how religious concepts are related to generosity, and these subjects were dropped from analysis. Five subjects (2 in the religious-prime condition, 2 in the control condition, and 1 in the secular-prime condition) mentioned religion in general, vague terms in their probe responses. We decided to retain these 5 subjects in our analyses. We found that neither excluding these 5 subjects nor including the 3 who were dropped had any effect on the final results. Furthermore, these 8 subjects who mentioned religion at all were distributed across all three conditions (in fact, the majority were in the control and secular-prime conditions). This suggests that their references to religion were not due to the priming procedure; more likely, a few subjects mentioned religion because the suspicion probes followed immediately after demographic questions, which included questions about self-reported religiosity.

These findings on suspicion are consistent with the literature showing that priming categories, goals, and emotions using the method employed in our study, as well as other, related methods, affects behavior, even for the vast majority of subjects who report no awareness of the prime (e.g., Bargh, Chen, & Burrows, 1996). Taken together, the facts point to the conclusion that our main findings cannot be explained away by the priming procedure and demand characteristics.

Discussion

This study demonstrates, then, that the prosocial effect of our religious prime is not limited to college students, but is in fact robust across a much more diverse sample. Moreover, given that a neutral prime was used in the control condition, and the suspicion probe revealed little reflective awareness of the religious nature of the prime, we can rule out the possibility that the effect of religious concepts on prosocial behavior was an artifact of the priming procedure itself or was a by-product of demand characteristics. Finally, we showed that implicit activation of concepts related to secular moral institutions restrained selfishness as much as did religious suggestion.

GENERAL DISCUSSION

God concepts, activated implicitly, increased prosocial behavior even when the behavior was anonymous and directed toward strangers. God concepts had as much effect in reducing selfishness as did concepts that activated a secular social contract, and the effect size was quite large. The results regarding how much God concepts affected atheists were, however, inconclusive. The first study demonstrated a clear effect for atheists, but this effect all but disappeared in the second study. Although further investigation is needed, we speculate that the inconsistency may have been due to our stricter definition of atheism in the second study. It is conceivable that avowed atheists, unlike other nonreligious people, doubt the existence of supernatural agents even at the implicit level. We leave these questions about atheism open for future investigation. In the meantime, we examine potential explanations for the effect we did find among the majority of our subjects.

Possible Explanations

Prosocial behavior can be influenced by increased positive or negative mood (Schaller & Cialdini, 1990), or by increased feelings of empathic concern (Eisenberg & Miller, 1987). It is conceivable that the religious primes increased prosocial behavior by acting via these mechanisms. However, in a follow-up study (Shariff & Norenzayan, 2006) in which we measured self-reported positive and negative affect (Watson, Clark, & Tellegen, 1988) and dispositional empathy (Davis, 1983) immediately after subjects were primed, we found no evidence for these mechanisms. Subjects who had received the religious prime reported neither increased positive or negative mood nor increased empathic concern.

Two possible theoretical explanations for the effect of the religious prime on prosocial behavior remain to be explored in the future. One is a behavioral-priming, or ideomotor-action, account based on the fact that the activation of perceptual-conceptual representations increases the likelihood of goals, plans, and motor behavior consistent with those representations (Bargh et al., 1996). Supernatural concepts such as "God" and "prophet" can refer to moral actors semantically and dynamically associated with acts of generosity and charitable giving. Irrespective of any attempt to manage their reputations, subjects may have automatically behaved more generously when these concepts were activated, much as subjects are more likely to interrupt a conversation when the trait construct "rude" is primed, or much as university students walk more slowly when the "elderly" stereotype is activated (Bargh et al., 1996).

Another possible explanation is that the religious prime aroused an imagined presence of supernatural watchers, and that this perception then increased prosocial behavior (for similar observations about supernatural concepts, see Bering, 2006, and Boyer, 2001). Although religions vary profoundly, central to all faiths is the idea of one or more omnipresent and omniscient moralizing agents who defy death, ignorance, and illusion; who demand costly sacrifice; and who arbitrate behavior in groups (Atran, 2002; Atran & Norenzayan, 2004; Norenzayan & Hansen, 2006). Generosity in cooperative games has been shown to be sensitive to even minor changes that compromise anonymity and activate reputational concerns (Haley & Fessler, 2005; Hoffman et al., 1994). Debates continue as to whether or not cooperative behaviors toward unrelated individuals, especially behaviors driven by passionate commitment, exist independent of short-term self-interest (e.g., Gintis, Bowles, Boyd, & Fehr, 2003). However, reputation management can go a long way in explaining the evolutionary stability of cooperative behavior between strangers, to the extent that selfish individuals are detected and subsequently excluded from future cooperative ventures.

A recent experiment (Haley & Fessler, 2005) found that even as subtle a cue as stylized eyespots on a computer background increased the amount of money that was offered in the dictator game. Similarly, an image of a pair of eyes increased money contributions to an "honesty box" used to collect money for drinks in a university lounge (Bateson, Nettle, & Roberts, 2006). If the mere presence of eyespots could increase generosity, it is very plausible that rousing belief in a supernatural watcher could produce similar effects, as was shown in an experiment by Bering et al. (2005) in which the belief that a dead graduate student's ghost resided in the testing room reduced cheating. In sum, we are suggesting that activation of God concepts, even outside of reflective awareness, matches the input conditions of an agency detector and, as a result, triggers this hyperactive tendency to infer the presence of an intentional watcher. This sense of being watched then activates reputational concerns, undermines the anonymity of the situation, and, as a result, curbs selfish behavior.

There is no reason why only one of these mechanisms need be responsible for the effect of God concepts on prosocial behavior. Religious sentiments have been culled and honed through hundreds of generations and may rely on multiple psychological mechanisms (Dennett, 2006), a possibility we leave open for exploration in future research.

Religion and the Origins of Civilization

There has been much speculation about why the emergence of religious iconography coincided with a rapid increase in population densities (Cauvin, 1999). It is possible—even likely—that early religions greatly facilitated population growth. Prior to around 12,000 years ago, group sizes remained small—limited by the threat of nonreciprocating defectors (Axelrod, 1984). A social group was restricted to genetically related individuals, bound by kin selection (Hamilton, 1964), and a handful of recognizable neighbors, bound by reciprocal altruism (Trivers, 1971). Theorists of religion, from Durkheim to Rappaport, have commonly attributed religion's socially cohesive effects to collective participation in costly ritual, rather than to belief in supernatural agents (see Sosis & Alcorta, 2003, for a discussion). However, in the present studies, we have found evidence that the invocation of supernatural agents may have played a central role. If the cultural spread of supernatural moralizing agents expanded the circle of cooperation to unrelated strangers, it may well have allowed small groups to grow into large-scale societies, from the early towns of Jericho and Ur to the metropolises of today.

One evolutionary explanation for our results invokes group selection. That is, ancestral societies with culturally widespread God concepts would have outcompeted societies without such concepts, given the cooperative advantage of believing groups (Wilson, 2002). However, group-selection accounts of religion, and altruistic behavior in general, although plausible in principle, face a number of well-known theoretical and empirical challenges (e.g., Atran, 2002). One does not have to appeal to group-selectionist arguments to explain why the likelihood of generosity increases when God concepts become cognitively accessible. As we have discussed, another plausible scenario centers on responsiveness to reputational concerns. These concerns—naturally selected because they ultimately maximized individual fitness in social groups (e.g., Bateson et al., 2006; Haley & Fessler, 2005)—could be activated by the perceived presence of any intentional, moralizing agents.

An Experimental Procedure to Measure the Effects of God Concepts

Religions are widespread elements of all societies and deeply affect the lives of most people in most societies. Yet scientific understanding of religion's impact on psychological processes remains poor. Implicit primes of concepts, goals, and affective states have been fruitfully used in social psychology in a wide range of domains (see Bargh & Chartrand, 1999). Similar causal and unobtrusive priming of God concepts has a number of potentially useful applications. This experimental procedure facilitates the measurement of the causal effect of specific religious concepts on people with a wide variety of explicit beliefs—theists and atheists alike, and everyone in between. Because priming operates largely outside of explicit awareness, subjects are less likely to respond to demand characteristics or to consciously revise their behaviors and beliefs in a priming paradigm than in a procedure with a manipulation that is more apparent to them. The priming technique can be readily and interestingly applied to study how religion affects prosocial behavior (Batson et al., 1993), moral intuitions (Cohen & Rozin, 2001), teleological reasoning (Kelemen, 2004), and prejudice (Allport & Ross, 1967). An experimental procedure activating religious concepts implicitly can be an important complement to other research designs, contributing to the growing efforts of cognitive and social scientists toward developing a natural science of religion.

Acknowledgments—We thank Sarah Allen and Courtney Edgar for their assistance and contributions. The Social Sciences and Humanities Research Council of Canada supported this research via a grant to Ara Norenzayan (410-2004-0197) and a fellowship to Azim F. Shariff (766-2005-0756).

REFERENCES

Allport, G.W., & Ross, J.M. (1967). Personal religious orientation and prejudice. *Journal of Personality and Social Psychology*, *5*, 432–443.

Atran, S. (2002). *In gods we trust: The evolutionary landscape of religion*. Oxford, England: Oxford University Press.

Atran, S., & Norenzayan, A. (2004). Religion's evolutionary landscape: Counterintuition, commitment, compassion, communion. *Behavioral and Brain Sciences*, *27*, 713–730.

Axelrod, R. (1984). *The evolution of cooperation*. New York: Basic Books.

Bargh, J.A., & Chartrand, T.L. (1999). The unbearable automaticity of being. *American Psychologist*, *54*, 462–479.

Bargh, J.A., Chen, M., & Burrows, L. (1996). Automaticity of social behavior: Direct effects of trait construct and stereotype activation on action. *Journal of Personality and Social Psychology*, *71*, 230–244.

Bateson, M., Nettle, D., & Roberts, G. (2006). Cues of being watched enhance cooperation in real-world setting. *Biology Letters*, *2*, 412–414.

Batson, C.D., Oleson, K.C., Weeks, S.P., Healy, S.P., Jennings, P., & Brown, T. (1989). Religious prosocial motivation: Is it altruistic or egoistic? *Journal of Personality and Social Psychology*, *57*, 873–884.

Batson, C.D., Schoenrade, P.A., & Ventis, L.W. (1993). *Religion and the individual: A social-psychological perspective*. Oxford, England: University Press.

Bering, J.M. (2003, August). *On reading symbolic random events: Children's causal reasoning about unexpected occurrences*. Paper presented at the Psychological and Cognitive Foundations of Religiosity Conference, Atlanta, GA.

Bering, J.M. (2006). The folk psychology of souls. *Behavioral and Brain Sciences, 29*, 453–462.

Bering, J.M., McLeod, K., & Shackelford, T.K. (2005). Reasoning about dead agents reveals possible adaptive trends. *Human Nature, 16*, 360–381.

Boyer, P. (2001). *Religion explained: The evolutionary origins of religious thought*. New York: Basic Books.

Cauvin, J. (1999). *The birth of the gods and the origins of agriculture* (T. Watkins, Trans.). Cambridge, England: Cambridge University Press.

Cohen, A.B., & Rozin, P. (2001). Religion and the morality of mentality. *Journal of Personality and Social Psychology, 81*, 697–710.

Davis, M.H. (1983). Measuring individual differences in empathy: Evidence for a multidimensional approach. *Journal of Personality and Social Psychology, 44*, 113–126.

Dennett, D.C. (2006). *Breaking the spell*. New York: Viking.

Eisenberg, N., & Miller, P.A. (1987). The relation of empathy to prosocial and related behaviors. *Psychological Bulletin, 101*, 91–119.

Fazio, R.H., & Olson, M.A. (2003). Implicit measures in social cognition research: Their meaning and use. *Annual Review of Psychology, 54*, 297–327.

Gintis, H., Bowles, S., Boyd, R., & Fehr, E. (2003). Explaining altruistic behavior in humans. *Evolution and Human Behavior, 24*, 153–172.

Haley, K.J., & Fessler, D.M.T. (2005). Nobody's watching? Subtle cues affect generosity in an anonymous economic game. *Evolution and Human Behavior, 26*, 245–256.

Hamilton, W.D. (1964). The evolution of social behavior. *Journal of Theoretical Biology, 7*, 1–52.

Henrich, J.R., Boyd, S., Bowles, H., Gintis, E., Fehr, C., Camerer, R., et al. (2005). "Economic man" in cross-cultural perspective: Ethnography and experiments from 15 small-scale societies. *Behavioral and Brain Sciences, 28*, 795–815.

Hoffman, E., McCabe, K., Shachat, K., & Smith, V. (1994). Preferences, property rights and anonymity in bargaining games. *Games and Economic Behavior, 7*, 346.

Irons, W. (1991). How did morality evolve? *Zygon: Journal of Religion and Science, 26*, 49–89.

Kelemen, D. (2004). Are children "intuitive theists"? Reasoning about purpose and design in nature. *Psychological Science, 15*, 295–301.

Norenzayan, A., & Hansen, I.G. (2006). Belief in supernatural agents in the face of death. *Personality and Social Psychology Bulletin, 32*, 174–187.

Rousseau, J.J. (1968). *The social contract, or principles of political right* (H.J. Tozer, Trans.). London: Penguin Classics. (Original work published 1762)

Schaller, M., & Cialdini, R.B. (1990). Happiness, sadness, and helping: A motivational integration. In E.T. Higgins & R.M. Sorrentino (Eds.), *Handbook of motivation and cognition: Foundations of social behavior* (Vol. 2, pp. 265–296). New York: Guilford Press.

Shariff, A., & Norenzayan, A. (2006). [The effect of religious priming on affect and attitudes]. Unpublished raw data.

Sosis, R., & Alcorta, C. (2003). Signaling, solidarity, and the sacred: The evolution of religious behavior. *Evolutionary Anthropology, 12*, 264–274.

Sosis, R., & Ruffle, B.J. (2004). Ideology, religion, and the evolution of cooperation: Field tests on Israeli kibbutzim. *Research in Economic Anthropology, 23*, 89–117.

Srull, T.K., & Wyer, R.S., Jr. (1979). The role of category accessibility in the interpretation of information about persons: Some determinants and implications. *Journal of Personality and Social Psychology, 37*, 1660–1672.

Trivers, R.L. (1971). The evolution of reciprocal altruism. *Quarterly Review of Biology, 46*, 35–57.

Voltaire. (1977). *The portable Voltaire* (B.R. Redman, Trans.). New York: Penguin Books. (Original work published 1727)

Watson, D., Clark, L.A., & Tellegen, A. (1988). Development and validation of brief measures of positive and negative affect: The PANAS scales. *Journal of Personality and Social Psychology, 54*, 1063–1070.

Wilson, D.S. (2002). *Darwin's cathedral*. Chicago: University of Chicago Press.

(Received 8/20/06; Revision accepted 9/13/06; Final materials received 10/13/06)

[17]

The Capacity for Religious Experience Is an Evolutionary Adaptation to Warfare

Allen D. MacNeill

> If we were forced to say in one word who God is and in another what the Bible is about, the answer would have to be: God is a *warrior*, and the Bible is about *victory*.
>
> —Jack Miles, *God, A Biography* (1995, p. 106)

Recent events have reinforced a pattern observable for at least the past 6 millennia: that religion and warfare are tightly, perhaps inextricably, intertwined. Even a cursory review of the history of wars and warlike conflicts indicates that religion has played a central role in these events. The common use of words and phrases such as "crusade," "holy war," and the now-infamous "jihad" point to the intimate connection between religion and warfare. From the most exalted "god-kings" to the lowliest "grunts" in the foxholes (where, as tradition tells us, there are no atheists), religion has both accompanied and facilitated warfare.

That religion and warfare are at some level related is virtually undeniable. What is less obvious at first glance is the quality of this relationship; is it causal, and if so, in which direction? There are at least three possibilities:

- Religions cause wars,
- Warfare promotes religion, or
- Both religion and warfare are causally linked to other, more general causative factors.

However, a more sophisticated analysis may indicate that religion and warfare are both cause and effect of each other. That is the thesis of this chapter: that the capacity for religious experience exists among humans primarily because it has facilitated warfare, which in turn reinforces the underlying causes of religion. In other words, the human capacities for religious experience and warfare have adapted to each other in a coevolutionary spiral that has made individual and group mass murder and suicide virtually inevitable, given prevailing ecological subsistence patterns.

The Capacity for Religion Is an Evolutionary Adaptation

In *The Descent of Man*, Darwin argued that humans do not have an innate instinct to believe in God:

> The belief in God has often been advanced as not only the greatest, but the most complete of all the distinctions between man and the lower animals. It is however impossible ... to maintain that this belief is innate or instinctive in man. (1871, p. 612)

He based this conclusion on the widespread observation that many human cultures do not include a belief in a deity that can be interpreted as being in any way conceptually similar to the monotheistic Judeo-Christian God. However, Darwin went on to point out that "a belief in all-pervading spiritual entities seems to be universal; and apparently follows from a considerable advance in man's reason, and from a still greater advance in his faculties of imagination, curiosity and wonder" (p. 612).

Implicit in Darwin's argument is the idea that only species-wide behavior patterns can legitimately be thought of as evolutionary adaptations. This is a common assumption among both ethologists and evolutionary psychologists, and in my opinion is at best misguided. I prefer G. C. Williams's definition of an adaptation: "An adaptation is any trait that enhances fitness and [has been] modified by selection to perform that role" (1966, p. 4). However, even this definition is somewhat muddied by the inclusion of the term "fitness." The working definition for "evolutionary adaptation" that will be used throughout the rest of this chapter is as follows:

- **An evolutionary adaptation** is any heritable phenotypic character whose frequency of appearance in a population is the result of increased reproductive success relative to alternative versions of that heritable phenotypic character.

Let us set aside for the moment the question of how such phenotypic characters are inherited. As we will see, this is not a trivial question, but one that in the long run does not fundamentally alter the argument I am about to make. Given this definition of evolutionary adaptation, it should be immediately clear why pan-specificity alone is a poor criterion for determining whether some character is an adaptation.

Implicit in this definition of evolutionary adaptation is the idea that there is some real (i.e., nontrivial) *variation* in the phenotypic characteristics present among the members of a population. Indeed, it was the recognition of the existence of such variation, and the insistence that this variation is the basis for natural selection, that was perhaps Darwin's most revolutionary discovery. Any trait that is an evolutionary adaptation will show some nontrivial variation in the expression of that trait, from individuals who express it to a very high degree, to individuals in whom its expression is virtually unnoticeable. What appears at first glance to be pan-specificity is actually the numerical preponderance of individuals whose expression of the trait is close to the population mean for that trait.

Is there other evidence that can be used to determine if a particular characteristic is an evolutionary adaptation? In addition to showing that the numerical preponderance of a particular phenotypic trait is the result of differential survival and reproduction, it may also be possible to link the phenotypic trait with an underlying anatomical and/or physiological substrate that is the efficient cause of the trait in question. For example, it is commonly accepted at present that the ability to speak and understand speech is an evolutionary adaptation in humans. This conclusion was originally based primarily on linguistic grounds (cf. Lenneberg, 1964, 1967; Chomsky, 1965) but has more recently been correlated with underlying neurological processes (Pinker & Bloom, 1999).

Another way of determining if a characteristic is an evolutionary adaptation is to correlate the population dynamics of the adaptation with its evolutionary environment of adaptation (Bowlby, 1969):

- **The evolutionary environment of adaptation** (*EEA*) is the ecological milieu under which a particular adaptation has arisen as the result of selection.

The concept of an EEA can be fruitfully employed when trying to determine whether a particular characteristic is an adaptation by attempting to show how the ecological circumstances prevalent in the EEA would have resulted in differential survival and reproduction. However, application of this technique is complicated by the fact that determination of the EEA of a given adaptation can be a somewhat circular process. Ideally, the circumstances of the EEA should be determined by means other than reference to a particular adaptation, followed by an analysis of the effects of the inferred EEA on the survival and reproduction of the organisms inhabiting it.

Final verification that a particular trait is indeed an evolutionary adaptation would require all of the foregoing, plus linking the appearance of the trait to an underlying gene or gene complex and showing that the frequency of the controlling gene(s) in the population in question has indeed been altered as the result of differential survival and reproduction. This is difficult to do even with very simple genetic traits, such as sickle-cell anemia. Furthermore, it may be that the causal connection between the underlying genes and the trait for which they code may be indirect at best. However, rather than abandon the concept of evolutionary adaptation altogether (as some have suggested; cf. Margulis, 1997), it may still be useful to apply Definition 1 (above) with four further qualifications:

- **Qualification 1:** An evolutionary adaptation will be expressed by most of the members of a given population, in a pattern that approximates a *normal distribution*;
- **Qualification 2:** An evolutionary adaptation can be correlated with *underlying anatomical and physiological structures*, which constitute the efficient (or proximate) cause of the evolution of the adaptation;

- **Qualification 3:** An evolutionary adaptation can be correlated with a preexisting *evolutionary environment of adaptation*, the circumstances of which can then be correlated with differential survival and reproduction; and
- **Qualification 4:** An evolutionary adaptation can be correlated with the presence and expression of an *underlying gene or gene complex*, which directly or indirectly causes and influences the expression of the phenotypic trait that constitutes the adaptation.

Given the foregoing, we can reframe Darwin's question thusly: Is *religion* an evolutionary adaptation? This question is similar to the question, "Is speaking English an adaptation?" Clearly, speaking English is not adaptive, any more than is speaking French or Tagalog. Given Definition 1 and the qualifications enumerated above, to assert that speaking English is an evolutionary adaptation would require that one verify that individuals who speak English survive and reproduce more often than individuals who speak some alternative language. Furthermore, their differential survival and reproduction must be shown to be causally related to their speaking English, and not to some other, related characteristic, such as the ability to speak, regardless of what language is spoken.

Here is the essential distinction: the *capacity* to speak English (or any other language, for that matter) is quite clearly an evolutionary adaptation (Pinker, 1994). That it is so is reinforced by the fact that there are specific circuits and regions in the human brain that are dedicated to the production and understanding of speech. Damage to these structures can severely limit or even completely destroy a person's ability to speak or understand spoken language (Penfield & Roberts, 1959). Furthermore, although every neurologically normal person can learn to speak and understand speech, there is the same kind of natural variation in this capacity that Darwin first pointed out as the basis for natural selection. That is, some individuals learn to speak and understand speech with great difficulty, others do so with great facility, while the vast majority of humanity muddles through with one "mother tongue." The point here is that the *capacity* for religious experience appears to have the same characteristics as the capacity for language. While it is pan-specific, there is considerable individual variation in the capacity for religious experience, with some individuals having very high capacity, others very low, and the average person somewhere in the middle (see Qualification 1 above).

Furthermore, there is accumulating evidence that there are underlying neurological structures that facilitate religious experience. The work of d'Aquili and Newberg (1999), Newberg and d'Aquili (2001), Persinger (1987), Ramachandran and Blakeslee (1998), and Saver and Rabin (1997) all purport to find correlates between religious experiences and specific brain structures and neurological processes. The point here is not to argue for the specific neurobiology underlying the particular states described by these researchers nor to argue that the neurological states they have studied comprise the whole of what we mean by the term "religion." Rather, the fact that some psychological states identified with religious experience have been correlated with specific neurological activity in specific structures in the human brain satisfies, at least in part, the criterion enumerated in Qualification 2 (above).

Finally, is there a "religion gene" that can be shown to correlate with the capacity for religious experience and whose frequency can be shown to vary in such a way as to approximate the patterns characteristic of an evolutionary adaptation? No, nor should we expect there to be one. Only in so-called "vulgar socio-biology" are there presumed to be single genes (or even gene complexes) that code for complex human behaviors such as the capacity for speech or religion. Rather, there are genes that code for the assembly, operation, and modification of "mental modules" that bring about these complex behavior patterns.

One way to avoid the whole morass of gene-behavior linkages is to employ what Stephen Emlen has called the "correlation approach" to behavioral ecology (1976). According to this method, one may be able to "interpret and partially predict the social structure of a species on the basis of a limited set of environmental or ecological variables ... [that] impose limits on the range of types of social organization that will be adaptive" (p. 736). According to this viewpoint, "[Species] faced with similar ecological 'problems' exhibit a predictable convergence in their 'solutions,' as shown in their social organizations" (p. 737). Following Emlen's lead, we can compare the types of religious experiences and patterns of religious behaviors exhibited by humans in different ecological subsistence patterns and at different times and places. In so doing, we may find some general patterns that will point to the underlying evolutionary dynamics influencing the development of the biological and cultural mechanisms producing those experiences and behaviors.

Adaptive characteristics do not increase in frequency monotonically in populations, nor are selective pressures usually limited to one or even a few parameters. Following the lead of Sewell Wright (1968–1978), it has been very common for evolutionary biologists to model the *adaptive landscape* for a given population or species (cf. Ridley, 1996, pp. 215–219). However, what is sometimes lost in diagrams such as these is the fact that there are individual organisms represented by nearly every point on the surface illustrated. That is, not all individuals in the population have risen to the adaptive peaks in the population, nor are all of them slipping down into the troughs of maladaptation. Perhaps it would be better to imagine each individual and its descendents as a boatlike cluster of adjacent points on this surface, tossed up and down by the vicissitudes of ecological change.

In this viewpoint, it may be easier to see that some changes in the environment are much less important than others and that some shifts in adaptive character may be much more significant than others. In particular, it is quite possible for a major change in one parameter in the environment to cause a corresponding change in the adaptive characteristics of the population, swamping any effects of smaller changes. In essence, what I am describing here is an evolutionary override of sorts, in which selection for one characteristic swamps selection for most or all other characteristics among the members of a population.

At this point, it appears likely that the capacity for religious experience has many of the characteristics of an evolutionary adaptation:

- the capacity for religious experience is pan-specific in humans, although there is considerable variation in this capacity, both within and between human groups,
- the capacity for religious experience has been correlated with underlying neurological structures and processes,
- the capacity for religious experience can be correlated with a known evolutionary environment of adaptation (i.e., intergroup warfare in agricultural societies, as will be discussed in more detail below), and
- although no underlying genetic mechanisms for the development of the capacity for religious experience are now known, the existence of consistent cross-cultural patterns of religious expression indicates that religious behavior is subject to evolutionary convergence in a manner analogous to other evolutionary adaptations.

The Capacity for Warfare Is Also an Evolutionary Adaptation

Now it is time to address the other half of the coevolutionary spiral, to wit: Is warfare (or, more precisely, the capacity for warfare) an evolutionary adaptation? Once again, Darwin was unequivocal on this subject. In *The Descent of Man*, he wrote,

> When two tribes of primeval man, living in the same country, came into competition, if (other circumstances being equal) the one tribe included a great number of courageous, sympathetic and faithful members, who were always ready to warn each other of danger, to aid and defend each other, this tribe would succeed better and conquer the other … . Thus the social and moral qualities would tend slowly to advance and be diffused throughout the world. (1871, p. 130)

Two things are immediately noticeable about this description: that Darwin assumes that the ability to be successful in warfare arises from courage, sympathy, and faithfulness within human groups and that the level at which selection is operating in the evolution of such qualities is the group, rather than the individual. As we shall soon see, neither of these assumptions is necessarily in accord with the evidence.

Before we can decide if warfare is adaptive, it is first necessary to define precisely what we mean by warfare. There appear to be at least three intergroup aggressive activities that are often referred to by the same name. It is important to my later argument that these be distinguished, and so here they are:

- **Raiding** (or *rustling*, as in cattle rustling) is an activity in which small groups of humans, virtually always men and almost always close kin groups, spontaneously and with relatively little planning or hierarchical organization, temporarily enter the recognized territory of a nearby group with the intention of forcibly obtaining resources, usually domesticated animals or women, or both. Not all members of a given kin group will necessarily participate in raiding, and all "warlike" activity and organization occurs immediately before, during, and after a raid. At all other times, the participants in such a raid are engaged in other domestic activities, generally unrelated to raiding.
- **Militia Warfare** is an activity in which somewhat larger groups of humans, again almost exclusively male but not necessarily close kin groups, band together periodically with some

planning and hierarchical organization, with the intention of either forcibly entering the recognized territory of a nearby group or defending against the forcible entry by similarly constituted raiding parties or militias from other groups. In militia warfare, most of the able-bodied males in a given social group will participate in some way, either in direct combat or combat support. However, once the immediate warlike activity has ended, the militia disbands and most, if not all, of its members turn to other tasks. An important characteristic that distinguishes raiding from militia warfare is the presence in the latter of generally recognized hierarchical ranks and specialized duties and training, a situation generally lacking in raiding/rustling.

- **Professional Warfare** is an activity in which relatively large groups of humans (i.e., armies), again almost exclusively male but usually not close kin groups (and often including noncombatant female auxiliaries), band together regularly or permanently with considerable planning and hierarchical organization, with the intention of either forcibly commandeering the recognized territory of a nearby group or defending against the forcible entry by similarly constituted armies from other groups. In professional warfare, many of the able-bodied males in a given social group will participate in some way, either in direct combat or combat support. Furthermore, regardless of when or if the immediate warlike activity has ended, the army continues to exist, and its members pursue specialized tasks within the military organization. An important characteristic that distinguishes raiding and militia warfare from professional warfare is the proliferation in the latter of strictly defined hierarchical ranks and specialized duties and the existence of a permanent professional class of warriors which includes almost all officers, but not necessarily all combatants (i.e., the grunts get to go home and take up other occupations after the war … assuming they survive).

Although there are many variations on these three themes (and an almost infinite gradation of one into the other), there are broad patterns of correlation between these three patterns across most societies. Furthermore, the three types of warlike organizations are generally correlated with ecological subsistence patterns. Raiding is most common among hunter-gatherers and pastoralists (i.e., people who raise domesticated animals as an important part of their subsistence) Militia warfare is more common among simple agriculturalists, especially those who live in widely dispersed villages and who depend primarily on domesticated crop plants for their subsistence. Professional warfare is most common among societies that are characterized by a combination of village agriculture and urban living. In particular, the maintenance of a professional army requires both large populations and a large surplus of food and other resources, as the members of the army themselves are no longer available for food production or distribution and must therefore be supported by the rest of the population.

According to Wallace, "[t]here are few, if any, societies that have not engaged in at least one war in their known history" (1968, p. 173). Indeed, there is reason to believe that warfare (or at least raiding) predates the evolution of the genus *Homo* and may not even be restricted to the order Primates. Jane Goodall (1986) describes behaviors among the chimpanzees (*Pan troglodytes*) of the Gombe preserve that are remarkably similar to the raiding behavior of humans in pastoral societies. Moving beyond the Primates, Hans Kruuk (1972, pp. 253–258) describes behaviors among the spotted hyaenas (*Crocuta crocuta*) of east Africa, which bear some resemblance to the behaviors described by Goodall.

Given that raiding and other forms of social aggression appear to be pan-specific, can they be considered to be adaptations? Or, to be more precise, is the *capacity* for social aggression, either offensive or defensive (or both), an evolutionary adaptation? I believe the answer to this question is yes. Clearly, there are neurological modules for aggressive behavior in humans and related primates. Primatologist Richard Wrangham (1999) has proposed that both chimpanzee and human males have a genetically influenced tendency to raid and kill members of neighboring groups whenever there is a state of intergroup hostility and one group can muster sufficient force to raid the other with relatively little fear of losses (1999, pp. 1–12). Wrangham and Peterson (1996) have pointed out that there are striking similarities between the raiding behavior of wild chimpanzees and the raiding behavior of the Yanomami (pp. 64–72). Chagnon (1990) has taken this argument further; in a discussion of the behavior of Yanomami warriors designated as *unokais* (a designation given to males that have undergone a ritual purification following a killing), Chagnon points out that *unokais* have a significantly higher reproductive success than non-*unokais*, as shown by statistical analysis of reproductive success at different ages. Males with the highest relative reproductive success are middle-aged men with children, a pattern that is repeated in many other societies (1988, pp. 985–992).

Tooby and Cosmides (1988) have argued that the capacity for raiding and warlike behavior shown by humans and other primates is based on an evolutionary "algorithm" in which the costs of warfare are balanced by the corresponding benefits to reproductive success (p. 5). I think we need to be clear that, in this context, "reproductive success" is used in the same sense as it is used by evolutionary biologists: that is, the net number of offspring produced by individuals performing different behaviors. We are not necessarily speaking of a kind of sexual selection for the capacity for warfare. Rather, we are referring simply to the number of offspring that survive in each behavioral cohort, for whatever reason. Ecological factors, such as the availability of food resources (especially proteins), the spatial and temporal distribution of such resources (e.g., dispersed and nondefensible versus clumped and defensible), the availability of specific tools and weapons (such as metal tools), and the number, size, and physical development of potential fighters, all play a part in the calculation of potential costs and benefits of warlike social behavior. In other words, the algorithm postulated by Tooby and Cosmides is a mental means of factoring in all of the various costs and benefits of alternative behaviors to determine which alternative will result in the most positive outcome.

At what level—individual or group—must such outcomes be positive for the capacity for warfare to be adaptive? Sober and Wilson (1998) have argued that cooperative behavior (i.e., "altruism") can evolve as the result of natural selection at the level of groups. Although, at first glance, it may not seem that warfare is altruistic. However, it clearly is, as individual members of a society engaged in warfare risk (and sometimes lose) their lives in defense of the group. Sober and Wilson do not specifically discuss warfare, but clearly it would qualify as a form of cooperative behavior. So, does the capacity for warfare evolve as the result of selection at the level of groups?

In the early 1960s, V. C. Wynne-Edwards used the concept of group selection as the basis for the explanation of nearly all of animal social behavior (1962). It was in response to Wynne-Edwards that G. C. Williams wrote *Adaptation and Natural Selection* (1966) in which he argued forcefully for the primary importance of selection at the level of individuals, rather than groups. Williams pointed out that any group of organisms in which reproductive success has been lowered by group processes (specifically by decreasing the number of offspring per individual by means of various mechanisms) is vulnerable to invasion and ultimate replacement by individuals who are not so constrained. His model for individual selection has been extended to the evolution of social behavior and cooperation by Hamilton (1964), Trivers (1971), Axelrod (1984), and Dawkins (1982).

Warfare is often thought of as an aberration, rather than a central characteristic of human sociality. However, even a brief review of human history should impress upon one that warfare has been a constant, if episodic, aspect of human social behavior. The point here is that, even if it does not happen regularly, warfare can have an effect on natural selection equivalent to—and in some cases greater than—a constant selective pressure. During periods in which warfare is not occurring, selection will result primarily from those sources of mortality and reduced reproductive success characteristic of peacetime society: disease, famine, competition for scarce resources, etc. However, during periods of warfare, these "everyday" forms of selection can be overwhelmed by the effects of warfare-specific changes in mortality and reproductive success. In other words, periods of warfare act like evolutionary "bottlenecks" selecting with greatly increased relative intensity for any physiological or behavioral characteristic that allows for differential survival and reproductive success.

Furthermore, it seems likely that these selective pressures will be exerted primarily at the level of individuals, rather than groups. To understand why, consider Sober and Wilson's definition of a group: "a set of individuals that influence each other's fitness with respect to a certain trait, but not the fitness of those outside the group" (1998, p. 92). To be consistent with standard practice in evolutionary theory, let fitness (symbolized by $\mathbf{w}$) be defined as the average per capita lifetime contribution of an individual of a particular genotype to the population after one or more generations, measured in number of offspring bearing that individual's genotype. If we apply this definition of fitness to Sober and Wilson's definition of a group, then each individual member of a group will have some fitness ($\mathbf{w}_i$), with the group fitness reducing to the sum of the fitnesses of the individuals that make up the group ($\mathbf{w}_G = \mathbf{w}_I + \mathbf{w}_j + \ldots \mathbf{w}_n$). Furthermore, let us assume that fitness is a function of some limited resource. In real terms, this resource might be food, or shelter, or access to mates, or some other factor that contributes directly or indirectly to survival or reproduction. Each individual can exploit some small fraction of the limited resource with the aggregate consumption of resources eventually reaching a maximum value (i.e., when each individual has maximized its fitness via the exploitation of that resource and the resource has been completely subdivided among the individuals in the group). Until that limit has been reached, competition between individuals in the group is relatively unimportant, and so individual fitness of each member of the group will not be limited by group membership.

Under such circumstances, there are at least three different ways in which intra-group cooperation could affect individual fitness:

- The fitness effect of group membership on individual fitness could be **negative**, compared with the fitness of each individual acting alone; that is, being a member of the group **detracts** from each individual's fitness, compared with acting alone; see Figure 10.1.

This is the situation described by Wynne-Edwards (1962) and criticized by G. C. Williams (1966). Under these conditions, the addition of each new member to a group *decreases* the average fitness of each member of the group, with the effect that such a group is constantly vulnerable to invasion and replacement by individuals who act entirely in their own interests. Given the relationship between group size and individual fitness, it is unlikely that this type of group selection would prevail under most natural conditions.

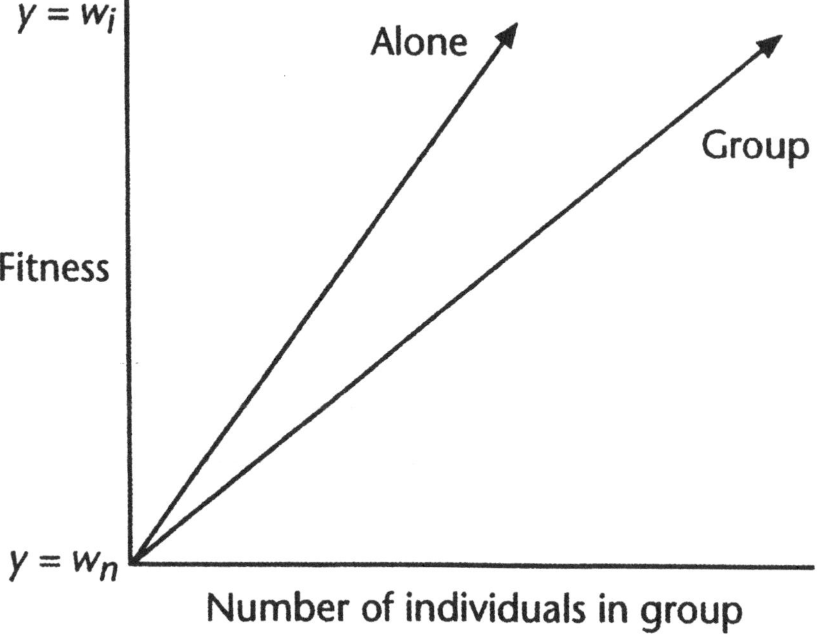

Figure 10.1
Negative Group Selection

- The fitness effect of group membership on individual fitness could be **positive**, compared with the fitness of each individual acting alone; that is, being a member of the group **adds** to each individual's fitness, compared with acting alone; see Figure 10.2.

This is the essentially the situation proposed by Sober and Wilson (1998). Under these conditions, the addition of each new member to a group *increases* the average fitness of each member of the group, with the overall effect that such a group becomes less vulnerable to invasion and replacement by "selfish" individuals as it grows larger. Unlike the situation with negative group fitness above, this type of group selection could easily evolve, as membership in the group clearly benefits individuals and vice versa.

- The fitness effect of group membership on individual fitness could be **negative** when the group size is below some critical value but could become **positive** as that critical group size is exceeded; see Figure 10.3.

Under such conditions, a group would have to reach a "critical group size" before the fitness benefits of intragroup cooperation would begin to be felt. This would seem to present a barrier to the evolution of such cooperation via group selection. However, the group might reach critical size for reasons unrelated to the activities

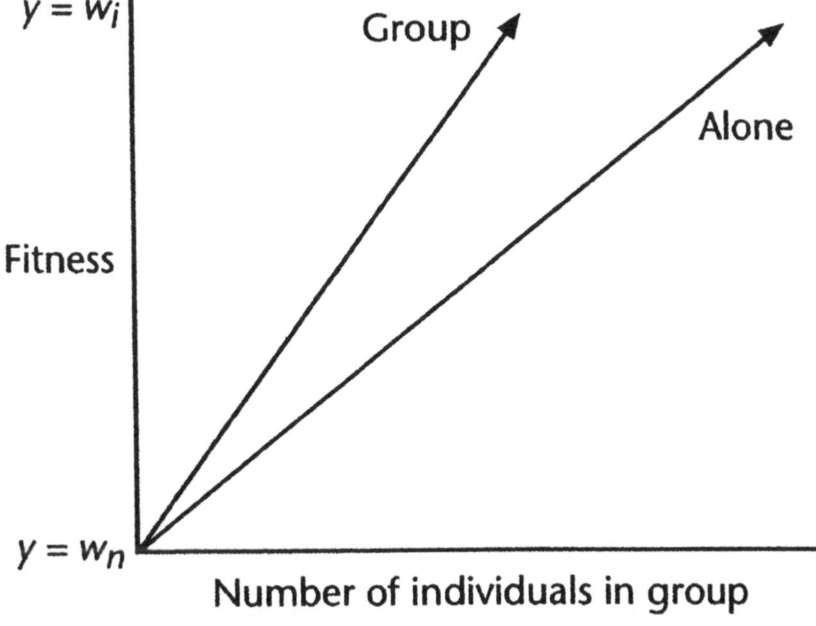

Figure 10.2
Positive Group Selection

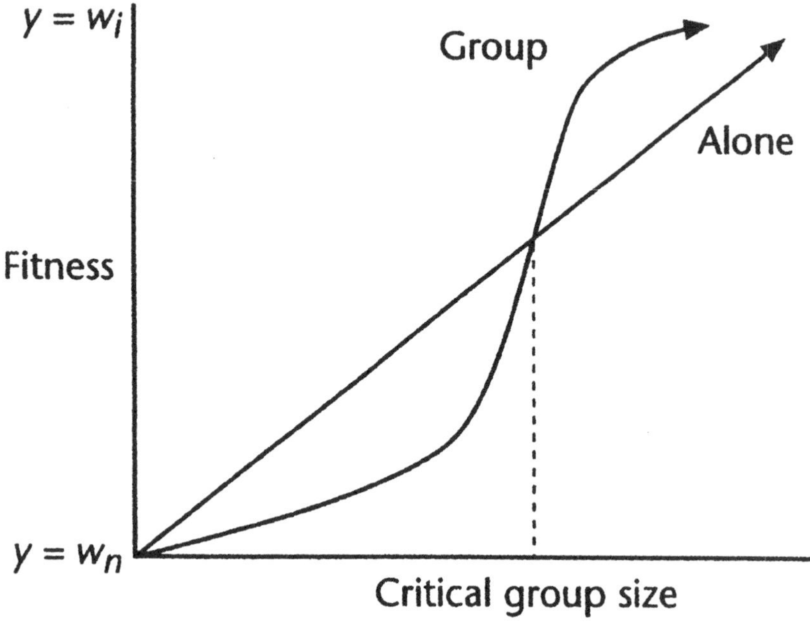

Figure 10.3
Variable Group Selection

resulting in reduced aggregate fitness. Once above the critical size, adding new members to the group would add to the fitness of each individual.

There are many circumstances in nature in which the kind of variable group fitness described above has been shown to exist. For example, in a study of the hunting behavior of wolves (*Canis lupus*) on Isle Royale, Mech (1970) found that the larger a wolf pack was, the better able it was to exploit larger prey (1970, p. 38). Intragroup cooperation is often essential to the success of hunting forays among social carnivores in general. Teleki (1973) found the same to be true for hunting behavior among wild chimpanzees. Indeed, chimpanzees exhibit unusually cooperative behavior when hunting, especially when their prey is other primates.

It is very likely that human warfare follows the pattern described in Figure 10.3. That is, there is a critical group size above which the effectiveness of warfare increases, as reflected in its effects on fitness. There are two primary reasons for this:

- As virtually any military commander would point out, the larger the military force, the more likely it is to prevail over its opponents. Calculation of the relative sizes and strengths of the opposing sides in any warlike interaction would be of crucial importance to the participants, regardless of rank. Therefore, it is likely that natural selection would have resulted in the evolution of mental algorithms that would facilitate such calculations under conditions of repeated or sustained warfare.

- As the size of a military force increases, the probability of injury or death to each individual member of the group generally decreases. One-on-one violent interactions between individual combatants are most likely to result in the injury or death of one or both combatants. As the number of combatants increases, the number of injuries and deaths per capita generally decreases (except in the case of modern technological warfare, where overwhelmingly powerful weapons can injure or kill huge numbers of combatants and noncombatants). This decrease in probability of injury and death with the size of a military force is another factor in the mental calculations performed by any potential participant in warfare.

Given the foregoing, it should now be clear that participation in warfare can have positive effects on both individual fitness and group fitness, when group fitness is measured as the aggregate fitness of the individuals making up the group. Participants in warfare—combatants and their supporters—can gain access to territory and to resources if they are on the winning side in a conflict. In particular, it is a well-known (but not often discussed) fact that the winners in virtually all warlike conflicts have greatly increased reproductive success compared with both the losers and nonparticipants in their own group. Wars and warlike interactions (including simple raids) are often followed by increases in birth rates among the winners. In particular, soldiers (i.e., combatants) are notorious for the commission of rape during war (Thornhill & Palmer, 2000). That this is the case has been recognized as far back as the founding myths of Western civilization: the legend of the rape of the Sabine women is based on an actual event in the early history of the city of Rome. Many authors have pointed out that access (including, of course, forcible access) to reproductive females is a constant theme in the genesis and prosecution of warfare. The Old Testament contains numerous examples of such forcible reproductive access, including several cases in which God specified which females (young, but not yet pregnant) were to be forcibly taken for reproductive purposes and which females (pregnant, old, or infertile) were to be killed, along with all males (including children).

Rape as a constant in warfare has of course continued to the present day (cf. Beevor, 2002). That rape would result in increased reproductive success on the part of soldiers is fairly obvious. What is not obvious is that this would also result in increased reproductive success on the part of the females being raped. So long as being raped does not result in injury or death, and so long as the person being raped is not subsequently harmed or placed under conditions of increased risk of harm, having been raped by a soldier would result in essentially the same increase in reproductive success as any other form of copulation. That the male Yanomami studied by Chagnon (1988) who participated in raids on neighboring groups in which females were forcibly abducted would have an increased reproductive success as a result has not been seriously questioned (except see Ferguson, 2001). What has not been systematically investigated are the effects of such abduction on the reproductive success of the females so abducted. In the absence of such data, the idea that such victims could indeed benefit (in a purely Darwinian sense) from the affects of warfare remains at present an interesting but untested hypothesis.

Being on the losing side in a warlike conflict need not be entirely negative for males either. From a Darwinian standpoint, what matters is reproductive success, not happiness or freedom from oppression. In ancient Rome, it was quite common for members of the conquered peoples to be pressed into slavery by the Romans. Although being a slave under such conditions might not be what one would have preferred, it was quite common for slaves to be allowed to marry and have children. Indeed, if the children of a slave became the property of the slave's owner, then there would have been a positive incentive for the slave owner to encourage the fecundity of his slaves. The point here is obviously not to endorse slavery but rather to point out that there are conditions under which the losers in a warlike conflict might benefit from participation in such conflict almost as much as the winners.

In sum, then, we may also conclude that the capacity for warfare, like the capacity for religious experience, has many of the characteristics of an evolutionary adaptation:

- the capacity for warfare is pan-specific in humans, although there is considerable variation in this capacity, both within and between human groups,
- although it has not yet been possible to correlate the capacity for warfare with underlying neurological structures and processes, there is ample evidence for a correlation between aggressive and violent behavior and the emotional control centers of the brain,
- the capacity for warfare can be correlated with known evolutionary environments of adaptation (raiding with hunting/gathering and pastoral agriculture, militia warfare with settled agriculture, and professional warfare with large-scale agriculture and urban culture), and
- although no underlying genetic mechanisms for the development of the capacity for warfare are now known, the existence of consistent cross-cultural patterns of group violence and coercion indicates that warlike behavior is subject to evolutionary convergence in a manner analogous to other evolutionary adaptations.

The Evolution of Religion: The Standard Model

Before turning to the crux of the argument, it is necessary to consider in more detail what the capacity for religious experience consists of. Recent work on the evolutionary dynamics of religion have converged on a "standard model" in which religions are treated as epiphenomena of human cognitive processes dealing with the detection of and reaction to agents, especially human agents, under conditions of stress anxiety and perceived threat. Boyer has proposed a comprehensive theory of the evolution of religion based on an underlying cognitive process whereby "Our minds are prepared [to give] us particular mental predispositions" (2001, p. 3). In particular, "evolution by natural selection gave us a particular kind of mind so that only particular kinds of religious notions can be acquired" (p. 4).

Boyer begins by asserting that "[r]eligion is about the existence and causal powers of nonobservable entities and agencies" (p. 7). He then proceeds to show that the common explanations for the origin of religion—explanations of puzzling physical and mental phenomena, explanations of evil and suffering, provision of comfort in times of adversity, and provision of the moral basis for social order—cannot be reduced to or included in an explanation of the evolutionary origin of the capacity for religious experience (pp. 5–12). Boyer then points out that "there is only a limited catalogue of possible supernatural beliefs" (p. 29). This is because "[t]he religious concepts we observe are relatively successful ones selected among many other variants" (p. 32). Therefore, "religion emerges ... in the selection of concepts and the selection of memories" (p. 33).

What are the criteria by which certain concepts are reinforced and others are lost? Following Sperber (1985), Boyer distinguishes between simple concepts and templates. The latter are large-scale concepts that subsume many smaller, simpler concepts, essentially by analogy. For example, the word "animal" designates a template, which is usually applied to any entity that is obviously alive (especially because it moves under its own power and with intentionality), eats things, reproduces, and has a general body plan that conforms to what most people would agree is an animal body plan. According to this model of mental classification, religious concepts are easily transmitted from person to person because they both conform to such templates in most respects but violate them in obvious and memorable ways: they "surprise people by describing things and events they could not possibly encounter in actual experience" (p. 55).

In this way, religious concepts are much more easily remembered and transmitted than nonreligious concepts:

> Some concepts ... connect with inference systems in the brain in a way that makes recall and communication very easy. Some concepts ... trigger our emotional programs in particular ways. Some concepts ... connect to our social mind. Some ... are represented in such a way that they soon become plausible and direct behavior. The ones that do *all* this are the religious ones we actually observe in human societies. They are most successful because they combine features relevant to a variety of mental systems. (Boyer, 2001, p. 50)

Central to Boyer's theory on the evolution of the capacity for religious experience is the concept of *agency*:

- **Agency** is that set of characteristics by which we infer the existence and action of an *agent*; that is, a living (or lifelike) entity whose behavior indicates that it has intentions and can act upon them. Agents are purposeful, and purposeful (i.e., teleological) action is the hallmark of agency.

Along with other cognitive and evolutionary psychologists, most notably Barrett (1996), Boyer asserts that the ability to detect agency has high selective value. Barrett points out that humans, like other potential prey animals, should have "hyperactive agency detectors," because any human who did not would be more likely to be injured or killed by a predator. Selection for ultrasensitive agency detectors would result in a tendency for such detectors to produce "false positives," that is, the tendency to infer the existence of agency in an entity in which it is absent.

Although Boyer seems to be on the right track, there is a strong implication throughout his work that the capacity for religious experience is an epiphenomenon that arises secondarily as the result of the action of agency detection and the increased mnemonic transmissability of concepts that violate cognitive templates. J. Donovan (1994, 2002) disagrees: for him, "religion has direct evolutionary advantages that have been directly selected. That benefit relates to the mitigation of existential anxiety with its roots in death awareness" (2002, p. 18). Donovan asserts that religion arises primarily as the result of the selectively positive effect of the reduction of anxiety arising as the result of the awareness of death. Donovan looked at the ability of "spiritual healers" to enter into possession trances, and he concluded that this is arguably a genetically based ability that has been selected as a belief-enhancing mechanism by which the palliative effect of religious participation can be rendered (1994, personal communication, March 19, 2003).

Atran incorporates Boyer's argument from cognitive processes into a more comprehensive selectionist explanation for the evolution of the capacity for religious belief and behavior. Atran agrees with Boyer that there are underlying cognitive (and therefore presumably neurological) processes by which certain types of beliefs can be spread with greatly increased ease and fidelity of transmission. However, he adds a social and political dimension to Boyer's argument, tying religion to the establishment and maintenance of social organization and political power. He quotes Irons (1996) to the effect that "[r]eligions in large-scale societies all show evidence of social dominance" (Atran, 2002, p. 103). Atran goes on to point out that religious rituals usually involve submissive displays, such as kneeling, bowing, prostration, hand spreading, and throat baring, which he likens to the submissive displays of subordinate nonhuman primates (p. 127). Taking this line of reasoning further, Atran points out that "[h]uman worship requires even dominant individuals to willingly submit to a higher moral authority in displays of costly, hard-to-fake commitment or risk losing the allegiance of their subordinates" (p. 127).

This is the heart of Atran's argument: that religion forms a kind of "social glue" that uses ritualized demonstrations of commitment to supernatural authority to encourage and even coerce individual adherence to group norms and goals. Atran's argument is essentially that all members of a society (i.e., a "group," in the parlance of group selection), from the most subordinate to the most dominant, benefit from the social cohesion and singularity of purpose that religion fosters. He states that "[t]he more a ruler sacrifices and suffers, the more the ruler earns respect and devotion" (p. 127). But clearly the same principle would apply to his subordinates, at any level: individual demonstrations of sacrifice and suffering (or at least the willingness to do so) on behalf of the group tend to encourage group solidarity.

There are two problems with this outlook: it assumes that costs and benefits are shared approximately equally throughout such groups, and it implicitly focuses on the group as the primary unit of selection. Atran is clearly aware of the first of these shortcomings. He refers to the classical Marxist "coercion argument" for the origin of religion, by which he means that "religion was [according to Marx] created by and for rulers to materially exploit the ruled, with ... secondary benefits to the oppressed masses of a low but constant level of material security" (as cited in Atran, p. 128). He then goes on to cite Diamond's theory that in large-scale societies (by which he presumably means settled agricultural societies with a mixed village and urban settlement pattern), the members of the ruling hierarchy (or "kleptocracy") gains the support of their subordinates by "constructing an ideology or religion justifying kleptocracy" (Diamond, 1997, p. 277). Diamond asserts that this reification of the supercision of the ruling hierarchy by religion represents a fundamental shift from the situation in bands and tribes (of hunter-gatherers and pastoralists), in which "supernatural beliefs ... did not serve to justify central authority" (p. 277).

Diamond concludes his discussion of the origin of religion by asserting that institutionalized religion confers two important benefits to centralized societies:

- shared ideology or religion helps solve the problem of how (genetically) unrelated individuals can cooperate, by providing a bond not based on (genetic) kinship, and
- religion gives people a motive, rather than genetic self-interest, for sacrificing their lives on the behalf of others (Diamond, p. 278).

I believe that both Atran and Diamond are on the right track, but their arguments are derailed by a common misapplication of selectionist thinking. Atran proposes what appears to be a relatively weak counterargument to the coercion argument, pointing out that religions can be liberating as well as oppressive (Atran, p. 129). While this is true in some cases, I believe it misses the point: if religion (or, more properly, the *capacity* for religious experience) is to evolve by natural selection, it must do so at the level of individuals in the context of specific ecological circumstances.

If Williams is correct about the nonexistence of group selection, then for the capacity for religion to evolve it must somehow increase *individual* reproductive success. In the context of small, relatively nonhierarchic bands or tribes of hunter-gatherers or nomadic pastoralists, it seems most likely that selection at the level of individuals would result in behaviors that would approximate those observed in hunting groups of primates and social carnivores. Although there are clearly recognized dominant individuals in such groups, all of the members of such groups clearly benefit from their membership in them. This is because there are circumstances in which groups of cooperative individuals can obtain resources that would be out of the reach of individuals acting on their own.

This same argument applies at all levels of social organization. For particular social processes to evolve by classical Darwinian selection, there must be some benefit that accrues to individuals from participating in such processes, a benefit that equals if not clearly supercedes the benefit to be gained from acting alone. That such benefits to individuals do result from highly organized social interactions in human and other animal societies is not in question. What is still to be decided here is whether such increases in individual fitness can be observed as the result of the capacity for religious experience, specifically in the context of warfare.

The Capacity for Religious Experience Has Evolved via Individual Selection Among Humans in the Context of Warfare

Here we come to the crux of my argument: that the capacity for religious experience is an adaptation that facilitates warfare. Let me begin by carefully defining the following terms:

- **The capacity for religious experience** is the capacity to formulate, communicate, and act on beliefs (that is, concepts, memories, and intentions or plans) that include reference to supernatural entities and processes. Like the capacity for language, such a capacity must be based on a corresponding neurological "hard wiring," although the dimensions (and limitations) of such neurological structures and processes await further investigation.
- **Religion** is not the specific content of the beliefs that arise from such a capacity. Rather, religion is the overall pattern of such beliefs, including concepts like omniscience, omnipotence, and omnipresence (on the part of supernatural deities), the existence of a soul that is separable from (and can live on after the death of) a physical body, and the existence of supernatural realms inaccessible to normal senses but accessible to deities and incorporeal entities such as souls.

Note that this definition of religion is not as inclusive as that used by Boyer, who includes not only the concepts and entities noted above, but also virtually all forms of "folk belief" (i.e., superstition) (2001). It seems likely to me that the origin of the capacities for both folk beliefs and religion has its roots in the same neurological substrate: a neurological mechanism that reduces anxiety in the face of stress induced by unknown, unpredictable, and presumably dangerous circumstances However, part of my thesis here is that true *religious* experience is a later development in the evolution of the human mind (and presumably the human nervous system), one that has evolved as the result of individual selection primarily in the context of warfare.

How, precisely, does the capacity for religious experience evolve in the context of warfare? Consider the decision that each potential combatant must make prior to participating in a raid, a battle, or an extended military campaign. This decision will include (but is certainly not limited to) the following:

- The probability that one will be seriously injured or killed in the raid, battle, or campaign,
- The possible consequences of *not* participating (e.g., everything from social disapprobation to summary execution),
- The probability that one will gain something (e.g., resources, social position, access to mates, etc.) as the result of one's participation, and
- The quality of such gains, especially when compared with the costs of nonparticipation.

It is important to note that the calculation of such costs and benefits need not be overtly conscious. Whether conscious or unconscious, the outcome of such a calculation would be either an increased or decreased motivation to participate in the impending conflict.

What happens during war? According to von Clausewitz,

> War is nothing but a duel on an extensive scale. If we would conceive as a unit the countless number of duels which make up a war, we shall do so best by supposing to ourselves two wrestlers. Each strives by physical force to compel the other to submit to his will: his first object is to throw his adversary, and thus to render him incapable of further resistance. (1832, p. 12)

War involves violent force, up to and including killing people. To participate in a war means to participate in an activity in which there is a significant probability that one will either kill other people or will be killed by them.

This means that any participant in warfare is faced with the possibility of painful and violent death as the result of such participation. Given this probability, if natural selection acts at the level of individuals, how can natural selection result in a propensity to participate in warfare? Clearly, either the probability that one will be killed must be perceived as low or the potential payoff from such participation must be perceived as high. If natural selection is to operate at the level of individuals, these two circumstances should ideally be obtained simultaneously,

Here is where the capacity for religious experience is crucial. By making possible the belief that a supernatural entity knows the outcome of all actions and can influence such outcomes, that one's "self" (i.e., "soul") is not tied to one's physical body, and that if one is killed in battle, one's essential self (i.e., soul) will go to a better "place" (e.g., heaven, valhalla, etc.) the capacity for religious experience can tip the balance toward participation in warfare. By doing so, the capacity for religious belief not only makes it possible for individuals to do what they might not otherwise be motivated to do, it also tends to tip the balance toward victory on the part of the religiously devout participant. This is because success in battle, and success in war, hinges on commitment: the more committed a military force is in battle, the more likely it is to win, all other things being equal. When two groups of approximately equal strength meet in battle, it is the group in which the individuals are more committed to victory (and less inhibited by the fear of injury or death) that is more likely to prevail. To give just one example, the battle cry and motto of the clan Neil has always been "*Buaidh na bas*!"—"Victory or death!"

Religions tell people what they most want to hear: that those agents and processes that they most fear have no ultimate power over them or pose no threat to themselves or the people they care about. In particular, by providing an intensely memorable, emotionally satisfying, and tension-releasing solution to the problem of mortality, religions make it possible for warriors to master their anxieties and do battle without emotional inhibitions. This makes them much more effective warriors, especially in the hand-to-hand combat that humans have fought throughout nearly all of our evolutionary history.

Consider the characteristics that are most often cited as central to religious experience. Newberg and d'Aquili have presented an integrated model of the neurobiological underpinnings of religious experience. They have pointed out that central to most religious experience is a sensation of awe, combined with "mildly pleasant sensations to feelings of ecstasy" (2001, p. 89). They have shown that such sensations can be induced by rhythmic chanting and body movements, combined with loud music and colorful visual displays, all of which produce a condition of sensory overload. This process then induces a neurological condition characterized by a sense of depersonalization and ecstatic union with one's surroundings.

This is precisely what happens as the result of military drill and training. It is no accident that humans preparing for war use exactly the same kinds of sensory stimuli described by Newberg and d'Aquili. They have tied such displays to religious activities and shown the deep similarities between religious rituals and secular ones: "patriotic rituals ... emphasize the 'sacredness' of a nation, or a cause, or even a flag ... turn[ing] a meaningful idea into a visceral experience" (p. 90). The two types of activities—religious rituals and patriotic rituals—use the same underlying neurological pathways and chemistry.

Religious experience is often equated with a state of mystical union with the supernatural. But what exactly does this mean, and in the context of this chapter, is there a connection between mystical experience and warfare? The answer is almost certainly yes. That combatants have had experiences that would be classified as mystical before, during, and after battle is a simple historical fact. The Scottish flag is based on just such an experience: the white crossed diagonal bars against a field of azure of the St. Andrew's cross is said to have appeared to King Hungus and his warriors during a battle against in the Saxons. Legend says this so encouraged the Scots and frightened their adversaries that a victory was won (Middlemass, 2000).

A common thread in all mystical experiences is a loss of the sense of self and a union with something larger than oneself (Newberg & d'Aquili, 2001, p. 101) Additionally, there is often a sense of submission to a higher power, in which one's personal desires and fears are subordinated to the purposes of that higher power. If that higher power were identified with the leaders of a military hierarchy, it is easy to see how such experiences could be used to increase one's loyalty and submission to that hierarchy.

It is likely that the same underlying neurological circuits that produce the sensations described by mystics also produce the sensations of fear, awe, and ecstasy that are experienced by combatants during the course of a battle. Like the evolutionary implications of rape, this is a topic that is rarely discussed outside of military circles but is a well-known phenomenon during battle. The noise and movement, the confusion and excitement, intensified tremendously by the imminence of injury and death combine to produce a state of massive arousal in the sympathetic nervous system of the combatant. This state of intense arousal is very similar to the state of arousal felt during copulation; indeed, some soldiers will candidly admit that during the heat of battle, they often experience a kind of sexual arousal, leading in some cases to ejaculation. This fusion of sensory and motor states in a condition of intense arousal, combined with a sensation of depersonalization, can easily produce in susceptible individuals a condition in which a kind of "blood lust" overwhelms most thoughts of self-preservation.

The Ultimate Sacrifice

Let us return to group selectionist arguments for the evolution of both religion and warfare. D. S. Wilson (2002) has proposed that the capacity for religion has evolved among humans as the result of selection at the level of groups, rather than individuals. Specifically, he argues that benefits that accrue to groups as the result of individual sacrifices can result in increased group fitness, and this can explain what is otherwise difficult to explain: religiously motivated behaviors (such as celibacy and self-sacrifice) that apparently lower individual fitness as they benefit the group.

At first glance, Wilson's argument seems compelling. Consider the most horrific manifestation of religious warfare: the suicide bomber. A person who blows himself or herself up in order to kill his or her opponents has lowered his or her individual fitness. Does this not mean that such behavior must be explainable only at the level of group selection? Not at all: the solution to this conundrum is implicit in the basic principles of population genetics. Recall that one of Darwin's requirements for evolution by natural selection was the existence of variation between the individuals in a population (1859, pp. 7–59). Variation within populations is a universal characteristic of life, an inevitable outcome of the imperfect mechanism of genetic replication. Therefore, it follows that if the capacity for religious experience is an evolutionary adaptation, then there will be variation between individuals in the degree to which they express such a capacity.

Furthermore, it is not necessarily true that when an individual sacrifices his or her life in the context of a struggle, the underlying genotype that induced that sacrifice will be eliminated by that act. Hamilton's principle of kin selection (1964) has already been mentioned as one mechanism, acting at the level of individuals (or, more precisely, at the level of genotypes), by which individual self-sacrifice can result in the increase in frequency of the genotype that facilitated such sacrifice. Trivers (1971) has proposed a mechanism by which apparently altruistic acts on the part of genetically unrelated individuals may evolve by means of reciprocal altruism.

Given these two mechanisms, all that is necessary for the capacity for religious behavior, including extreme forms of self-sacrifice, to evolve is that as the result of such behaviors, the tendency (and ability) to perform them would be propagated throughout a population. The removal of some individuals as the result of suicide would merely lower the frequency of such tendencies and abilities in the population, not eliminate them altogether. If by making the ultimate sacrifice, an individual who shares his or her genotype with those who benefit by that sacrifice will, at the level of his or her genes, become more common over time (E. O. Wilson, 1975, p. 4).

To the Winner Go the Spoils

Let us now consider the flip side of war: the benefits that accrue to the winners of warlike conflicts. Given the mechanisms of kin selection, one can see how warfare and the religious beliefs that facilitate it might evolve among the closely related kin groups that constitute the raiding parties characteristic of hunting/gathering and pastoral peoples. It is also possible to construct an explanation for militia warfare and professional warfare on the basis of a blend of kin selection and reciprocal altruism. However, a closer examination of the spoils of war make such explanations relatively unnecessary.

Betzig (1986) performed a cross-cultural analysis of the correlation between despotism and reproductive success in 186 different cultures. Her conclusion was that

> [n]ot only are men regularly able to win conflicts of interest more polygynous, but the degree of their polygyny is predictable from the degree of bias with which the conflicts are resolved. Despotism, defined as an exercised right to murder arbitrarily and with impunity, virtually invariably coincides with the greatest degree of polygyny, and presumably, with a correspondingly high degree of differential reproduction. (Betzig, 1986, p. 88)

In other words, males who most successfully use violence and murder as a means of influencing the actions of others have historically had the most offspring. In the context of warfare, this means that the winners of a battle, or even more so, of a war will pass on to their offspring whatever traits facilitated their victory, including the capacity to believe in a supernatural force that guides their destiny and protects them in battle. The effects of such capacities are not trivial; as Betzig points out, the differences between the reproductive success of the winners of violent conflicts and the losers is measured in orders of magnitude. As noted earlier, wars are bottlenecks through which only a relative few may pass, but which reward those who do with immensely increased reproductive success.

Putting all of this together, it appears likely that the capacity for religious experience and the capacity for warfare have constituted a coevolutionary spiral that has intensified with the transitions from a hunting/gathering existence through subsistence agriculture to the evolution of the modern nation-state. As pointed out earlier, there is a correlation between the type of intergroup violence and the ecological context within which that violence occurs. Generally speaking, raiding/rustling is correlated with hunting/gathering and pastoralism, militia warfare with village agriculture, and professional warfare with urban society and the nation-state. There is a corresponding progression in the basic form of religious experience and practice: animism is most common among hunter-gatherers, while polytheism is more common among agriculturalists, and monotheism is most common in societies organized as nation-states. This is not to say there are no exceptions to this correlation. However, the fact that such a correlation can even be made points to the underlying ecological dynamics driving the evolution of subsistence patterns, patterns of warfare, and types of religious experience.

Genes, Memes, or Both?

It is extremely unlikely that any human behavior (or the behavior of any animal with a nervous system complex enough to allow learning) is the result of the expression of any single gene. On the contrary, it is almost universally accepted among evolutionary psychologists that all behaviors show a blend of innate and learned components. What is interesting to ethologists is not the question of "how much," but rather the much simpler question of "how"?

One answer that has been suggested is that there are two different carriers of information that can be transmitted among humans: genes and memes. According to Dawkins, a meme is "a unit of cultural transmission" corresponding to things like "tunes, ideas, catch-phrases, clothes fashions, ways of making pots or of building arches" (1976, p. 206). Dawkins even addressed the possibility that God Himself might be a meme:

> Consider the idea of God .... What is it about the idea of a god which give it its stability and penetrance in the cultural environment? The survival value of the god meme in the meme pool results from its great psychological appeal. It provides a superficially plausible answer to deep and troubling questions about existence. It suggests that injustices in this world may be rectified in the next .... God exists, if only in the form of a meme with high survival value, or infective power, in the environment provided by human culture. (p. 207)

Is all of religion simply a meme, or more precisely, a "meme complex"? And does the answer to this question tell us anything about the connection between the capacity for religion and warfare? There are at least three hypotheses for the mode of transmission of the capacity for religious experience:

- **Hypothesis 1:** The capacity for religious experience might be almost entirely innate; that is, it arises almost entirely out of "hard-wired" neural circuits in the human brain, which produce the sensations, thoughts, and behaviors that we call religious.
- **Hypothesis 2:** The capacity for religious experience might be almost entirely learned; that is, it arises almost entirely from concepts (i.e., "memes") that are transmitted from person to person via purely linguistic means, and without any underlying neurological predisposition to their acquisition.
- **Hypothesis 3:** The capacity for religious experience might arise from a combination of innate predispositions and learning; that is, like many animal behaviors, the capacity for religious experience might be the result of an innate predisposition to learn particular memes.

Both Boyer's and Atran's theories of the origin of religion are closest to the third hypothesis. From the foregoing analysis, it should also be clear that my own hypothesis for the origin of the capacity for religious experience is closest to hypothesis 3. However, unlike Boyer and Atran, I have proposed that the specific context within which the human nervous system has evolved has been persistent, albeit episodic, warfare.

A common objection to the hypothesis that the capacity for religious experience is an evolutionary adaptation is that there has been insufficient time for natural selection to produce the vast diversity in religious experiences and practices that exists in our species. I think there are two responses to this objection. First, although the diversity of religious beliefs and practices is quite surprising at first glance, this diversity is neither unlimited nor devoid of general trends. For example, virtually all religions include supernatural entities. However, the class of actual supernatural entities is not unlimited. Indeed, most supernatural entities bear a strong resemblance to humans, although with some qualities that humans are not observed to possess, such as the ability to fly, pass through walls, hear other's thoughts, etc. Furthermore, the qualities of most deities are remarkably similar to those attributed to kings, priests, and military leaders, although to a greater extent and with fewer "human" limitations. The global pantheon is overpopulated with warrior gods, and this overpopulation is not accidental.

Furthermore, there are circumstances under which selection can produce a dramatically accelerated rate of evolutionary change. Lumsden and Wilson (1981, 1983) describe this kind of evolutionary change as "autocatalytic gene-culture coevolution" (1981, p. 11) According to their theory, genes prescribe, not specific behaviors, but rather epigenetic rules of development by which minds are assembled (1983, p. 117). The mind then grows by incorporating parts of the culture (i.e., memes) already in existence. Culture, therefore, is created constantly from the combined decisions and innovations of all of the members of society. Most importantly, some individuals possess genetically inherited epigenetic rules that enable them to survive and reproduce better than other individuals. Consequently, the more successful epigenetic rules spread through the population, along with the genes that encode them. In other words, culture is created and shaped by biological processes, while those same biological processes are simultaneously altered in response to further cultural change. Genes and memes coevolve, with each change in one catalyzing a corresponding change in the other (1983, pp. 117–118).

The primary reason for the accelerated rate of evolution that results from gene/meme coevolution is the alternation between the temporal modes of the two types of evolution. If one conceives of time as passing along a vertical axis, then genetic transmission is almost entirely vertical. That is, genes are passed from parents to offspring. Genetic transmission also involves a very low mutation rate, relative to memetic evolution. Memetic transmission, by contrast, is both vertical and horizontal. That is, memes can be transmitted between contemporaries, as well as between parents and offspring. Furthermore, as Boyer has pointed out, the mutation rate of memes is immensely higher than that of genes. "Cultural memes undergo mutation, recombination, and selection inside the individual mind every bit as much and as often (in fact probably more so and more often than) during transmission between minds" (Boyer, 2001, p. 39).

Combining the concept of gene/meme coevolution with the episodic nature of selection during warfare, it appears that the evolution of the capacity for religious experience evolves via a kind of bootstrap effect. Each change in the underlying neurological capacity for religious experience is followed by a corresponding change in the conceptual (i.e., "memetic") structure of the religions that are produced as a result of that capacity. This, in turn, sets the stage for further selection at the level of genes, as individuals with particular religious meme complexes succeed (or fail). Stir warfare into the mix, including the tremendous assymetries in reproductive success described by Betzig (1986), and it appears likely that a substantial fraction of the whole of what we call "religion" is the result of gene/meme coevolution in the context of intergroup warfare.

New Directions in the Evolution of Religion and Warfare

Given the current state of our knowledge of the underlying neurobiology of religious experience, the foregoing amounts to little more than a tantalizing hypothesis for the evolution of the capacity for religious experience. However, it suggests some avenues of investigation that would help to clarify the relationships between the capacity for religious experience and warfare. For example, it would be very interesting to know whether and to what extent religious experience and concomitant beliefs are reinforced by participation in warfare, and whether there is a positive or negative effect on such experiences and beliefs as the result of being on the winning or losing side in a warlike conflict. Collection of what would essentially be natural history data on the prevalence, spread, or disappearance of religious experiences or beliefs in the context of warfare vs. peacetime would help to determine both rates of change and possible mechanisms of spread or extinction. Empirical studies using controlled test populations could also shed light on the connections between religious experiences and beliefs and stress and perceptions of potential threat. Finally, and most importantly, detailed demographic analysis of reproductive success and religious beliefs, especially as they relate to a history of warfare, might find the kinds of correlations suggested here.

In closing, it seems likely that throughout the history of our species warfare has contributed significantly to the evolution of the capacity for religious experience, which has in turn facilitated warfare. Intergroup warfare can be adaptive whenever resources are concentrated, predictable, and defensible. Agriculture and industrial/urban subsistence patterns have facilitated warfare but have also steadily increased its costs. High technology warfare, especially when waged using weapons of mass destruction, has greatly increased the costs of warfare without appreciably increasing its benefits. In an age when the decisions of a single military leader can unleash nuclear annihilation, warfare is clearly maladaptive. As a consequence, it may be desirable to eliminate, or at least redirect, our capacity for warfare. However, if the deep evolutionary connections between the capacities for religion and warfare that I have proposed do, in fact, exist, this may mean redirecting (or possibly eliminating) the capacity for religious experience. Only time will tell, and only God (if He exists) knows how much time we have left.

Note

This chapter originally appeared under the same title in *Evolution and Cognition*, volume 10, number 1, 2004, pages 43–60. Used by permission.

References

Atran, S. (2002). *In gods we trust: The evolutionary landscape of religion.* Oxford, England: Oxford University Press.

Axelrod, R. (1984). *The evolution of cooperation.* New York: Basic Books.

Barrett, J. L. (1996). *Anthropomorphism, intentional agents, and conceptualizing God.* Unpublished Ph.D. dissertation, Cornell University, Ithaca, New York.

Beevor, A. (2002). *The fall of Berlin 1945.* New York: Viking.

Betzig, L. (1986). *Despotism and differential reproduction: A Darwinian view of history.* New York: Aldine.

Bowlby, J. (1969). *Attachment.* New York: Basic Books.

Boyer, P. (2001). *Religion explained: The evolutionary origins of religious thought.* New York: Basic Books.

Chagnon, N. (1988). Life histories, blood revenge, and warfare in a tribal population. *Science, 239,* 985–992.

Chagnon, N. (1990). Reproductive and somatic conflicts of interest in the genesis of violence and warfare among tribesmen. In J. Haas (Ed.), *The anthropology of war* (pp. 77–104). Cambridge, England: Cambridge University Press.

Chomsky, N. (1965). *Aspects of the theory of syntax.* Cambridge, MA: MIT Press.

d'Aquili, E. G., & Newberg, A. B. (1999). *The mystical mind: Probing the biology of religious experience.* Minneapolis, MN: Fortress.

Darwin, C. R. (1859). *On the origin of species by means of natural selection, or the preservation of favoured races in the struggle for life.* London: Murray.

Darwin, C. R. (1871). *The descent of man and selection in relation to sex.* London: Murray.

Dawkins, R. (1976). *The selfish gene.* Oxford, England: Oxford University Press.

Dawkins, R. (1982). *The extended phenotype.* San Francisco: Freeman.

Diamond, J. (1997). *Guns, germs, and steel: The fates of human societies.* New York: W. W. Norton.

Donovan, J. (1994). Multiple personality, hypnosis, and possession trance. *Yearbook of Cross-Cultural Medicine and Psychotherapy,* pp. 99–112.

Donovan, J. (2002). Implicit religion and the curvilinear relationship between religion and death anxiety. *Implicit Religion, 5*(1), 17–28.

Emlen, S. T. (1976). An alternative case for sociobiology. *Science, 192,* 736–738.

Ferguson, B. (2001). Materialist, cultural, and biological theories on why Yanomami make war. *Anthropological Theory, 1*(1), 99–116.

Goodall, J. (1986). *The Chimpanzees of Gombe.* Cambridge, MA: Belknap.

Hamilton, W. D. (1964). The genetical theory of social behavior. *Journal of Theoretical Biology, 12*(1), 1–52.

Irons, W. (1996). Morality, religion, and human nature. In W. M. Richardson and W. Wildman (Eds.), *Religion and science.* New York: Routledge.

Kruuk, H. (1972). *The spotted hyena: A study of predation and social behavior.* Chicago: University of Chicago Press.

Lenneberg, E. H. (1964). A biological perspective on language. In E. H. Lenneberg (Ed.), *New directions in the study of language* (pp. 65–88). New York: Wiley.

Lenneberg, E. H. (1967). *Biological foundations of language.* New York: Wiley.

Lumsden, C. J., & Wilson, E. O. (1981). *Genes, mind, and culture.* Cambridge, MA: Harvard University Press.

Lumsden, C. J., & Wilson, E. O. (1983). *Promethean fire: Reflections on the origin of mind.* Cambridge, MA: Harvard University Press.

Margulis, L. (1997). Big trouble in biology: Physiological autopoiesis versus mechanistic neo-Darwinism. In L. Margulis and D. Sagan (Eds.), *Slanted truth: Essays on Gaia, symbiosis, and evolution* (pp. 265–282). New York: Springer-Verlag.

Mech, L. D. (1970). *The wolf: The ecology and behavior of an endangered species.* Garden City, NY: Natural History Press.

Middlemas, T. (2000). Legendary origin of the [Scottish] flag. [Electronic Version] VisualNet. Retrieved March 30, 2003, from http://www.fotw.ca/flags/gb-scotl.html#leg

Miles, J. (1995). *God, A biography.* New York: Knopf.

Newberg, A. B., & d'Aquili, E. G. (2001). *Why God won't go away: Brain science and the biology of belief.* New York: Ballantine.

Penfield, W., & Roberts, L. (1959). *Speech and brain mechanisms.* Princeton: Princeton University Press.

Persinger, M. A. (1987). *Neuropsychological bases of God beliefs.* New York: Praeger.

Pinker, S. (1994). *The language instinct.* New York: Morrow.

Pinker, S., & Bloom, P. (1999). Natural language and natural selection. In J. H. Barkow, L. Cosmides, & J. Tooby (Eds.), *The adapted mind: Evolutionary psychology and the generation of culture* (pp. 451–493). Oxford, England: Oxford University Press.

Ramachandran, V. S., & Blakeslee, S. (1998). *Phantoms in the brain: Probing the mysteries of the human mind.* New York: Morrow.

Ridley, M. (1996). *Evolution* (2nd ed.). Cambridge, MA: Blackwell Science.

Saver, J. L., & Rabin, J. (1997). The neural substrates of religious experience. *Journal of Neuropsychiatry, 9*(3), 498–510.

Sober, E., & Wilson, D. S. (1998). *Unto others: The evolution and psychology of unselfish behavior.* Cambridge, MA: Harvard University Press.

Sperber, D. (1985). Anthropology and psychology: Towards an epidemiology of representations. *Man, 12,* 73–89.

Teleki, G. (1973). *The predatory behavior of wild chimpanzees.* Lewisburg, PA: Bucknell University Press.

Thornhill, R., & Palmer, C. T. (2000). *A natural history of rape: Biological bases of sexual coercion.* Cambridge, MA: MIT Press.

Tooby, J., & Cosmides, L. (1988). *The evolution of war and its cognitive foundations.* Institute for Evolutionary Studies Technical Report No. 88-1.

Trivers, R. L. (1971). The evolution of reciprocal altruism. *Quarterly Review of Biology, 46*(4), 35–57.

Von Clausewitz, C. (1832). *On war* (J. J. Graham, Trans.) Berlin: Dümmlers Verlag [Electronic version] Retrieved March 28, 2003, from http://www.clausewitz.com/CWZHOME/On_War/ONWARTOC.html

Wallace, A. (1968). Psychological preparations for war. In M. Freud, M. Harris, & R. Murphy (Eds.), *War.* New York: Natural History Press.

Williams, G. C. (1966). *Adaptation and natural selection.* Princeton, NJ: Princeton University Press.

Wilson, D. S. (2002). *Darwin's cathedral: Evolution, religion, and the nature of society.* Chicago: University of Chicago Press.

Wilson, E. O. (1975). *Sociobiology: The new synthesis.* Cambridge, MA: Belknap.

Wrangham, R. (1999). Evolution of coalitionary killing. *Yearbook of Physical Anthropology,* 1–29.

Wrangham, R., & Peterson, D. (1996). *Demonic males: Apes and the origins of human violence.* Boston: Houghton Mifflin.

Wright, S. (1968–1978). *Evolution and genetics of populations* (Vols. I–IV). Chicago: University of Chicago Press.

Wynne-Edwards, V. C. (1962). *Animal dispersion in relation to social behavior.* Edinburgh, Scotland: Oliver and Lloyd.

[18]

Ritual, Emotion, and Sacred Symbols

The Evolution of Religion as an Adaptive Complex

Candace S. Alcorta and Richard Sosis
University of Connecticut

This paper considers religion in relation to four recurrent traits: belief systems incorporating supernatural agents and counterintuitive concepts, communal ritual, separation of the sacred and the profane, and adolescence as a preferred developmental period for religious transmission. These co-occurring traits are viewed as an adaptive complex that offers clues to the evolution of religion from its nonhuman ritual roots. We consider the critical element differentiating religious from nonhuman ritual to be the conditioned association of emotion and abstract symbols. We propose neurophysiological mechanisms underlying such associations and argue that the brain plasticity of human adolescence constitutes an "experience expectant" developmental period for ritual conditioning of sacred symbols. We suggest that such symbols evolved to solve an ecological problem by extending communication and coordination of social relations across time and space.

KEY WORDS Adolescence; Costly signals; Emotion; Neuropsychology; Religion; Ritual; Symbolic thought

The evolution of religion and its possible adaptive function have been the subject of considerable recent investigation by a wide array of researchers with diverse theoretical and methodological approaches. Cognitive scientists and evolutionary psychologists have been prominent among these researchers (Atran 2002; Barrett 2000; Bering 2005; Boyer 2001; Bulbulia 2004a, 2004b; Guthrie 1993; Kirkpatrick 1999; Mithen 1996, 1999). They have primarily studied religion in terms of beliefs, uncovering the psychological mechanisms that produce supernatural agents in all cultures. With the notable exceptions of Bering (2005) and Bulbulia

Received August 11, 2004; accepted October 28, 2004; final version received December 7, 2004.

Address all correspondence to Candace S. Alcorta, Department of Anthropology U-2176, University of Connecticut, Storrs, CT 06269-2176. Email: candace.alcorta@uconn.edu

(2004a), these researchers have concluded that religion constitutes a by-product of cognitive adaptations selected for "more mundane" survival functions. Evolutionary anthropologists have also revitalized studies of religion over the past two decades (see Sosis and Alcorta 2003). In contrast to the cognitive scientists, however, these researchers have tended to focus on religious behaviors rather than beliefs. The primary debate among these investigators has centered on the relative importance of group selection and individual selection in the evolution of religious systems (Cronk 1994a, 1994b; Rappaport 1994; Sosis 2003a; Sosis and Alcorta 2003; Wilson 2002). Drawing on both ethological studies and a rich theoretical legacy beginning with Durkheim (1969), evolutionary anthropologists have proposed that religious behaviors constitute costly signals that contribute to social cohesion (Cronk 1994a; Irons 1996a, 1996b, 2001; Sosis 2003b). These theorists situate religious ritual within a broader, nonhuman evolutionary continuum related to socially adaptive behaviors. Costly signaling theory has received empirical support from the research of Sosis and colleagues (Sosis 2000; Sosis and Bressler 2003; Sosis and Ruffle 2003, 2004), whose work has demonstrated a significant and positive association between participation in religious ritual and enhanced cooperation. However, these researchers have yet to examine how the high levels of cooperation observed within religious communities (e.g., Sosis and Bressler 2003; Sosis and Ruffle 2003) translate into individual fitness gains.

Although not guided by evolutionary analyses, the cumulative findings of a third body of research that has emerged over the past two decades does provide evidence of individual benefits for religious practitioners. This work has been conducted by sociologists, epidemiologists, psychologists, and physicians, and has explored the health impacts of religion on adherents (Hummer et al. 1999; Levin 1994, 1996; Matthews et al. 1998; Murphy et al. 2000). Accumulating findings from this body of research show significant positive associations between religious participation and individual health. These studies demonstrate decreased mental and physical health risks, faster recovery times for a wide variety of disorders, and greater longevity for those who regularly attend weekly Western religious services, even when social and lifestyle confounds are controlled (Hummer et al. 1999; Matthews et al. 1998; Murphy et al. 2000). In association with ongoing neurophysiological research (Austin 1998; McNamara 2001, 2002; Newberg et al. 2001; Saver and Rabin 1997; Winkelman 1986, 1992, 2000), these findings suggest proximate mechanisms by which religious participation may impact psychoneuroimmunological systems and, thus, individual fitness.

FOUR FEATURES OF RELIGION

These various approaches to religion have provided significant insights, but individually each is insufficient for an evolutionary understanding of religion. A synthesis that encompasses religion's cross-culturally recurrent features and captures that which differentiates the religious from the secular is required. We propose that religion may best be understood as an evolved complex of traits incorporating cognitive, affective, behavioral, and developmental elements selected to solve an adaptive problem. Here we focus on four cross-culturally recurrent features of all religions that we consider to be integral components of this complex. These are:

- Belief in supernatural agents and counterintuitive concepts;
- Communal participation in costly ritual;
- Separation of the sacred and the profane; and
- Importance of adolescence as the life history phase most appropriate for the transmission of religious beliefs and values.

These four elements emerge and reemerge throughout the anthropological and sociological literature and encompass cognitive, behavioral, affective, and developmental aspects of religious systems across a wide variety of cultures (Douglas 1966; Durkheim 1969; Eliade 1958, 1959; Malinowski 1948; Rappaport 1999; Turner 1967, 1969; Tylor 1871). Although each trait may be variably expressed across different socioecological systems, their recurrence in societies as diverse as totemic Arunta hunter-gatherers and Protestant American industrialists suggests that they constitute basic elements of religion.

In this paper we examine each of these traits in relation to an evolutionary theory of religion as an evolved mechanism for social cooperation. We posit that the critical element in the differentiation of religious from nonhuman ritual was the emergence of emotionally charged symbols. Drawing on the seminal insights of Durkheim (1969), Turner (1967, 1969), and Rappaport (1999), we propose proximate mechanisms by which religious ritual serves to invest stimuli with motivational meaning. The brain plasticity of extended human adolescence is examined as an "experience expectant" developmental period for the emotional valencing of emergent symbolic systems. Following Richerson and Boyd (1998) we conclude that the symbolic systems of religious ritual in early human populations solved an ecological problem by fostering cooperation and extending the communication and coordination of social relations across time and space.

Supernatural Agents and Counterintuitive Concepts

Belief in supernatural agents may be the most commonly offered definition of religion (see Sosis and Alcorta 2003). Durkheim (1969) was the first to propose that supernatural agents represent the reification of society itself and function to maintain social order. Although Durkheim's reification of society as a causal explanation for religion has largely fallen into disfavor, his observation that the type of agent represented in a society's religion reflects the social organization of that society has been subsequently supported by the work of Wallace (1966) and the cross-cultural analyses of Swanson (1960).

More recently, Guthrie (1993), and other cognitive scientists (Atran 2002; Barrett 2000; Boyer 2001; Kirkpatrick 1999; Pinker 1997) have reexamined the supernatural beliefs of religious systems and have concluded that such beliefs are merely a "byproduct of numerous, domain-specific psychological mechanisms that evolved to solve other (mundane) adaptive problems" (Kirkpatrick 1999:6). Rejecting any adaptive function of religious beliefs per se, these researchers view the conceptual foundations of religion as deriving from categories related to "folkmechanics, folkbiology, (and) folkpsychology" (Atran and Norenzayan 2004). Supernatural agents, similar to moving dots on computer screens or faces in the clouds, are simply the result of innate releasing mechanisms of agency detection modules evolved to respond to animate, and therefore potentially dangerous, entities (Atran and Norenzayan 2004). Likewise, the attribution of intentionality to supernatural agents is viewed as the application of folkpsychology mental modules evolved in response to complex human social interactions. For many cognitive scientists, supernatural agents, as well as religious beliefs in general, constitute little more than "mental module misapplications."

Anthropological and psychological evidence, however, suggests that supernatural agents of religious belief systems not only engage, but also modify, evolved mental modules. Moreover, they do so in socioecologically specific and developmentally patterned ways. Although agency detection modules probably do give rise to the human ability to imagine a broad array of supernatural agents, those that populate individual religions are neither random nor interchangeable. Whether supernatural agents are envisioned as totemic spirits, ancestral ghosts, or hierarchical gods is very much dependent upon the socioecological context in which they occur (Durkheim 1969; Swanson 1960; Wallace 1966). The types of religious practitioners present, as well as the nature of religious practices performed in a society, have been shown to be significantly correlated with measures of social complexity and integration (Bourguignon 1976; Winkelman 1986, 1992). The shamanic use of trance to communicate with totemic ancestors found among the Athapaskan hunter-gatherers of the Arctic would be very familiar to the desert-dwelling Arunta hunter-gatherers of Australia. Likewise, the presence of priests and hierarchical gods typifies religions of state-level agricultural societies from the Maya of Mexico to the Ashanti of Africa. Cross-cultural statistical research by Swanson (1960) and subsequent analyses by Roes and Raymond (2003) have shown that the presence of moralizing gods "who tell people what they should and should not do" is significantly and positively related to group size, social stratification, environmental resource levels, and extent of external conflict.

The supernatural beings of all these religious belief systems engage evolved mental modules of agency and intentionality, as noted by cognitive scientists. This, however, does not preclude the possibility that religion is an evolved adaptation. As we have argued elsewhere (Sosis and Alcorta 2004), evolution is opportunistic and necessarily co-opts existing traits to solve novel ecological problems. It is the modification of these traits through natural selection that constitutes evolution. The question to be posed, therefore, is not "Does religion incorporate preexistent mental modules?" Instead, the relevant question is whether there exists evidence of adaptation of those modules to solve ecological challenges. Recent experimental work by developmental psychologists suggests that the answer to this question is "yes." The supernatural agents of religious belief systems incorporate attributes of agency and intentionality, but they also possess an additional attribute not shared with natural category agents. In contrast to natural category agents, the supernatural agents of religious belief systems are "full access strategic agents" (Boyer 2001). They are "envisioned as possessing knowledge of socially strategic information, having unlimited perceptual access to socially maligned behaviors that occur in private and therefore outside the perceptual boundaries of everyday human agents" (Bering 2005:419). Moreover, accumulating research indicates that humans exhibit a developmental predisposition to believe in such socially omniscient supernatural agents, appearing in early childhood and diminishing in adulthood (Bering 2005; Bering and Bjorklund 2004). Cross-cultural studies conducted with children between the ages of 3 and 12 indicate that young children possess an "intuitive theism" (Kelemen 2004) that differentiates this social omniscience of supernatural agents from the fallible knowledge of natural social agents (Bering 2005). As the child's theory of mind develops, parents and other natural agents are increasingly viewed as limited in their perceptual knowledge. Supernatural agents, however, not only remain socially omniscient, but are viewed by children in late childhood as agents capable of acting on such knowledge. This developmental predisposition to believe in socially omniscient and declarative supernatural agents contrasts with evolved mental modules of folkpsychology for natural categories. It also goes far beyond natural agency-detection modules to encompass socially strategic agents with behaviorally motivating characteristics.

Supernatural agents of religious belief systems also diverge from evolved mental modules for natural ontological categories (e.g., animate/inanimate; people/animals) in another significant way. Such agents do not uphold natural categories; they violate them. Totemic animals that can talk, dead ancestors who demand sacrificial offerings and visit the living, and incorporeal gods capable of being in all places at all times violate basic premises of natural ontological categories. Yet, these exceedingly unnatural constructs comprise powerful religious schema that elicit deep devotion and belief across traditional and contemporary cultures alike. If religious beliefs are merely by-products of mental modules evolved to deal with the "natural world," why do such beliefs consistently violate the basic cognitive schema from which they are presumed to derive?

In addressing this question, a number of cognitive scientists have noted that the counterintuitive concepts that characterize religious beliefs are both attention-arresting and memorable (Atran 2002; Boyer 2001; Kirkpatrick 1999). Experimental tests validate these observations (Atran and Norenzyan 2004; Boyer and Ramble 2001). Counterintuitive concepts, such as bleeding statues and virgin births, do grab attention. Atran and Norenzayan (2004) note, however, that the efficacy of counterintuitive concepts in engaging attention, improving recall, and promoting transmission is highly dependent upon the broader context within which these concepts are framed. Comparing belief sets with intuitive and counterintuitive concepts, they found that the specific profile of the counterintuitive/intuitive concepts most frequently encountered in religious belief systems achieved the "highest rate of delayed recall and lowest rate of memory degradation over time" (Atran and Norenzyan 2004:723). Thus, the counterintuitive beliefs of religious systems not only violate natural ontological categories, they do so in a specifically patterned way that renders them maximally memorable and maximally transmissible. This suggests selection for such concepts.

Counterintuitive concepts have yet another important feature of significance for social groups. In addition to their mnemonic efficacy, they comprise almost unbreakable "codes" for the uninitiated. Most language distortions occur within ontological categories (Bartlett [1932] as reported in Atran and Norenzayan 2004). When distortions do cross ontological boundaries, they are most common from counterintuitive to intuitive concepts; distortions occurring from intuitive to counterintuitive concepts are extremely rare. For example, it would be much more likely for a listener to modify "talking horse" to "walking horse" than the converse. These findings indicate that counterintuitive concepts are not readily generated on the basis of intuitive concepts, and they suggest that the chances of spontaneously re-creating a preexistent counterintuitive concept are exceedingly low. This probability is lowered even further by embedding multiple counterintuitive concepts within religious belief sets. By incorporating counterintuitive concepts within belief systems, religion creates reliable costly signals that are difficult to "fake." They must be learned, and since such learning has been orally transmitted throughout the vast majority of human evolution, this also implies participation in religious ritual. As a result, religious belief systems serve as both costly and reliable signals of group membership.

Finally, the irrationality of counterintuitive concepts contributes to their efficacy as honest signals of commitment to a group who share that belief (Lee Cronk, personal communication 2002). Within a pluralistic context, adherents who propound counterintuitive beliefs risk censure. Early Christian belief in the resurrection of Christ constituted a potent signal to Romans, Jews, and other Christians. Only individuals knowledgeable about the religious tenets of Roman Catholicism would conceive of the transmutation of wine to blood, and only those initiated into the faith through the emotional conditioning of those tenets would truly believe that such a transmutation occurs during the sacrament of Communion. Through the eyes of nonadherents, such beliefs may be viewed as extraordinary and irrational. Such perceptions contribute to both the costliness and the effectiveness of religious signals.

In summary, neither the content nor the structure of religious belief systems supports the assertion that such beliefs constitute epiphenomenal "by-products." Although supernatural agents engage mental modules of agency and intentionality that evolved in response to "mundane" selection pressures, they modify these modules in specific and developmentally patterned ways. Cross-culturally, supernatural agents are integral elements of religious beliefs and they consistently reflect significant socioecological relations within their respective cultures. The agents of religious belief are not natural category agents, as would be predicted if they were simply by-products of mental modules evolved to deal with such agents. They are instead counterintuitive agents that not only modify natural agency module parameters, but do so in consistently patterned and behaviorally significant ways. A developmental propensity to believe that such agents are not only intentional, but also socially omniscient, is indicated by accumulating experimental evidence (Bering and Bjorklund 2004). Although the predisposition to believe in such supernatural agents appears to be innate, the development of such beliefs is dependent upon cultural transmission. Religious cognitive schema exhibit structural elements that maximize transmission through the incorporation of minimally counterintuitive concepts that engage attention, promote recall, and insure exclusivity.

These features of religious belief systems provide ontogenetic lability for the construction of socially relevant moral systems across diverse ecologies, and they do so within a structure that is maximally transmissible and minimally invasive. Bering (2005:430) notes that "children are simultaneously immersed in unique cultural environments where morality is chiefly determined by socioecological conditions. Although there is likely a common 'moral grammar' underlying all children's development in this domain, the moral particulates of any given society are given shape by the demands of local environments." Bulbulia, too, argues that religion, like language, exhibits an innate grammar in which "development consists of fixing labels to preexisting cognitive structures" (J. Bulbulia, personal communication 2004). For both Bulbulia (2004a, 2004b) and Bering (2005), the idea of socially omniscient supernatural agency is a central component of this system. These researchers view the adaptive value of such agents to be the maintenance of group cooperation and cohesion across a broad spectrum of socioecologies. Atran, likewise, acknowledges religion's use of supernatural agents in "maintaining the cooperative trust of actors and the trustworthiness of communication by sanctifying the actual order of mutual understandings and social relations" but asserts that "religion has no evolutionary function per se"(2002:278–279). In contrast, we argue that religion's ability to promote cooperation is its evolutionary function, and that the costliness of religious ritual bears a direct relationship to the nature of the collective action problems faced. When individual costs are high, but the potential benefits of cooperation are great, costly religious ritual provides a reliable mechanism for minimizing free-riding and maximizing cooperation. We consider the cognitive schema of religious systems to be a fundamental evolved element in ensuring such cooperation. Both the ontogenetic and structural features of religious belief systems suggest evolved features. Yet, we also maintain that religious belief systems in isolation are incapable of "sanctifying the actual order of mutual understandings and social relations" (Atran 2002:278). It is certainly possible to be cognizant of religious beliefs without subscribing to them, as any schoolchild who has ever studied Greek mythology can attest. In order for religious beliefs to sanctify social relations, they must first themselves be sanctified. This is achieved through ritual.

Communal Participation in Costly Ritual

The pivotal role of communal ritual in religion has been noted by numerous researchers (Bloch 1989; Bourguignon 1973, 1976; Durkheim 1969; Eliade 1958, 1959; McCauley 2001; Rappaport 1999; Turner 1967, 1969). The formality, patterning, repetition, and rhythm of religious ritual have direct parallels in nonhuman ritualized display (Laughlin and McManus 1979; Lorenz 1965; Rappaport 1999; Rogers and Kaplan 2000; Smith 1979). In animal species such displays have evolved to serve intra- and inter-specific communication functions (Dugatkin 1997; Lorenz 1965; Rogers and Kaplan 2000; Rowe 1999). On its most basic level, nonhuman ritual constitutes "a process by which behavior specialized to be informative becomes differentiated from behavior that is informative only incidentally to its other functions" (Smith 1979:54). Ritualized displays represent but one end of a continuum in animal signaling systems that also includes simple indexical signals. Signal costs appear to be driven by both competition and receiver selection. Under conditions of ambiguity, or when signals can readily be faked, costlier signals may evolve to improve signal reliability. Ritualized displays are among the costliest of animal signals in terms of time, energy, and somatic resources required of the signaler. Zahavi (1975, 1981) has argued that such costly signals provide honest information for receiver assessment since only those who are sufficiently fit can bear the costs of such displays (Johnstone 2000; Zahavi and Zahavi 1997). Empirical research supports this hypothesis (Johnstone 2000; Zahavai and Zahavi 1997). Laboratory experiments indicate that the costliness of ritualized display is driven by receiver selection for reliable signals (Rowe 1999). The formality, sequence, repetition, and patterning that increase both time and energy costs of ritual also improve the ability of the receiver to assess the reliability of the message transmitted. These elements alert and focus attention, enhance memory, and promote associational learning (Rowe 1999). They neurophysiologically "prime" both the sender and receiver for action (Lewis and Gower 1980; Rogers and Kaplan 2000; Tinbergen 1965). The type of action that results is dependent both upon the receiver's assessment of the sender and upon the encoded "action releasers" embedded within the ritual display (Lewis and Gower 1980; Tinbergen 1965).

Animal signals and signal responses show considerable ontogenetic and socioecological malleability (Ball 1999; Lewis and Gower 1980; Marler 1999; Rogers and Kaplan 2000; Wingfield et al. 1999). Although some species-specific signals, such as the pecking response of herring-gull chicks to red dots, constitute relatively fixed, environmentally stable action-response sequences (Lewis and Gower 1980; Tinbergen 1965), others incorporate individually variant and ontogenetically learned patterns, as seen in the male courtship songs of various bird species (Ball 1999; Marler 1999). Signals of some species, including the esthetic nest constructions of male bowerbirds (Dissanayake 1995) and the friendship greeting rituals of baboons (Watanabe and Smuts 1999) show considerable malleability and high proportions of environmentally variable behaviors. The continuum of simple to complex ritual signals clearly encompasses a broad range of "fixed" and "learned" elements.

Ritual signals communicate important information regarding the condition, status, and intent of the sender. The intensity of plumage coloration in birds, the pitch of croaking in frogs, and the stotting height of springboks constitute indexical signals that provide information regarding parasite load, size, and agility, respectively (Krebs and Davies 1984; Rogers and Kaplan 2000). Such signals may also convey information regarding intent. In many species, intent signals frequently involve the transference of behaviors from their original context to a ritual context. The incorporation of food begging displays in bird courtship rituals, and presentation of the ano-genital area by subordinate primates to dominants, both represent signals that have been emancipated from their original feeding and copulating behaviors and transferred to new contexts of courtship and social hierarchies. In both instances, affiliative responses eliciting approach behaviors are associated with the original function of the signal and the transferred signal intent (Lewis and Gower 1980). Transference of these signals from their original contexts to ritual communicates the intent of the sender by evoking the autonomic and neurophysiological state associated with the signal's origins. The incorporation of these intent signals in ritual derives from their preexisting motivational characteristics (Laughlin and McManus 1979).

Religious ritual, like nonhuman ritualized displays, is demarcated from ordinary behaviors and is composed of the same structural elements (Rappaport 1999). Formality, patterning, sequencing, and repetition are basic components of religious ritual, and signals of condition, status, and intent constitute "action releasers" embedded within that structure. Pan-human social signals of dominance and submission, such as bowing and prostration, are prominent components of religious ritual worldwide (Atran 2002; Bloch 1989; Boyer 2001; Leach 1966; Rappaport 1999). As in nonhuman ritual, these signals convey information regarding status and intent. Religious ritual also incorporates indexical and iconic signals. Masks, statues, and other "agent" representations are prominent elements in religious ritual across cultures. They engage innate mental modules evolved for mundane functions and potentiate human predispositions to autonomically respond to specific classes of stimuli, including animate agents and angry faces (LeDoux 2002). Incorporation of evocative, grotesque, and dissonant features further intensify such responses. Like the signals of nonhuman ritual, the signals of religious ritual clearly elicit neurophysiological responses in participants and influence the nature of social interaction (Lewis and Gower 1980; Reichert 2000; Rogers and Kaplan 2000; Sapolsky 1999). In contrast to nonhuman ritual, however, iconic, indexical, and ontogenetic signals are not the primary encoded elements of human religious ritual. The fundamental elements of human religious ritual are, instead, abstract symbols devoid of inherent emotional or cognitive meaning. Words such as "Allah," the geometric designs of Australian Dreamtime paintings, and religious beliefs do not, in and of themselves, elicit any innate or ontogenetically derived neurophysiological response. Although, like language, religious systems across cultures appear to share a "deep structural grammar" that has an ontogenetic basis, the specific symbols embedded within that syntax are shaped by historical and socioecological parameters. In contrast to the signals of animal ritual, the meaning of abstract religious symbols must be created, both cognitively and emotionally. This important difference between nonhuman ritual and human religious systems not only requires that the abstract symbols of religious ritual be learned; it additionally requires that the emotional and behavioral significance of these symbols be learned as well. Whereas animal ritual elicits behavior through encoded signals, religious ritual elicits behavior to encode symbols. The creation of these symbols provides ritual tools for the shaping of social behaviors across space and time. Sosis (2003b) has argued that ritual participation generates belief among performers. He examined the psychological mechanisms underlying this process. Here we extend this argument to explain the interrelationship between emotions, symbols, and the sacred, and describe the neurological underpinnings of how ritual participation impacts belief.

Separation of the Sacred and the Profane

Religious ritual is universally used to define the sacred and to separate it from the profane (Douglas 1966; Durkheim 1969; Eliade 1959; Rappaport 1999). As noted by Rappaport (1999), ritual does not merely identify that which is sacred; it *creates* the sacred. Holy water is not simply water that has been discovered to be holy, or water that has been rationally demonstrated to have special qualities. It is, rather, water that has been *transformed* through ritual. For adherents who have participated in sanctifying rituals, the cognitive schema associated with that which has been sanctified differs from that of the profane. For Christians, profane water conjures associations of chemical structure and mundane uses; holy water, however, evokes associations of baptismal ritual and spiritual cleansing. Of greater importance from a behavioral perspective, the emotional significance of holy and profane water is quite distinct. Not only is it inappropriate to treat holy water as one treats profane water, it is emotionally repugnant. Although sacred and profane things are cognitively distinguished by adherents, the critical distinction between the sacred and the profane is the emotional charging associated with sacred things.

This distinction in emotional valence is created through participation in religious ritual. Sacred symbols have distinct cognitive schema, but their sanctity derives from their emotional meaning. It is the emotional significance of the sacred that underlies "faith," and it is ritual participation that invests the sacred with emotional meaning. The creation of religious symbols from abstract objects, and the imbuing of these symbols with attributions of "awe," "purity," and "danger" (Douglas 1966), are consistent and critical features of religious ritual everywhere (Douglas 1966; Durkheim 1969; Rappaport 1999; Turner 1969). Why is this so?

Emotions Motivate Behavior. Accumulating research indicates that emotions constitute evolved adaptations that weight decisions and influence actions. Emotions "rapidly organize the responses of different biological systems including facial expression, muscular tonus, voice, autonomic nervous system activity, and endocrine activity" (Levenson 1994:123) in order to prepare the organism for appropriate response to salient sensory stimuli. The ability of emotions to "alter attention, shift certain behaviors upward in response hierarchies, and activate relevant associative networks in memory" (Levenson 1994:123) directly impacts individual fitness. Since emotions are generated from limbic cortices that are outside conscious control, they are difficult to "fake" (Ekman et al. 1983). They, therefore, provide reliable communication signals among conspecifics. EEG patterns for simulated and real emotions are not the same, nor are the motor control areas for an emotion-related movement sequence and a voluntary act (Damasio 1994, 1998; Ekman and Davidson 1993). The somatic markers of emotion, including such things as pulse rate, skin conductance, pupil dilation, and facial expressions, differ from those under voluntary control. Emotionally motivated smiles engage different muscles from "Duchenne smiles," as do emotionally motivated frowns (Ekman 2003). As a result, emotions constitute powerful and honest cues of state and intent (Ekman 2003; Ekman et al. 1983).

Emotions may be elicited by sensory stimuli both internal and external to the organism. Predators, passing thoughts, and pulse rate are all capable of evoking emotional response. The emotional processing and appraisal of these stimuli engage widespread and complex cortical and subcortical systems within the brain. Initial unconscious processing of stimuli occurs in subcortical structures of the brain, including the basal ganglia, the amygdala, and the hypothalamus. This "first pass" level of processing appears to incorporate a superordinate division based on positive/approach and negative/withdrawal ratings of stimuli (Cacioppo et al. 2002).

Positive Stimuli Activate the Dopaminergic Reward System. The dopaminergic reward system constitutes "an emotional system that has evolved to motivate forward locomotion and search behavior as a means of approaching and acquiring rewarding goals" (Depue et al. 2002:1071). This system originates in the ventral tegmental area of the midbrain and projects to the nucleus accumbens of the ventral striatum. Its activation triggers the release of dopamine (DA), a neuromodulator which functions as a reward for the organism (Davidson and Irwin 2002). Stimuli intrinsic to somatic and reproductive success, such as food and sex, activate dopamine neurons within this system, and initiate goal-seeking behaviors. The potentiation of dopaminergic neurons induces a positive motivational state in the organism and simultaneously increases stimuli salience and locomotor activity (Pearson 1990). "Activation of this system has been shown to function as a reward, and animals will perform an arbitrary operant in order to self-administer stimulation of this pathway" (Pearson 1990:503). Drugs of addiction potentiate this system, as do subjectively rated "pleasurable" activities (Cacioppo et al. 2002). Repeated potentiation of this system transfers "the ability to phasically activate DA transmission from incentive stimuli intrinsic to the goal to incentive stimuli extrinsic to it" (DiChiara 1995:95). This results in the assignment of a positive affective valence to stimuli perceived under that state (DiChiara 1995). Such "incentive learning" creates associational neural networks that link stimuli associated with rewarding experiences to behavioral motivators, thereby investing previously neutral stimuli with positive valence. For former drug addicts, the paraphernalia, settings, and even neighborhoods associated with drug use constitute such incentive stimuli capable of activating mesolimbic neural networks, as revealed through brain imaging studies (DiChiara 1995).

Negative Stimuli Activate the Amygdala. The amygdala is a subcortical collection of specialized nuclei located beneath the temporal cortex. A central function of the amygdala is the rapid appraisal of potentially dangerous and threatening stimuli. Activation of the amygdala initiates a cascade of specific neuroendocrine events that prepare the organism to respond quickly to threats and danger. These responses appear to be "hard wired" in the nervous system (LeDoux 2002). Animals with lesioned or removed amygdala lack a fear response, even when placed in highly dangerous situations (LeDoux 1996).

In humans, the amygdala is also pivotal in initiating fear responses. Humans exhibit an innate predisposition to negatively valence potentially harmful and threatening stimuli, including animate objects and angry or fearful faces. There is considerable evidence that such stimuli elicit a greater response than positive stimuli, particularly in relation to action tendencies (Ito et al. 2002). The amygdala also processes human facial cues in relation to social judgments of trust. This processing occurs both consciously and unconsciously by the left and right amygdala, respectively (Adolphs 1999, 2002a, 2002b; Adolphs et al. 1998; Dolan 2000; Morris et al. 1998; Oram and Richmond 1999).

While specific stimuli innately activate the amygdala, it is also possible for neutral stimuli to acquire negative valence through classical and contextual conditioning. Previously neutral stimuli that are present or otherwise associated with a negatively valenced stimulus that activates the amygdala may subsequently initiate such response themselves. Once such conditioning occurs, it is difficult to reverse. Extinction of such conditioning "is not a process of memory erasure, [but rather] involves cortical inhibition of indelible, amygdala-mediated memories" (LeDoux 2002:404). As a result of both the negativity bias in information processing and the indelible nature of emotional memory, amygdala-conditioned stimuli constitute powerful long-term elicitors of emotional response.

The amygdala is highly interconnected with sensory, motor, and autonomic output systems. These interconnections "provide an anatomical basis for adaptive responses to stimuli" (Dolan 2000:1117). Interconnections with the hypothalamus ensure rapid somatic responses to stimuli through a cascade of neuroendocrine events.

These events prepare the organism for behavioral response and provide feedback information regarding body state to the amygdala. Reciprocal interconnections with the nuclear basalis ensure amygdalar participation in cortical arousal and selective attention. Direct interconnections of the amygdala with the hippocampal formation allow affective modifications of spatial behavior (Cacioppo et al. 2002; Cardinal et al. 2002; Damasio 1994, 1998; LeDoux 2002). Specific reciprocal projections from the amygdala to other emotional processing regions, including the ventral striatum and brainstem nuclei, provide an important link between positive and negative affective systems (Dolan 2000; LeDoux 2002; Rolls 1998). It is, however, the direct interconnections between the amygdala and the prefrontal cortex that are of particular significance for human social and symbolic systems (Deacon 1997; Groenewegen and Uylings 2000; Rolls 1998).

The Prefrontal Cortex Plays a Critical Role in Decision Making. McNamara (2001, 2002) has convincingly argued that the self-responsibility, impulse control, and morality which religions seek to instill in adherents are frontal lobe functions. Ongoing research supports the pivotal role of the prefrontal cortex in social judgment and impulse control, as well as symbolic thought (Deacon 1997; Dehaene and Changuex 2000; McNamara 2001, 2002; Rolls 1998). The orbitofrontal (OFC) region of the prefrontal cortex is the area of the brain activated in anticipation of rewards and punishments. Injuries to this brain area affect the delicate calculus of personal interest, environmental contingencies, and social judgments that motivate and guide individual behaviors within a social group (Dehaene and Changeux 2000; Rolls 1998). The valuation of behavioral alternatives, particularly in relation to social behaviors, appears to be processed in the OFC. Impairments to this area correlate highly with socially inappropriate or disinhibited behavior (Anderson et al. 2002; Damasio 1994; Kolb et al. 2004).

The behavioral deficits of OFC impaired patients are also seen in individuals who have intact prefrontal cortices and intact amygdala but lack interconnections between the two (Damasio 1994; LeDoux 1996, 2002). These individuals perform well on abstract reasoning tasks but are unable to apply such reasoning to personal decision making (Damasio 1994; LeDoux 1996, 2002). The loss of emotion typical of OFC impaired patients is also a characteristic of these disconnect patients. For these individuals, the affective cues required for valuation of predicted outcomes are absent. In the absence of emotional input from the basolateral amygdala, the OFC lacks valuation information necessary for the prediction of reward/punishment outcomes. Recent laboratory experiments conducted by Schoenbaum and colleagues demonstrate that both the orbitofrontal cortex and the basolateral amygdala are "critical for integrating the incentive value of outcomes with predictive cues to guide behavior" (Schoenbaum et al. 2003:855). It is through the emotional inputs of the amgydala that "otherwise neutral cues acquire motivational significance or value through association with biologically significant events" (Schoenbaum et al. 2003:863).

Ritual, Emotion, and Sanctification. Religious rituals are biologically significant events. Ongoing research with ritual participants engaged in meditation and trance demonstrate changes in brain wave patterns, heart and pulse rate, skin conductance, and other autonomic functions (Austin 1998; Davidson 1976; Kasamatsu and Hirai 1966; MacLean et al. 1997; Mandel 1980; Newberg et al. 2001; Winkelman 2000). Meditation also alters neuroendocrine levels, including testosterone, growth hormone, and cortisol (MacLean et al. 1997). Although little research has been conducted on the neurophysiological effects of less intense religious participation, there is mounting evidence that participation in weekly Western religious services may impact blood pressure (Brown 2000; Dressler and Bindon 2000), adolescent testosterone levels (Halpern et al. 1994), and other neurophysiological systems (Levin 1994, 1996; Matthews et al. 1998; Murphy et al. 2000). Experiments suggest that some of these neurophysiological changes may be associated with the "rhythmic drivers" that characterize human religious ritual.

Music Is a Universal Feature of Religious Ritual. Human and nonhuman ritual share basic structural components of formality, pattern, sequence, and repetition. Human religious ritual further amplifies and intensifies these elements through the incorporation of "rhythmic drivers." Described by Bloch as "distinguishing marks of ritual" (1989:21), these elements, including music, chanting, and dance, constitute recurrent and important components of religious ritual across cultures. Although Bloch derived these features from ethnographies of traditional societies, the recent survey of U.S. congregations conducted by Chaves et al. (1999) found music to be a consistent feature of contemporary U.S. religious services, as well. Even in the most ritually constrained religions, music remains a key consistent feature (Atran 2002). Not only is music an important component of religious ritual, across traditional cultures it is inseparable from it (Becker 2001).

Music has important neurophysiological effects. As a "rhythmic driver," it impacts autonomic functions and synchronizes "internal biophysiological oscillators to external auditory rhythms" (Scherer and Zentner 2001:372). The coupling of respiration and other body rhythms to these drivers affects a wide array of physiological processes, including brain wave patterns, pulse rate, and diastolic blood pressure (Gellhorn and Kiely 1972; Lex 1979; Mandel 1980; Neher 1962; Walter and Walter 1949). This "coupling effect" has been shown to be present in humans at a very early age (Scherer and Zentner 2001). Music amplifies and intensifies this effect through the use of instruments, or "tools," thereby providing a means of synchronizing individual body rhythms within a group. Recent work by Levenson (2003) has shown that synchronized autonomic functions, including such things as pulse rate, heart contractility, and skin conductance, are positively and significantly associated with measures of empathy. The prominent role of music in religious ritual promotes such empathy.

Music also has demonstrated effects on measures of stress and immunocompetence. A significant negative correlation between exposure to "relaxing" music and salivary cortisol levels was found in experiments conducted by Khalfa et al. (2003). Other research has demonstrated significant positive correlations between music and immunocompetence, as measured by salivary immunoglobin A (SIgA), with active participation correlating most highly with immunocompetence and no music exposure correlating the least (Hirokawa and Ohira 2003; Kuhn 2003). These associations between music and measures of stress and health may be mediated by music's ability to alter autonomic functions and evoke emotions. The capacity of music to alter skin temperature, muscle tension, cardiovascular function, respiration, norepinephrine, and brain wave patterns all have subjectively reported "emotion inducing effects" (Hirokawa and Ohira 2003; Scherer and Zentner 2001). The contour, rhythm, consonance/dissonance, and expectancy within a musical structure contribute to both the intensity and valence of the experienced emotion (Hirokawa and Ohira 2003; Scherer and Zentner 2001; Sloboda and Juslin 2001). Studies of subliminal facial expression demonstrate that musically induced physiological changes closely correspond with both involuntary facial expressions of emotion and subjectively described emotions evoked by particular types of music (Krumhansl 1997).

The capacity of music to entrain autonomic states and evoke congruent emotions in listeners provides the basis for creating and synchronizing motivational states in ritual participants. Although the communal songs and vocalizations of nonhuman species, including birds, whales, and wolves, may also function in social accommodation, only human music is capable of amplifying, intensifying, and modifying these effects through the use of "tools." The externalization of auditory signal production through the use of musical instruments fundamentally alters the signal/signaler relationship. The signal produced through the use of musical instruments is no longer indexical of either the signaler's state or condition. Two warriors can sound like twenty through the use of drums. Moreover, discrete sounds produced with musical instruments can be manipulated and juxtaposed to create emotionally evocative signals independent of the musician's state. Like the phonemes, words, and sentences of language, the use of musical instruments to produce sounds permits the combining of such sounds to create emotionally meaningful signals. These, in turn, can be arranged and rearranged within encompassing musical structures. The formality, sequence, pattern, and repetition of such musical structures themselves elicit emotional response through their instantiation of ritual. Music thereby creates an emotive "proto-symbolic" system capable of abstracting both the signals and structure of ritual. This abstraction and instantiation of ritual through music may well have established the foundation for symbolic thought in human evolution. It certainly provided a tool for the evocation of communal emotions across time and space.

Religious Ritual Evokes Both Positive and Negative Emotions. Cross-culturally, the emotion most frequently evoked by music in religious ritual is happiness (Becker 2001). In its most intense version, this may reach ecstasy. Such extreme joy "almost by definition involves a sense of the sacred" (Becker 2001:145) and is not unlike that attained through use of various psychoactive drugs. Such drugs also constitute prominent elements in many religious systems. These components of religious ritual activate noradrenergic, serotonergic, and dopaminergic systems in the brain that heighten attention, enhance mood, and increase sociability (Regan 2001). These components of religious ritual elicit positive emotional responses in participants and engage the brain's dopaminergic reward system.

There are also numerous elements of religious ritual that evoke fear and pain rather than happiness and joy (Douglas 1966; Eliade 1959; Glucklich 2001; Turner 1967, 1969). Many ritual settings, including caves, caverns, and cathedrals, arouse vigilance by altering sensory perception through unpredictable illumination. Grotesque masks, bleeding statues, and fearsome icons engage innate "agency" modules that initiate emotional responses to danger and threat. Physical and mental ordeals inflict suffering and alter autonomic states. Vengeful gods and demons mete out punishment and demand painful sacrifices. Such negative stimuli comprise central elements of many religious systems and are particularly prevalent within the context of rites of passage (Eliade 1958, 1959; Glucklich 2001; Turner 1969). In contrast to the positive affect induced by ecstatic religious ritual, these components of ritual initiate responses related to fear and danger and evoke intense negative emotions in ritual participants.

The ability of religious ritual to elicit both positive and negative emotional responses in participants provides the substrate for the creation of motivational communal symbols. Through processes of incentive learning, as well as classical and contextual conditioning, the objects, places, and beliefs of religious ritual are invested with emotional significance. The rhythmic drivers of ritual contribute to such conditioning through their "kindling effects." Research on temporal lobe syndrome patients has shown that repeated neuronal firing of the amygdala can result in the conditioned association of arbitrary stimuli with heightened emotional significance (Bear 1979; Bear et al. 1981; Damasio 1994; Geschwind 1979). The increased religiosity characteristic of some temporal lobe epileptics has been attributed to this kindling effect (Bear 1979; Saver and Rabin 1997). Rhythmic environmental stimuli, including both music (Peretz 2001) and rapid, flashing lights, contribute to the rapid neuronal firing that results in such kindling (LeDoux 2002). Temporal lobe patients have a low threshold for such firing. EEG recordings have shown that the driving effects of ritual, such as music, drumming, and dancing, are capable of altering neuronal firing patterns in nonclinical populations, as well (Lex 1979; Neher 1962; Walter and Walter 1949). Elements of religious ritual that increase neuronal firing rates prime ritual participants for the conditioned association of symbols and emotions, both positive and negative, and create communal conditions for investing religious stimuli with these emotions (DiChiara 1995). The "ecstasy" achieved through the music and movement of Sufi dancing is transferred to the religious poetry with which it is associated. Likewise, ingestion of peyote by the Huichol Indians with its potentiation of the dopaminergic reward system provides a neurophysiological basis for investing the communal Peyote Hunt itself with sacred significance (Myerhoff 1974). The negative emotional responses elicited by shadowed cathedrals, fearsome masks, and painful ordeals are heightened by drumming, music, and chants. The emotions thereby elicited and intensified become conditionally associated with the gods, ghosts, and demons that populate religious belief systems. Such symbols are not inherently pleasurable, but they are motivationally powerful and emotionally indelible. The use of communal ritual to invest previously neutral stimuli with deep emotional significance creates a shared symbolic system that subsequently valences individual choices and motivates behavior (Dehaene and Changeux 2000).

Most Religions Incorporate Both Positive and Negative Elements. The extent to which positive and negative elements are emphasized varies considerably both across the rituals within a given religion and among religions. Whether religious ritual predominantly incorporates positively or negatively valenced symbols appears to be correlated with both the political characteristics of the group and the risk-to-benefit ratio of their cooperative endeavors. We anticipate that when collective action issues are predominantly problems of coordination with few potential costs to individuals, positively valenced rituals will serve to promote affiliative cooperation. Such rituals engender empathy among participants and conditionally associate religious symbols with internal reward systems through incentive learning. When the predominant collective action issues faced by a group involve high individual costs but potentially great collective benefits, however, we expect increases in the costliness of religious ritual through the incorporation of negatively valenced stimuli to deter free riders. Since negatively valenced components of ritual are motivationally more powerful than positive stimuli, they provide a more reliable emotionally anchored mechanism for the subordination of immediate individual interests to cooperative group goals. In societies lacking a central political authority with police powers capable of subordinating individual interests to those of the group, intense and negatively valenced religious rituals address the inherent free-rider problems of collective action. The prominence of negatively valenced elements in religions associated with large, socially stratified, preindustrial societies (Roes and Raymond 2003) underscores this "policing" role of religion in motivating cooperation when a central secular authority is weak (Paige and Paige 1981). This is particularly pronounced in adolescent rites of passage in such societies (Eliade 1958; Glucklich 2001; Turner 1969). The incorporation of painful and dangerous elements in such rites is positively and significantly correlated with the incidence of warfare in preindustrial societies (Sosis et al., n.d.). These highly charged negative ritual experiences not only bond initiates, they also motivate intense cooperation and obedience under conditions of high individual risk and low central authority. The less powerful in such societies bear a larger share of the fitness costs of such subordination, but they may still gain greater benefits as members of a successful cooperative group than they would otherwise realize.

Yet, even when religious systems emphasize negatively valenced symbols, the use of ritual to invest such symbols with emotional meaning necessarily incorporates positively valenced components that benefit ritual participants both psychologically and politically. Powerfully valenced symbols that motivate behavioral choices reduce cognitive dissonance, particularly under conditions of socioecological stress. Research by Bradshaw (2003) indicates that in contemporary Western societies, weekly worship attendance results in relatively greater decreases in psychological distress for the socioeconomically disadvantaged. The positive correlation of music and immunocompetence, and its inverse correlation with stress, suggests that ritual participation may differentially benefit group members facing the highest stress loads.

At the same time, joint participation in costly ritual creates empowering conditions. Ritual not only promotes more efficient and effective group functioning for politically and socially sanctioned endeavors, it simultaneously creates motivationally coordinated coalitions that can surmount existing in-group/out-group boundaries and provide a mechanism for social and political change (Bourguignon 1973). The Protestant Reformation of the sixteenth century, the role of African-American churches in the U.S. Civil Rights Movement, the contemporary importance of Pentecostalism in Latin America, and messianic movements in general, all illustrate the important role of religion in creating cooperative coalitions that have been instrumental in transforming existent social and political relationships.

Adolescence and Religion

Adolescent Rites of Passage. Adolescent rites of passage are one of the most consistent features of religions across cultures (Bettleheim 1962; Brown 1975; Lutkehaus and Roscoe 1995; Paige and Paige 1981; van Gennep 1960). In some societies, such as the Yamana and Halakwulup of Tierra del Fuego, such rites traditionally consisted of little more than oral transmission of sacred knowledge from elder to youth (Eliade 1958). In other cultures, such as the Ndembu and the Elema, pubertal initiation rites involved "kidnapping" of adolescents, months of sequestered seclusion, and ritual ordeals that included dietary restrictions, sleep deprivation, physical pain, and genital mutilation (Eliade 1958; Glucklich 2001; Paige and Paige 1981; Turner 1969; van Gennep 1960). In modern societies, adolescence also constitutes an important developmental period for religious training (Atran 2002; Elkin 1999; Regnerus et al. 2003). Although the intensity and duration of adolescent rites of passage vary from culture to culture, all share a common structure (Turner 1969; van Gennep 1960), as well as a common emphasis on the evocation of emotion and its association with symbols in the teaching of sacred things (Eliade 1958; Turner 1967, 1969).

The expressed purpose of rites of passage is to *initiate* particular categories of a society's adolescents into "the sacred." Initiates not only learn the sacred, they live it. The social and psychological death, transformation, and rebirth of the individual achieved through these rites not only train initiates, but transform them as well

(Turner 1967, 1969). Initiates enter as children but leave as adults invested with both social and reproductive rights, as well as the responsibilities entailed therein.

Through rites of passage initiates learn what things constitute the sacred. This requires the development of new cognitive schema for previously mundane things, whether words, images, or objects, involving the generation of new neural associative networks. More importantly, however, initiates directly experience the sacred. The separation; sleep and food deprivation; exposure to novel, dangerous, and terrifying stimuli; and subjection to physical and mental ordeals that are frequently an integral part of such rites evoke autonomic and emotional responses in initiates. Rites of passage purposefully engage unconscious emotional processes, as well as conscious cognitive mechanisms. The conditioned association of such emotions as fear and awe with symbolic cognitive schema achieved through these rites results in the sanctification of those symbols, whether places, artifacts, or beliefs. Because such symbols are deeply associated with emotions engendered through ritual, they take on motivational force. When such rites are simultaneously experienced by groups of individuals, the conditioned association of evoked emotions with specific cognitive schema creates a cultural community bound in motivation, as well as belief.

Adolescent Brain Development. Adolescence may constitute a neurophysiologically sensitive developmental period for the learning of abstract concepts and the conditioned association of such concepts with emotions (Kolb et al. 1998; Kwon and Lawson 2000; Plant 2002; Spear 2000). The human brain demonstrates great plasticity during development. Infancy, childhood, adolescence, and adulthood are marked by differentiated growth patterns in various brain cortices and nuclei (Casey et al. 2000; Giedd et al. 1999; Keshavan et al. 2002; Kolb and Whishaw 1998; Kolb et al. 1998; Kwon and Lawson 2000; LeDoux 2002; Plant 2002; Sowell et al. 1999; Spear 2000; Walker and Bollini 2002). The differential patterns of brain growth across the life course create sensitive periods for particular types of learning (Greenough 1986). Early childhood language acquisition is an example of such "experience expectant" learning (Pinker 1997). We propose that adolescence constitutes a second critical period of "experience expectancy" for the learning of emotionally valenced symbolic systems.

The Adolescent Brain Does Not Mature Uniformly. Whereas the preadolescent brain grows through an increase in cortical gray matter, during adolescence synaptic pruning eliminates as much as one-half of the number of cortical synapses per neuron (Spear 2000). Synapse elimination does not occur uniformly throughout the human cortex, however. Frontal and parietal lobes follow a similar developmental trajectory, with increases in gray matter up to a maximum occurring at 12.1 and 11.8 years, respectively, for males and 11.0 and 10.2 years, respectively, for females, followed by a decline, resulting in a net decrease in volume across adolescence. The growth of temporal lobe gray matter has also been found to be nonlinear, with maximum size reached at 16.5 years for males and 16.7 years for females, and slight declines thereafter (Giedd et al. 1999; Sowell et al. 1999). Both frontal and temporal lobe maturation occurs late in development and is completed in early adulthood (Keshavan et al. 2002; Sowell et al. 1999). This heterochronous adolescent loss of cortical gray matter is accompanied by increased volume in amygdalar and hippocampal nuclei. Concurrent changes in white matter density facilitate the propagation of electrical signals and increase the speed of neural transmission (Keshavan et al. 2002; Walker and Bollini 2002).

These changes streamline brain function by eliminating irrelevant interconnections and enhancing those that remain. This ontogenetic sculpting of the brain results from differential activation of specific neurons on the basis of experience in the accommodation of environmental needs (Greenough and Black 1991; Kolb and Whishaw 1998; Kolb et al. 1998; LeDoux 2002). Kolb and colleagues note that "experience can alter different parts of neurons differently (and) . . . changes in synaptic organization are correlated with changes in behavior" (1998:156). As a result, "the environment or activities of the teenager may guide selective synapse elimination during adolescence" (Giedd et al. 1999:863). Emotionally evocative experiences that occur during adolescence may, therefore, actually shape neural networks in the maturing brain. This is particularly true for brain areas such as the temporal lobes and prefrontal cortices undergoing maturation.

The maturation of the prefrontal cortex that occurs during adolescence has important implications for abstract reasoning abilities and symbolic thought. The prefrontal cortex (PFC) is "essential for such functions as response inhibition, emotional regulation, planning and organization" (Sowell et al. 1999:860). The interconnectivity of the PFC with nearly all other brain regions uniquely situates this cortical structure in its ability to associate diverse stimuli (Groenewegen and Uylings 2000; Robbins 2000; Rolls 1998). Maturation of the prefrontal cortex during adolescence provides the neurophysiological substrate for social cognition, abstract reasoning, and symbolic thought (Adolphs 2002a; Deacon 1997; Robbins 2000).

There Is a Shift in the Dopaminergic Reward System during Adolescence. Significant changes in neurotransmitter systems occur during adolescence. Receptors for dopamine, serotonin, acetycholine and GABA (γ-aminobutyric acid) are pruned from their preadolescent over-production, and limbic areas, including the hippocampus, also undergo pruning of excitatory receptors. Hippocampal receptors for endogenous cannabinoids peak during adolescence at higher than adult levels (Spear 2000). Studies by Carlson et al. (2002) demonstrate increased long-term potentiation as a result of endocannabinoid production, suggesting enhanced memory functions during this period. Concurrent with the decline in excitatory neurotransmitter receptors during adolescence, a shift in dopamine balance from mesolimbic to mesocortical regions occurs. This shift impacts reward learning and has significant behavioral implications (Schultz et al. 2002). Dopamine inhibitory input to the prefrontal cortex is greatest during adolescence, whereas dopamine activity in the anterior cingulate cortex and other subcortical regions, including the amygdala, is lowest. While dopamine activity in the anterior cingulate cortex is under inhibitory control of the amygdalar dopamine system, the amygdala is, in turn, tonically inhibited by prefrontal cortex activity. According to Walker and Bollini, "the enhancement of neuronal connection between the cortex and limbic regions may play a role in the integration of emotional behaviors with cognitive processes" (2002:18) during this time. The shifting dominance of amygdalar dopamine projections from anterior cingulate cortex to the prefrontal cortex during adolescence impacts both conditioned associations and the intrinsic reward system. In addition to cortical maturation during adolescence, MRI studies have shown differences in the activity of the amygdala in adolescents, as compared with adults. Human adolescents exhibited "greater brain activity in the amygdala than in the frontal lobe when engaged in a task requiring the subjects to identify emotional state from facial expressions, while adults conversely exhibited greater activation in frontal lobe than amygdala when engaged in the same task" (Spear 2000:440).

Adolescent Changes in Brain Function Have Important Implications for Learning and Behavior. The concurrent maturation of the temporal lobe and amygdala are relevant to facial recognition and social judgments (Adolphs et al. 1998). Studies indicate that the amygdala mediates judgment of other people's social behavior, particularly with regard to approachability and trustworthiness (Adolphs 2002a, 2002b; Cardinal et al. 2002). The shift in the dopaminergic reward system from mesolimbic to mesocortical dominance that occurs during adolescence provides a unique developmental window for the conditioned association of abstract symbols with intensely experienced emotions and for the integration of these associations with both social interactions and symbolic thought. Heightened adolescent sensitivity to stressors amplifies this process (Spear 2000). The synaptogenesis and neurotransmitter shifts occurring during adolescence intensify the impacts of environmental stimuli experienced during this developmental phase. This is particularly true for the late-maturing frontal and temporal cortices, and for such limbic nuclei as the amygdala and the hippocampus. The specific changes occurring in the adolescent brain render this a particularly sensitive developmental period in relation to social, emotional, and symbolic stimuli. These are precisely the type of stimuli of greatest importance in adolescent rites of passage.

Adolescent Rites of Passage Bombard Initiates with Environmental Stimuli that Engage Prefrontal, Temporal, and Limbic Functions. The ritual components of these rites optimize stimulus impacts while amplifying the kindling effects of the stimuli through rhythmic drivers, including music, chanting, and dance, all of which may be particularly salient for adolescents. Intensification of the stimuli through sleep and food deprivation, fear, physical ordeals, and drugs can be expected to increase the neurophysiological impacts in terms of memory, reward learning, and emotional charging of stimuli. The "breaking down" of initiates during the liminal phase of adolescent rites of passage engenders a common autonomic state among initiates. The empathy and shared emotional charging experienced in rites of passage valence the cognitive schema associated with sacred things.

Not all schemata constitute equal candidates for sanctification. Accumulating evidence suggests a developmental propensity for schema incorporating socially omniscient and declarative supernatural agents (Bering 2005). Moreover, schema of religious belief systems exhibit consistent structural features. Dichotomies, inversions, and counterintuitive concepts are consistent elements of this structure (Atran 2002; Atran and Norenzayan 2004; Boyer 2001; Boyer and Ramble 2001; Levi-Strauss 1963). The cognitive schema of religious systems also consistently incorporate the salient socioecological features of the society in which they occur, yet they do so while transcending the momentary, individual, and specific attributes of those features. Both the abstraction of social relations and their transformation into eternal truths are the hallmarks of religious schema (Rappaport 1999). These schema leave no outwardly visible signs but instead carve their indelible mark on the very minds of initiates. Through adolescent rites of passage, "the abstract is made alive and concrete by the living substance of men and women" (Rappaport 1999:148).

THE EVOLUTION OF RELIGION

Many recent evolutionary studies define religion in terms of cognition, focusing on the beliefs rather than the behaviors of religious systems. From a cross-cultural perspective, however, it is ritual that lies at the heart of all religions (Durkheim 1969; Eliade 1958, 1959; Rappaport 1999; Turner 1967, 1969), and it is participation in ritual that creates believers (Sosis 2003b). In the absence of ritual indoctrination and practice, religious beliefs lack both emotional salience and motivational force.

Ritual in nonhuman species functions to communicate social information and to coordinate social behaviors through the use of species-specific signals evolved to elicit neurophysiological responses in participants (Dugatkin 1997; Rogers and Kaplan 2000; Rowe 1999; Wingfield et al. 1999). Although ritual displays may be costly in terms of time, energy, and somatic expenditures, they provide information to participants that can impact individual fitness. By providing reliable signals, ritual allows accurate assessment of conspecific condition and intent (Zahavi and Zahavi 1997). It also "primes" participants for social interaction. Ritual winners reap resource and mating advantages; losers, however, also benefit from reductions in conflict achieved through ritual. Within the context of social groups, ritual further functions to decrease individual stress through the stabilization of social organization (Sapolsky 1999) and provides a means of facilitating both group fission/fusion and the coordination of group activities (Dugatkin 1997; Goodall 1986; Laughlin and McManus 1979; Rogers and Kaplan 2000). The pre-hunt ritual of wolves represents such coordination, and the friendship rituals of chimps and baboons have been observed to facilitate cooperative alliances that force changes in troop hierarchies (Goodall 1986; Watanabe and Smuts 1999). Among human groups, these same functions are apparent in the rituals of both sports and politics.

Religious ritual, too, functions to communicate and coordinate social behaviors and does so through the elicitation of neurophysiological responses. Participation in religious ritual results in empirically demonstrated effects on both cooperation (Sosis and Bressler 2003; Sosis and Ruffle 2003, 2004) and individual health and longevity (Hummer et al. 1999; Matthews et al. 1998; Murphy et al. 2000). Like the ritualized displays of nonhuman species, religious ritual is positively associated with decreased stress and improved immunological function (Murphy et al. 2000). Bradshaw (2003) has further found that decreases in psychological distress associated with participation in religious ritual may be particularly relevant for the relatively deprived. Under conditions of inequality, religious ritual may, thus, confer direct fitness benefits for participants while simultaneously providing a mechanism for cooperative action for political change. The parallels between nonhuman and religious ritual extend, as well, to the use of religious ritual in the reintegration of social groups across cultures, and the coordination of group endeavors. Nonhuman and religious ritual clearly share important structural and functional elements selected for their adaptive value in social communication (Rowe 1999). The two are, however, separated by a critical distinction. While nonhuman ritual encodes signals as neurophysiological primes for behavior, religious ritual encodes symbols created through the ritual process itself.

Although it is impossible to retrace ritual's evolution to a symbolic signaling system, the "distinguishing marks of ritual"—chanting, music, and dance—may provide important clues. As discussed above, all religions incorporate music in some form, and in most it is a dominant element. Music is uniquely adapted to instantiate the structure of ritual precisely because it incorporates the formality, sequencing, patterning, and repetition that define ritual. As a result, it is able to elicit the neurophysiological responses associated with such ritual in the absence of ritual behaviors. Music's direct impacts on autonomic function, its ability to enhance immunocompetence (Kuhn 2002), and its role in entraining ritual participants may all have led to its selection as a fundamental component of early hominid ritual. Ultimately, however, the most important evolutionary consequence of music may well have been its "proto-symbolic" attributes. The ability of music to abstract and codify ritual meaning over time and space may have been the critical first step toward symbolic thought. The introduction of such a symbolic ritual system introduced a new type of cognition in hominid evolution. The use of ritual to create associational neural networks linking symbolic, social, and affective systems provided social groups with a highly flexible tool for motivating individual behavior, forging inter-group alliances, and discriminating between friends and enemies. Individuals within such groups would have realized fitness benefits resulting from inter-alliance sharing of patchily distributed resources, as well as enhanced cooperation for in-group ventures, including hunting and warfare.

When symbolic behavior emerged in human evolution remains unknown. Some researchers argue for the emergence of symbolic culture in early archaic populations (Bednarik 1995; Hayden 1993; Marshack 1990); others maintain that symbolic thought appeared in early *Homo sapiens sapiens* prior to migration out of Africa (Henshilwood et al. 2001; McBrearty and Brooks 2000; Watts 1999). Still others argue for a "big bang" theory of symbolic culture first appearing approximately 50,000 years ago in western European populations (Mithen 1996). All, however, associate the emergence of symbolic systems with ritual. Mithen notes that "the very first art we possess appears to be intimately associated with religious ideas by containing images of what are likely to be supernatural beings" (Mithen 1996:155). Watts (1999) also argues for a ritual origin of symbolic systems but maintains that such systems emerged some 100,000 years earlier than posited by Mithen. Watts (1999) argues that the ubiquitous presence of red ochre pigments at numerous African MSA (Middle Stone Age) sites indicates ritual activity. Noting a jump in ochre presence over time, he concludes that "the preoccupation with redness clearly indicates that ochre was primarily used for signalling" (1999:128) and argues that "the habitual nature of such behaviour from the MSA2b onwards strongly suggests that the signalling was symbolic rather than solely indexical or iconic" (1999:137). The MSA Blombos Cave excavations of Henshilwood and colleagues lend further support for symbolic behaviors in African MSA populations. These researchers recovered twenty-eight bone tools dated ca. 70,000 years ago exhibiting "formal" techniques of bone tool manufacture, as well as ochre pencils and objects bearing geometric designs. They note that "bone tools are . . . only one element of a range of techniques used at BBC during the MSA to produce practical and/or symbolic artefacts indicative of a complex technological society" (2001:668). The occurrence of pigment processing at numerous MSA sites, as well as the notching and incising of ochre, bone, and ostrich shell, are also interpreted by McBrearty and Brooks as evidence of symbolic behavior. These researchers note that "Despite the relatively small number of excavated MSA sites, the quantity and quality of evidence for symbolic behavior . . . far exceeds that known for the European Middle Paleolithic where the site sample is more than ten times greater" (2000:531).

The irregularly patterned and increasing use of red ochre pigment by African MSA/LSA populations suggests that ritual was of variant but increasing importance in human social groups throughout this period. The widespread occurrence of red ochre pigments has been interpreted by Dunbar (1999) as evidence of "badging." He argues that red ochre badging increased during the African MSA in order to mark and identify group members when both the size and the number of groups were increasing. Yet, Dunbar notes that "external badges encounter a common problem . . . they are easy to fake" (1999:202).

If, however, red ochre badging is viewed within a broader context of ritual, as Watts and others (Knight et al. 1995) have interpreted it to be, then both the costs and the reliability of these badges increase, as well. Participation in ritual entails time and energy costs which may deter free riders (Irons 2001; Sosis 2003b). More importantly, participation in communal ritual provides the context for the creation and internalization of communally shared motivators. The use of ritual to emotionally charge badges and other selected artifacts would have added to the costs of such badges, but would have significantly increased their reliability as signals of motivational intent. This ritual transformation of signal badges to emotionally charged and positively valenced symbols of social relationships may have served to facilitate the creation of alliances under conditions of resource scarcity and conflict (Hayden 1987). The red ochre, beadwork, bone incising and regional stone- and bone-working styles evident in the archaeological record of the African MSA between 250,000 and 50,000 BP all indicate an increasing importance of ritual, an intensification of costly signals, and the emergence of symbolic systems specific to social groups. The emergence of dance, music, and even language may have their roots in this intensification process. Why did these changes occur during the MSA?

There is evidence of increasing population, increasing use of a fission/fusion social organization, and shared use of patchy resources within an environment of overlapping group ranges throughout the MSA. McBrearty and Brooks (2000) report that MSA sites in Africa are more numerous than those of the Acheulian and are found in previously uninhabited zones, suggesting both the need and the ability of MSA populations to exploit a wider range of habitats. Moreover, these sites provide evidence of deliberate foresight and planning in cooperative hunting strategies (Chase 1989), specialized tool use (McBrearty and Brooks 2000; Shea 1988), and in the transport of both water and materials across long distances (Deacon 1989). The development of technologies such as ostrich eggshell containers that permitted the transport of critical resources such as water opened up previously uninhabitable areas (Watts 1999). The appearance of blades, as well as retouched stone and bone points, indicates increasing technological sophistication, as well. McBrearty and Brooks (2000) have interpreted the diversification of MSA toolkits and the varying proportions of different artifact classes at different sites as evidence of regional tradition differences, as well as differences in extractive activities. These authors present compelling arguments for continuing intensification and scheduling of resource use throughout the African MSA into the LSA. Evidence of both selective, tactical hunting of large game and intensifying use of aquatic and small-scale resources is cited, as well as proliferation and geographic extension of trade networks. This intensification of extractive and hunting technologies, as well as expansion into previously unexploited habitats and increasing territorial sizes during the African MSA, have been viewed by McBrearty and Brooks (2000) as evidence of both population growth and environmental degradation.

The picture that emerges from the accumulating archaeological record for the African MSA is one of population growth, geographic dispersion, and technological intensification and specialization. Tactical hunting strategies for large game emerged. Simultaneously, the irregular distribution of critical items, such as water, and the regional distribution of other prized resources, such as obsidian, introduced increased inter-group interaction and competition for utilization of these patchy resources. The ecological context of human groups in the African MSA suggests that the nature of hominid social groups underwent change during this period. Larger group sizes punctuated by seasonal fission/fusion, and the creation and maintenance of alliances in response to resource irregularity, are indicated. An increased reliance on cooperative subsistence strategies, including large game hunting and joint utilization of dispersed water sources, as well as increased competition between groups for patchy resources, can be surmised from the archaeological record. Red ochre pigments and decoratively incised stone and bonework suggest that these changes were accompanied by increases in ritual and the emergence of an abstract symbolic system.

It is likely that the incorporation of rhythmic drivers in human ritual preceded these developments. The drumming and "proto-dances" of chimpanzees suggest that precedents of music, chanting, and dance existed in common ancestral hominoids (Goodall 1986). Such behaviors may have originated as communication signals. The ability of these drivers to enhance positive affect would have rendered rhythmic ritual a useful tool in the reintegration of fissioned groups and in the creation of inter-group alliances. The use of rhythmic ritual to invest artifacts with symbolic, emotionally valenced meaning would have provided dispersed groups with a tangible and motivational symbol of the abstract social relationships codified through the ritual process. With increasing resource competition, however, there would also be increasing need to differentiate and cohesify groups in order to more efficiently and effectively extract and defend resources. These conditions would further promulgate in-group specialization and stratification. Under such conditions, negatively valenced religious symbols would assume increasing importance owing to both their greater motivational force and their signaling efficacy (Johnson and Kruger 2004).

In contrast to the indexical signals of animal ritual, which elicit congruent motivational states within an immediate time and space, the symbols of religious ritual afforded early humans a means of engendering congruent motivational states across space and time. And, although signals elicit neurophysiological responses that permit social interaction in the here and now, symbols extend the horizon of those responses to future activities, as well. Religious symbols, thus, provided tools for creating cooperative coalitions across time. In doing so, they introduced a new level of cognition and social organization in human evolution.

CONCLUSION

Religion is an important and unique human adaptation defined by four recurrent traits: belief systems incorporating supernatural agents and counterintuitive concepts, communal ritual, separation of the sacred and the profane, and adolescence as a preferred developmental period for religious transmission. Although the specific expression of each of these traits varies across cultures in socioecologically patterned ways, the belief systems and communal rituals of all religions share common structural elements that maximize retention, transmission, and affective engagement. The roots of these structural elements can be found in nonhuman ritual where they serve to neurophysiologically prime participants and ensure reliable communication. Religion's incorporation of music, chanting, and dance intensifies such priming and extends the impacts of ritual beyond dyadic interactions. Music constitutes an abstract representation of ritual that can be recreated across time and space to evoke the emotions elicited by ritual. Human use of ritual to conditionally associate emotion and abstractions creates the sacred; it also lies at the heart of symbolic thought. The brain plasticity of human adolescence offers a unique developmental window for the creation of sacred symbols. Such symbols represent powerful tools for motivating behaviors and promoting in-group cooperation. Although religion evolved to solve an ecological problem by promoting group communication and cooperation across space and time, the symbols it created laid the foundation for a new adaptive niche in human evolution.

A number of empirically testable hypotheses emerge from this view of religion. We have posited that the adaptive function of religion is to ensure cooperation when individuals can achieve net benefits through collective action, and we have proposed that ritual serves to engender such cooperation through the motivational valencing of symbols. If so, religious ritual should be most pronounced within groups of individuals who are not genetically related and are pursuing high-cost cooperative endeavors, and least pronounced among kin groups pursuing individualistic subsistence strategies. Significant associations between ritual intensity, positive and negative symbolic valence, and age of initiation should also exist among these variables. We expect to find the highest intensity of ritual in groups encompassing unrelated individuals who must engage in intermittent, high-risk, cooperative endeavors, such as external warfare or long-term sharing of scarce and patchy resources. In contrast, the lowest levels of religious ritual should occur among non-cooperating groups of kin. We would further expect to find permanent, highly charged religious symbolic systems in non-kin groups engaged in high risk or widely dispersed cooperative endeavors. Based on emotion theory, we expect ritual systems to incorporate more negative affect in the emotional conditioning of symbols under conditions of large group size and political inequality. We have argued that adolescence constitutes an experience expectant period for the emotional valencing of symbols. We, therefore, anticipate adolescent rites of passage to be most intense and prolonged among unrelated adolescents in societies engaging in high-risk, cooperative activities. In addition, there should be a positive association between the duration/intensity of adolescent rites of passage and concomitant changes in both brain response patterns to religious symbols and individual cooperative behaviors. Music should be a particularly powerful elicitor of such responses.

Numerous research questions remain. If adolescence is an "experience expectant" period for the emotional valencing of symbolic systems, is adolescent development dependent on such valencing? In the absence of religious ritual, how is such valencing achieved? Does ritual participation impact adolescent health and behavior? Do adolescent rites of passage measurably alter neurotransmitter and endocrine levels? Are there gender and/or status differences in the neurophysiological effects of ritual? Can we empirically demonstrate autonomic congruence in ritual participants? If so, is such congruence significantly associated with perceived empathy and increased cooperation? To what extent do the various components of ritual impact emotional charging of symbolic stimuli? Can we define socioecological parameters associated with the positive and negative emotional charging of religious symbols? Does the developmental propensity to believe in socially omniscient supernatural agents peak in adolescence? Are such agents a necessary component of symbolically charged belief systems, or can such systems instead achieve cooperation through the emotional charging of unfalsifiable non-agent schema, such as "liberty" and "freedom"? Finally, if religion is an evolved adaptation for cooperation, can humanity achieve such cooperation in its absence? This is among the most salient questions facing the world today. The answer must begin with a better understanding of religion as a specifically human adaptation.

Signals are necessarily bound to the moment; symbols, however, have existence and meaning that extend beyond the immediate to link the past, present, and future. They, thus, lay the foundation for creating and identifying groups, but also for motivating cooperation among the individuals within these groups across both space and time (Rappaport 1999). Far from being an evolutionary by-product, religion represents a critical adaptive complex evolved in response to ecological challenges faced by early human populations. Individual fitness benefits resulted both from participation in ritual itself, and from the cooperative activities it enabled. The use of music-based ritual to imbue group signals with emotional and motivational meaning gave impetus to a new system of social communication and a new level of human cognition.

The authors would like to thank Joseph Bulbulia, Marc Dichter, James Dow, Patrick McNamara, Eric Smith, Robert Storey, and two anonymous reviewers for their helpful comments on an earlier draft of this paper.

Candace Alcorta is currently a doctoral student in the Department of Anthropology at the University of Connecticut. Her research interests include the behavioral ecology and evolution of religion, and the interrelationship between cultural and neurophysiological systems. She is currently conducting research on adolescent religious participation, stress, and health.

Richard Sosis is an associate professor in the Department of Anthropology at the University of Connecticut and a senior lecturer in the Department of Sociology and Anthropology at The Hebrew University of Jerusalem. His current research interests include the evolution of cooperation, utopian societies, and the behavioral ecology of religion. He has conducted fieldwork on Ifaluk Atoll in the Federated States of Micronesia and is currently pursuing various projects in Israel aimed at understanding the benefits and costs associated with religious behavior.

REFERENCES CITED

Adolphs, R.
1999 Neural Systems for Recognizing Emotions in Humans. In *The Design of Animal Communication*, M. Hauser and M. Konishi, eds. Pp. 187–212. Cambridge: MIT Press.

2002a Social Cognition and the Human Brain. In *Foundations in Social Neuroscience,* J. T. Cacioppo, G. G. Berntson, R. Adolphs, et al., eds. Pp. 313–332. Cambridge: MIT Press.
2002b Trust in the Brain. *Nature Neuroscience* 5:192–193.

Adolphs, R., D. Tranel, and A. R. Damasio, eds.
1998 The Human Amygdala in Social Judgment. *Nature* 393:470–474.

Anderson, S. W., A. Bechara, H. Damasio, D. Tranel, and A. R. Damasio
2002 Impairment of Social and Moral Behavior Related to Early Damage in Human Prefrontal Cortex. In *Foundations of Social Neuroscience,* J. T. Cacioppo, G. G. Berntson, R Adolphs, et al., eds. Pp. 333–343. Cambridge: MIT Press.

Atran, S.
2002 *In Gods We Trust: The Evolutionary Landscape of Religion.* Oxford: Oxford University Press.

Atran, S., and A. Norenzayan
2004 Religion's Evolutionary Landscape: Counterintuition, Commitment, Compassion, Communion. *Behavioral and Brain Sciences* 27:713–730.

Austin, J. H.
1998 *Zen and the Brain.* Cambridge: MIT Press.

Ball, G. F.
1999 The Neuroendocrine Basis of Seasonal Changes in Vocal Behavior among Songbirds. In *The Design of Animal Communication,* M. D. Hauser and M. Konishi, eds. Pp. 213–254. Cambridge: MIT Press.

Barrett, J. L.
2000 Exploring the Natural Foundation of Religion. *Trends in Cognitive Science* 4:29–34.

Bartlett, F.
1932 *Remembering.* Cambridge: Cambridge University Press.

Bear, D.
1979 Temporal Lobe Epilepsy: A Syndrome of Sensory-Limbic Hyperconnectionism. *Cortex* 15:357–384.

Bear, D., L. Schenk, and H. Benson
1981 Increased Autonomic Responses to Neutral and Emotional Stimuli in Patients with Temporal Lobe Epilepsy. *American Journal of Psychiatry* 138:843–845.

Becker, J.
2001 Anthropological Perspectives on Music and Emotion. In *Music and Emotion,* P. Juslin and J. Sloboda, eds. Pp. 135–160. Oxford: Oxford University Press.

Bednarik, R. G.
1995 Concept-Mediated Marking in the Lower Paleolithic. *Current Anthropology* 36:605–634.

Bering, J. M.
2005 The Evolutionary History of an Illusion: Religious Causal Beliefs in Children and Adults. In Origins of the Social Mind: Evolutionary Psychology and Child Development, B. Ellis and D. Bjorklund, eds. Pp. 411–437. New York: Guilford Press.

Bering, J. M., and D. F. Bjorklund
2004 The Natural Emergence of Reasoning about the Afterlife as a Developmental Regularity. *Developmental Psychology* 40:217–233.

Bettleheim, B.
1962 *Symbolic Wounds: Puberty Rites and the Envious Male.* New York: Collier.

Bloch, M.
1989 *Ritual, History, and Power.* London: Athlone Press.

Bourguignon, E., ed.
1973 *Religion, Altered States of Consciousness, and Social Change.* Columbus: Ohio State University Press.
1976 *Possession.* San Francisco: Chandler and Sharpe.

Boyer, P.
2001 *Religion Explained: The Evolutionary Origins of Religious Thought.* New York: Basic Books.

Boyer, P., and C. Ramble
2001 Cognitive Templates for Religious Concepts: Cross-Cultural Evidence for Recall of Counterintuitive Representations. *Cognitive Science* 25:535–564.

Bradshaw, M.
2003 Religion as Compensation for Deprivation: The Interactive Influence of Religion and SES on Psychological Distress. Paper presented at annual meeting of the Society for the Scientific Study of Religion, October, Norfolk, Virginia.

Brown, J.
1975 Adolescent Initiation Rites: Recent Interpretations. In *Studies in Adolescence,* R. E. Grinder, ed. Pp. 40–52. New York: MacMillan.

Brown, C. M.
2000 Exploring the Role of Religiosity in Hypertension Management among African Americans. *Journal of Health Care for the Poor and Underserved* 11:19–32.

Bulbulia, J.
2004a Religious Costs as Adaptations that Signal Altruistic Intention. *Evolution and Cognition* 10:19–42.
2004b The Cognitive and Evolutionary Psychology of Religion. *Biology and Philosophy* 19:655–686.

Cacioppo, J. T., W. L. Gardner, and G. G. Berntson
2002 The Affect System Has Parallel and Integrative Processing Components: Form Follows Function. In *Foundations in Social Neuroscience,* J. T. Caccioppo, G. G. Berntson, R. Adolphs, et al., eds. Pp. 493–522. Cambridge: MIT Press.

Cardinal, R., J. Parkinson, J. Hall, and B. Everitt
2002 Emotion and Motivation: The Role of the Amygdala, Ventral Striatum and Prefrontal Cortex. *Neuroscience and Biobehavioral Reviews* 26:321–352.

Carlson, G., Y. Wang, and B. E. Alger
2002 Endocannabinoids Facilitate the Induction of LTP in the Hippocampus. *Nature Neuroscience* 5:723–724.

Casey, G. J., J. N. Giedd, and K. N. Thomas
2000 Structural and Functional Brain Development and Its Relation to Cognitive Development. *Biological Psychology* 54:241–257.

Chase, P. G.
1989 How Different Was Middle Paleolithic Subsistence? A Zooarchaeological Perspective on the Middle to Upper Paleolithic Transition. In *The Human Revolution: Behavioral and Biological Perspectives on the Origins of Modern Humans,* P. Mellars and C. P. Stringer, eds. Pp. 321–327. Edinburgh: Edinburgh University Press.

Chaves, M., M. E. Konieczny, K. Beyerlein, and E. Barman
1999 The National Congregations Study: Background, Methods, and Selected Results. *Journal for the Scientific Study of Religion* 38:458–476.

Cronk, L.
1994a Evolutionary Theories of Morality and the Manipulative Use of Signals. *Zygon* 29:32–58.
1994b The Use of Moralistic Statements in Social Manipulation: A Reply to Roy A. Rappaport. *Zygon* 29:351–355.

Damasio, A. R.
1994 *Descartes' Error: Emotion, Reason, and the Human Brain.* New York: Avon Books.
1998 The Somatic Marker Hypothesis and the Possible Function of the Prefrontal Cortex. In *The Prefrontal Cortex,* A. C. Roberts, T. W. Robbins, and J. Weiskrantz, eds. Pp. 36–50. New York: Oxford University Press.

Davidson, J. R.
1976 The Physiology of Meditation and Mystical States of Consciousness. *Perspectives in Biology and Medicine* (Spring):345–379.

Davidson, R.
1994 On Emotion, Mood, and Related Affective Constructs. In *The Nature of Emotion,* P. Ekman and R. J. Davidson, eds. Pp. 51–55. New York: Oxford University Press.

Davidson, R., and W. Irwin
2002 The Functional Neuroanatomy of Emotion and Affective Style. In *Foundations in Social Neuroscience,* J. T. Caccioppo, G. G. Berntson, R. Adolphs, et al., eds. Pp. 473–490. Cambridge: MIT Press.

Deacon, H. J.
1989 Late Pleistocene Paleoecology and Archaeology in the Southern Cape. In *The Human Revolution: Behavioral and Biological Perspectives on the Origins of Modern Humans,* P. Mellars and C. P. Stringer, eds. Pp. 547–564. Edinburgh: Edinburgh University Press.

Deacon, T.
1997 *The Symbolic Species.* New York: W.W. Norton.

Dehaene, S., and J. P. Changeux
2000 Reward-Dependent Learning in Neuronal Networks for Planning and Decision-Making. In *Cognition, Emotion, and Autonomic Responses: The Integrative Role of the Prefrontal Cortex and Limbic Structures,* H. B. M. Uylings, C. G. van Eden, J. P. D. deBruin, et al., eds. Pp. 219–230. Elsevier: New York.

Depue, R. A., M. Luciana, P. Arbisi, P. Collins, and A. Leon
2002 Dopamine and the Structure of Personality: Relation of Agonist-Induced Dopamine Activity to Positive Emotionality. In *Foundations in Social Neuroscience,* J. T. Caccioppo, G. G. Berntson, R. Adolphs, et al., eds. Pp. 1071–1092. Cambridge: MIT Press.

DiChiara, G.
1995 The Role of Dopamine in Drug Abuse Viewed from the Perspective of Its Role in Motivation. *Drug and Alcohol Dependence* 38:95–137.

Dissinayake, E.
1995 Homo aestheticus: *Where Art Comes From and Why.* Seattle: University of Washington Press.

Dolan, R.
2000 Emotional Processing in the Human Brain Revealed through Functional Neuroimaging. In *The New Cognitive Neurosciences,* second ed., M. Gazzaniga, ed. Pp. 1115–1132. Cambridge: MIT Press.

Douglas, M.
1966 *Purity and Danger.* Frederick A. Praeger: New York.

Dressler, W. W., and J. R. Bindon
2000 The Health Consequences of Cultural Consonance: Cultural Dimensions of Lifestyle, Social Support and Arterial Blood Pressure in an African American Community. *American Anthropologist* 102:244–260.

Dugatkin, L. A.
1997 *Cooperation among Animals: An Evolutionary Perspective.* New York: Oxford University Press.

Dunbar, R.
1999 Culture, Honesty and the Freerider Problem. In *The Evolution of Culture,* R. Dunbar, C. Knight, and C. Power, eds. Pp. 194–213. New Brunswick: Rutgers University Press.

Durkheim, E.
1969 *The Elementary Forms of the Religious Life.* New York: Free Press. (Originally published in 1915)

Ekman, P.
2003 Darwin, Deception and Facial Expression. In *Emotions Inside Out,* P. Ekman, J. J. Campos, R. J. Davidson, and F. B. M. de Waal, eds. Pp. 205–221. Annals of the New York Academy of Sciences 1000.

Ekman, P., and R. Davidson
1993 Voluntary Smiling Changes Regional Brain Activity. *Psychological Science* 4:342–345.

Ekman, P., R. W. Levenson, and W. V. Friesen
1983 Autonomic Nervous System Activity Distinguishes among Emotions. *Science* 22:1208–1210.

Eliade, M.
1958 *Rites and Symbols of Initiation: The Mysteries of Birth and Rebirth.* Dallas: Spring Publications.
1959 *The Sacred and the Profane: The Nature of Religion.* New York: Harcourt Brace Jovanovich.

Elkin, D.
1999 Religious Development in Adolescence. *Journal of Adolescence* 22:291–295.

Gellhorn, E., and W. F. Kiely
1972 Mystical States of Consciousness: Neurophysiological and Clinical Aspects. *Journal of Nervous and Mental Disease* 154:399–405.

Geschwind, N.
1979 Behavioral Changes in Temporal Lobe Epilepsy. *Psychological Medicine* 9:217–219.

Giedd, J. N., J. Blumenthal, N. O. Jeffries, F. X. Catellanos, H. Liu, A. Zijdenbos, T. Paus, A. C. Evans, and J. L. Rapoport
1999 Brain Development during Childhood and Adolescence: A Longitudinal MRI Study. *Nature Neuroscience* 2:861–863.

Glucklich, A.
2001 *Sacred Pain.* New York: Oxford University Press.

Goodall, J.
1986 *The Chimpanzees of Gombe: Patterns of Behavior.* Cambridge: Harvard University Press.

Greenough, W. T.
1986 What's Special about Development? Thoughts on the Bases of Experience Sensitive Synaptic Plasticity. In *Developmental Neuropsychobiology,* W. T. Greenough and J. M. Juraska, eds. Pp. 387–408. New York: Academic Press.

Greenough, W. T., and J. Black
1991 Induction of Brain Structure by Experience: Substrate for Cognitive Development. In *Minnesota Symposia on Child Psychology,* Vol. 24, M. R. Gunnar and C. A. Nelson, eds. Pp. 155–200. Hillsdale, N. J.: Lawrence Erlbaum.

Groenewegen, H. J., and H. B. M. Uylings
2000 The Prefrontal Cortex and the Integration of Sensory, Limbic, and Autonomic Information. In *Cognition, Emotion, and Autonomic Responses: The Integrative Role of the Prefrontal Cortex and Limbic Structures,* H. B. M. Uylings, C. G. van Eden, J. P. D. de Bruin, et al., eds. Pp. 3–28. New York: Elsevier.

Guthrie, S. E.
1993 *Faces in the Clouds: A New Theory of Religion.* New York: Oxford University Press.

Halpern, C. T., J. R. Udry, and B. Campbell
1994 Testosterone and Religiosity as Predictors of Sexual Attitudes and Activity among Adolescent Males: A Biosocial Model. *Journal of Biosocial Science* 26:217–234.

Hayden, B.
1987 Alliances and Ritual Ecstasy: Human Responses to Resource Stress. *Journal for the Scientific Study of Religion* 26:81–91.
1993 The Cultural Capacities of Neanderthals: A Review and Re-evaluation. *Journal of Human Evolution* 24:113–146.

Henshilwood, C. S., F. d'Errico, C. W. Marean, R. G. Milo, and R. Yates
2001 An Early Bone Tool Industry from the Middle Stone Age at Blombos Cave, South Africa: Implications for the Origins of Modern Human Behaviour, Symbolism, and Language. *Journal of Human Evolution* 41:631–678.

Hirokawa, E. and H. Ohira
2003 The Effects of Music Listening After a Stressful Task on Immune Functions, Neuroendocrine Responses, and Emotional States in College Students. *Journal of Music Therapy* 40:189–211.

Hummer, R. A., R. G. Rogers, C. B. Narn, and C. G. Ellison
1999 Religious Involvement and U.S. Adult Mortality. *Demography* 36:273–285.

Irons, W.
1996a In Our Own Self-Image: The Evolution of Morality, Deception and Religion. *Skeptic* 4:50–61.
1996b Morality as an Evolved Adaptation. In *Investigating the Biological Foundations of Morality,* J. P. Hurd, ed. Pp. 1–34. Lewiston, Idaho: Edwin Mellon Press.
2001 Religion as a Hard-to-Fake Sign of Commitment. In *Evolution and the Capacity for Commitment,* R. Nesse, ed. Pp. 292–309. New York: Russell Sage Foundation.

Ito, T. A., J. T. Larsen, N. K. Smith, and J. T. Cacioppo
2002 Negative Information Weighs More Heavily on the Brain: The Negativity Bias in Evaluative Categorizations. In *Foundations in Social Neuroscience,* J. T. Cacioppo, G. G. Berntson, R. Adolphs, et al., eds. Pp. 575–598. Cambridge: MIT Press.

Johnson, D. D. P., and O. Kruger
2004 The Good of Wrath: Supernatural Punishment and the Evolution of Cooperation. *Political Theology* 5:159–176.
Johnstone, R. A.
2000 Game Theory and Communication. In *Game Theory and Animal Behavior,* L. A. Dugatkin and H. K. Reeve, eds. Pp. 94–117. New York: Oxford University Press.
Kasamatsu, A., and T. Hirai
1966 An Electroencephalographic Study on the Zen Meditation. *Folio Psychiatrica & Neurologica Japonica* 20:315–336.
Kelemen, D.
2004 Are Children "Intuitive Theists"? Reasoning about Purpose and Design in Nature. *Psychological Science* 15:295–301.
Keshavan, M. S., V. A. Diwadkar, M. DeBellis, E. Dick, R. Kotwal, D. R. Rosenberg, J. A. Sweeney, N. Minshew, and J. W. Pettegrew
2002 Development of the Corpus Callosum in Childhood, Adolescence and Early Adulthood. *Life Science* 70:1909–1922.
Khalfa, S., S. D. Bella, M. Roy, I. Peretz, and S. J. Lupien
2003 Effects of Relaxing Music on Salivary Cortisol Level after Psychological Stress. In *The Neurosciences and Music,* G. Avanzini, C. Faienza, D. Minciacchi, L. Lopez, and M. Majno, eds. Pp. 374–376. Annals of the New York Academy of Sciences 999.
Kirkpatrick, L. A.
1999 Toward an Evolutionary Psychology of Religion and Personality. *Journal of Personality* 67:921–951.
Knight, C., C. Power, and I. Watts
1995 The Human Symbolic Revolution: A Darwinian Account. *Cambridge Archaeological Journal* 5:75–114.
Kolb, B., and I. Q. Whishaw
1998 Brain Plasticity and Behavior. *Annual Review of Psychology* 49:43–64.
Kolb, B., M. Forgie, R. Gibb, G. Gorny, and S. Rontree
1998 Age, Experience and the Changing Brain. *Neuroscience and Biobehavioral Reviews* 22:143–159.
Kolb, B., S. Pellis, and T. E. Robinson
2004 Plasticity and Functions of the Orbital Frontal Cortex. *Brain and Cognition* 55:104–115.
Krebs, J. R., and R. Dawkins
1984 Animal Signals: Mind-Reading and Manipulation. In *Behavioural Ecology: An Evolutionary Approach,* second ed., J. R. Krebs and N. B. Davies, eds. Pp. 380–402. Oxford: Blackwell Scientific.
Krumhansl, C. L.
1997 An Exploratory Study of Musical Emotions and Psychophysiology. *Canadian Journal of Experimental Psychology* 51:336–353.
Kuhn, D.
2002 The Effects of Active and Passive Participation in Musical Activity on the Immune System as Measured by Salivary Immunoglobulin A (SIgA). *Journal of Music Therapy* 39:30–39.
Kwon, Y. J., and A. E. Lawson
2000 Linking Brain Growth with the Development of Scientific Reasoning Ability and Conceptual Change during Adolescence. *Journal of Research in Science Teaching* 37:44–62.
Laughlin, C. D., Jr., and J. McManus
1979 Mammalian Ritual. In *The Spectrum of Ritual,* E. G. d'Aquili, C. D. Laughlin, and J. McManus, eds. Pp. 80–116. New York: Columbia University Press.
Leach, E. R.
1966 Ritualisation in Man in Relation to Conceptual and Social Development. In *A Discussion on Ritualisation of Behaviour in Animals and Man,* J. Huxley, ed. Philosophical Transactions of the Royal Society, B:403–408.
LeDoux, J. E.
1996 *The Emotional Brain.* New York: Simon and Schuster.
2002 Emotion: Clues from the Brain. In *Foundations in Social Neuroscience,* J. T. Cacioppo, G. G. Berntson, R. Adolphs, et al., eds. Pp. 389–410. Cambridge: MIT Press.

Levenson, R. W.
1994 Human Emotions: A Functional View. In *The Nature of Emotion,* P. Ekman and R. J. Davidson, eds. Pp. 123–126. New York: Oxford University Press.
2003 Blood, Sweat and Fears: The Autonomic Architecture of Emotion. In *Emotions Inside Out,* P. Ekman, J. J. Campos, R. J. Davidson, and F. B. M. de Waal, eds. Pp. 348–366. Annals of the New York Academy of Sciences 1000.
Levi-Strauss, C.
1963 *Structural Anthropology.* New York: Basic Books.
Levin, J. S.
1994 Religion and Health: Is There an Association, Is It Valid, and Is It Causal? *Social Science in Medicine* 38:1475–1482.
1996 How Religion Influences Morbidity and Health: Reflections on Natural History, Salutogenesis, and Host Resistance. *Social Science and Medicine* 43:849–864.
Lewis, D. B., and D. M. Gower
1980 *Biology of Communication.* New York: John Wiley and Sons.
Lex, B. W.
1979 The Neurobiology of Ritual Trance. In *The Spectrum of Ritual,* E. G. d'Aquili, C. D. Laughlin, Jr., and J. McManus, eds. Pp. 117–151. New York: Columbia University Press.
Lorenz, K.
1965 *Evolution and Modification of Behavior.* Chicago: University of Chicago Press.
Lutkehaus, N. C., and P. B. Roscoe, eds.
1995 *Gender Rituals: Female Initiation in Melanesia.* New York: Routledge.
MacLean, C. R. K., K. G. Walton, S. R. Wenneberg, D. K. Levitsky, J. P. Mandarino, R. Wziri, S. L. Hillis, and R. H. Schneider
1997 Effects of the Transcendental Meditation Program on Adaptive Mechanisms: Changes in Hormone Levels and Responses to Stress after 4 Months of Practice. *Psychoneuroendocrinology* 22:277–295.
Malinowski, B.
1948 *Magic, Science, Religion and Other Essays.* Garden City, New Jersey: Doubleday.
Mandel, A.
1980 Toward a Psychobiology of Transcendence: God in the Brain. In *The Psychobiology of Consciousness,* J. Davidson and R. Davidson, eds. Pp. 379–464. New York: Plenum Press.
Marler, P.
1999 On Innateness: Are Sparrow Songs "Learned" or "Innate"? In *The Design of Animal Communication,* M. D. Hauser and M. Konishi, eds. Pp. 293–318. Cambridge: MIT Press.
Marshack, A.
1990 Evolution of the Human Capacity: The Symbolic Evidence. *Yearbook of Physical Anthropology* 32:1–34.
Matthews, D. A., M. E. McCullough, D. B. Larson, H. G. Koenig, J. P. Swyers, and M. G. Milano
1998 Religious Commitment and Health Status. *Archives of Family Medicine* 7:118–124.
McBrearty, S., and A. Brooks
2000 The Revolution That Wasn't: A New Interpretation of the Origin of Modern Human Behavior. *Journal of Human Evolution* 39:453–563.
McCauley, R. N.
2001 Ritual, Memory and Emotion: Comparing Two Cognitive Hypotheses. In *Religion in Mind,* J. Andresen, ed. Pp. 115–140. Cambridge: Cambridge University Press.
McNamara, P.
2001 Religion and the Frontal Lobes. In *Religion in Mind,* J. Andresen, ed. Pp. 237–256. Cambridge: Cambridge University Press.
2002 The Motivational Origins of Religious Practices. *Zygon* 37:143–160.
Meyerhoff, B.
1974 *Peyote Hunt: The Sacred Journey of the Huichol Indians.* Ithaca: Cornell University Press.
Mithen, S.
1996 *The Prehistory of the Mind.* London: Thames & Hudson.
1999 Symbolism and the Supernatural. In *The Evolution of Culture,* R. Dunbar, C. Knight, and C. Power, eds. Pp. 147–172. New Brunswick: Rutgers University Press.

Morris, J. S., A. Ohman, and R. J. Dolan
1998 Conscious and Unconscious Emotional Learning in the Human Amygdala. *Nature* 393:467–470.

Murphy, P. E., J. W. Ciarrocchi, R. L. Piedmont, S. Cheston, and M. Peyrot
2000 The Relation of Religious Belief and Practices, Depression, and Hopelessness in Persons with Clinical Depression. *Journal of Consulting and Clinical Psychology* 68:1102–1106.

Neher, A.
1962 A Physiological Explanation of Unusual Behavior in Ceremonies Involving Drums. *Human Biology* 34:151–161.

Newberg, A. B., E. G. d'Aquili, and V. Rause
2001 *Why God Won't Go Away.* New York: Ballantine Books.

Oram, M. W., and B. J. Richmond
1999 I See a Face—A Happy Face. *Nature Neuroscience* 2:856–858.

Paige, K. E., and J. M. Paige
1981 *The Politics of Reproductive Ritual.* Los Angeles: University of California Press.

Pearson, J.
1990 Neurotransmission in Brain Regions Associated with Emotion and Autonomic Control. In *An Introduction to Neurotransmission in Health and Disease,* P. Riederer, N. Koppe, and J. Pearson, eds. Pp. 253–263. New York: Oxford University Press.

Peretz, I.
2001 Brain Specialization for Music: New Evidence from Congenital Amusia. In *The Biological Foundations of Music,* R. Zatorre and I. Peretz, eds. Pp. 153–165. Annals of the New York Academy of Sciences 930.

Pinker, S.
1997 *How the Mind Works.* New York: Norton.

Plant, T. M.
2002 Neurophysiology of Puberty. *Journal of Adolescent Health* 31:185–191.

Rappaport, R. A.
1994 On the Evolution of Morality and Religion: A Response to Lee Cronk. *Zygon* 29:331–349.
1999 *Ritual and Religion in the Making of Humanity.* London: Cambridge University Press.

Regan, C.
2001 *Intoxicating Minds.* New York: Columbia University Press.

Regnerus, M., C. Smith, and M. Fritsch
2003 *Religion in the Lives of American Adolescents: A Review of the Literature.* Research Report of the National Study of Youth and Religion, No. 3. Chapel Hill: University of North Carolina.

Reichert, S. E.
2000 Game Theory and Animal Contests. In *Game Theory and Animal Behavior,* L. A. Dugatkin and H. K. Reeve, eds. Pp. 64–93. New York: Oxford University Press.

Richerson, P., and R. Boyd
1998 The Evolution of Human Ultra-Sociality. In *Indoctrinability, Ideology and Warfare: Evolutionary Perspectives,* I. Eibl-Eibisfeldt and F. Salter, eds. Pp. 71–95. New York: Berghahn Books.

Robbins, T. W.
2000 The Arousal to Cognition: The Integrative Position of the Prefrontal Cortex. In *Cognition, Emotion and Autonomic Responses: The Integrative Role of the Prefrontal Cortex,* H. B. M. Uylings, C. G. van Eden, J. P. C. de Bruin, M. G. P. Feenstra, and C. M. A. Pennartz, eds. Pp. 469–483. New York: Elsevier.

Roes, F. L., and M. Raymond
2003 Belief in Moralizing Gods. *Evolution and Human Behavior* 24:126–135.

Rogers, L. J., and G. Kaplan
2000 *Songs, Roars, and Rituals.* Cambridge: Harvard University Press.

Rolls, E. T.
1998 The Orbitofrontal Cortex. In *The Prefrontal Cortex,* A. C. Roberts, T. W. Robbins, and L. Weiskrantz, eds. Pp. 67–86. New York: Oxford University Press.

Rowe, C.
1999 Receiver Psychology and the Evolution of Multi-Component Signals. *Animal Behaviour* 58:921–931.

Sapolsky, R.
1999 Hormonal Correlates of Personality and Social Contexts: From Non-human to Human Primates. In *Hormones, Health and Behavior,* C. Panter-Brick and C. Worthman, eds. Pp. 18–46. Cambridge: Cambridge University Press.

Saver, J. L., and J. Rabin
1997 The Neural Substrates of Religious Experience. *Journal of Neuropsychiatry* 9:498–510.

Scherer, K. R., and M. R. Zentner
2001 Emotional Effects of Music: Production Rules. In *Music and Emotion,* P. Juslin and J. Sloboda, eds. Pp. 361–392. Oxford: Oxford University Press.

Schoenbaum, G., B. Setlow, M. P. Saddoris, and M. Gallagher
2003 Encoding Predicted Outcome and Acquired Value in Orbitofrontal Cortex during Cue Sampling Depends upon Input from Basolateral Amygdala. *Neuron* 39:855–867.

Schultz, W., P. Dayan, and P. R. Montague
2002 A Neural Substrate of Prediction and Reward. In *Foundations in Social Neuroscience,* J. T. Cacioppo, G. G. Berntson, R. Adolphs, et al., eds. Pp. 313–332. Cambridge: MIT Press.

Shea, J.
1988 Spear Points from the Middle Paleolithic of the Levant. *Journal of Field Archaeology* 15:441–450.

Sloboda, J., and P. Juslin
2001 Psychological Perspectives on Music and Emotion. In *Music and Emotion,* P. Juslin and J. Sloboda, eds. Pp. 71–104. Oxford: Oxford University Press.

Smith, J. W.
1979 Ritual and the Ethology of Communicating. In *The Spectrum of Ritual,* E. G. d'Aquili, C. D. Laughlin, Jr., and J. McManus, eds. Pp. 51–79. New York: Columbia University Press.

Sosis, R.
2000 Religion and Intragroup Cooperation: Preliminary Results of a Comparative Analysis of Utopian Communities. *Cross-Cultural Research* 34:70–87.
2003a Review of "Darwin's Cathedral: Evolution, Religion and the Nature of Society" by David Sloan Wilson. *Evolution and Human Behavior* 24:137–143.
2003b Why Aren't We All Hutterites? Costly Signaling Theory and Religious Behavior. *Human Nature* 14:91–127.

Sosis, R., and C. S. Alcorta
2003 Signaling, Solidarity and the Sacred: The Evolution of Religious Behavior. *Evolutionary Anthropology* 12:264–274.
2004 Is Religion Adaptive? *Behavioral and Brain Sciences* 27:749–750.

Sosis, R., and E. Bressler
2003 Cooperation and Commune Longevity: A Test of the Costly Signaling Theory of Religion. *Cross-Cultural Research* 37:211–239.

Sosis, R., and B. Ruffle
2003 Religious Ritual and Cooperation: Testing for a Relationship on Israeli Religious and Secular Kibbutzim. *Current Anthropology* 44:713–722.
2004 Ideology, Religion, and the Evolution of Cooperation: Field Experiments on Israeli Kibbutzim. *Research in Economic Anthropology* 23:87–115.

Sosis, R., H. Kress, and J. Boster
n.d. Scars for War: Evaluating Alternative Signaling Explanations for Cross-Cultural Variance in Ritual Costs. Ms. in the authors' possession.

Sowell, E. R., P. M. Thompson, C. J. Holmes, T. L. Jernigan, and A. W. Toga
1999 In Vivo Evidence for Post-Adolescent Brain Maturation in Frontal and Striatal Regions. *Nature Neuroscience* 2:859–861.

Spear, L. P.
2000 The Adolescent Brain and Age-Related Behavioral Manifestations. *Neuroscience and Biobehavioral Reviews* 24:417–463.

Swanson, G. E.
1960 *The Birth of the Gods: The Origin of Primitive Beliefs.* Ann Arbor: University of Michigan Press.

Tinbergen, N.
1965 *The Study of Instinct.* Oxford: Clarendon Press.
Turner, V.
1967 *The Forest of Symbols.* New York: Cornell University Press.
1969 *The Ritual Process.* Chicago: Aldine.
Tylor, E. B.
1871 *Primitive Culture.* London: Murray.
van Gennep, A.
1960 *The Rites of Passage.* Chicago: University of Chicago Press. (Originally published in 1909)
Walker, E., and A. M. Bollini
2002 Pubertal Neurodevelopment and the Emergence of Psychotic Symptoms. *Schizophrenia Research* 54:17–23.
Wallace, A. F. C.
1966 *Religion: An Anthropological View.* New York: Random House.
Walter, V. J., and W. G. Walter
1949 The Central Effects of Rhythmic Sensory Stimulation. *Electroencephalography and Clinical Neurophysiology* 1:57–86.
Watanabe, J. M., and B. B. Smuts
1999 Explaining Religion without Explaining It Away: Trust, Truth, and the Evolution of Cooperation in Roy A. Rappaport's "The Obvious Aspects of Ritual." *American Anthropologist* 101:98–112.
Watts, I.
1999 The Origin of Symbolic Culture. In *The Evolution of Culture,* R. Dunbar, C. Knight, and C. Power, eds. Pp. 113–146. New Brunswick: Rutgers University Press.
Wilson, D. S.
2002 *Darwin's Cathedral.* Chicago: Chicago University Press.
Wingfield, J. C., J. D. Jacobs, K. Soma, D. L. Maney, K. Hunt, D. Wisti-Peterson, S. Meddle, M. Ramenofsky, and K. Sullivan
1999 Testosterone, Aggression, and Communication: Ecological Bases of Endocrine Phenomena. In *The Biology of Communication,* M. Konishi and M. D. Hauser, eds. Pp. 255–284. Cambridge: MIT Press.
Winkelman, M.
1986 Trance States: A Theoretical Model and Cross-Cultural Analysis. *Ethos* 14:174–203.
1992 *Shamans, Priests and Witches: A Cross-Cultural Study of Magico-religious Practitioners.* Anthropological Research Papers No. 44. Tempe: Arizona State University.
2000 *Shamanism: The Neural Ecology of Consciousness and Healing.* Westport, Conn.: Bergin

Part V
Intelligent Design

[19]

Intelligent Design: The Original Version

FRANCISCO J. AYALA

Abstract *William Paley* (Natural Theology, *1802) developed the argument-from-design. The complex structure of the human eye evinces that it was designed by an intelligent Creator. The argument is based on the irreducible complexity ("relation") of multiple interacting parts, all necessary for function. Paley adduces a wealth of biological examples leading to the same conclusion; his knowledge of the biology of his time was profound and extensive. Charles Darwin's* Origin of Species *is an extended argument demonstrating that the "design" of organisms can be explained by natural selection. Moreover, the dysfunctions, defects, waste, and cruelty that prevail in the living world are incompatible with a benevolent and omnipotent Creator. They come about by a process that incorporates chance and necessity, mutation and natural selection. In addition to science, there are other ways of knowing, such as art, literature, philosophy, and religion. Matters of value, meaning, and purpose transcend science.*

Key words: Evolution; Natural selection; Intelligent design; Paley; Darwin

In his *Natural Theology* of 1802, the English theologian William Paley advanced the "argument from design". The living world, he argues, provides compelling evidence of being designed by an omniscient and omnipotent Creator. Paley's first example is the human eye that he compares with a telescope: they are both made upon the same principles and bear a complete resemblance to one another, in their configuration, position of the lenses, and effectiveness in bringing each pencil of light to a point at the right distance from the lens. Could, he asks, these attributes be in the eye without purpose? "There cannot be design without designer; contrivance, without a contriver."

The argument-from-design is elaborated by Paley with greater cogency and more extensive knowledge of biological detail than by any other author before or since. Paley brings in all sorts of biological knowledge, from the geographic distribution of species to the interactions between predators and their prey, the interactions between the sexes, the camel's stomach and the woodpecker's tong, the compound eyes of insects and the spider's web. He explores the possibility of a sort of "natural selection": organisms may have come about by chance in an endless multiplicity of forms; those now in existence are those that happened to be functionally organized because they are the only ones able to survive and reproduce. Paley's evidence against chance derives from a notion akin to what some contemporary authors have named "irreducible complexity," that he calls "relation": the presence of a great variety of parts interacting with each other to produce an effect, which cannot be accomplished if any of the parts is missing.

10 **Theology and Science**

Charles Darwin read and enjoyed *Natural Theology* while a student at Cambridge University and found the argument compelling, but this would change later. I propose that the motivating objective of Darwin's *Origin of Species* was to provide a solution to Paley's problem; namely, to demonstrate how his discovery of natural selection would account for the design of organisms, without the need to recourse to supernatural agencies. As Darwin saw it, if his explanation were correct, biological evolution would follow; organisms would have changed over time and diversified, in response to a diversity of conditions in different places and at different times. Darwin, therefore, assembled evidence for evolution, because the occurrence of evolution corroborates his explanation of design as a result of natural selection.

The interaction between chance processes, such as genetic mutation and recombination, with natural selection yields a creative process which generates novelty (new sorts of organisms) and adaptation. The organisms appear to be designed to live in their environments, and their parts appear to be designed to fulfill certain functions, as a consequence of the incremental, step-by-step dialogue and barter between chance and natural selection, exercised over eons of time. But the process is haphazard, imperfections are pervasive, and the immense majority of species become extinct. The defective and dysfunctional design of organisms amounts to an argument-from-imperfection for the origin of organisms by natural processes.

I conclude with a brief statement pointing out that science is not the only way to acquire knowledge about the natural world. A scientific view of the world is by itself hopelessly incomplete, because there remain questions of value, purpose, and meaning that are outside science's realm, while they may be approached by other ways of knowing such as art and literature, philosophical reflection, and religious inspiration.

William Paley's intelligent design

The English clergyman and author William Paley (1743–1805) was intensely committed to the abolition of the slave trade and had become by the 1780s a much sought after public lecturer against slavery. He was also an influential writer of works on Christian philosophy, ethics, and theocracy. *The Principles of Moral and Political Philosophy* (1785) and *A View of the Evidence of Christianity* (1794) earned him prestige and well-endowed ecclesiastical benefices, which allowed him a comfortable life. Illness forced him in 1800 to give up his public speaking career, which provided ample time to study science, particularly biology, and write *Natural Theology; or, Evidences of the Existence and Attributes of the Deity* (1802), the book by which he has become best known to posterity and which would greatly influence Darwin. With *Natural Theology,* Paley sought to update John Ray's *Wisdom of God Manifested in the Works of the Creation* (1691), taking advantage of one century of additional scientific knowledge.

Paley's keystone claim is that: "There cannot be design without a designer; contrivance, without a contriver; order, without choice; . . . means suitable to an end, and executing their office in accomplishing that end, without the end ever having been contemplated."[1] *Natural Theology* is a sustained argument manifesting the obvious design of humans and their parts, as well as the design of all sorts of organisms, in themselves, and in their relations to one another and to their environment. There are chapters dedicated to the complex design of the human eye; to the human frame, which displays a precise mechanical arrangement of bones, cartilage, and joints; to the circulation of the blood and the disposition of blood vessels; to the comparative anatomy of humans and animals; to the digestive system, kidneys, urethras, and bladder; to the wings of birds and the fins of fish; and much more. For 352 pages, *Natural Theology* conveys Paley's expertise: extensive and accurate biological knowledge, as detailed and precise as it was available in the year 1800. After detailing the precise organization and exquisite functionality of each biological object or process, Paley draws again and again the same conclusion, that only an omniscient and omnipotent Deity could account for these marvels of mechanical perfection, purpose, and functionality, and for the enormous diversity of inventions that they entail.

Paley's first model example is the human eye, in chapter three, "Application of the Argument." I will quote him at some length, for there is no better way to display his knowledge of the anatomy of the eye or his skill of argumentation.

> I know no better method of introducing so large a subject, than that of comparing a single thing with a single thing: an eye, for example, with a telescope. As far as the examination of the instrument goes, there is precisely the same proof that the eye was made for vision as there is that the telescope was made for assisting it. They are made upon the same principles; both being adjusted to the laws by which the transmission and refraction of rays of light are regulated. . . . For instance, these laws require, in order to produce the same effect, that the rays of light, in passing from water into the eye, should be refracted by a more convex surface than when it passes out of air into the eye. Accordingly we find that the eye of a fish, in that part of it called the crystalline lens, is much rounder than the eye of terrestrial animals. What plainer manifestation of design can there be than this difference? What could a mathematical instrument maker have done more to show his knowledge of [t]his principle, his application of that knowledge, his suiting of his means to his end . . . to testify counsel, choice, consideration, purpose?[2]

It is worthwhile to follow Paley's argument further:

> The lenses of the telescopes and the humors of the eye bear a complete resemblance to one another, in their figure, their position, and in their power over the rays of light, namely, in bringing each pencil to a point at the right distance from the lens; namely, in the eye, at the exact place where the membrane is spread to receive it. How is it possible, under circumstances of such close affinity, and under the operation of equal evidence, to exclude contrivance from the one, yet to acknowledge the proof of contrivance having been employed, as the plainest and clearest of all propositions, in the other?[3]

He brings in, to his argument's advantage, the issue of dioptric distortion:

> In dioptric telescopes there is an imperfection of this nature. Pencils of light, in passing through glass lenses, are separated into different colors, thereby tinging the object, especially the edges of it, as if it were viewed through a prism. To correct this

> inconvenience has been long a desideratum in the art. At last it came into the mind of a sagacious optician, to inquire how this matter was managed in the eye, in which there was exactly the same difficulty to contend with as in the telescope. His observation taught him that in the eye the evil was cured by combining lenses composed of different substances, that is, of substances which possessed different refracting powers. Our artist borrowed thence his hint, and produced a correction of the defect by imitating, in glasses made from different materials, the effects of the different humors through which the rays of light pass before they reach the bottom of the eye. Could this be in the eye without purpose, which suggested to the optician the only effectual means of attaining that purpose?[4]

The functional anatomy of the eye is, later on, summarized as follows:

> [We marvel] knowing as we do what an eye comprehends, namely, that it should have consisted, first, of a series of transparent lenses—very different, even in their substance, from the opaque materials of which the rest of the body is, in general at least, composed, and with which the whole of its surface, this single portion of it excepted, is covered: secondly, of a black cloth or canvas—the only membrane in the body which is black—spread out behind these lenses, so as to receive the image formed by pencils of light transmitted through them; and placed at the precise geometrical distance at which, and at which alone, a distinct image could be formed, namely, at the concourse of the refracted rays: thirdly, of a large nerve communicating between this membrane and the brain; without which, the action of light upon the membrane, however modified by the organ, would be lost to the purposes of sensation.[5]

Could the eye have come about without design or preconceived purpose, as a result of chance? Paley had set the argument against chance, in the very first paragraph of *Natural Theology*, arguing rhetorically by analogy:

> In crossing a heath, suppose I pitched my foot against a *stone*, and were asked how the stone came to be there, I might possibly answer, that for any thing I knew to the contrary it had lain there for ever; nor would it, perhaps, be very easy to show the absurdity of this answer. But suppose I had found a *watch* upon the ground, and it should be inquired how the watch happened to be in that place, I should hardly think of the answer which I had before given, that for any thing I knew the watch might have always been there. Yet why should not this answer serve for the watch as well as for the stone; why is it not as admissible in the second case as in the first? For this reason, and for no other, namely, that when we come to inspect the watch, we perceive—what we could not discover in the stone—that its several parts are framed and put together for a purpose, *e.g.* that they are so formed and adjusted as to produce motion, and that motion so regulated as to point out the hour of the day; that if the different parts had been differently shaped from what they are, or placed after any other manner or in any other order than that in which they are placed, either no motion at all would have been carried on in the machine, or none which would have answered the use that is now served by it.[6]

The strength of the argument against chance derives, Paley tells us, from what he names "relation," a notion akin to what contemporary anti-evolutionists have named "irreducible complexity" (and that some of them have given themselves credit for its discovery). This is how Paley formulates the argument:

> When several different parts contribute to one effect, or, which is the same thing, when an effect is produced by the joint action of different instruments, the fitness of such parts or instruments to one another for the purpose of producing, by their united action, the effect, is what I call *relation*; and wherever this is observed in the works of nature or of man, it appears to me to carry along with it decisive evidence of understanding, intention, art . . . all depending upon the motions within, all upon the system of intermediate actions.[7]

A remarkable example of complex parts, fit together so that they cannot function one without the other, are the sexes, "manifestly made for each other . . . subsisting, like the clearest relations of art, in different individuals, unequivocal, inexplicable without design."[8]

The outcomes of chance do not exhibit relation among the parts or, as we might say, organized complexity:

> the question is, whether a useful or imitative conformation be the product of chance . . . Universal experience is against it. What does chance ever do for us? In the human body, for instance, chance, that is, the operation of causes without design, may produce a wen, a wart, a mole, a pimple, but never an eye. Among inanimate substances, a clod, a pebble, a liquid drop might be; but never was a watch, a telescope, an organized body of any kind, answering a valuable purpose by a complicated mechanism, the effect of chance. In no assignable instance has such a thing existed without intention somewhere.[9]

Paley considers and rejects an interesting hypothesis that would combine chance and natural selection:

> The hypothesis teaches, that every possible variety of being hath, at one time or other, found its way into existence—by what cause or in what manner is not said—and that those which were badly formed perished . . . The hypothesis, indeed, is hardly deserving of the consideration which we have given to it. What should we think of a man who, because we had never ourselves seen watches, telescopes, stocking-mills, steam-engines, etc., made, knew not how they were made, nor could prove by testimony when they were made, or by whom, would have us believe that these machines . . . derive [their curious structures] from no other origin than this; namely, that a mass of metals and other materials having run, when melted, into all possible figures, and combined themselves in all possible forms and shapes and proportions, these things which we see are what were left from the incident . . . I cannot distinguish the hypothesis, as applied to the works of nature, from this solution, which no one would accept as applied to a collection of machines.[10]

This hypothesis is reminiscent of the philosopher of classic Greece Empedocles' account of the origin of complex entities, such as animals and plants, and it seems similar to the model of chance and selection rejected by some contemporary anti-evolutionists, but it has nothing significant in common with Darwin's theory of natural selection, an incremental process that incorporates adaptive changes, one small step at a time, in response to the environmental circumstances of the organisms. But Darwin's theory had not yet been formulated and one can only credit Paley for a serious exploration of the subject, carrying it as far as the biology of his time made it possible.

14 **Theology and Science**

Paley's natural theology fails, even in his time, when seeking an account of imperfections, defects, pain and cruelty that would be consistent with his notion of the Creator. Chapter 23 is entitled "Of the Personality of the Deity" and it would surprise many by its well meaning, if naïve, arrogance, as Paley seems convinced that he can determine God's "personality." This is how the chapter starts:

> Contrivance, if established, appears to me to prove ... the *personality* [Paley's emphasis] of the Deity, as distinguished from what is sometimes called nature, sometimes called a principle ... Now, that which can contrive, which can design, must be a person. These capacities constitute personality, for they imply consciousness and thought.... The acts of a mind prove the existence of a mind; and in whatever a mind resides, is a person. The seat of intellect is a person.[11]

Paley proceeds, in the ensuing chapter to set "the natural attributes of the Deity," namely, omnipotence, omniscience, omnipresence, eternity, self-existence, necessary existence and spirituality—all these Paley infers from the observation of natural processes!

Paley raises the question of organs or parts seemingly unnecessary or superfluous. He considers two possible states of affairs: "in some instances the operation, in others the use, is unknown."[12] Examples of the first kind include the lungs of animals, which we know to be necessary for survival, although we are not "acquainted with the action of the air upon the blood, or in what manner that action is communicated by the lungs."[13] He cites the lymphatic system as a second example of this kind. Instances "may be numerous; for they will be so in proportion to our ignorance. ... Every improvement of knowledge diminishes their number."[14] Examples of organs with unknown use include the spleen, which seems not to be necessary for "it has been extracted from dogs without any sensible injury to their vital functions." But it may well be the case that the part serves some unknown function, even if not necessary for survival in the short run. In any case, he adds, "superfluous parts do not negative the reasoning which we instituted concerning those parts which are useful."[15]

This last comment seems to me remarkable in that it is so unconvincing and so inconsistent with Paley's conceptual framework. Yet this is his general explanation for nature's imperfections: "Irregularities and imperfections are of little or no weight ... but they are to be taken in conjunction with the unexceptionable evidences which we possess of skill, power, and benevolence displayed in other instances."[16] But if functional design manifests an intelligent designer, why should not deficiencies indicate that the designer is less than omniscient, or less than omnipotent, or less than omnivolent? Paley cannot have it both ways. Moreover, we know that some deficiencies are not just imperfections, but they are outright dysfunctional, jeopardizing the very function the organ or part is supposed to serve. This is a matter to which I shall return below. We now know, of course, that the explanation for dysfunction and imperfection is natural selection, which can account for design and functionality, but does not achieve any sort of perfection, nor is it omniscient or omnipotent.

I am filled with amazement and respect for Paley's extensive and profound biological knowledge. He discusses the air-bladder of fish, the fang of vipers, the claw of herons, the camel's stomach, the woodpecker's tongue, the elephant's proboscis, the hook in the bat's wing, the spider's web, the compound eyes of insects, and their metamorphosis, the glowworm, univalve and bivalve mollusks, seed dispersal, and on and on, with accuracy and as much detail as known to the best biologists of his time.

Paley's textbooks were part of the canon at the University of Cambridge for nearly half a century after his death and thus were read by Charles Darwin, who was an undergraduate student there between 1827 and 1831, with profit and "much delight." Darwin writes in his *Autobiography*:

> In order to pass the B.A. examination, it was also necessary to get up Paley's *Evidences of Christianity*, and his *Moral Philosophy*. This was done in a thorough manner, and I am convinced that I could have written out the whole of the *Evidences* with perfect correctness, but not of course in the clear language of Paley. The logic of this book and, as I may add, of his *Natural Theology*, gave me as much delight as did Euclid. The careful study of these works, without attempting to learn any part by rote, was the only part of the academic course which, as I then felt and as I still believe, was of the least use to me in the education of my mind. I did not at that time trouble myself about Paley's premises; and taking these on trust, I was charmed and convinced by the long line of argumentation.[17]

William Paley was not alone in Britain in the first half of the nineteenth century. The *Bridgewater Treatises*, published between 1833 and 1840, were written by eminent scientists and philosophers to set forth "the Power, Wisdom, and Goodness of God as manifested in the Creation." The complex functional organization of the human hand was one example elaborated as incontrovertible evidence that the hand had been designed by the same omniscient Power that had created the world. The treatises are marvels of biological knowledge and insight, even though the line of argumentation seems derivative by comparison to Paley's *Natural Theology*.

The emergence of modern science: Copernicus and Darwin

There is a priggish version of the history of the ideas that sees a parallel between Copernicus' and Darwin's monumental intellectual contributions, which are said to have eventuated two revolutions. According to this version, the Copernican Revolution consisted in displacing the Earth from its previously accepted locus as the center of the universe, moving it to a subordinate place as one more planet revolving around the sun. In congruous manner, this version affirms, the Darwinian Revolution consisted in displacing humans from their position as the center of life on Earth, with all other species created for the purpose of humankind, and placing humans instead as one species among many in the living world, so that humans are related to chimpanzees, gorillas, and other species by shared common ancestry. Copernicus had accomplished his revolution with the heliocentric theory of the solar system; Darwin's achievement emerged from his theory of organic evolution.

I will proffer that this version of the two revolutions is inadequate. What it says is true, but it misses what is most important about these two intellectual revolutions, namely, that they ushered in the beginning of science in the modern sense of the word. These two revolutions may jointly be seen as the one Scientific Revolution, with two stages, the Copernican and the Darwinian.

Darwin is deservedly given credit for the theory of biological evolution, because he accumulated evidence demonstrating that organisms evolve; and he discovered the process, natural selection, by which they evolve their functional organization. But the import of *The Origin of Species* is that it completed the Copernican Revolution, initiated three centuries earlier, and thereby radically changed our conception of the universe and the place of humankind in it.

The Copernican Revolution was launched with the publication in 1543, the year of Nicolaus Copernicus' death, of his *De revolutionibus orbium celestium* (*On the Revolutions of the Celestial Spheres*), and bloomed with the publication in 1687 of Isaac Newton's *Philosophiae naturalis principia mathematica* (*The Mathematical Principles of Natural Philosophy*). The discoveries of Copernicus, Kepler, Galileo, Newton, and others, in the sixteenth and seventeenth centuries, had gradually ushered in a conception of the universe as matter in motion governed by natural laws. It was shown that the earth is not the center of the universe, but a small planet rotating around an average star; that the universe is immense in space and in time; and that the motions of the planets around the sun can be explained by the same simple laws that account for the motion of physical objects on our planet. (Laws such as $f = m \times a$, *force = mass* $\times$ *acceleration*, or the inverse-square law of attraction, $f = g(m_1 m_2)/r^2$.) These and other discoveries greatly expanded human knowledge, but the conceptual revolution they brought about was more fundamental yet: a commitment to the postulate that the universe obeys immanent laws that account for natural phenomena. The workings of the universe were brought into the realm of science: explanation through natural laws. Physical phenomena could be accounted for whenever the causes were adequately known.

Darwin completed the Copernican Revolution by drawing out for biology the ultimate conclusion of the notion of nature as a lawful system of matter in motion. The adaptations and diversity of organisms, the origin of novel and highly organized forms, the origin of mankind itself, could now be explained by an orderly process of change governed by natural laws.

The origin of organisms and their marvelous adaptations were attributed, before Darwin, to the design of an omniscient Creator. God had created the birds and bees, the fish and corals, the trees in the forest, and best of all, humans. God had given human beings eyes so that they might see; and God had provided fish with gills to breathe in water. Paley, like the authors of the *Bridgewater Treatises* and many other philosophers and theologians, argued that the functional design of organisms manifests the existence of an all-wise Creator. Wherever there is design, there is a designer; the existence of a watch evinces the existence of a watchmaker.

The advances of physical science had driven humankind's conception of the universe to a split-personality state of affairs, which persisted well into the mid-nineteenth century. Scientific explanations, derived from natural laws, dominated the world of nonliving matter, on the earth as well as in the heavens. Supernatural explanations, depending on the unfathomable deeds of the Creator, accounted for the origin and configuration of living creatures—the most diversified, complex, and interesting realities of the world.

It was Darwin's genius to resolve this conceptual schizophrenia. Darwin completed the Copernican Revolution by drawing out for biology the notion of nature as a lawful system of matter in motion that human reason can explain without recourse to extra-natural agencies.

The conundrum faced by Darwin can hardly be overestimated. The strength of the argument-from-design to demonstrate the role of the Creator was easily set forth. Wherever there is function or design, we look for its author. Paley had belabored this argument with great skill and profusion of detail. It was Darwin's greatest accomplishment to show that the complex organization and functionality of living beings can be explained as the result of a natural process, natural selection, without any need to resort to a Creator or other external agent. The origin and adaptation of organisms in their profusion and wondrous variations were thus brought into the realm of science.

Darwin accepted that organisms are "designed" for certain purposes, i.e. they are functionally organized. Organisms are adapted to certain ways of life and their parts are adapted to perform certain functions. Fish are adapted to live in water, kidneys are designed to regulate the composition of blood, and the human hand is made for grasping. But Darwin went on to provide a natural explanation of the design. The seemingly purposeful aspects of living beings could now be explained, like the phenomena of the inanimate world, by the methods of science, as the result of natural laws manifested in natural processes.

Darwin's discovery

The central argument of the theory of natural selection is summarized by Darwin in *The Origin of Species* as follows:

> As more individuals are produced than can possibly survive, there must in every case be a struggle for existence, either one individual with another of the same species, or with the individuals of distinct species, or with the physical conditions of life. . . . Can it, then, be thought improbable, seeing that variations useful to man have undoubtedly occurred, that other variations useful in some way to each being in the great and complex battle of life, should sometimes occur in the course of thousands of generations? If such do occur, can we doubt (remembering that more individuals are born than can possibly survive) that individuals having any advantage, however slight, over others, would have the best chance of surviving and of procreating their kind? On the other hand, we may feel sure that any variation in the least degree injurious would be rigidly destroyed. This preservation of favorable variations and the rejection of injurious variations, I call Natural Selection.[18]

Darwin's argument addresses the same issues as Paley's: how to account for the adaptive configuration of organisms, the obvious "design" of their parts to fulfill certain functions. Darwin argues that hereditary adaptive variations ("variations useful in some way to each being") occasionally appear, and that these are likely to increase the reproductive chances of their carriers. The success of pigeon fanciers and animal breeders clearly evinces the occasional occurrence of useful hereditary variations. Over the generations, favorable variations will be preserved, multiplied, and conjoined; injurious ones will be eliminated. In one place, Darwin adds: "I can see no limit to this power [natural selection] in slowly and beautifully *adapting* each form to the most complex relations of life."[19] Natural selection was proposed by Darwin primarily to account for the adaptive organization, or "design," of living beings; it is a process that preserves and promotes adaptation. Evolutionary change through time and evolutionary diversification (multiplication of species) are not directly promoted by natural selection (hence, the so-called evolutionary stasis emphasized by the theory of punctuated equilibrium), but they often ensue as by-products of natural selection fostering adaptation.

There is a possible reading of Darwin's *Origin of Species* that sees it, first and foremost, as a sustained effort to solve Paley's problem within a scientific explanatory framework. It is, indeed, how I interpret Darwin's masterpiece. The Introduction and Chapters I to VIII explain how natural selection accounts for the adaptations and behaviors of organisms, their "design." The extended argument starts in Chapter I, where Darwin describes the successful selection of domestic plants and animals and, with considerable detail, the success of pigeon fanciers seeking exotic sports. This evidence manifests what selection can accomplish using spontaneous variations beneficial to man. The ensuing chapters extend the argument to variations propagated by natural selection (i.e. reproductive success) for the benefit of the organisms, rather than by artificial selection for traits desirable to humans. Organisms exhibit design, but it is not "intelligent design," imposed by God as a supreme engineer, but the result of natural selection promoting the adaptation of organisms to their environments. Organisms exhibit complexity, but it is not irreducible complexity emerged all of a sudden in its current elaboration, but has arisen gradually and cumulatively, step by step, promoted by the adaptive success of individuals with incrementally more complex elaborations.

If Darwin's explanation of the adaptive organization of living beings is correct, evolution necessarily follows, as organisms become adapted to different environments and to the changing conditions of all environments, and as hereditary variations become available at a particular time that improve the organisms' chances of survival and reproduction. The *Origin*'s evidence for biological evolution is central to Darwin's explanation of "design," because his explanation postulates the occurrence of biological evolution, which he, therefore, seeks to demonstrate in most of the remainder of the book (Chapters IX–XIII), returning to the original theme in the concluding Chapter XIV. In the last paragraph of the *Origin*, Darwin eloquently returns, indeed, to the dominant theme of adaptation or design:

> It is interesting to contemplate an entangled bank, clothed with many plants of many kinds, with birds singing on the bushes, with various insects flitting about, and with worms crawling through the damp earth, and to reflect that these

elaborately constructed forms, *so different* from each other, and dependent on each other *in so complex a manner*, have all been produced by laws acting around us. . . . Thus, from the war of nature, from famine and death, the most exalted object which we are capable of conceiving, namely, the production of the higher animals, directly follows. There is grandeur in this view of life, with its several powers, having been originally breathed into a few forms or into one; and that, whilst this planet has gone cycling on according to the fixed law of gravity, from so simple a beginning *endless forms most beautiful and most wonderful* have been, and are being, evolved.[20] (Emphasis added.)

Natural selection as a "design" process

The modern understanding of the principle of natural selection is formulated in genetic and statistical terms as differential reproduction. Natural selection implies that some genes and genetic combinations are transmitted to the following generations with a higher probability than their alternates. Such genetic units will become more common in subsequent generations and their alternates less common. Natural selection is a statistical bias in the relative rate of reproduction of alternative genetic units.

Natural selection does not operate in the manner of Paley's unaccepted hypothesis, acting on randomly formed organisms, allowing the functional ones to survive while the great majority dies. Natural selection does not operate, either, as a sieve that retains the rarely arising useful genes and lets go the more frequently arising harmful mutants; at least, not only. Natural selection acts in the filtering way of a sieve, but it is much more than a purely negative process, for it is able to generate novelty by increasing the probability of otherwise extremely improbable genetic combinations. Natural selection is thus a creative process. It does not create the entities upon which it operates, but it produces adaptive (functional) genetic combinations that could not have existed otherwise.

The creative role of natural selection must not be understood in the sense of the absolute creation that traditional Christian theology predicates of the divine act by which the universe was brought into being *ex nihilo*, or in the manner of creation in which Paley assumes God, the supreme engineer, had created the adaptations of organisms. Natural selection may rather be compared to a painter who creates a picture by mixing and distributing pigments in various ways over the canvas. The canvas and the pigments are not created by the artist but the painting is. It is inconceivable that a random combination of the pigments might result in the orderly whole that is the final work of art, say Leonardo's *Mona Lisa*. In the same way, the combination of genetic units which carries the hereditary information responsible for the formation of the vertebrate eye could have never been produced by a random process such as mutation. Not even if we allow for the three billion years plus during which life has existed on earth. The complicated anatomy of the eye like the exact functioning of the kidney are the result of a nonrandom process—natural selection.

How natural selection, a purely material process, can generate novelty in the form of accumulated hereditary information may be illustrated by the following example. Some strains of the colon bacterium, *Escherichia coli,* in order to be able to reproduce in a culture medium, require that a certain substance, the amino acid histidine, be provided in the medium. When a few such bacteria are added to ten cubic centimeters of liquid culture medium, they multiply rapidly and produce between 20 and 30 billion bacteria in a few hours. Spontaneous mutations to streptomycin resistance occur in normal (i.e. sensitive) bacteria at rates of the order of one in one hundred million (1×10^{-8}) cells. In the bacterial culture we expect between 200 and 300 bacteria to be resistant to streptomycin due to spontaneous mutation. If a proper concentration of the antibiotic is added to the culture, only the resistant cells survive. The two or three hundred surviving bacteria will start reproducing, however, and allowing one or two days for the necessary number of cell divisions, 20 billion or so bacteria are produced, all resistant to streptomycin. Among cells requiring histidine as a growth factor, spontaneous mutants able to reproduce in the absence of histidine arise at rates of about four in one hundred million (4×10^{-8}) bacteria. The streptomycin resistant cells may now be transferred to a culture with streptomycin but with no histidine. Most of them will not be able to reproduce, but about 1000 will and will start reproducing until the available medium is saturated.

Natural selection has produced in two steps bacterial cells resistant to streptomycin and not requiring histidine for growth. The probability of the two mutational events happening in the same bacterium is of about four in ten million billion ($1 \times 10^{-8} \times 4 \times 10^{-8} = 4 \times 10^{-16}$) cells. An event of such low probability is unlikely to occur even in a large laboratory culture of bacterial cells. With natural selection, cells having both properties are the common result.

Critics have sometimes alleged as evidence against Darwin's theory of evolution examples showing that random processes cannot yield meaningful, organized outcomes. It is thus pointed out that a series of monkeys randomly striking letters on a typewriter would never write *The Origin of Species*, even if we allow for millions of years and many generations of monkeys pounding at typewriters.

This criticism would be valid if evolution would depend only on random processes. But natural selection is a non-random process that promotes adaptation by selecting combinations that "make sense," i.e., that are useful to the organisms. The analogy of the monkeys would be more appropriate if a process existed by which, first, meaningful words would be chosen every time they appeared on the typewriter; and then we would also have typewriters with previously selected words rather than just letters in the keys, and again there would be a process to select meaningful sentences every time they appeared in this second typewriter. If every time words such as "the," "origin," "species," and so on, appeared in the first kind of typewriter, they each became a key in the second kind of typewriter, meaningful sentences would occasionally be produced in this second typewriter. If such sentences became incorporated into keys of a third type of typewriter, in which meaningful paragraphs were selected whenever they appeared, it is clear that pages and even chapters "making sense" would eventually be produced. The end product would be an irreducibly complex text.

We need not carry the analogy too far, since the analogy is not fully satisfactory, but the point is clear. Evolution is not the outcome of purely random processes, but rather there is a "selecting" process, which picks up adaptive combinations because these reproduce more effectively and thus become established in populations. These adaptive combinations constitute, in turn, new levels of organization upon which the mutation (random) plus selection (non-random or directional) process again operates. The complexity of organization of animals and plants is "irreducible" to simpler components in one or very few steps, but not thorough the millions and millions of generations and the multiplicity of steps and levels made possible by eons of time.

The critical point is that evolution by natural selection is an incremental process, operating over eons of time and yielding organisms better able to survive and reproduce than others, which typically differ from one another at any one time only in small ways; for example, the difference between having or lacking an enzyme able to catalyze the synthesis of the amino acid histidine. Notice also that increased complexity is not a necessary outcome of natural selection, although such increases occur from time to time, so that, although rare, they are very conspicuous over eons of time. Increased complexity is not a necessary consequence of evolution by natural selection, but rather emerges occasionally as a matter of statistical bias. The longest living organisms on earth are the microscopic bacteria, which have continuously existed on our planet for 3.5 billion years and yet those now living exhibit no greater complexity than their old time ancestors. More complex organisms came about much later, without the elimination of their simpler relatives. For example, the primates appeared on earth some 50 million years ago and our species, *Homo sapiens*, came about two hundred thousand years ago.

As illustrated by the bacterial example, natural selection produces combinations of genes that would otherwise be highly improbable because natural selection proceeds stepwise. The vertebrate eye did not appear suddenly in all its present perfection. Its formation required the appropriate integration of many genetic units, and thus the eye could not have resulted from random processes alone, nor did it come about suddenly or in a few steps. The ancestors of today's vertebrates had for more than half a billion years some kind of organs sensitive to light. Perception of light, and later vision, were important for these organisms' survival and reproductive success. Accordingly, natural selection favored genes and gene combinations increasing the functional efficiency of the eye. Such genetic units gradually accumulated, eventually leading to the highly complex and efficient vertebrate eye. Natural selection can account for the rise and spread of genetic constitutions and, therefore, of types of organisms, that would never have existed under the uncontrolled action of random mutation. In this sense, natural selection is a creative process, although it does not create the raw materials—the genes—upon which it acts.[21]

Design and chance

There is an important respect in which an artist makes a poor analogy of natural selection. A painter has a preconception of what he wants to paint and will consciously modify the painting so that it represents what he wants. Natural selection has no foresight, nor does it operate according to some preconceived plan. Rather it is a purely natural process resulting from the interacting properties of physicochemical and biological entities. Natural selection is simply a consequence of the differential multiplication of living beings, as pointed out. It has some appearance of purposefulness because it is conditioned by the environment: which organisms reproduce more effectively depends on what variations they possess that are useful in the place and at the time where the organisms live. But natural selection does not anticipate the environments of the future; drastic environmental changes may be insuperable to organisms that were previously thriving. Species extinction is the common outcome of the evolutionary process. The species existing today represent the balance between the origin of new species and their eventual extinction. More than 99% of all species that ever lived on earth have become extinct without issue. These may have been more than one billion species; the available inventory of living species has identified and described less than two million out of some ten million estimated to be now in existence.

The team of typing monkeys is also a bad analogy of evolution by natural selection, because it assumes that there is "somebody" who selects letter combinations and word combinations that make sense. In evolution there is no one selecting adaptive combinations. These select themselves because they multiply more effectively than less adaptive ones.

There is a sense in which the analogy of the typing monkeys is better than the analogy of the artist, at least if we assume that no particular statement was to be obtained from the monkeys' typing endeavors, but just any statements making sense. Natural selection does not strive to produce predetermined kinds of organisms, but only organisms that are adapted to their present environments. Which characteristics will be selected depends on which variations happen to be present at a given time in a given place. This in turn depends on the random process of mutation, as well as on the previous history of the organisms (i.e., on the genetic make-up they have as a consequence of their previous evolution). Natural selection is an opportunistic process. The variables determining in what direction it will go are the environment, the preexisting constitution of the organisms, and the randomly arising mutations.

Thus, adaptation to a given environment may occur in a variety of different ways. An example may be taken from the adaptations of plant life to the desert climate. The fundamental adaptation is to the condition of dryness, which involves the danger of desiccation. During a major part of the year, sometimes for several years in succession, there is no rain. Plants have accomplished the urgent necessity of saving water in different ways. Cacti have transformed their leaves into spines, having made their stems into barrels storing a reserve of water; photosynthesis is performed on the surface of the stem instead of in the leaves. Other plants have no leaves during the dry season, but after it rains they burst into leaves and flowers and produce seeds. Ephemeral plants germinate from seeds, grow, flower, and produce seeds—all within the space of the few weeks while rainwater is available; the rest of the year the seeds lie quiescent in the soil.

The opportunistic character of natural selection is also well evidenced by the phenomenon of adaptive radiation. The evolution of drosophila flies in Hawaii is a relatively recent adaptive radiation. There are about 1500 drosophila species in the world. Approximately 500 of them have evolved in the Hawaiian archipelago, which has a small land area, about one twenty-fifth the size of California. Moreover, the morphological, ecological, and behavioral diversity of Hawaiian drosophila exceeds that of drosophila in the rest of the world. There are more than 1000 species of land snails in Hawaii, all of which have evolved in the archipelago. There are 72 bird species, all of which but one exist nowhere else.

Why should have such explosive evolution have occurred in Hawaii? The overabundance of drosophila flies there contrasts with the absence of many other insects. The ancestors of Hawaiian drosophila reached the archipelago before other groups of insects did, and thus they found a multitude of unexploited opportunities for living. They responded by a rapid adaptive radiation; although they are all derived from a single colonizing species, they adapted to the diversity of opportunities available in diverse places or at different times by developing appropriate adaptations, which range broadly from one to another species. The geographic remoteness of the Hawaiian archipelago seems, in any case, a more reasonable explanation for these explosions of diversity of a few kinds of organisms than assuming an inordinate preference on the part of the Creator for providing the archipelago with numerous drosophila, but not with other insects, or a peculiar distaste for creating land mammals in Hawaii, since none existed there until introduced by humans.

The process of natural selection can explain the adaptive organization of organisms, as well as their diversity and evolution, as a consequence of their adaptation to the multifarious and ever changing conditions of life. The fossil record shows that life has evolved in a haphazard fashion. The radiations, expansions, relays of one form by another, occasional but irregular trends, and the ever-present extinctions, are best explained by natural selection of organisms subject to the vagaries of genetic mutation and environmental challenge. The scientific account of these events does not necessitate recourse to a pre-ordained plan, whether imprinted from without by an omniscient and all-powerful designer, or resulting from some immanent force driving the process towards definite outcomes. Biological evolution differs from a painting or an artifact in that it is not the outcome of preconceived design.

Natural selection accounts for the "design" of organisms, because adaptive variations tend to increase the probability of survival and reproduction of their carriers at the expense of maladaptive, or less adaptive, variations. The arguments of Paley against the incredible improbability of chance accounts of the adaptations of organisms are well taken as far as they go. But not Paley and not any other author before Darwin, was able to discern that there is a natural process (namely, natural selection) that is not random, but rather is oriented and able to generate order or "create." The traits that organisms acquire in their evolutionary histories are not fortuitous but determined by their functional utility to the organisms, designed as it were to serve their life needs.

Chance is, nevertheless, an integral part of the evolutionary process. The mutations that yield the hereditary variations available to natural selection arise at random, independently of whether they are beneficial or harmful to their carriers. But this random process (as well as others that come to play in the great theatre of life) is counteracted by natural selection, which preserves what is useful and eliminates the harmful. Without hereditary mutation, evolution could not happen because there would be no variations that could be differentially conveyed from one to another generation. But without natural selection, the mutation process would yield disorganization and extinction because most mutations are disadvantageous. Mutation and selection have jointly driven the marvelous process that starting from microscopic organisms has yielded orchids, birds, and humans.

The theory of evolution conveys chance and necessity jointly intricated in the stuff of life; randomness and determinism interlocked in a natural process that has spurted the most complex, diverse, and beautiful entities in the universe: the organisms that populate the earth, including humans who think and love, endowed with free will and creative powers, and able to analyze the process of evolution itself that brought them into existence. This is Darwin's fundamental discovery, that there is a process that is creative though not conscious. And this is the conceptual revolution that Darwin completed: that everything in nature, including the 'design' of living organisms, can be accounted for as the result of natural processes governed by natural laws. This is nothing if not a fundamental vision that has forever changed how mankind perceives itself and its place in the universe.

The 'fact' of evolution

The biological disciplines provide overwhelming evidence that organisms are related by common descent with modification: paleontology, comparative anatomy, biogeography, embryology, biochemistry, molecular genetics, and others. The idea first emerged from observations of graded changes in the succession of fossil remains found in a sequence of layered rocks, as well as numerous remains of kinds of organisms no longer in existence. The layers have a cumulative thickness of tens of kilometers that represent up to 3.5 billion years of geological time. The general sequence of fossils from bottom upward in layered rocks had been recognized before Darwin proposed that the succession of biological forms strongly implied evolution. The farther back into the past one looked, the less the fossils resembled recent forms, the more the various lineages merged, and the broader the implications of a common ancestry for organisms presently quite diverse, such as fish, reptiles, and mammals.

Although gaps in the paleontological record remain now, many have been filled by the research of paleontologists since Darwin's time. Millions of fossil organisms found in well-dated rock sequences represent a succession of forms through time and manifest many evolutionary transitions. Microbial life of the simplest type (i.e. procaryotes, which are cells whose genetic matter is not bound by a nuclear membrane) was already in existence more than three billion years ago. The oldest evidence of more complex organisms (i.e. eukaryotic cells with their genetic matter enclosed in a chamber known as the nucleus) has been discovered in flinty rocks approximately 1.4 billion years old. More advanced forms, like algae, fungi, higher plants, and a great variety of animals have been found only in younger geological strata.

The sequence of observed forms and the fact that all (except the procaryotes) are constructed from the same basic cellular type; strongly imply that all these major categories of life (including animals, plants, algae, and fungi) have a common ancestry in the first eukaryotic cells. Moreover, there have been so many discoveries of intermediate forms between fish and amphibians, between amphibians and reptiles, between reptiles and mammals, and so on, that it is often difficult to identify categorically along the line when the transition occurs from one to another particular genus or, more generally, from one to another kind of organism. Nearly all fossils can be regarded as intermediates in some sense; they are life forms that come between ancestral forms that preceded them and those that followed.

Inferences about common descent derived from paleontology have been reinforced by comparative anatomy. The skeletons of humans, dogs, whales, and bats are strikingly similar, despite the different ways of life led by these animals and the diversity of environments in which they have flourished. The correspondence, bone by bone, can be observed in every part of the body, including the limbs: a person writes, a dog runs, a whale swims, and a bat flies with structures built of the same bones organized in the same pattern. Structures that manifest great similarity in their composition and configuration are called "homologous," and are best explained by common descent from a kind of organism that already exhibited the same composition and configuration, but so that modifications followed that made the structures suitable to the way of life of the descendants. Comparative anatomists investigate such homologies, not only in bone structure, but in other parts of the body as well, working out degrees of relationships from degrees of similarity.

The mammalian ear and jaw offer an example in which paleontology and comparative anatomy combine to show common ancestry through transitional stages. The lower jaws of mammals contain only one bone, whereas those of reptiles have several. The additional bones in the reptile jaw are homologous with bones now found in the mammalian ear. What function could these bones have had, either in the mandible or in the ear, during intermediate stages? Paleontologists have discovered two transitional forms of mammal-like reptiles (Therapsida) with a double jaw-joint—one joint composed of the bones that persist in the mammalian jaw, the other consisting of the quadrate and articular bones that eventually became the hammer and anvil of the mammalian ear. The complex structure of the jaw of the Therapsida made possible the gradual evolution of some of its bones into a different function, while the remainder retained the jaw function. Similar examples are numerous.

Other biological disciplines that manifest biological evolution include embryology and biogeography, already known in Darwin's time, as well as more recently developed disciplines, such as biochemistry, genetics, and comparative ethology. It is not my intention here to review the evidence forthcoming from these biological disciplines, because it is readily available in numerous textbooks and treatises. However, I do want to point out that the most encompassing as well as detailed evidence comes from molecular biology, a recent discipline that emerged in the second half of the twentieth century. It is the most encompassing evidence because the most diverse kinds of organisms can all be compared in many different respects at once, from the lowly bacteria and the microscopic protozoa to the multicellular plants, fungi, and animals visible to the human eye. Molecular biology is remarkable in that organisms encompass thousands of genes and proteins, each one of which can be evaluated as an independent test of the evolutionary relationships among any particular organisms. Moreover, the evidence can readily be quantified. The possibility exists today of determining the evolutionary history of any group of organisms with as much detail as wanted. Only the limitations of human or other resources stand in the way of reconstructing the grand panorama of the evolution of all life, from the microscopic creatures of 3.5 thousand million years ago to the microorganisms, animals, and plants of today.

The proteins and nucleic acids that are essential to the makeup of all organisms are informational macromolecules that retain a record of their evolutionary history. The evolutionary information is contained in the linear sequence of their component elements in much the same way as semantic information is contained in the sequence of letters of an English sentence. This evolutionary information is so detailed that it not only makes it possible to reconstruct the phylogenetic topology, or evolutionary relationships of parentage among organisms, but also opens up the possibility of timing the events in that history, even those that occurred in the remote past of life's history. The information is quantifiable because the number of units that differ between organisms is readily established when the sequences of the component units are obtained for a given protein or gene. There is very little that comparative anatomy can say about the relative similarity of organisms as diverse as yeasts, pine trees, and human beings, but there are homologous macromolecules that can be compared among all three.

Nucleic acids (such as DNA, or deoxyribonucleic acid, which embodies the hereditary information) and proteins are linear molecules made up of units, called nucleotides in the case of nucleic acids, amino acids in the case of proteins. Evolution typically occurs by the substitution of some of these units gradually, one at a time, so that the number of differences between two organisms is an indication of the recency of their common ancestry, in a similar way as the distance between two cars reflects how long they have been traveling in opposite directions.

The theory of evolution encompasses three issues: (i) the fact of evolution; that is, that organisms are related by common descent with modification; (ii) evolutionary history; that is, the time when lineages split from one another and the changes that occur in each lineage; and (iii) the mechanisms or processes by which evolutionary change occurs.

The fact of evolution is the most fundamental issue, and one established with utmost certainty. Most biologists agree that the evolutionary origin of organisms is today a scientific conclusion established with the kind of certainty attributable to such scientific concepts as the roundness of the earth, the motions of the planets, and the molecular composition of matter. This degree of certainty beyond reasonable doubt is what is implied when biologists say that evolution is a "fact"; the evolutionary origin of organisms is accepted by the immense majority of biologists.

The theory of evolution seeks to ascertain the evolutionary relationships between particular organisms and the events of evolutionary history (the second issue above). Many conclusions of evolutionary history are well established; for example, that the chimpanzee and gorilla are more closely related to humans than is any of those three species to the baboon or other monkeys. Other matters are less certain and still others—such as precisely when life originated on earth or when multi-cellular animals, plants, and fungi first appeared—remain largely unresolved. But uncertainty about these issues does not cast doubt on the fact of evolution. Similarly we do not know all the details about the configuration of the Sierra Nevada Mountains, but that is not reason to doubt that the mountains exist.

Some anti-evolutionists argue that the theory of evolution is only that, a theory and not a fact. Science relies on observation, replication, and experimentation but, they say, nobody has seen the origin of life or the evolution of species, nor have these events been replicated in the laboratory or by experiment.

This argument ignores that when scientists talk about the theory of evolution, they use the word "theory" differently from its use in ordinary language. In everyday English, a theory is an imperfect fact, as in "I have a theory as to where Osama bin Laden is hiding." In science, however, a theory is based on and incorporates a body of knowledge. According to the theory of evolution, organisms are related by common descent. There is a multiplicity of species because organisms change from generation to generation, and different lineages change in different ways. Species that share a recent ancestor are, therefore, more similar than those with more remote ancestors. Thus, humans and chimpanzees are, in configuration and genetic make-up, more similar to each other than they are to baboons or to elephants. That evolution has occurred is, in ordinary language, a fact.

How is this factual claim compatible with the accepted view that science relies on observation, replication, and experimentation since nobody has observed the evolution of species, much less replicated it by experiment? What scientists observe are not the concepts or general conclusions of theories, but their consequences. Copernicus heliocentric theory affirms that the earth revolves around the sun. Nobody has observed this phenomenon, but we accept it because of numerous confirmations of its predicted consequences. We accept that matter is made of atoms, even though nobody has seen them, because of corroborating observations and experiments in physics and chemistry. The same applies with the theory of evolution. For example, the claim that humans and chimpanzees are more closely related to each other than they are to baboons leads to the prediction that the DNA is more similar between humans and chimps than between chimps and baboons. To test this prediction, scientists select a particular gene, examine its DNA structure in each species, and thus corroborate the inference. Experiments of this kind are replicated in a variety of ways to gain further confidence in the conclusion. So it is for myriad predictions and inferences between all sorts of organisms.

Defects, deficiencies, and dysfunctions

I pointed out earlier the unsatisfactory answer that Paley advances to account for the imperfections of organisms. With respect to organs with unknown functions, he points out, correctly, that this may be a matter of our limited knowledge; a function may eventually be discovered. His suggestion is that, for example, the function of the lungs or of the spleen might eventually be discovered, as indeed it has been the case for these two organs. With respect to actual irregularities or imperfections, he makes the unsatisfactory and, to me, unexpected claim that "they are of little or no weight . . . when taken in conjunction . . . with the unexceptionable evidences that we possess of skill, power, and benevolence displayed in other instances." He adds: " . . . apparent blemishes . . . ought to be referred to some cause, though we be ignorant of it."[22]

One of the recent authors who have reformulated Paley's argument-from-design responds to the critics who point out the imperfections of organisms in the following way.

> The most basic problem is that the argument [against intelligent design] demands perfection at all. Clearly, designers who have the ability to make better designs do not necessarily do so. . . . I do not give my children the best, fanciest toys because I don't want to spoil them, and because I want them to learn the value of a dollar. The argument from imperfection overlooks the possibility that the designer might have multiple motives, with engineering excellence oftentimes relegated to a secondary role. . . . Another problem with the argument from imperfection is that it critically depends on psychoanalysis of the unidentified designer. Yet the reasons that a designer would or would not do anything are virtually impossible to know unless the designer tells you specifically what those reasons are.[23]

So, God may have had his reasons for not designing organisms as perfect as they could have been.

A problem with this explanation is that it destroys intelligent design as a scientific hypothesis, because it provides it with an empirically impenetrable shield.[24] If we cannot reject intelligent design because the designer may have reasons that we could not possibly ascertain, there would seem to be no way to test intelligent design by drawing out predictions logically derived from the hypothesis that are expected to be observed in the world of experience. Intelligent design as an explanation for the adaptations of organisms could be (natural) theology, as Paley would have it, but, whatever it is, it is not a scientific hypothesis.

I would argue, moreover, that is not good theology either, because it leads to conclusions about the nature of the designer quite different from those of omniscience, omnipotence, and omnibenevolence that Paley had inferred as the attributes of the Creator. It is not only those organisms and their parts that are less than perfect, but also that deficiencies and dysfunctions are pervasive, evidencing defective design. Consider the human jaw. We have too many teeth for the jaw's size, so that wisdom teeth need to be removed and orthodontists make a decent living straightening the others. Would we want to blame God for such defective design? A human engineer could have done better. Evolution gives a good account of this imperfection. Brain size increased over time in our ancestors, and the remodeling of the skull to fit the larger brain entailed a reduction of the jaw. Evolution responds to the organisms' needs through natural selection, not by optimal design but by tinkering, as it were, by slowly modifying existing structures. Consider now the birth canal of women, much too narrow for easy passage of the infant's head, so that thousands upon thousands of babies die during delivery. Surely we do not want to blame God for this defective design or for the children's deaths. Science makes it understandable, a consequence of the evolutionary enlargement of our brain. Females of other animals do not experience this difficulty. Theologians in the past struggled with the issue of dysfunction because they thought it had to be attributed to God's design. Science, much to the relief of many theologians, provides an explanation that convincingly attributes defects, deformities and dysfunctions to natural causes.

One more example: why are our arms and our legs, which are used for such different functions, made of the same materials, the same bones, muscles and nerves, all arranged in the same overall pattern? Evolution makes sense of the anomaly. Our remote ancestors' forelimbs were legs. After our ancestors became bipedal and started using their forelimbs for functions other than walking, these became gradually modified, but retaining their original composition and arrangement. Engineers start with raw materials and a design suited for a particular purpose; evolution can only modify what is already there. An engineer, who would design cars and airplanes, or wings and wheels, using the same materials arranged in a similar pattern, would surely be fired.[25]

The defective design of organisms could be attributed to the gods of the ancient Greeks, Romans, and Egyptians, who fought with one another, made blunders, and were clumsy in their endeavors. But, in my view, it is not compatible with special action by the omniscient and omnipotent God of Judaism, Christianity, and Islam.[26]

Powers and limits of science

I will add a final comment that should not be necessary, but probably is, owing to the hubris of some scientists and the pusillanimity of some believers. Science is a wondrously successful way of knowing. Science seeks explanations of the natural world by formulating explanations based on observation and experimentation that are subject to the possibility of rejection or corroboration, by cycles upon cycles of additional observations and experimentation. A scientific explanation is tested by ascertaining whether or not predictions about the world of experience derived from the explanation agree with what is later observed.

30 Theology and Science

Science as a mode of inquiry into the nature of the universe has been successful and of great consequence. Witness the proliferation of science academic departments in universities and other research institutions, the enormous budgets that the body politic and the private sector willingly commit to scientific research, and its economic impact. The Office of Management and the Budget (OMB) of the US government has estimated that 50% percent of all economic growth in the United States since the Second World War can directly be attributed to scientific knowledge and technical advances. The technology derived from scientific knowledge pervades our lives: the high-rise buildings of our cities, thruways and long span-bridges, rockets that bring men to the moon, telephones that provide instant communication across continents, computers that perform complex calculations in millionths of a second, vaccines and drugs that keep bacterial parasites at bay, gene therapies that replace DNA in defective cells. All these remarkable achievements bear witness to the validity of the scientific knowledge from which they originated.

Scientific knowledge is also remarkable in the way it emerges by way of consensus and agreement among scientists and in the way new knowledge builds upon past accomplishments rather than starting anew with each generation or each new practitioner. Surely scientists disagree with each other on many matters; but these are issues not yet settled, and the points of disagreement generally do not bring into question previous knowledge. Modern scientists do not challenge that atoms exist, or that there is a universe with a myriad stars, or that heredity is encased in the DNA.

What I want to add is something that seems rather obvious to me: science is a way of knowing, but it is not the only way. Knowledge also derives from other sources, such as common sense, artistic and religious experience, and philosophical reflection. The validity of the knowledge acquired by non-scientific modes of inquiry can be simply established by pointing out that science (in the modern sense of the word) dawned in the sixteenth century, but mankind had for centuries built cities and roads, brought forth political institutions and sophisticated codes of law, advanced profound philosophies and value systems, and created magnificent plastic art, as well as music and literature. We thus learn about ourselves and about the world in which we live and we also benefit from products of this non-scientific knowledge. We learn about the human predicament reading Shakespeare's *King Lear*, watching a Rembrandt *Self-portrait*, and listening to Tchaikovsky's *Symphonie Pathétique* or Elton John's *Candle in the Wind*. The crops we harvest and the animals we husband emerged millennia before science's dawn from practices set down by farmers in the Middle East, Andean Sierras, and Mayan plateaus.

It is not my intention to belabor the extraordinary fruits of nonscientific modes of inquiry. But I have set forth the view that nothing in the world of nature escapes the scientific mode of knowledge, and that we owe this universality to Darwin's revolution. Here I wish simply to state that successful as it is, and universally encompassing as its subject is, a scientific view of the world is hopelessly incomplete. There are matters of value, meaning, and purpose that are outside science's scope. Even when we have a satisfying scientific understanding of a natural object or process, we are still missing matters that may well be thought by many to be of equal or greater import. Scientific knowledge may enrich esthetic and moral perceptions, and illuminate the significance of life and the world, but these are matters outside science's realm.

Notes

1 William Paley, *Natural Theology* (New York: American Tract Society), 15–16. I will cite pages following this American edition, which is undated, but seems to have been printed in the late nineteenth century.
2 Ibid., 20–21.
3 Ibid., 22.
4 Ibid., 22–23.
5 Ibid., 48.
6 Ibid., 1.
7 Ibid., 175–176.
8 Ibid., 180.
9 Ibid., 49.
10 Ibid., 51.
11 Ibid., 265.
12 Ibid., 46.
13 Ibid.
14 Ibid., 47.
15 Ibid.
16 Ibid., 46.
17 Sir Francis Darwin, ed., *Charles Darwin's Autobiography* (New York, 1961), 34–35.
18 Darwin's quote (as well as the two that follow) is from *On The Origin of Species*, a facsimile of the first edition of 1859 (New York: Atheneum, 1967), 63, 80–81.
19 Darwin, *Origin*, 469.
20 Darwin, *Origin*, 489–490.
21 A common objection posed to the account I have sketched of how natural selection gives rise to otherwise improbable features, is that some postulated transitions, for example, from a leg to a wing, cannot be adaptive. The answer to this kind of objection is well known to evolutionists. For example, there are rodents, primates, and other living animals that exhibit modified legs used for both running and gliding. The fossil record famously includes the reptile *Archaeopteryx* and many other intermediates showing limbs incipiently transformed into wings endowed with feathers. One other example is described later in this article; namely, the transition involving bones that make up the lower jaw of reptiles but later evolved into bones now found in the mammalian ear. What possible function could a bone have, either in the mandible or in the ear, during the intermediate stages?
22 Paley, *Natural Theology*, 46.
23 Michael J. Behe, *Darwin's Black Box. The Biochemical Challenge to Evolution* (New York: Touchstone, Simon & Schuster, 1996), 223.
24 Robert T. Pennock, ed., *Intelligent Design Creationism and Its Critics. Philosophical, Theological, and Scientific Perspectives* (Cambridge, MA and London: MIT, 2001), 249. The implications of this point with respect of the teaching of evolution in the schools have been drawn in the public arena. In *The Washington Times*, 21 March 2002, US Senator Edward Kennedy, who has publicly supported the teaching of alternate scientific theories when there is diversity of opinion among scientists, writes that "intelligent design is not a genuine scientific theory and, therefore, has no place in the curriculum of our nation's public school science classes."

32 Theology and Science

25 Examples of deficiencies and dysfunctions in all sorts of organisms can be endlessly multiplied, reflecting the opportunistic, tinkerer-like character of natural selection, rather than intelligent design. The world of organisms also abounds in characteristics that might be called "oddities," as well as those that have been characterized as "cruelties," an apposite qualifier if the cruel behaviors were designed outcomes of a being holding on to human or higher standards of morality. But the cruelties of biological nature are only metaphoric cruelties when applied to the outcomes of natural selection. Examples of cruelty involve not only the familiar predators (say, a chimpanzee) tearing apart their prey (say, a small monkey held alive by a chimpanzee biting large flesh morsels from the screaming monkey), or parasites destroying the functional organs of their hosts, but also, and very abundantly, between organisms of the same species, even between individuals of different sexes in association with their mating. A well-known example is the female praying mantis that devours the male after coitus is completed. Less familiar is that, if she gets the opportunity, the female will eat the head of the male *before* mating, which thrashes the headless male mantis into spasms of sexual frenzy that allow the female to connect his genitalia with hers (S. E. Lawrence, "Sexual cannibalism in the praying mantis, *Mantis religiosa*: A field study," *Animal Behaviour* 43 (1992): 569–583. See also, M. A. Elgar, "Sexual cannibalism in spiders and other invertebrates," *Cannibalism: Ecology and Evolution among Diverse Taxa*, eds. M. A. Elgar and B. J. Crespi (Oxford: Oxford University, 1992.)) In some midges (tiny flies), the female captures the male as if he were any other prey and with the tip of her proboscis she injects into his head her spittle that starts digesting the male's innards that are then sucked by the female; partly protected from digestion are the relatively intact male organs that break off inside the female and fertilize her (J. A. Downes, "Feeding and mating in the insectivorous Ceratopogoninae (Diptera)," *Memoirs of the Entomological Society of Canada* 104 (1978): 1–62.) Male cannibalism is known in dozens of species, particularly spiders and scorpions. Diverse sorts of oddities associated with mating behavior are described in the delightful, but accurate and documented, book by Olivia Judson, *Dr. Tatiana's Sex Advice to All Creation* (New York: Holt, 2002).

26 With a somewhat more strident tone, the distinguished American philosopher of biology, David Hull, has made the same point: "What kind of God can one infer from the sort of phenomena epitomized by the species on Darwin's Galapagos Islands? The evolutionary process is rife with happenstance, contingency, incredible waste, death, pain and horror . . . Whatever the God implied by evolutionary theory and the data of natural selection may be like, he is not the Protestant God of waste not, want not. He is also not the loving God who cares about his productions. He is not even the awful God pictured in the Book of Job. The God of the Galapagos is careless, wasteful, indifferent, and almost diabolical. He is certainly not the sort of God to whom anyone would be inclined to pray" (David L. Hull, "God of the Galapagos," *Nature* 352 (1992): 485–486).

Francisco J. Ayala is the Donald Bren Professor of Biological Sciences and Professor of Philosophy at the University of California, Irvine. He has been President and Chairman of the Board of the American Association for the Advancement of Science. From 1994 to 2001, he was a member of the US President's Committee of Advisors on Science and Technology. He has published more than 750 articles and is author or editor of 18 books, which include *Variation and Evolution in Plants and Microorganisms* (2000), *Genetics and The Origin of Species* (1997), *Tempo and Mode in Evolution* (1995) and *Modern Genetics* (2nd ed, 1984). He is recipient of the 2002 National Medal of Science, the highest scientific honor in the United States.

[20]

Intelligent design and probability reasoning

ELLIOTT SOBER*
Philosophy Department, University of Wisconsin, Madison, WI 53706, USA; Department of Philosophy, Logic, and Scientific Method, London School of Economics, London WC2A 2AE, England (E-mail: ersober@facstaff.wisc.edu)

Philosophers schooled in the rules of deductive logic often feel that they can find their way when reasoning about probabilities by using the idea that probability arguments are approximations of deductively valid arguments. In a deductively valid argument, the premisses necessitate the conclusion; in a strong probability argument, the premisses confer a high probability on the conclusion. As a probability argument is strengthened, the probability of the conclusion, conditional on the premisses, increases; in the limit, the premisses confer a probability of unity on the conclusion. Deductive validity thus seems to be the limit case of strong probability arguments.

There is nothing wrong with this idea, though it does require refinement.[1] However, there is a distinct though closely related thought that can lead one very much astray. This is the idea that for each deductively valid form of argument, there exists a strong probabilistic argument that has roughly the same form. Granted, this principle is vague as stated, but nonetheless I think it plays a heuristic role for many philosophers (and nonphilosophers also). I want to explain why there are fundamental reasons why this heuristic is not to be trusted.

I'll begin with an example in which the heuristic does no harm. *Modus ponens* has the following logical form:

(MP) If X then Y
X
———————
Y

A probabilistic analog of *modus ponens* can be constructed as follows:

* My thanks to Branden Fitelson, Alan Hajek, and Terry Sullivan for helpful discussion.

Pr(Y|X) is high
X
p [========= (where p is high)
Y

"Pr(Y|X)" represents conditional probability (the probability of Y given X) and is standardly defined as Pr(Y&X)/Pr(X). The double line separating premisses and conclusion is meant to indicate that the argument is not deductively valid. The letter "p" that labels this line denotes the probability that the premisses confer on the conclusion.

With some tinkering this pattern of reasoning can be turned into a respectable form of argumentation. My preference is to turn it into a deductively valid argument in which a claim about the probability of Y is deduced. A first step in that direction might be the following:

Pr(Y|X) is high
X
———————
Pr(Y) is high

However, this is unsatisfactory as it stands. It is perfectly possible for Y to have a high probability conditional on X, but a low probability unconditionally; Even though it is very probable that the roulette wheel ball landed double-zero on the last spin, given that your honest and visually acute friend told you that this is what happened, it is still unconditionally improbable that the ball landed double-zero. The way forward is to time-index the probability functions:

(Prob-MP) Pr_{t1}(Y|X) is high
X
———————
Pr_{t2}(Y) is high

We are to imagine that an agent at time t_1 assigns a high value to Pr(Y|X). The agent then learns that X is true; this means that the probability assignment needs to be updated. If X is the total evidence that the agent acquires about Y in the temporal interval separating t_1 and t_2, then he or she should assign Y a high probability at time t_2. This is nothing other than the Principle of Conditionalization[2] applied so as to respect the Principle of Total Evidence. I didn't mention either of these in my formulation of (Prob-MP), so a fuller statement of this form of argument should go as follows:

$Pr_{t1}(Y|X)$ is high
X is the total evidence that the agent acquires between t_1 and t_2
Updating proceeds by conditionalization

$Pr_{t2}(Y)$ is high

If (Prob-MP) is ok, what is wrong with the heuristic idea that deductively valid arguments have analogs that are probabilistically strong? We need look no farther than *modus tollens*:

(MT) If X then Y
not-Y

---

not-X

If we construct a probabilistic analog of (MT), and assume both the Principle of Conditionalization and the Principle of Total Evidence, we obtain:

(Prob-MT) $Pr_{t1}(Y|X)$ is high
not-Y

Pr_{t2}(not-X) is high

In other words, if a theory X says that Y is very probable, and we learn that Y fails to obtain, then we should conclude that the theory is probably false. Here is an equivalent formulation:

Pr_{t1}(not-Y|X) is low
not-Y

Pr_{t2}(not-X) is high

If a theory X says that something probably won't occur, but it does, then the theory is probably false.

It is easy to find counterexamples to this principle. You draw from a deck of cards. You know that if the deck is normal and the draw occurs at random, then the probability is only 1/52 that you'll obtain the seven of hearts. Suppose you *do* draw this card. You can't conclude just from this that it is improbable that the deck is normal and the draw was at random.

This example makes it seem obvious that there is no probabilistic analog of *modus tollens*. However, this feeling of obviousness can fade when we look at other examples in which the relevant probability is far less than 1/52. Consider the following argument proposed by the biologist Richard Dawkins.[3] He is considering what a respectable theory of the origin of life on earth is permitted to say was the probability that life would evolve from nonliving materials:

> . . . there are some levels of sheer luck, not only too great for puny human imaginations, but too great to be allowed in our hard-headed calculations about the origin of life. But . . . how great a level of luck, how much of a miracle, *are* we allowed to postulate? . . . The answer to our question . . . depends upon whether our planet is the only one that has life, or whether life abounds all around the universe.
>
> . . . the maximum amount of luck that we are allowed to assume, before we reject a particular theory of the origin of life, has odds of one in N, where N is the number of suitable planets in the universe. There is a lot hidden in that word 'suitable' but let us put an upper limit of 1 in 100 billion billion for the maximum amount of luck that this argument entitles us to assume.

Since there are approximately 100 billion billion planets in the universe, Dawkins thinks that we can reject any theory of the origin of life on earth that says that the probability of that event was less than 1/100 billion billion:

Pr(life evolved on earth | theory T) < 1/100 billion billion
Life evolved on earth
―――――――――――――
Theory T is false

One curious feature of this argument is Dawkins' choice of a lower bound. Why is the number of planets relevant? Perhaps Dawkins is thinking that if α is the *frequency* of life-bearing planets among "suitable" planets (i.e., planets on which it is possible for life to evolve), then the true *probability* of life's evolving on earth must also be α. There is a mistake here, which we can identify by examining how actual frequency and probability are related. With small sample size, it is perfectly possible for these quantities to have very different values; consider a fair coin that is tossed three times and then destroyed. However, Dawkins is obviously thinking that the sample size is very large, and here he is right that the actual frequency provides a good estimate of the probability. It is interesting that Dawkins tells us to reject a theory if the probability it assigns is too *low*, but why doesn't he also say that we should reject it if the probability it assigns is too *high*? The reason, presumably, is that we cannot rule out the possibility that our planet was not just suitable but *highly conducive* to the evolution of life. However, this point cuts both ways. Although α is the *average* probability that a suitable planet will have life evolve, different suitable planets still might have different probabilities; some planets may have values that are greater than α while others may have values that are lower. Dawkins' lower bound assumes *a priori* that the earth was above average; this is a mistake that might be termed the "Lake Woebegone Fallacy."[4]

There's a general reason why no probabilistic version of *modus tollens* is to be had. Theories that make good probabilistic predictions about lots of events will typically say that the *conjunction* of those events has a very low probability. Even if $Pr(E_1|T)$, $Pr(E_2|T)$, . . ., $Pr(E_n|T)$ are each high (but less than unity), $Pr(E_1 \& E_2 \& \ldots \& E_n|T)$ will be very low, if the E_i's are sufficiently numerous and are probabilistically independent of each other, conditional on T. Consider a roulette wheel in which we distinguish only double-zero and not-double-zero as possible outcomes. A perfectly satisfactory theory of this device might say that the probability of double-zero is 1/38 and the probability of not-double-zero is 37/38 on each spin. Suppose we spin the wheel 3800 times and obtain a sequence of outcomes in which there are 100 double zero's. The probability of this exact sequence of outcomes is $(1/38)^{100}(37/38)^{3700}$, which is a tiny number. The fact that the theory assigns this outcome a very low probability hardly suffices to reject the theory.[5]

The accompanying table depicts the asymmetry between *modus ponens* and *modus tollens* for which I have argued. I assume that the riders concerning the Principle of Conditionalization and the Principle of Total Evidence are in place. There is a "smooth Transition" between probabilistic and deductive *modus ponens*; the minor premiss ("X") either ensures that Y is true, or makes Y very probable, depending on how the major premiss is formulated. In contrast, there is a radical discontinuity between probabilistic and deductive *modus tollens*. The minor premiss ("not-Y") guarantees that X is false in the one case, but has no implications whatever about the probability of X in the other.

Given that probabilistic *modus tollens* is invalid, there is a fallback position that we should consider. Perhaps if a theory says that an event is very improbable, but the event happens anyway, then the event counts as *evidence against* the theory. The event doesn't allow you to conclude that the theory is false, nor even that it has a low probability, but maybe the event lowers whatever probability you had assigned the theory before:

Table 1. Although *modus ponens* has a probabilistic analog, *modus tollens* does not

	Deductive		Probabilistic	
Modus Ponens	If X then Y X ——— Y	**VALID**	$Pr_{t1}(Y\|X)$ is high X ——— $Pr_{t2}(Y)$ is high	**VALID**
Modus Tollens	If X then Y not-Y ——— not-X	**VALID**	$Pr_{t1}(Y\|X)$ is high not-Y ——— $Pr_{t2}(X)$ is low	**INVALID**

$Pr_{t1}(Y|X)$ is low
Y
———————
$Pr_{t2}(X)$ is lower than $Pr_{t1}(X)$

Whereas (Prob-MT) allows you to draw a conclusion about the *absolute value* of X's probability (it is *low*), the present proposal is that your conclusion should merely be *comparative* (X's probability is *lower* than it was before). This principle also is wrong, as a nice example from the statistician Richard Royall[6] illustrates: Suppose I send my valet to bring me one of my urns. I want to test the hypothesis that the urn he returns with contains 2% white balls. I draw a ball and find that it is white. Is this evidence against the hypothesis? It may not be. Suppose I have only two urns – one of them is as described, while the other contains 0.0001% white balls. In this instance, drawing a white ball is evidence *in favor* of the hypothesis, not evidence *against*.

Royall's example brings out an important feature of the concept of evidence. To say whether an observation is evidence for or against a hypothesis, we have to know what the other hypotheses are that we should consider. The evidence relation is to be understood in terms of the idea of *discrimination*.[7] E is evidence for or against the hypothesis H_1 only relative to an alternative hypothesis H_2. For evidence to be evidence, it must discriminate between the competing hypotheses. Another way to put this point is by saying that the evidence relation is ternary, not binary. The right concept to consider is "E favors H_1 over H_2," not "E is evidence for (or against) H." It needs to be understood that this thesis is restricted to hypotheses that do not deductively entail observations; if H entails E, and E fails to obtain, then E rules out (and hence disconfirms) H.

The idea that evidence is essentially a comparative concept is often associated with the *Law of Likelihood*:[8]

> Evidence E favors hypothesis H_1 over hypothesis H_2 if and only if $Pr(E|H_1) > Pr(E|H_2)$.

Notice that the absolute values of $Pr(E|H_1)$ and $Pr(E|H_2)$ don't matter here; all that matters is how they compare. The Likelihood Principle does not tell you what to believe nor does it even indicate which hypothesis has the higher probability of being true. It merely assesses the weight of the evidence at hand.

I say that the thesis that evidence is comparative is often "associated with" the Likelihood Principle, not that the two are essentially connected. There are two reasons for this. The first is that the Likelihood Principle, taken at its word, does not rule out the possibility that one can talk about the evidence for or against a given hypothesis without reference to alternative hypotheses. True, advocates of "likelihoodism"[9] have endorsed the Likelihood Principle and have also insisted that evidence is essentially comparative, but this just shows that likelihoodism goes beyond the letter of the Law of Likelihood. The second reason is that there are theories of evidence that depart from the dictates of likelihoodism but nonetheless agree that the evidence relation is ternary rather than binary. For example, standard Neyman-Pearson statistical theory (interpreted evidentially) tells one how to deal with the possibility of both "type 1" and "type 2" errors, and this entails that *two* hypotheses are being assessed,[10] not just *one*.

I don't want to give the impression that the comparative conception of evidence is universally endorsed in science. Unfortunately, there is a statistical methodology that is sometimes used that purports to assess how evidence bears on a single hypothesis. This is the theory, due to R.A. Fisher, of significance testing. The story of the valet and the two urns already suggests what is wrong with this approach, but let me add another example to help flesh out the picture a bit. Consider the hypothesis that a coin is fair. If the coin is tossed a large number of times (say, 1000 times), there will be 2^{1000} possible sequences of heads and tails that might occur. If the hypothesis that the coin is fair is true, then each of these exact sequences has the same tiny probability (namely $(\frac{1}{2})^{1000}$) of occurring. Yet, it seems utterly wrong to say that each outcome would count as evidence against the hypothesis.[11] To make sense of what it means to test this hypothesis about the coin, we need to say what the alternative hypotheses are; if the alternative hypothesis one wishes to consider is that the coin is strongly biased in favor of heads, then sequences in which there are large numbers of tails count as evidence *in favor* of the hypothesis that the coin is fair; but if the alternative one wishes to consider is that the coin is strongly *biased against* heads, precisely the opposite interpretation of that observation would be correct. Fisher remarked that when a theory says that what one has observed is very improbable, that one's conclusion should take the form of a "... simple disjunction. Either an exceptionally rare chance has occurred, or the theory ... is not true."[12] There is nothing wrong with this point; the mistake is to think that this disjunction entails that one has obtained evidence against the theory.

This is not the place to present a systematic critique of Fisherian significance testing (under the interpretation of the method that equates the improbability of E if H is true with the strength of the evidence against H). That critique has been developed in several places already,[13] and no adequate response has been provided. So let us take stock. The first conclusion is that there is no probabilistic analog of *modus tollens*. This should be uncontroversial. Separate from this thesis about arguments that draw conclusions about the *probabilities* of hypotheses is the thesis I have defended about *evidence: Assessing whether an observation counts as evidence for or against a hypothesis must consider alternative hypotheses and what they predict about the observation.* As noted above, this comparative thesis is restricted to hypotheses that don't deductively entail observational claims, but merely confer probabilities on them. In discussing the Law of Likelihood, I mentioned that this principle does not tell you which hypotheses to accept or reject. However, if the acceptance or rejection of hypotheses requires the accumulation of evidence *pro* and *con*, then the comparative principle just stated provides a simple but important lesson about acceptance.[14]

These points about probability reasoning allow us to identify the central deficiency in the Intelligent Design (ID) movement. "Intelligent design" is the label that Michael Behe, William Dembski, and Philip Johnson prefer so that their position will not be confused with old-fashioned creationism.[15] The term "creationism" suggests the idea of *special* creation – a denial of the claim that all life on earth is genealogically related; ID theorists don't endorse the idea of special creation – that each species (or "basic kind of organism") was separately created by an intelligent designer. Rather, their beef with evolutionary theory concerns the power of natural selection to produce complex adaptations. Behe is the pointman here, arguing that traits that exhibit "irreducible complexity" pose an in-principle difficulty for evolutionary theory and, indeed, for any theory that limits itself to mindless natural processes. The vertebrate eye, for example, exhibits irreducible complexity because all of its many parts must be arranged *just so* if the eye is to perform the function of allowing the organism to see. For this reason, Behe's argument isn't different in form from Paley's.[16] The novelty in Behe's presentation consists in his choice of examples. Behe thinks that basic features of biochemistry, such as the machinery that drives the bacterial flagellum and the mechanisms that get blood to coagulate, are irreducibly complex. Just as earlier creationists complained that an organism would gain no benefit from having 10% of an eye or wing, Behe argues that having 10% of the clotting process would be useless.

There are a number of philosophical and scientific objections that might be considered in connection with Behe's argument. Before I move on to my main complaint, I need to mention the fact that Behe equivocates between the process of gradual natural selection, taken on its own, and "evolutionary processes" construed more broadly. "Darwinian gradualism," taken in its strict sense, requires the steady accumulation of modifications, each conferring a small benefit. This means that 10% of a wing has to represent an advantage compared with 9%, if this type of selection is to transform a population from one in which all individuals have no wings at all to one in which all have 100% of a wing.[17] On the other hand, it is important to recognize that evolutionary theory countenances many processes additional to that of pure Darwinian gradualism. For example, since the theory is probabilistic, it is perfectly possible for a population to move from each individual's having 9% of a wing to each individual's possessing 10%, even if the latter state represents no selective advantage. This is called "random genetic drift." My point here is not that this transition is *probable*, but that it is *possible*, according to the theory. Behe is correct that the pure process of Darwinian gradualism cannot lead a wing to evolve if the fitnesses are those described in lines 2 or 3 in Figure 1, and that the monotonic increase depicted in line 1 is required. However, he concludes from this that "evolutionary theory" cannot explain the emergence of traits whose fitnesses conform to lines 2 and 3; this does not follow and it is not correct.

The objection to Behe's argument that I want to focus on here concerns the type of reasoning he employs against evolutionary theory and in favor of the hypothesis of intelligent design. Behe repeatedly vacillates between using a deductive and a probabilistic *modus tollens* against evolutionary theory. The vacillation sometimes occurs on the same page. Consider the following passage:

> ... I have shown why many biochemical systems cannot be built up by natural selection working on mutations: no direct, gradual route exists to these irreducibly complex systems ... There is no magic point of irreducible complexity at which Darwinism is logically impossible. But the hurdles for gradualism become higher and higher as structures are more complex, more interdependent (p. 203).

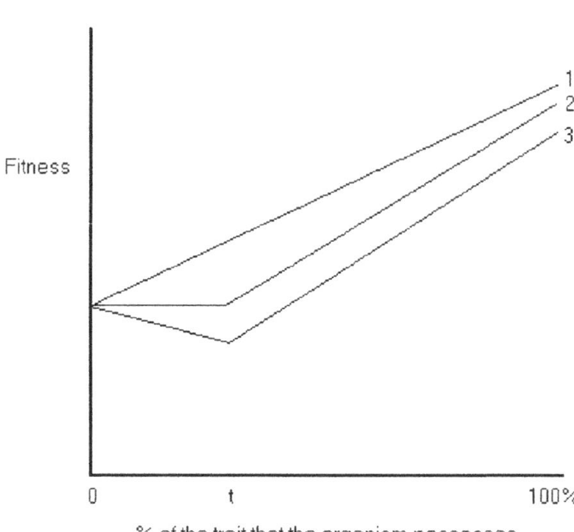

Figure 1. What are the fitness consequences of having *n*% of a wing or eye, as opposed to having $(n - 1)$%? According to line 1, each small increase represents an increase in fitness. According to line 2, having more and more of the trait makes no difference in fitness until a threshold (*t*) is crossed. Line 3 also depicts a threshold effect, but here having more of the wing or eye is deleterious, not neutral, until the threshold is crossed. Evolution via the pure process of Darwinian gradualism requires the monotonic increase that line 1 exhibits, and cannot occur if the fitnesses are those represented by lines 2 or 3. However, evolutionary theory countenances processes additional to that of "pure Darwinian gradualism," so, in fact, the theory say that it *is* possible for the trait to evolve under all three scenarios.

Behe's first sentence says that irreducible complexity *cannot arise* by Darwinian processes; however, the next two assert, more modestly, that irreducibly complex features are *improbable* on the Darwinian model and that they become more improbable the more complex they are. I hope it is clear from what I've said earlier why this shift is important. If evolutionary theory really did have the deductive consequence that organisms *cannot* have features that are irreducibly complex, then that theory would have to be false, if such features exist. But what if the theory merely entailed that irreducibly complex features are very improbable? Would the existence of such features show that the theory is improbable? Would it follow that the theory is disconfirmed by those observations? Would it follow that these features provide evidence in favor of intelligent design? The answers to all these questions are the same – *no*. There is no probabilistic analog of *modus tollens*.

In addition to rejecting evolutionary explanations, Behe advances the positive thesis that the biochemical systems he describes in loving detail "were designed by an intelligent agent" (p. 204). However, for these details to favor intelligent design over mindless evolution, we must know how probable those details are under each hypothesis. This is the point of the Law of Likelihood. Behe asserts that these details are very improbable according to evolutionary theory, but how probable are they according to the hypothesis of intelligent design? It is here that we encounter a great silence. Behe and other ID theorists spend a great deal of time criticizing evolutionary theory, but they don't take even the first steps towards formulating an alternative theory of their own that confers probabilities on what we observe. If an intelligent designer built the vertebrate eye ,or the bacterial flagellum, or the biochemical cascade that causes blood to clot, what is the probability that these devices would have the features we observe? The answer is simple – *we do not know*. We lack knowledge of what this putative designer's intentions would be if he set his mind to constructing structures that perform these functions.

The sad fact about ID theory is that there is no such theory. Behe argues that evolutionary theory entails that adaptive complexity is very improbable, Johnson rails against the dogmatism of scientists who rule out *a priori* the possibility of supernatural explanation, and Dembski tries to construct an epistemology in which it is possible to gain evidence for the hypothesis of design without ever having to know what, if anything, that hypothesis predicts. A lot goes wrong in each of these efforts,[18] but notice what is not even on the list.

Intelligent design theorists may feel that they have already stated their theory. If the *existence* of the vertebrate eye is what one wishes to explain, their hypothesis is that an intelligent designer constructed the vertebrate eye. If it is the *characteristics* of the vertebrate eye (the fact that it has features $F_1, F_2, \ldots, F_n$), rather than its mere existence, that one wants to explain, their hypothesis is that an intelligent designer constructed the vertebrate eye with the intention that it have features $F_1, F_2, \ldots, F_n$ and that this designer had the ability to bring his plan to fruition. Notice that both of these formulations of the hypothesis of intelligent design simply build into that hypothesis the observations whose explanation we seek. The problem with this strategy is that the same game can be played by the other side. If the evolutionary hypothesis is formulated by saying "evolution by natural selection produced the vertebrate eye" or by saying that "evolution by natural selection endowed the eye with features $F_1, F_2, \ldots, F_n$," then it too entails the observations.

To avoid trivializing the problem in this way, we should formulate the observations so that they are *not* built into the hypotheses we want to test. This can be achieved by organizing the problem as follows:

(O) The vertebrate eye has features $F_1, F_2, \ldots, F_n$.

(ID) The vertebrate eye was created by an intelligent designer.

(ENS) The vertebrate eye was the result of evolution by natural selection.

Behe claims that (O) has a low probability according to the (ENS) hypothesis. My complaint is that we do not know what the probability of (O) is according to (ID). If an intelligent designer made the eye, perhaps he would have been loathe to give it the features we observe. Or perhaps he would have aimed at producing those very characteristics.[19] The single sentence stated in (ID) does not a theory make. This problem is not solved by simply *inventing* assumptions about the putative designer's goals and abilities; what is needed is information about the putative designer(s) that is independently attested. Without that information, the theory makes no predictions about the eye or about the other examples of "irreducible complexity" that Behe discusses. And without those predictions, the intelligent design movement can provide no evidence against the evolutionary hypothesis.

After concluding that evolutionary theory cannot explain adaptations that are irreducibly complex, Behe briefly broaches the subject of whether some "as-yet-undiscovered natural process" might be the explanation. Here is his analysis:

> No one would be foolish enough to categorically deny the possibility ... [however] if there is such a process, no one has a clue how it would work. Further, it would go against all human experience, like postulating that a natural process might explain computers ... In the face of the massive evidence we do have for biochemical design, ignoring that evidence in the name of a phantom process would be to play the role of the detectives who ignore an elephant (pp. 203–204).

Notice that Behe claims that there is "massive evidence for biochemical design," but what *is* that evidence? It seems to consist of two facts, or alleged facts – that evolutionary theory says that irreducibly complex adaptations have low probabilities and that no one has yet formulated any other theory restricted to mindless natural processes that could be the explanation. However, if the comparative principle about evidence stated earlier is correct, this "evidence" is no evidence at all.

After evolutionary theory and "as-yet-undiscovered natural process[es]" are swept from the field, Behe immediately concludes that the biological mechanisms whose details he has described

> ... were designed by an intelligent agent. We can be as confident of our conclusion for these cases as we are of the conclusions that a mousetrap

> was designed, or that Mt. Rushmore or an Elvis poster were designed . . . Our ability to be confident of the design of the cilium or intracellular transport rests on the same principles as our ability to be confident of the design of anything: the ordering of separate components to achieve an identifiable function that depends sharply on those components (p. 204).

Behe is right that the nonbiological examples he cites favor hypotheses of intelligent design over hypotheses that postulate strictly mindless natural processes, but he is wrong about the reason and wrong to think that biochemical adaptations can be assimilated to the same pattern. In the case of mousetraps, Mount Rushmore, and Elvis posters, we are confident about intelligent design because we have strong evidence for *human* intelligent design. We know that all of these objects are just the sorts of things that human beings are apt to make. The probability of their having the features we observe, on the hypothesis that they were made by intelligent human designers, is fairly large, whereas the probability of their having those features, if they originated by chance, is low. The likelihood inference is unproblematic. But the probability that the bacterial flagellum would have the features we observe, or that the mechanism for blood clotting would have its observed features, if *human beings* somehow made those devices, is very very low. ID theorists therefore are led to consider possible *non*human designers – indeed, possible designers who are *supernatural.* Some of these *possibilia* would, if they existed, have goals and abilities that would make it highly probable that these devices have the features we observe; others would not. Averaging over all these possibilities, what is the probability that the device will have the features we observe if it was made by some possible intelligent designer or other? We do not know, even approximately.

Behe would like to be able to identify an observable feature of natural objects that *could not exist* if those objects were produced by strictly mindless processes and that therefore *must* be due to intelligent design (natural or supernatural). There is no such property. It is not *impossible* for irreducibly complex functional features to arise by the evolutionary process of natural selection, which is *not* a random process.[20] Indeed, it isn't even impossible for them to arise by a purely random chance process. This is the simple point made vivid by thinking about monkeys and typewriters and of particles whirling in the void. The next step is to think about the properties that an object *probably will* have if it is made by an intelligent designer and *probably won't* have if it isn't. The problem here is that there are many kinds of possible intelligent designers, and many kinds of possible mindless processes. Is there a property that a natural object probably will have, no matter what sort of possible intelligent designer made it? I am confident that the answer to this question is *no*. Is there a property that it probably won't have, no matter what sort of possible mindless process made it? As for this second question, here I am in agreement with Behe – *we really don't know*. But ignorance does not constitute a reason to reject the possibility that what we observe is due to mindless natural processes that we have not yet considered and conclude that what we observe must be due to intelligent design.

My critique of the intelligent design movement has been based on the comparative principle I stated about evidence – *to say whether an observation counts as evidence against evolutionary theory and in favor of the hypothesis of intelligent design, one must know what each predicts about the observation*. I have challenged intelligent design theorists to produce a theory that has implications about the detailed examples of "irreducible complexity" that Behe describes. However, there is another response that intelligent design theorists might contemplate. This is to deny the comparative principle itself. Dembski has seized this horn of the dilemma.[21] If he succeeds in developing an epistemology of this sort (so far he has not), the way will be paved for an unprecedented result in the history of science – the rejection of a logically consistent theory that confers probabilities on observations, but does not entail them, and its replacement by another, without its needing to be said what the replacing theory predicts.

Notes

1. One needed refinement is that the number of mutually exclusive and collectively exhaustive propositions be finite. When this fails, a probability of unity is not the same as necessity, and a probability of zero is not the same as impossibility. If I randomly choose a fraction that is between 0 and 1, the probability that I'll choose 13/345 is zero, but it isn't impossible that I'll choose that number.
2. The principle of conditionalization assumes that acquiring evidence involves becoming certain that various propositions are true. If we are never entitled to be certain about the truth values of observation reports, then a new rule for updating is needed. This is supplied by the idea of Jeffrey-conditionalization; see Richard Jeffreys, *The Logic of Decision* (Chicago: University of Chicago Press, 1983). This point does not invalidate the claim that (Prob-MP) is correct; it merely points to a limit on its applicability.
3. See Richard Dawkins, *The Blind Watchmaker* (New York: Norton, 1986), pp. 144–146.
4. See Elliott Sober, "The Design Argument." In W. Mann (ed.), *Blackwell Companion to the Philosophy of Religion* (Oxford: Blackwell, 2003). Also available at the following URL: http://philosophy.wisc.edu/sober.
5. This conclusion cannot be evaded by saying that the theory entails that obtaining approximately 100 double zero's in 3800 spins is highly probable. The Principle of Total Evidence says that we have to use *all* of the evidence available in evaluating theories, not just part. Theories also entail that tautologies will be true, and tautologies are part of every data set, but this is no reason to set aside the total evidence and focus just on the tautologies that the evidence entails.

6. See Richard Royall, *Statistical Evidence – A Likelihood Paradigm* (Boca Raton, FL: Chapman and Hall, 1997), p. 67.
7. Elliott Sober, "Testability," *Proceedings and Addresses of the APA* 73 (1999): 47–76. Also available at the following URL: http://philosophy.wisc.edu/sober.
8. See Ian Hacking, *The Logic of Statistical Inference* (Cambridge: Cambridge University Press, 1965), Anthony Edwards, *Likelihood* (Cambridge: Cambridge University Press, 1972), and Richard Royall, *op cit.*
9. Anthony Edwards, *op cit*, Richard Royall, *op cit*, and Elliott Sober, *op cit.*
10. See Richard Royall, *op cit*, chapter 2 for discussion.
11. See Ian Hacking, *op cit*, p. 85. For some alternative formulations of the Fisherian idea, and objections, see Richard Royall, *op cit*, chapter 3.
12. R.A. Fisher, *Statistical Methods and Scientific Inference* (New York: Hafner. 1959, 2nd edition), p. 39.
13. See Ian Hacking, *op cit*, Anthony Edwards, *op cit*, and Richard Royall, *op cit.*
14. The present point about the comparative character of evidence can be connected with the earlier argument about probabilistic *modus tollens* by considering the fact that every deductively invalid argument can be turned into a valid argument by adding premisses. How can this be done in the case of (Prob-MT)? Bayes' Theorem says that $Pr(H|O) = Pr(O|H)Pr(H)/Pr(O)$. In consequence, the following argument is valid (assuming, as before, that O is the total evidence and that updating proceeds by conditionalization):

    $Pr_{t1}(O|H)$ is low
    O
    $Pr_{t1}(H) \leq Pr_{t1}(O)$
    ―――――――――――――
    $Pr_{t2}(H)$ is low

    The new third premiss is equivalent to

    $$Pr_{t1}(H)[1 - Pr_{t1}(O|H)] \leq Pr_{t1}(O|\text{not-}H)Pr_{t1}(\text{not-}H).$$

    Notice that the prior probability and the likelihood of the *alternative* hypothesis (not-H) enters into this formula. The lesson, again, is this: *if you want to argue that H is improbable, based on the fact that H says that what you observe is very improbable, you must have additional information about how probable the observations would be if H were false.*
15. Michael Behe, *Darwin's Black Box* (New York: Free Press, 1996); William Dembski, *The Design Inference* (Cambridge: Cambridge University Press, 1998); Philip Johnson, *Darwin on Trial* (Downers Grove, IL: Intervarsity, 1991).
16. William Paley, *Natural Theology, or, Evidences of the Existence and Attributes of the Deity, Collected from the Appearances of Nature* (London: Rivington, 1802).
17. This problem is apparently more pressing when it comes to wings than it is with respect to eyes. Eyes come in various forms, some far more rudimentary than the vertebrate eye, and it is not at all difficult to see how all these various forms might provide an adaptive advantage; see Richard Dawkins, *Climbing Mount Improbable* (New York: W.W. Norton, 1996), chapter 5. The wing is more puzzling, since 5% of a wing provides no lift at all; being able to fly is a threshold effect. One part of the solution to this problem is to see that the rudimentary beginnings of wings can serve other functions, and that once wing evolution is under way, wings can continue to evolve because they facilitate flight.

This is more than just speculation; J. Kingsolver and M. Koehl (in "Aerodynamics, Thermoregulation, and the Evolution of Insect Wings – Differential Scaling and Evolutionary Change." *Evolution* 39 (1985) 488–504) provide empirical evidence for the claim that insect wings began evolving as devices for regulating temperature and then continued to evolve as devices for flying.

18. For a critical evaluation of Dembski's epistemology, see Branden Fitelson, Elliot Sober, and Christopher Stephens, "How Not to Detect Design – Critical Notice of W. Dembski's *The Design Inference*." *Philosophy of Science* 66 (1999) 472–488, reprinted in R. Pennock (ed), *Intelligent Design Creationism and its Critics* (Cambridge, MA: MIT Press, 2001), pp. 597–616. Although Dembski tries to build on the (flawed) foundation of Fisherian significance testing, his proposals go far beyond Fisher's. For one thing, Fisher's method applies to the problem of testing a *specific* chance hypothesis, whereas Dembski thinks that he can "sweep from the field" the entire class of *all* chance hypotheses. Another novelty in Dembski's approach is his use of ideas from complexity theory.
19. In fact, if the ID hypothesis says that *some* intelligent designer produced the effects cited, then one must consider different possible intelligent designers, weigh the probability that they were the ones involved, and assess the probability of the outcome if they were doing the work.
20. A random process is one that has a large number of equiprobable outcomes. The whole point of natural selection is that some outcomes are vastly more probable than others. Selection is a probabilistic process, but not all probabilistic processes are random.
21. William Dembski, *op cit.* Since Behe (pp. 285–286) praises Dembski's epistemological insights, he presumably would embrace this response.

[21]

THE INTELLIGENT-DESIGN MOVEMENT: SCIENCE OR IDEOLOGY?

by Gregory R. Peterson

Abstract. The past decade has seen the rise of a new wave of criticism of evolutionary biology, led by claims that it should be replaced by a new science of intelligent design. While the general question of inferring design may fairly be considered worthy of attention, claims that intelligent-design theory (IDT) constitutes a biological science are highly problematic. This article briefly summarizes the assertions made about IDT as a biological science and indicates why they do not stand up to analysis. While claiming that IDT is a biological science, its advocates have failed to actually produce a research program that merits serious attention. As such, it is clear that IDT is more driven by ideological considerations than by attention to actual scientific research.

Keywords: Michael Behe; demarcation; William Dembski; evolutionary theory; intelligent-design theory; Imre Lakatos.

The past decade has seen the rise of a new movement seeking to create a science of intelligent design. Explicitly theological in character, intelligent-design theorists have sought to reintroduce the notion of divine design as a scientific hypothesis to be considered alongside and in place of naturalistic accounts of cosmic and biological origins and change. To say the least, such claims have been far from uncontroversial, having produced considerable polemic and, within some quarters, heated academic debate.

Gregory R. Peterson is Assistant Professor of Religion and codirector of the Global Institute at Thiel College, Greenville, PA 16125; e-mail gpeterso@thiel.edu. A version of this paper was originally presented as "Will Intelligent Design Ever Be a Science?" at the conference "Intelligent Design: Science and Theology in Consonance?" in Fredericton, New Brunswick, September 2000.

Much of this debate has centered on the validity of the design inference (Dembski 1998a), the viability of specific examples or models of design (Behe 1996), the role of methodological naturalism in science (Moreland 1994b), and the proper relationship between science and theology (Dembski 1999).

Although of some significance, these issues are in many ways tangential to the central and most controversial question: Is intelligent design truly a scientific research program? Because of the wide-ranging assertions made by intelligent-design advocates, this question is difficult to answer. Intelligent design is said to apply to physical cosmology, biochemistry, human evolution, and cryptography, among other areas. The central area of contention, however, has been biology and the theory of biological evolution, and it is here that intelligent-design theory (IDT) has most emphatically staked its claim, insisting that current evolutionary theory is incomplete and that the development of intelligent design within biology will stimulate a revolution in thinking about issues of origin and speciation. According to its proponents, IDT stands to revolutionize biology.

But has IDT really developed a scientific program to compete with evolution and natural selection? At best, the answer is not yet. One can argue that IDT is still quite young, but it has not engaged in the kind of behavior appropriate for a rising research program. Rather, it has largely pursued a strategy of populist persuasion, by and large eschewing the kinds of activities normal to scientific development. This, combined with the meagerness of its scientific claims and agenda, suggests that not only is intelligent design not yet a scientific research program, there seems little reason to believe that it will ever constitute such. In their pursuit to discredit evolutionary biology and to portray intelligent design as a scientific theory, ID advocates have obscured much of interest and importance in thinking about the relationship of God and world. By radically polarizing and politicizing the science-and-religion dialogue, advocates of IDT stand to reverse two decades' worth of constructive dialogue and to reinvigorate the fractious ghosts of religion-science conflict and legal action.

The Central Claim: Intelligent Design as a Biological Science

According to one view, theories of intelligent design are as old as the philosophical tradition in the West. Aristotle's inclusion of final cause in his analysis of the physical and biological worlds and the Stoics' inference of the existence of God from biological complexity point to the early existence of arguments regarding purpose and design (cf. Cicero, *De Natura Deorum*, Book II:12–14). While the modern ID movement draws from this historical well, its primary affinity is with the more scientifically (some might say scientistically) minded design arguments of the eighteenth and nineteenth centuries. Indeed, William Dembski gives a positive evaluation of these early efforts at the same time that he acknowledges weaknesses (1999). Like these scientists from an earlier era, the modern ID movement seeks to explain biological and other forms of physical complexity in terms of the actions of a divine intelligent agent. Furthermore, they claim that such an explanation is scientific in character and therefore should be funded and taught in the same manner as other scientific theories.

The great difference between modern proponents of IDT and their predecessors is, to put it succinctly, one hundred years of evolutionary theory. Almost any example of IDT literature reveals that two wars are being fought. On one hand, ID theorists are resolutely opposed to completely naturalist explanations of biological complexity, with special antagonism reserved for the modern neo-Darwinian synthesis. On the other hand, ID theorists offer the hypothesis of intelligent design as a superior alternative, whose rejection is (allegedly) based on the secularist dogmatism of modern scientists. For IDT these two positions are closely linked. The central task of IDT is, in short, to reintroduce God into the equations of science. To do this, theorists must also counter four hundred years of history that has moved in precisely the opposite direction.

Consequently, opposition to methodological naturalism is central to ID rhetoric. According to the common construal of this principle, methodological naturalism is seen as an underlying principle of natural science. Methodological naturalism prohibits reference to God or other supernatural entities in any scientific explanation. If an atheist were struck by lightning after taking the Lord's name in vain, a vindictive theist might attribute it to God's will. A scientist, however, would look for natural causes.

ID theorists reject the principle of methodological naturalism, claiming that it is a form of implicit atheism and that it has failed to account for basic features of biological function and history (Johnson 1991; Moreland 1994b). Instead, they argue for a theistic science. According to J. P. Moreland,

> Theistic science can be considered a research program . . . that, among other things, is based on two propositions:
>
> 1. God, conceived of as a personal, transcendent agent of great power and intelligence, has through direct, primary agent causation and indirect, secondary causation created and designed the world for a purpose and has directly intervened in the course of its development at various times. . . .
> 2. The commitment expressed in proposition 1 can appropriately enter into the very fabric of the practice of science and the utilization of scientific methodology. (Moreland 1994b, 41–42).

Moreland's second point is, of course, the more important. Theistic science asserts not only that traditional theological doctrines such as *creatio ex nihilo* (creation from nothing) are true but that some theological claims are actually scientific in character and should be a part of appropriate scientific inquiry. In fact, any true account of such issues as the origin of life, according to the hypothesis of theistic science, necessarily invokes the concept of God—or at least a designer.

While theistic science may be an eventual goal, the claim is that ID is a present reality, is a scientific hypothesis that successfully explains the occurrence of irreducible (or specified) complexity in (among other places) biological organisms. Consequently, IDT is in conflict with evolutionary and particularly Darwinian accounts of the origin of life and the origin of species. The scientific character and status of IDT is emphasized.

> What has emerged is a new program for scientific research known as intelligent design. Within biology intelligent design is a theory of biological origins and development. Its fundamental claim is that intelligent causes are necessary to explain the complex, information-rich structures of biology and that these causes are empirically detectable. To say intelligent causes are empirically detectable is to say there exist well-defined methods that, on the basis of observational features of the world, are capable of reliably distinguishing intelligent causes from undirected natural causes. (Dembski 1998b, 16)

The framework for IDT comes almost entirely from Dembski and Michael Behe. Dembski, a mathematician and philosopher at Baylor University, has developed theoretical and mathematical grounds for detecting design (1998a). Behe, a biochemist at Lehigh University, has been primarily responsible for developing examples from molecular biology that seem to be unexplainable by standard evolutionary explanation (1996). Fairly clearly, IDT conceives of itself as an ambitious project, providing not just an alternative scientific account for biological origins and specified complexity but an account that breaks down the wall between theology and science. Indeed, if Dembski and Behe are correct, IDT would truly be the most significant scientific theory ever, for it would in essence prove the existence of God. If vindicated, the names of Dembski and Behe will be remembered alongside those of Newton and Einstein. I suspect, however, that the truth lies elsewhere.

The Nature of Science

In bidding for scientific status, IDT advocates raise the question of demarcation. That is, by what criteria can we determine what is and is not science? Unfortunately, this has been one of the more difficult questions to answer, with the result that a number of formulations over the course of the past century have met with different levels of satisfactoriness. Among these, the earliest criterion is that set by the logical positivists, who argued that scientific (and indeed all) truth claims are those that can be verified. The verification principle, however, quickly ran into several problems, even when applied to the domain of science, for an experiment does not confirm a hypothesis as much as it disconfirms others. Karl Popper, therefore, asserted that scientific hypotheses were characterized by their falsifiability.

Physics may be considered a science, but psychoanalysis and Marxism, according to Popper's standard, fail (1959; 1971).

Although Popper's criterion remains useful in a heuristic way, it has since been significantly qualified. Most notably, Thomas Kuhn's work (1962) utilized the history of science to show that science rarely proceeded in such a neat and tidy fashion. Rather, science was often characterized by prolonged periods of research based on a paradigm—an undergirding set of theories, formulas, and exemplars that indicated acceptable areas of inquiry and prescribed the kinds of answers one was likely to obtain thereby. Paradigms, in Kuhn's analysis, are highly resistant to falsification, and it is only when a succession of paradigm defeats builds up that a crisis occurs, spawning competing and incompatible theories. Eventually, one such theory proves successful, establishing a new paradigm, and the cycle repeats.

While Kuhn's account of scientific change became widely popular, many saw it (somewhat wrongly, in my view) as too irrational in character. Of the successors to Kuhn's approach, one that has been most influential is that of Imre Lakatos (1970), particularly in theological circles (see Murphy 1990; Peterson 1998). Lakatos argued that science is characterized by competing research programs. A research program is characterized by a set of hard-core and sometimes unverifiable theoretical claims and commitments. Through the development of auxiliary hypotheses, testable consequences and elucidations of the theory are devised (positive heuristic), accompanied by needed ad hoc hypotheses and arguments that shore up weak areas of the theory. A research program that is progressive has its hypotheses repeatedly confirmed, discovers unexpected novel facts, is able to explain the phenomena accounted for by competitor theories, and furthermore explains phenomena that competing theories cannot explain. By contrast, a degenerating program will either do none of these or, on the whole, do them poorly. Lakatos acknowledges the complex communal and historical element of scientific research and at the same time provides a demarcation criterion. Progressive research programs can be considered scientific in character. Degenerating research programs, however, eventually reach a point at which they cease to be scientific. In Lakatos's analysis, however, demarcation is not clean-cut. Programs can experience brief periods of degeneration, only to rebound later. In comparison to Popper's clear falsification criteria, Lakatos's criteria are much more general and difficult to analyze. A number of factors must be considered before the success of a program can be determined and, consequently, an element of human judgment at the expense of algorithmic certainty appears. For this reason, it is not merely a matter of judging whether a particular research program is scientific but also of determining whether it is good science or bad science.

Despite these vagaries, Lakatos's approach appears to attain at least a minimum standard of science and, more generally, of all empirical query.

It is nicely consistent with a variety of historical examples, such as the transition from Ptolemaic to Copernican cosmology as well as the more recent advent of plate tectonics in geology. It also accounts for such prolonged issues as the dinosaur-extinction debate, which involved more than one discipline and several years of sifting through relevant data (Glen 1994) and provided a way of thinking about such "fringe" sciences as creation science, parapsychology, and astrology. In theory, one could announce the arrival of astrological science. We could even be charitable in evaluating it in the early years. But eventually it would have to produce. If no truly testable form of the theory emerged, if it continually appealed to ad hoc hypotheses and after-the-fact adjustments to failed predictions, there would be little reason to consider it seriously.

By affirming a broadly Lakatosian approach, I necessarily concede the philosophical point that science is not necessarily limited by methodological naturalism. In theory, one could design a theistic or metanatural research program that would be scientific in character. However, I emphasize the words *in theory*. While a Lakatosian approach may provide the minimum standard for science, it does not by itself mention all of the relevant criteria used in much of scientific practice. Quantification and experimentation, although not always attainable, are certainly desirable. Generally speaking, no science is completely quantifiable, and the natural sciences experience a range of levels of quantification. A similar observation may be made about experimentation. Furthermore, a good scientific program is often able to tie in claims with a variety of other existing programs in neighboring fields. With reference to plate tectonics, continental drift eventually helped thinking about species diversity and extinction events.

In the present case, I would add two further criteria that may be of particular importance in the current debate. A mature science is based on, among other things, a well-formed intellectual framework and accompanying hypotheses. That is, a well-formed scientific research program has clear claims and clearly testable goals, and it will be able to give clear interpretations of data when that information arrives. The theory of punctuated equilibrium has suffered in large part because of this very problem. While punctuated equilibrium in some form may be true, Stephen Jay Gould and Niles Eldredge have been criticized for never providing the kind of criteria to make it an unambiguous and clearly defined hypothesis (Gould and Eldredge 1972; Somit and Peterson 1989). Even more pertinent than this, however, is a second, albeit not completely scientific, criterion. Extraordinary claims require extraordinary evidence. Certainly, the claim of IDT is extraordinary. Not only are its theorists arguing that some organisms are intelligently designed; they are arguing that they are intelligently designed by God—an extraordinary claim of the first order. Little wonder, then, that the great majority of those in the scientific community have treated IDT with extreme skepticism.

While questions of demarcation are tricky, they are nevertheless necessary in making practical decisions about what counts as knowledge and, consequently, what should be taught in schools and funded by the government and granting institutions. By asserting that IDT is a science, advocates of IDT might genuinely appeal to a Lakatosian understanding to justify their status. Certainly, questions of detecting design and the legitimacy of inferring the existence of a designer are interesting ones, and they deserve some attention. Certainly, the question of whether one can infer God as a designer has a long intellectual tradition behind it and is still debated among philosophers, theologians, and some physical scientists. If this was all that IDT advocates proposed, their position would be much less controversial. But they have been insistent that IDT is primarily a *biological* science. It is here that the real trouble begins and where even a broadly Lakatosian approach to science works against them.

Intelligent Design as a Biological Hypothesis

According to its proponents, IDT is a scientific research program. Unlike many other branches of science, IDT is not focused on a limited temporal or spatial domain. It is not simply a science of biological organisms or subatomic particles. Rather, its theorists argue, IDT is relevant to all origin issues, from cosmology to biochemistry to human evolution. Despite this, however, advocates consistently claim that it is about the biological sciences that IDT is most concerned and toward which much of its attention is directed. Dembski states clearly that "the focus of intelligent design movement is on biology" (1999, 14). Behe's work clearly centers on biochemistry. Of the twenty-five articles contained in Moreland's *The Creation Hypothesis* (1994a) and Dembski's *Mere Creation* (1998a), at least eleven are centrally concerned with biology. In these articles, furthermore, IDT is not limited to the realm of biochemistry but includes accounts of human evolution, phylum diversity, and altruism. The claim, therefore, is that IDT has wide-ranging implications for the study of biology, presumably extending across a number of subdisciplines.

The biological science of IDT presumes, at least in theory, several propositions. First, IDT assumes the existence of what may be called "deep time," scientific evidence for the antiquity of the universe (on the order of 13 billion years) as well as the antiquity of Earth (roughly 4.5 billion years) and of life (roughly 4 billion years). For instance, Behe states, "For the record, I have no reason to doubt that the universe is the billions of years old that physicists say it is. Further, I find the idea of common descent (that all organisms share a common ancestor) fairly convincing, and have no particular reason to doubt it" (1996, 5). Similar affirmations have been made by Dembski and others in the IDT movement. Second, as the quotation from Behe implies, IDT does not claim that evolution never occurs or that natural selection does not play a role in the origin of some species. Thus, while IDT proponents spend a great deal of their time attacking Darwinism specifically and evolution in general, we are presumably to understand this not as complete refutation of Darwinian and evolutionary processes but as a critique of their misapplication to biological phenomena that are best viewed in terms of intelligent design.

It should be observed, however, that the commitment to both of these propositions seem rather tepid. ID theorists, for instance, make virtually no effort to distinguish themselves from the even more dubious young-Earth creation-science movement; indeed, it is unclear into which camp individuals like Moreland fall. Whereas muting these claims may be a way of showing evangelical Christian solidarity, it does little to further IDT's agenda in scientific circles, where the two movements are easily confused. Furthermore, such vitriol is heaped upon evolutionary explanations that it is not clear whether, in truth, IDT proponents accept any evolutionary account of biological organisms. I will choose to take IDT's support of deep time, evolution, and Darwinism at their word, but I maintain that their public presentation on these issues is highly problematic, leading one to suspect that they believe there to be no real support for these at all.

Two other propositions, however, form the core of IDT, and these are the most important for the scientific status of the movement. First, Dembski (1998a) has devised a logical apparatus for detecting design. According to Dembski, design can be detected through the elimination of chance and law hypotheses and that this "design filter" has a potentially wide range of applications, from cryptography and SETI (search for extraterrestrial intelligence) to cosmology and biology. I will not evaluate here the validity of Dembski's apparatus. In truth, the design filter on first glance captures much of the intuition behind the notion of design, and his packaging it as a logical, algorithmic structure is certainly thought provoking. At the same time, it is not clear to me that the design filter is unproblematic. Brandon Fitelson, Christopher Stephens, and Elliott Sober (1999), for instance, provide a rather stinging analysis of several central aspects of the design filter. Notoriously problematic is the purely negative character of the filter. Design is detected not by any positive characteristic such as the presence of iridium in a sedimentary layer that reveals a large meteor impact but rather by elimination of all the alternatives. According to Fitelson, Stephens, and Sober, this gives design a privileged status. Design can never be falsified, only confirmed. Furthermore, Dembski claims that his filter avoids the need for determining prior probabilities (that is, more or less, the need for knowing the likelihood of the hypothesis before being able to determine the likelihood of the hypothesis), something that he touts as a superior feature of his filter. Yet such prior probabilities must be implicit in any assessment of chance and law hypotheses. Both of these issues are generally relevant to the scientific status of intelligent design.

A second key proposition of IDT is that designed structures can be detected in the natural world. A designed structure is recognized by its irreducible complexity (Behe), denoted by Dembski as specified complexity. For Behe, biochemistry is the key science in finding irreducible complexity. Biochemistry is "Darwin's black box," the assumed substratum upon which all evolution must depend. Structures as diverse as the bacterium flagellum, blood-clotting systems, and the eye (yet again!) are all identified as irreducibly complex structures, unexplainable by appeals either to chance or to natural selection. Having proven unable to provide an empirically adequate account of transitional forms that could explain these structures, Behe concludes that they are irreducibly complex. Because they are irreducibly complex, they are products of design, and after briefly considering alternatives he ascertains that the only conceivable designer in these cases is a divine being, God.

Given these four propositions (deep time, limited role for evolution, a method for detecting design, and the applicability of the method to biology), it is worth commenting on what kind of theory of biology IDT presents. Despite the rhetoric to the contrary, IDT is not a competitor to evolutionary accounts of biological origins (which is outside evolutionary theory proper anyway) and the origins of species. Rather, it is a *modification* of such accounts. That is, IDT is a form of evolutionary theory. Common descent is accepted (at least by Behe), with the caveat that there are organisms whose normal progression of descent is modified by an intelligent designer. These design modifications may have strong phenotypic implications in multicellular organisms, but the locus of ID is at the biochemical level. Such design features, furthermore, are saltations. That is, they are sudden changes in the structure of the organism, creating a new "hopeful monster" that presumably would still have to meet the criteria of natural selection in order to survive.

Thus, IDT provides a saltationist account of evolution, the innovation being the claim that many important, adaptive saltations cannot happen by chance but must be the consequence of design. Put in this light, the impact of IDT when applied to biology seem less startling than at first glance. Saltationist theories of evolution have been proposed before. In fact, Lynn Margulis's account of the bacterial origin of mitochondria in eukaryotic cells might be regarded as a saltationist account of a whole branch of life that nevertheless falls within the domain of natural selection. The difference of IDT, of course, lies in the *mechanism* of saltation. For ID theorists, some saltations result from neither chance nor natural selection. Thus, IDT proposes a radical theory to account for, presumably, a scattered set of problematic phenomena within a broader evolutionary framework.

ID AS SCIENCE: AN ASSESSMENT

Given this description, we may then move to the central point by asking, Is IDT truly scientific in character? Does it warrant our attention in the same way that Newtonian mechanics, thermodynamics, or even evolutionary theory does? Among the several ways to address this issue, three are particularly relevant. First, is IDT well formed? Second, does IDT truly provide an account that competes with evolutionary claims? Third, is IDT progressive in a scientific sense?

It is certainly the case that IDT asks well-formed questions. IDT is concerned with the origin and development of complex, biological structures. This concern is in no way unique to IDT but is shared by a number of competing programs. Such an issue is especially acute at the point of the origins of life, and IDT theorists are right to see the RNA world hypothesis and Stuart Kauffman's (1995) claims about self-organizing systems as competing theories attempting to explain the same phenomena. If either of these proved correct, clearly this particular claim of IDT would be falsified.

It is not obvious, however, whether IDT provides a well-formed answer. ID theorists claim to be able to detect instances of specified complexity and then go on to attribute such phenomena to the work of intelligent design. Yet, this point is largely unargued; indeed, Dembski admits that specified complexity is not *necessarily* a sign of intelligent design (1998a, 9). As Howard Van Till observes (1999), this renders the central claim of IDT a bit circular. Specified complexity reveals intelligent design. Intelligent design is whatever produces specified complexity.

More problematic, however, is what is supposed to count as intelligence. ID theorists rely on a largely intuitive concept of intelligence. Intelligence, after all, is a human quality. According to Dembski (1998a, 62), the importance of intelligence is "directed contingency, or what we call choice." Although insightful in some ways, this definition is hardly conventional among psychologists and cognitive scientists, who in many cases avoid the term *intelligence* altogether. Intelligent actions can include a variety of things. The ability to follow an algorithm was, until the advent of the computer, one popular criterion. Trial-and-error learning is invoked in some quarters, whereas others might focus on the category of creative insight.

Furthermore, there are different kinds of intelligences in the world. Computers can now play an unbeatable game of chess, and robots are currently capable of self-navigation and problem solving. More recent computer programs and, some would argue, human intelligences as well use a kind of Darwinian problem-solving approach, choosing the best out of competing hypotheses, which are "selected" for further consideration. Chimpanzees produce tools, and some are capable of symbolic communication. We might regard termite mounds as intelligently designed, but they are designed by termites, hardly the conventional image of an intelligent designer. If such lowly creatures as termites can produce termite mounds, however, can even lowlier, lawlike (and one might say algorithmic-like) processes produce termites? Or must we presume that all creatures that construct artifacts are products of intelligent design and, therefore, are not truly intelligent but have only what philosophers call "derived intentionality"? If that were the case, would not human beings themselves be instances of derived intentionality and not truly a suitable example of original, intelligent agents?

Intelligent-design theorists have consistently resisted even reasoned speculations on the nature of the designer, claiming that this lies beyond the scope of IDT as a science. One can only regard this as one of the strangest examples of a lack of curiosity that exists in the scientific world. After all, if true, IDT has essentially proved the existence of a kind of being previously undemonstrated by science. ID theorists assert that, although irreducible complexity can suggest the existence of a creator, it unfortunately tells us nothing about the character or nature of the creator. Such an assertion, however, is not highly convincing. Archaeologists, after all, specialize in analyzing obviously designed artifacts constructed by human beings throughout history. And while the intent and character of the designers in some cases are not clear, archaeologists spend much of their time constructing hypotheses about the designers and their culture from these selfsame artifacts. It is unclear why biological artifacts should be different.

This is more than an idle point. An intelligent-design theory that refuses to say anything about the designer is either confused or incoherent. Such a theory also negatively affects the scientific character of the discipline, for it essentially disallows the formulation of any hypotheses that might in fact be testable in a positive way. Without a theological science to accompany its biological science, IDT retains a purely negative approach that explains by not explaining.

Does IDT truly provide a competing account to evolutionary claims? Again, the evidence is weak at best, not least because IDT does not truly address many of the central issues that evolutionary theory was designed to address. It is noteworthy to point out that Darwin's magnum opus was not titled *The Origin of Life* but *The Origin of Species*. Darwin wrote not because he was interested in biochemistry but because he was interested in species variation and, in particular, the fact that species varied in interesting and nonrandom ways. Often IDT critics concentrate not on much of the modern scientific character of evolutionary biology but on the slogans that, arguably, are not part of evolutionary biology but inform the way that evolutionary biology is done. Phrases such as "survival of the fittest" and "natural selection" are necessarily vague and amorphous. They and others provide not so much scientific theory as a metatheory, informing particular scientific practices and providing a heuristic for investigating natural phenomena.

For instance, ID theorists do not provide explanations for island biogeography. Islands typically host a percentage of endemic (unique) species out of all proportion to their size. Moreover, the species represent phyla and orders that one might expect to migrate by sea: birds, snakes, snails, small lizards, and the like. Furthermore, it is clear that the existing or recently extinct (largely because of human activity) examples of these species are not the species that originally migrated. This fact is most obvious in the case of the many examples of large, flightless birds that were found on islands such as Hawaii and New Zealand, which would obviously have been incapable of migration on their own. However, it also can be seen in the variations between the islands themselves, with the variation increasing as the distance between islands or island chains increases. Study of island biogeography was much furthered by Robert Macarthur and Edward O. Wilson's (1967) analysis that suggested a correlation between island size and species diversity.

One might think that the challenge posed by these observations is simply the presence of all these different species. If this were the case, an ID theorist could simply respond that islands have many unique examples of design and would, perhaps whimsically, observe off the scientific record that God has a fondness for islands. This, however, is not the challenge. The challenge is the *patterns* of diversity that are found on islands. A history of the island in terms of evolution and natural selection is consistent with, and to a certain extent predictive of, the existence of large, flightless birds; high rates of endemic lizards and snails; and increased diversity on large islands. IDT, however, not only does not say anything about these patterns; it seems unable to. Any account that IDT could give would necessarily be ad hoc. Why should we expect in advance a greater rate of endemic species on islands? Why should we expect these species to appear to be descended from likely migratory species? Why should we expect to see apparently evolutionary patterns of migration across islands?

While one can only rarely give specific lines of descent for any individual species, evolutionary theory provides a framework in which such diversity is understood and even expected. More important, general terms such as "natural selection" are used to advance more specific hypotheses. "Natural selection" is most cogent in the more specific theoretical claim of allopatric speciation, which suggests that new species are typically formed by small, founder groups that separate from existing populations and gradually grow in diversity. The theory of allopatric speciation is consistent with and even predictive of the observation of higher numbers of endemic species in geographies that do not allow a great deal of transportation, such as islands, insulated valleys, underground lakes, and isolated geothermal vents. Macarthur and Wilson's theory would be an even more specific cashing out of the implications of the framework of natural selection, even though it is not presented in strictly historical terms.

This argument is applicable as well to the more general trends of natural history. In short, life starts small and gets big. Whereas ID theorists seem willing to accept common descent (meaning, presumably, that God works with the existing biological materials at any given time), IDT gives no account of why we should expect fish to appear first or why all dinosaurs appear much later. It does not explain why Australia became dominated by marsupials such as kangaroos (in carnivorous and herbivorous varieties) and wallabies. It is not enough to say that God works with existing biological materials, a claim that implicitly breaks the ban on describing the designer's intentions. An intelligent designer of divine proportions could presumably implement quite radical saltations. The problem is that natural history exhibits patterns that ID is unable or unwilling to explain, patterns that are consistent with (and to a certain extent predicted by) evolutionary approaches.

Of course, ID theorists could simply concede these broad swaths of inquiry to evolutionary theory, maintaining all the while that within the rather broadly evolutionary world there nevertheless exist specific instances of design. Despite the frequent rhetoric to the contrary, perhaps ID theorists are really claiming that ID is the exception rather than the rule, that ID is relatively rare but important nonetheless. Such a position, at least in theory, is still potentially significant—not least because God is attached to the other end of the equation—but it hardly shakes the foundations of biology. Rather, IDT turns out to be an attempt to explain isolated phenomena that are currently unexplainable and which, as far as the IDT literature reveals, lack the comparative research to establish whether evolutionary hypotheses are even plausible.

The third, and perhaps most important, question remains. Is IDT progressive? That is, does IDT show the hallmarks of a nascent scientific research program? Are there serious parallels between the emergence of IDT and, say, relativity theory or plate tectonics? Sadly, the answer appears to be no. This is in part due to the surprisingly small domain of explanation that IDT has allowed itself. According to its proponents, IDT explains instances of irreducible complexity. That's it. It does not explain how they came to be. It does not explain why they came to be. It does not even explain when they came to be.

Yet, a truly scientific research program should be bursting with questions and working out innumerable implications. After all, IDT asserts not only that we have a reliable design filter but also that we can use this filter to accurately detect design in biological organisms. If this were the case, numerous lines of reasoning could be followed up, even given the dramatic strictures that ID imposes on itself. One interesting question would be, How frequently does design occur? Given that we can detect design and given that not all organisms are designed, it would be interesting to determine what percentage of organisms exhibit intelligent design. Such data, in turn, could lead to further, more specific questions. Do we find design to be more prevalent in some lines of descent than others? Do humans exhibit more design features than ostriches do? Does God have a fondness for the millions of species of insects, or are they simply the product of natural selection? Are human pathogens such as AIDS, cholera, and malaria intelligently designed? Would not these be interesting questions?

Surprisingly, ID theorists are not pursuing any of these questions. Instead, at the end of his article "Intelligent Design Theory as a Tool for Analyzing Biochemical Systems" Behe (describing the payoff of IDT for studying biology) states the following:

> So what difference does intelligent design theory make to the way we practice science? I believe it is this: a scientist no longer has to go to enormous lengths to shoehorn complex, interactive systems into a naturalistic scenario. . . . We should remain open to the possibility that further analysis will show our conclusion was wrong, but we should not be timid about reaching a conclusion of design and building on it. (Behe 1998, 194)

In other words, the payoff of ID is not increased scientific knowledge but rather relief from the attempt to answer unanswerable questions. Again, the lack of curiosity about the theorists' own hypothesis is remarkable. Instead of engaging in further questions and research, IDT allows us to stop asking questions. Instead of trying to formulate sophisticated scientific questions about design, ID theorists seem to consider this a matter of faith that is not amenable to rational discussion.

This lack of development alone is compounded by a lack of curiosity about the implications of related aspects of the theory that are connected to but outside the realm of biology. After all, how does design occur? Is it a single saltation or a series of them? Does it happen in a single organism or over a series of organisms? More important, what are the mechanisms of ID? How does the design occur? Any change in biochemical structure and information content would suggest an energy expenditure, presumably from outside the natural realm, implying a violation of the law of energy conservation. Does IDT imply this, and if so, how would it work?

These are relatively minor problems when compared to a more central issue. ID theorists have not carried out the agenda of their own program. Dembski's admonition at the end of *The Design Inference* is noteworthy:

> The fact is that the design inference does not yield design all that easily, especially if probabilistic resources are sufficiently generous. It is simply not the case that unusual and striking coincidences automatically generate design as the conclusion of a design inference. There is a calculation to be performed. *Do the calculation. Take the numbers seriously. See if the underlying probabilities really are small enough to yield design.* (Dembski 1998a, 228)

Remarkably, this is the one thing that ID theorists have not done. Nowhere does Behe use Dembski's design-inference apparatus to "do the numbers" and arrive at a solid conclusion of design for such structures as the bacterium flagellum. Even more to the point, Behe shows no inclination to do so; he seems satisfied to give brief examples of complexity in publications aimed at a nonscientific readership. No other ID theorist, to date, has taken up this challenge. Furthermore, ID scientists have not published their views in peer-reviewed scientific publications, so in the majority of cases this work has not been properly evaluated by appropriate professionals.

This last observation is perhaps the most serious. A scientific research program rises and falls on its publication track record, yet IDT does not seem to have even a research agenda that will lead to the kind of data that can be fairly evaluated in a peer-review process. Not only is there no scientific progress; there seems to be no intent to achieve scientific progress. It is not only that IDT does not meet the maximal criteria of science; it is hard to see how it even meets the minimum, Lakatosian criteria of science. While we can certainly characterize IDT as a research program, there seems to be little reason to regard it as a scientific research program, let alone a progressive one. IDT has not worked out the implications of its own claims, has not clearly elucidated its relation to evolutionary biology, has not provided specific scientific data, and has not submitted these for peer review. Furthermore, since it does not seek to explain many of the same phenomena that concern evolutionary biology, it is unclear to what extent IDT constitutes a competing hypothesis. At best, it is a science waiting in the wings.

Concluding Comments

Of course, it could be that IDT publications will soon reverse this situation. After all, Behe's *Darwin's Black Box* is only five years old, far too short a time to get a research program off the ground. And, it is often admonished, we should be charitable to new research programs. True enough, but it is strange that this new research is not reflected in current publications. Surely, if ID theorists had such plans and workings, its advocates would want to advertise them!

Interestingly enough, however, ID seems more inclined to move in the opposite direction, from scientific research to polemical debate. Indeed, the publication trajectory of IDT is revealing. Dembski followed his technical monograph with two books aimed at conservative Christian markets and a receptive conservative Christian audience. The content of these popular works is revealing. In the two anthologies mentioned earlier (Moreland 1994a; Dembski 1998c), as I previously observed, eleven of the twenty-five articles concern biology. Depending on how you define the topics, fourteen of these articles deal with philosophical attacks on methodological naturalism, claims about design "in theory," or an engagement of philosophical and theological issues. Of those that deal with biology, the vast majority of virtually every article deals not with intelligent design but with why Darwinism is wrong in the particular instance cited. Little that is specific is said about ID in these articles beyond the general claim that it provides a better account of the phenomena than Darwinism does. Dembski's second monograph (1999) abandons the scientific arena for a largely philosophical account that attempts to bridge the gap between science and theology. Science appears to play only a secondary role in IDT literature. The primary concerns are philosophical and theological in character. Indeed, IDT seems to be exactly what it accuses its opponents of being: an ideological agenda masquerading as science.

There is nothing wrong with ideological and philosophical agendas, at least in principle. Indeed, many would share the concerns of ID theorists about naturalism as a worldview. Many scientists agree that the origin of life is, to say the least, immensely puzzling. When an ideology masquerades as something else, however, it becomes deceptive in character. The issues that IDT addresses are serious and worthy of consideration. The approach, however, is misguided and will result in only more painful public encounters at the expense of the religious traditions its proponents seek to defend.

REFERENCES

Behe, Michael J. 1996. *Darwin's Black Box: The Biochemical Challenge to Evolution.* New York: Touchstone.

———. 1998. "Intelligent Design Theory as a Tool for Analyzing Biochemical Systems." In *Mere Creation: Science, Faith, and Intelligent Design*, ed. William Dembski. Downer's Grove, Ill.: Intervarsity Press.

Cicero, Marcus Tullus. n.d. *De Naturum Deorum*, Book II:12–14.

Dembski, William. 1998a. *The Design Inference: Eliminating Chance through Small Probabilities.* New York: Cambridge Univ. Press.

———. 1998b. "Introduction: Mere Creation." In *Mere Creation: Science, Faith, and Intelligent Design*, ed. William Dembski. Downer's Grove, Ill.: Intervarsity Press.

———, ed. 1998c. *Mere Creation: Science, Faith, and Intelligent Design.* Downer's Grove, Ill.: Intervarsity Press.

———. 1999. *Intelligent Design: The Bridge between Science and Theology.* Downer's Grove, Ill.: Intervarsity Press.

Fitelson, Brandon, Christopher Stephens, and Elliott Sober. 1999. "How Not to Detect Design—Critical Notice: William Dembski, *The Design Inference.*" *Philosophy of Science* 66:472–88).

Glen, William, ed. 1994. *The Mass-Extinction Debates: How Science Works in a Crisis.* Stanford: Stanford Univ. Press.

Gould, Stephen Jay, and Niles Eldredge. 1972. "Punctuated Equilibria: An Alternative to Phyletic Gradualism." In *Models in Paleobiology*, ed. Thomas J. M. Schopf. San Francisco: Freeman, Cooper.

Johnson, Phillip. 1991. *Darwin on Trial.* Washington, D.C.: Regnery Gateway.

Kauffman, Stuart. 1995. *At Home in the Universe: The Search for the Laws of Self-Organization and Complexity.* New York: Oxford Univ. Press.

Kuhn, Thomas. 1962. *The Structure of Scientific Revolutions.* Chicago: Univ. of Chicago Press.

Lakatos, Imre. 1970. "Falsification and the Methodology of Scientific Research Programmes." In *Criticism and the Growth of Knowledge*, ed. Imre Lakatos and Alan Musgrave, 91–196. Cambridge: Cambridge Univ. Press.

Macarthur, Robert, and Edward O. Wilson. 1967. *Theory of Island Biogeography.* Princeton, N.J.: Princeton Univ. Press.

Moreland, J. P., ed. 1994a. *The Creation Hypothesis: Scientific Evidence for an Intelligent Designer.* Downer's Grove, Ill.: Intervarsity Press.

———. 1994b. "Theistic Science and Methodological Naturalism." In *The Creation Hypothesis: Scientific Evidence for an Intelligent Designer*, ed. J. P. Moreland, 41–66. Downer's Grove, Ill.: Intervarsity Press.

Murphy, Nancey. 1990. *Theology in the Age of Scientific Reasoning.* Ithaca, N.Y.: Cornell Univ. Press.

Peterson, Gregory. 1998. "The Scientific Status of Theology: Imre Lakatos, Method and Demarcation." *Perspectives in Science and Christian Faith* 50:22–31.

Popper, Karl. 1959. *The Logic of Scientific Discovery.* New York: Basic Books.

———. 1971. *The Open Society and Its Enemies.* 2 vols. Princeton, N.J.: Princeton Univ. Press.

Somit, Albert, and Steven A. Peterson, eds. 1989. *The Dynamics of Evolution: The Punctuated Equilibrium Debate in the Natural and Social Sciences.* Ithaca N.Y.: Cornell Univ. Press.

Van Till, Howard. 1999. "Does 'Intelligent Design' Have a Chance? An Essay Review." *Zygon: Journal of Religion and Science* 34 (December): 667–75.

[22]

The Advantages of Theft over Toil: The Design Inference and Arguing from Ignorance

JOHN S. WILKINS
History and Philosophy of Science
The University of Melbourne
PO Box 542, Somerville 3912
Australia
E-mail: wilkins@wehi.edu.au

WESLEY R. ELSBERRY
Department of Wildlife and Fisheries Sciences
Texas A&M University 3027 Macaulay Street
San Diego, CA 92106
USA
E-mail: welsberr@inia.cls.org

Abstract. Intelligent design theorist William Dembski has proposed an "explanatory filter" for distinguishing between events due to chance, lawful regularity or design. We show that if Dembski's filter were adopted as a scientific heuristic, some classical developments in science would not be rational, and that Dembski's assertion that the filter reliably identifies rarefied design requires ignoring the state of background knowledge. If background information changes even slightly, the filter's conclusion will vary wildly. Dembski fails to overcome Hume's objections to arguments from design.

Key words: Bayesian inference, Darwin, Dembski, intelligent design, natural selection

Sam Spade enters his office to find "Fingers" Finagle, a reformed safecracker, standing in front of his open safe holding the priceless artifact the Cretan Sparrow that Spade was looking after for a client. "Fingers" insists he did not crack the safe, but merely spun the combination dial a few times idly, and it opened by itself. Spade knows from the promotional literature that came with the safe when he bought it at the Chump end-of-season sale that it has over 10 billion (10^{10}) possible combinations, and that only one of these will open it. Moreover, he knows that the dial must be turned in alternating directions, not – as "Fingers" claims he did – in the same direction repeatedly. What does Spade know about this situation? Is the safe open by design, or by accident? William Dembski (1998) thinks he can answer this question definitively.

712

Dembski has proposed an "explanatory filter" (EF) which, he claims, enables us to reliably distinguish events that are due to regularities, those that are due to chance, and those that are due to design. Such a filter is needed, he believes, to determine the reason for cases like Spade's safe, the discrimination of signals by the Search for Extra-Terrestrial Intelligence project (SETI) that are due to intelligent senders from those that are caused by ordinary phenomena like quasars, and most critically, whether all or some aspects of the biological world are due to accident or design. In other words, Dembski's filter is a reworking of Paley's design inference (DI) in the forensic manner of identifying the "guilty parties".

We will argue that Dembski's filter fails to achieve what it is claimed to do, and that were it to be adopted as a scientific heuristic, it would inhibit the course of science from even addressing phenomena that are not currently explicable. Further, the filter is a counsel of epistemic despair, grounded not on the inherent intractability of some classes of phenomena, but on the transient lacunae in current knowledge. Finally, we will argue that design is not the "default" explanation when all other explanations have been exhausted, but is another form of causal regularity that may be adduced to explain the probability of an effect being high, and which depends on a set of background theories and knowledge claims about designers.

Spade's immediate intuition is that "Fingers" has indeed burgled the safe, but Spade is no philosopher and he knows it. He has, however, read *The Design Inference* by the detective theoretician, Dembski, and so he applies the filter to the case in hand (literally, since he has "Fingers" by the collar as he works through the filter on the whiteboard).

The EF is represented as a decision chart (p. 37):

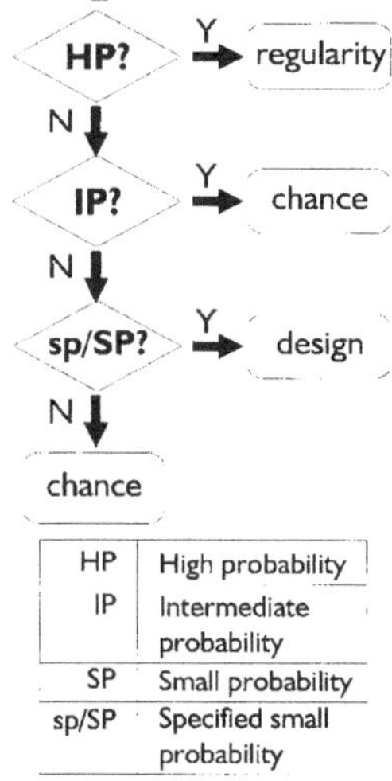

HP events are explained as causal regularities. If it is very likely that an event would turn out as it did, then it is explained as a regularity. IP events are events which occur frequently enough to fall within some deviation of a normal distribution, and which are sufficiently explained by being between those extremes. The rolling of a "snake eyes" in a dice game is an IP event, as is the once-in-a-million lottery win. SP events come in two flavours: specified and unspecified. Unspecified events of small probability do not call for explication. An array of stones thrown will have *some* pattern, but there is no need to explain exactly that pattern, unless the specifiable likelihood of a pattern is so small that its attainment calls for some account. If an array of stones spells out a pattern that welcomes travellers to Wales by British Rail, then *that* requires explanation; to wit, that the stones were placed there by an employee of British Rail, by design. The minuscule probability that a contextually significant message in English would occur by chance is ruled out by the specified complexity of that sentence. This Dembski calls the Law of Small Probabilities – *specified events of small probability do not occur by chance.*[1]

Spade, though not given to deep reflection, nevertheless studied statistics at the Institute of Forensic Studies, and so he wishes to be thorough. He traverses the filter step-by-step.

E: the safe door is open.

HP? No, the door regularly remains locked without intervention, and "Fingers" did not know the combination.

IP? No, there is no significant chance that random spinning of the dial would happen on the combination. Even had "Fingers" chanced to spin the dial the right directions – an IP event – the chance is one in ten billion (10^{-10}) that he would have happened on the combination. The chance is effectively zero, using the Law of Small Probabilities.

SP? Yes, the event has a very small probability.

sp/SP? Yes, the prior probabilities are exactly specified in addition to being very small.

Conclusion: "Fingers" opened the safe by design, not by accident.

"Fingers" is duly charged and arraigned for burglary. He engages the renowned deep thinking lawyer, Abby Macleal, and she defends him with skill. Before we get to the courtroom scene, however, let us go back in time, over a century, to the musings of a young naturalist.

This naturalist – call him Charles – is on a voyage of discovery. He has read his Paley; indeed, he might almost have written out Paley's *Evidences* with perfect correctness by memory. Although he has not heard of Dembski's filter, he knows the logic: whatever cannot be accounted for by natural law or chance must be the result of design. Young Charles encounters some pattern of the distribution and form of a class of organisms – let us suppose they are tortoises – on an isolated archipelago and the nearest large continent. Each island has a unique tortoise most similar to the autochthon of the neighbouring island and the species of the island closest to the continent is most similar to that continent's species. On the basis of the biological theories then current, he knows that there is no known process that can account for this pattern. It is so marked that one can draw a tree diagram from the continental form to the islands, and it will match a diagram showing the similarity of each form to the others. What should Charles *rationally* infer from this? Let us assume for comparative purposes that Charles is in possession of the filter; he will therefore reason like this:

E:	Species are distributed such that morphological distance closely matches geographic distance.
HP?	No, there is no regularity that makes this distribution highly probable.
IP?	No, the likelihood of such a distribution is extremely low.
SP?	Yes, it is a very small probability (made even smaller as more variables are taken into account).
sp/SP?	Yes, the problem is (more or less) specified.
Conclusion:	The tortoises have the biogeographic distribution and formal distribution they do by design.

By Dembski's framework, Rational Charles should have ascribed the tortoises' situation to intelligent agency, and his subsequent research *should* have been directed to identifying that agency, perhaps by building balsa rafts to test the likelihood that continental sailors might have taken varieties now extinct on the continent and placed them each to an island according to some plan. An even more parsimonious explanation, and one more agreeable to the Rev. Paley's natural theology, might be that a single agent had created them *in situ*, along a plan of locating similar species adjacent to each other, which has the added virtue of explaining a large number of similar distributions known throughout the world, as Alfred, a later young voyager, was to note.

Unfortunately for the progress of rational science, *Actual* Charles is not rational in this manner. He infers that some unknown process accounts for this distribution as a regularity, instead of inferring design. He irrationally conjectures that all the variants are modified descendants of the continental species, and that the morphological and geographical trees are evidence of a family tree of species evolution; and thus the theory of common descent is born. Charles is, rightly, castigated by his friends for irrationality and lack of scientific rigor. His leap to an unknown process is unwarranted, as is his subsequent search for a mechanism to account for it. Were his ideas to be accepted, perhaps out of fashion or irreligion, science would be put back for more than a century until Dembski came along to put it right.

Lest this seem to be a parody of Dembski's views, consider his treatment of the evolution versus creation debate and the origins of life. Dembski (wrongly) conflates the two, treating the origins of life as a test case for the validity of evolutionary theory (it isn't – even if the major groups of living organisms had separate origins, or were created by an agent, their subsequent history could and would have an explanation in terms of "undesigned" evolution). Creationists – the actual ones that do reject evolutionary theories in the way that Rational Charles should have in the 1830s – challenge what Dembski putatively does not, that species share common ancestors with their closest relatives and that natural selection accounts for adaptation. As an adjunct to their arguments, they also, along with Dembski, give credence to the "calculations" of the probability that prebiotic processes would spontaneously form the building blocks of life (the LIFE event), especially of genetic molecules, that various authors have given. Dembski addresses the arguments Stuart Kauffman (and others) use to block the design inference (Kauffman 1993, 1995) by presenting this argument:

Premise 1: LIFE has occurred.

Premise 2: LIFE is specified.

Premise 3: If LIFE is due to chance then LIFE has small probability.

Premise 4: Specified events of small probability do not occur by chance (the Law of Small Probabilities).

Premise 5: LIFE is not due to a regularity.

Premise 6: LIFE is due to regularity, chance, or design (the filter).

Conclusion: LIFE is due to design.

Dawkins' argued (1986: 139, 145–146) that there are a lot of "planetary years" available for LIFE to originate because there are a very large number of planets in the universe in which LIFE might have occurred and a lot of time available on each, Dembski says "... because Dawkins never assigns an exact probability to LIFE, he never settles whether LIFE is a high probability event and *thus could legitimately be attributed to a regularity*" (p. 58, italics added). Therefore, he says, we may infer that Dawkins accepts Premise 5! But what Dawkins actually says is that the improbability of life occurring had better not exceed the probability that it arose by chance on any one of the available number of planets on which it might have done. This sets a minimum bound to the probability of life, and Dawkins says that on (then) current knowledge, he doesn't know how probable life is. For all he knows, life is indeed due to a regularity. Kauffman's work on the dynamics of autocatalytic polymer sets supports the notion that the *upper* bound to the probability of life occurring is very high indeed, and life is to be "expected" in appropriate conditions.

716

Dembski's comment? This is a "commitment". The implication is that it is a mere belief or act of faith on Kauffman's part. In fact, it is considerably more than that, and the real problem for origins of life researchers is not to find a possible scenario, but to decide which of a growing number of them holds the most promise, or which combination. But Dembski's filter makes it unnecessary even to try.[2]

So, let us return now to the courtroom drama in time to hear Abby Macleal rebut prosecutor Pearl E. Mason's case. Abby calls retired Chump engineer Lachlan (Locky) Smith to the stand, and elicits from him the information that the Chump safe Spade owns has an inherent design flaw. If the tumbler is spun five times or more, centrifugal force will cause the lock to spontaneously open. Spade suddenly realizes why he got it so cheap. "Fingers" is acquitted, and initiates civil action for mental anguish and loss of reputation. Clearly, the background information has changed the probability assignments. At the time Spade found "Fingers" at the open safe, he was in possession of one set of background information, B_i. The probability of the event E requiring explanation led to a design inference. After Smith's testimony, a different set of background information, B_j, comes into play, and so the filter now delivers a "regularity" assignment to E. Suppose, though, that Smith had delivered yet another background set, B_k, by testifying that the model in question only actually used two of the five cylinders in the lock. Given that there are 100 possible numbers that might match the successful open state for each cylinder, the probability of a random opening is now 10^{-4}, which is a much higher probability, given the number of Chumps of that model in use in the Naked City (particularly after Chump's massive sell-off of that model to clear the faulty stock). *Now* the same filter delivers us a chance explanation given B_k. The point is that Dembski's filter is supposed to regulate rational explanation, especially in science, and yet it is highly sensitive to the current state of knowledge. One single difference of information can change the inference from design to regularity to chance. This goes to the claim that Dembski's explanatory filter reliably finds design. Reliability, Dembski tells us, is the property that once an event is found to have the property of "design", no further knowledge will cause the event to be considered to have the property of "regularity" or "chance". What the filter lacks that real-world design inferences already have is a "Don't know" decision. If we can say of a problem that it is currently intractable or there is insufficient information to give a regularity or chance explanation *now*, then the Filter tells us we must ascribe it to design if it is specifiable. But it can be specifiable without the knowledge required to rule out regularity or chance explanations. This is clearly a god of the gaps stance, and it can have only one purpose: to block further investigation into these problems.

Supposing we do insert a "don't-know" branch: where should it go? There is an ambiguity in Dembski's treatment of his argumentative framework. The Explanatory Filter is written about as if it describes a process of analysis, but Dembski's further argumentation is cast in terms of a first-order logical calculus. In a process, we would come to a "don't-know" conclusion after some evaluation of alternatives, but in a logical framework, there is no temporal dependency. We will here ignore the demands of process and concentrate on the logic. As Dembski's filter eliminates hypotheses from high probability to low probability, clearly an inability to assign a probability in the first place makes the decision the first branch point. So if, on B_i, the probability of E is undecidable, that needs to be worked out first:

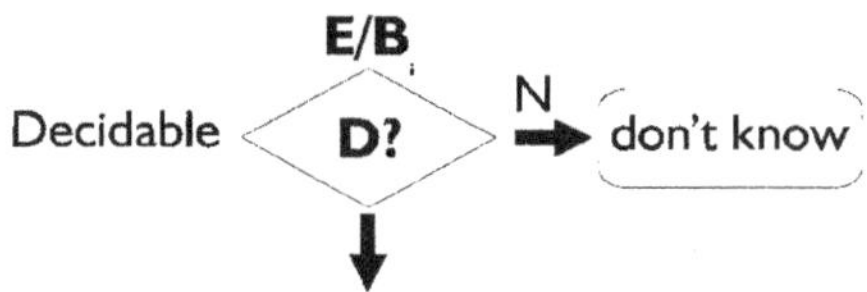

Undecidable probabilities lead us to a blocking of the inference at all. No further inferences can be drawn, and no design is required to explain any event for which there is no assignment. However, even if E *is* decidable on B_i, that in no way licences the expectation that on B_j or B_k those probabilities will remain fixed. For example, when Dawkins wrote in 1986, the state of knowledge about prebiotic chemical reactions was sparse; the range of possible RNA codes and molecular alternatives was not properly understood. As knowledge has grown, our estimate of the probability that *some* ribonucleotides, or perhaps ribonucleoproteins, or even polyaminoacids, might enter into protobiotic autocatalytic cycles has become much higher. Some even think that in a geologically short time after the cooling of the earth's surface, with the right conditions (themselves now expected to be of reasonably high probabilities on earth) life is almost certain to arise. Perhaps, then, we need another branching at each decision, leading to "Don't-know-*yet*". As Dembski's probabilities are Bayesian assignments made on the basis of a set of prior knowledge and default hypotheses, this seems to be a perfectly reasonable move. However, it has one glaring problem – it blocks *any* inferences of design, and that is too much. There are well attested cases of design in the world: we humans do things by design all the time. So an explanatory filter had better not exclude design altogether. How can it be included here? When is a design inference legitimate?

The problem with a simple conclusion that something is designed, is its lack of informativeness. If you tell me that skirnobs are designed but nothing else about them, then how much do I actually know about skirnobs? Of a

718

single skirnob, what can I say? Unless I already know a fair bit about the aims and intentions of skirnob designers, nothing is added to my knowledge of skirnobs by saying that it is designed. I do not know if a skirnob is a good skirnob, fulfilling the design criteria for skirnobs, or not. I do not know how typical that skirnob is of skirnobs in general, or what any of the properties of skirnobs are. I may as well say that skirnobs are "gzorply muffnordled",[3] for all it tells me. But if I know the nature of the designer, or of the class of things the designer is a member of, then I know something about skirnobs, and I can make some inductive generalizations to the properties of other skirnobs.

The way we find out such things about designers is to observe and interact, and if we can, converse, with them. In this way we can build up a model of the capacities and dispositions of designers. Experience tells me that a modernist architect will use certain materials to certain effect. Lacking any information about modernist architects leaves me none the wiser knowing that an architect is modernist (in contrast to other architects). Once we have such knowledge of designers, though, what we can say about them is that they generate *regularities of outcomes*. We know, for example, what the function of the Antikythera Device, a clockwork bronze assembly found in an ancient Greek shipwreck, was because we know the kinds of organisms that made it, we know the scientific, religious and navigational interests they had, we know about gears, and we know what they knew about the apparent motions of the heavens. Hence we can infer that the Antikythera Device is an astrolabe, used for open sea navigation by the stars, or a calendrical calculator, or both (de Solla Price 1974). But suppose it was found by interstellar visitors long after humans went extinct. What would *they* know about it? Unless they had similar interest and needs to ourselves, or were already able to reconstruct from other contexts what human needs and interests were, for all they know it might be the excrusion of some living organism (which, in a sense, it is), just like a sand dollar. It might never occur to them to compare it to the apparent motion of the heavens from earth circa 500 BCE.

So a revision to Dembski's filter is required beyond the first "Don't-know" branch. This sort of knowledge of designers is gained empirically, and is just another kind of regularity assignment. Because we know what these designers do to some degree of accuracy, we can assess the likelihood that E would occur, whether it is the creation of skirnobs or the Antikythera Device. That knowledge makes E a HP event, and so the filter short-circuits at the next branch and gives a design inference relative to a background knowledge set B_i available at time t. So now there appears to be *two* kinds of design – the ordinary kind based on a knowledge of the behavior of designers, and a "rarefied" design, based on an inference from ignorance, both of the possible causes of regularities and of the nature of the designer.[4]

Dembski (1999) critiques "evolutionary algorithms" (an 'everything and the kitchen sink' category in Dembski's idiosyncratic usage) as being incapable of yielding specified complexity, and thus actual design. The way in which this is done by Dembski bears upon our point that ordinary design is just another kind of regularity. First, recall how Dembski finds the complexity of an event, which is to use the probability of that event given a chance hypothesis as an estimate of the complexity. When Dembski gives an example involving known agent causation, such as a Visa card number or a phone number, the complexity is assigned on the basis that these numbers are drawn by chance. Yet we know that if a Visa credit card application is accepted, or the phone company accepts an application for phone service, then the likelihood that one's number will be assigned is very close to 1. However, when Dembski calculates the complexity of an event where he knows that an evolutionary algorithm is the proximal cause, the use of a chance hypothesis for comparison is eschewed. Instead, Dembski tells us that the complexity in this case is simply the likelihood that the evolutionary algorithm itself would yield the event. This probability is often very close to 1, and thus Dembski introduces the phrase "probability amplifier" to apply to evolutionary algorithms. But evolutionary algorithms are not the only probability amplifiers around; known intelligent agents are just as much probability amplifiers in Dembski's sense of the phrase. Dembski's inconsistency in handling complexity measurement in these two cases can be resolved in two ways. First, Dembski could consistently measure complexity by reference to the *difficulty* of the problem as measured by the length of bit strings, as discussed in Chapter 4 of *The Design Inference*. When this measure is deployed, the putative causal hypothesis has no bearing upon the complexity assigned to the problem. This would mean that evolutionary algorithms, among other things, could clearly be responsible for events that are classed as "design" in his original Explanatory Filter. This would occur when an evolutionary algorithm is *a priori* excluded as being a sufficient causative explanation for the event in question. Design would not be capable of distinguishing agent causation, as Dembski has so far claimed, from causation by natural processes.

The second way that the inconsistency can be resolved is as we have already indicated, by recognizing a distinction between ordinary design and rarefied design. For those events where our background information includes information about how agents or processes produce events of high probability, we would assign those to the HP category and explain them with reference to regularity. This would indeed preserve a place for a class of rarefied design in the Explanatory Filter. However, Dembski earlier argued that design indicated agent causation because his Explanatory Filter captures

720

our usual means of recognizing design, and so it would only apply to the class of ordinary design, not the desired rarefied design. It is only by the attempt to inconsistently treat agent causation as a privileged hypothesis that Dembski can (erroneously) claim that ordinary design and rarefied design share a node on the Explanatory Filter.

Where does this leave us? Dembski's filter is no longer looking so appealing. Now there are many possible reasons to accept alternative explanations to rarefied design because of uncertainty and on the basis of different background information. The filter now has many points at which elimination of alternative explanations to regularity and chance through to a rarefied design inference is blocked due to uncertainty and on the basis of different background information. In fact, it now looks rather like this:

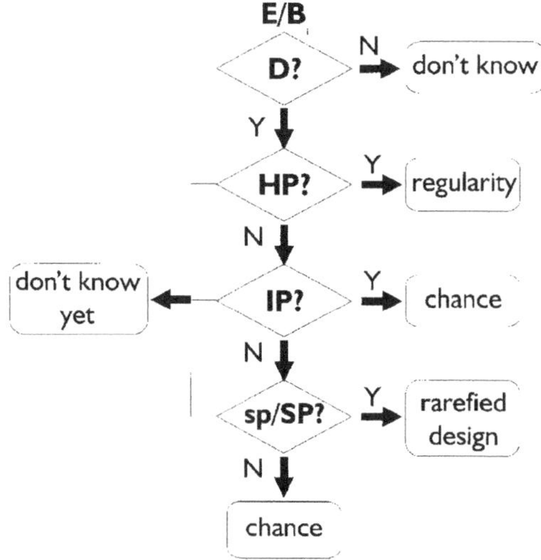

So far from leaving us satisfied with a rarefied design inference, as distinct from an ordinary design inference, in cases where we cannot give a satisfactory explanation in terms of causal regularities or chance, we now have to weigh carefully the reliability of our expected probability assignments given $B_{i...n}$, and the likelihood that any one of these will apply. Maybe Actual Charles' inferences were not so irrational after all. Perhaps Historical Charles (Darwin's) inferences were actually a rational bet based on a confidence that B would change. Similarly the work of Kauffman, Szathmáry (1997), Wächtershäuser (1997), and many others working on the origins of life rests on the assumptions of an as-yet-unknown causal mechanism or process is a rational bet, despite the lack of specified probability assignments. Only further work and time will tell, and the matter cannot be determined a priori on the basis of Dembski's filter or anything like it.

So instead of design being the penultimate default hypothesis in the decision tree, rarefied design becomes, at best, a tenuous conclusion to draw. There is an in-principle difference between rarefied and ordinary design inferences, based on the background knowledge available about ordinary, but not rarefied, design agencies. Rarefied design inferences tell us nothing that can be inductively generalized. Consequently, analogies between artifacts of ordinary design, which are the result of causal regularities of (known) designers, and the "artifacts" of rarefied design do not hold (as Philo noted in Hume's *Dialogues Concerning Natural Religion*, Book V). Sober (1993) argues that Hume got the Teleological Argument wrong – instead of the teleological argument being an analogical inductive inference, he says, Paley's argument (which actually came after Hume's criticisms, but it is unclear whether Paley had read Hume's *Dialogues*) is actually an *abductive* argument to the best explanation. But for the reasons given here, we think Sober is wrong. One cannot make a substantive case for rarefied design *except* through analogy, and Hume showed this fails since no ordinary designer works to produce the sorts of artifacts that living things would be if they were artifacts. So, if analogy is not used, no explanatory power inheres in rarefied design inferences. Of course, this does not preclude ordinary design inferences of the origin of life, for example (cf. Pennock 1998 on the Raelians, a group who believe life on earth was seeded by aliens), but the evidence that would make this a viable explanation is lacking, especially when compared to regularity explanations in terms of known chemistry, physics and geology.

Indeed, we might even conclude that the specified small probability of rarefied design is itself an artifact of our prior expectations. As our background knowledge changes and grows (due to the "irrational" inferences of people like Actual Charles), so too do the specifications, and sp/SP can become HP or IP. Why is there a rarefied design option in the filter at all? Dembski has not dealt with such Humean objections. His *a priori* expectation is that events of specified small probability (relative to whichever specification) do not happen by themselves through chance or regularity, and hence require some other "explanation". But if this is merely a statement about our expectations, and we already require a "don't know" or "don't know yet" option in our filter, why are we ever forced to a rarefied design conclusion? Surely we can content ourselves with regularities, chance and "don't know" explanations. Such overreaching inferences as a rarefied design inference carry a heavy metaphysical burden, and the onus is on the proponents of such an *a priori* assumption to justify it. Otherwise, "don't know" is adequate to the empirical task where background information is equivocal. This leads us to a truncated filter of the form of the following flow diagram:

722

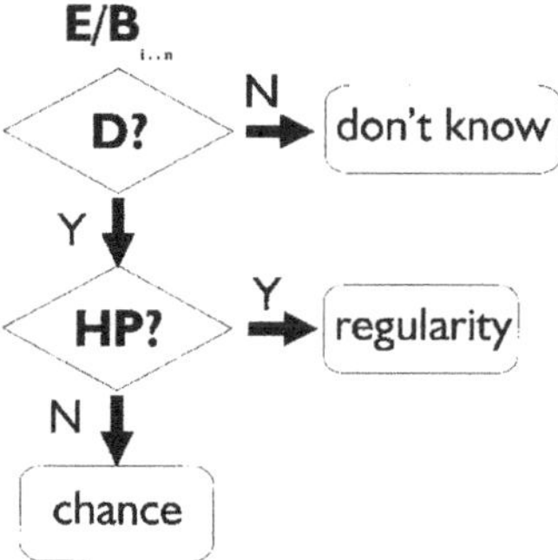

If B_i is replaced, then the explanation of E may change, and this is as it should be. Science is not a process of deduction from fixed axioms and *a priori* background specifications, or it is only transiently. We may expect that some background assumptions are harder to change than others, but they are all revisable in the light of better knowledge. The need for a rarefied design inference which does not offer scope for inductive inferences or add any predictive value beyond that already gained from B and HP/IP assignments is not apparent. One might cynically suppose that it is there not for heuristic and explanatory reasons but as a way to confer legitimacy upon natural theology and teleology.

If science is to be possible given a fallibilistic account of knowledge and if the knowledge it generates depends on empirical rather than innate rational information, then no rarefied design inference is needed, and all inferences are sensitive to the current state of knowledge. *A priori* assignments such as Dembski's filter requires make the *human* enterprise of discovery through trial and error impossible except where the metaphysical commitments of scientists and the broader society are unthreatened. On Dembski's account, any hypothesis – for example that humans are descended from apes – that offended public metaphysics would never get a chance to be considered and developed, since it would be deemed unnecessary before it could be. Naturally, once it *had* been incorporated into the B_i of the present, it could not be rejected except on evidentiary grounds by a DI-proponent, but neither would it have been incorporated in the first place if the filter were widely accepted. In effect, the DI filter is a counsel of heuristic despair, or perhaps of hope that some matters are protected from a scientific investigation that is not dependent upon final causes. Like all such accounts from Paley onwards, Dembski's filter has all the advantages of theft over honest toil, as "Fingers" knew well when he deliberately made use of the common knowledge of all safecrackers in the Naked City that the Chump safes had an inherent design flaw. It is a pity that the court didn't also have that bit of background knowledge in their own design inference.

Notes

[1] Dembski 1998: 218. Cf. Also p. 52.

[2] On pp. 105–106, Dembski discusses the problem of assigning a measure of the complexity of a problem needed to assign the probability estimate of E. He says, "I offer no algorithm for assigning complexities nor theory for how complexities change over time. For practical purposes it suffices that a community of discourse cans settle on a fixed estimate of difficulty." In effect, design inferences are necessarily fixed to the community's discourse at a single time. See Fitelson et al. (1999) for a discussion of Dembski's implicit use of background information.

[3] This wonderful phrase is due to Howard Van Till.

[4] Dembski's design is not necessarily the result of intelligent agency, he says (p. 9), but if it is not, design is just a residue of the EF. The EF classifies events, not causes. NS appears to be well-suited to produce events that land in the design bin. Causal regularities that account for design-like features (Kitcher 1998) such as self-organisation or natural selection are covered by the regularity option. This is the problem with Dembski's use of these terms. They lead to confusing events and causes. As Dembski notes, events caused by intelligent agents can be classified in any of the three bins of his EF. We would extend this to say that events due to NS can also be found in multiple bins, including "design", and the analogy he draws is explicitly with output of intelligent agency. The reason we can "reverse engineer" the "design" of extinct organisms is because we know a lot about the ecology, biology and chemistry of organisms from modern examples. But we can do none of this with rarefied design.

References

Dawkins, R.: 1986, *The Blind Watchmaker*, Longman Scientific and Technical, Harlow.

Dembski, W.A.: 1998, *The Design Inference: Eliminating Chance through Small Probabilities*, Cambridge University Press, Cambridge/New York.

Dembski, W.A.: 1999, Explaining specified complexity. Meta-Views #139, http://listserv.omni-list.com/scripts/wa.exe?A2=ind99&L=metaviews&D=1.

Fitelson, B., Stephens, C. and Sober, E.: 1999, 'How Not to Detect Design – Critical Notice: William A. Dembski, The Design Inference', *Philosophy of Science* **66**(3), 472–488.

Kauffman, S.A.: 1993, *The Origins of Order: Self-Organization and Selection in Evolution*, Oxford University Press, New York.

Kauffman, S.A.: 1995, *At Home in the Universe: The Search for Laws of Self-Organization and Complexity*. Oxford University Press, New York.

Kitcher, P.: 1998, 'Function and Design', in C. Allen, M. Bekoff and G. Lauder (eds), *Nature's Purposes: Analyses of Function and Design in Biology*, MIT Press: A Bradford Book, Cambridge, MA.

Pennock, R.T.: 1999, *Tower of Babel: The Evidence Against the New Creationism*, MIT Press, Cambridge, MA.

Sober, E.: 1993, *Philosophy of Biology*, Dimensions of Philosophy Series, Westview Press, Boulder, CO.

de Solla Price, D.: 1974, 'Gears from the Greeks: The Antikythera Mechanism – a Calendar Computer from ca. 80 BC', *Transactions of the American Philosophical Society* **64**(7), 1–70.

724

Szathmáry, E.: 1997, 'Origins of Life. The First Two Billion Years', *Nature* **387**(6634), 662–663.

Wächtershäuser, G.: 1997, 'The Origin of Life and Its Methodological Challenge', *Journal of Theoretical Biology* **187**(4), 483–494.

Part VI
Outcomes and Implications

[23]

EVOLUTION AND THE MEANING OF LIFE

by William Grey

Abstract. The last century has witnessed a succession of revolutionary transformations in the discipline of biology. However, the rapid expansion of our understanding of life and its nature has had curiously little impact on the way that questions about life and its significance have been discussed by philosophers. This paper explores the answers that biology provides to central questions about our existence, and it examines why the substitution of causal explanations for teleological ones appears natural and satisfying in the case of physical theory but meets widespread resistance in the case of biology.

Keywords: biology and values; human purpose; meaning of life.

To talk about the "meaning of life" is to introduce a set of issues which are raised, often facetiously, either by philosophers as an exemplary instance of fruitless inquiry or by nonphilosophers as an illustration of the impotence of the discipline to come to grips with the important issues to which they imagine philosophers should address themselves. The fact is, however, that over the last few years an increasing number of philosophers have confronted this rather vaguely delineated set of issues. Two comparatively recent anthologies entitled *The Meaning of Life*, one edited by Steven Sanders and David Cheney (1980), the other by E. D. Klemke (1981), have appeared. Also, Richard Routley (now Sylvan) and Nicholas Griffin (1982) have produced a discussion paper addressing this topic.

My indebtedness to these and other writers is extensive, as will be indicated in what follows. The main difference between my approach to the problems and that of the majority of the other writers to whom I

William Grey has taught philosophy at the Australian National University, Canberra, and Temple University, Philadelphia. His major philosophical interests include environmental philosophy and metaphysics. This paper was presented to the Fullarton Club, Bryn Mawr College, Philadelphia on 16 April 1983. William Grey is at present working for the Department of Industry, Technology, and Commerce in Canberra. His address is 12 Chowne Street, Campbell, A.C.T., Australia 2601.

shall refer lies in the injection of considerations derived from the discipline of biology. It would certainly be an exaggeration to suggest that philosophers are altogether ignorant of biology or, conversely, that biologists are unmindful of some of the far-reaching implications of discoveries in their discipline. Biological thought nevertheless has had curiously little impact on the way that philosophers have discussed, and continue to discuss, some important aspects of life and its significance. I hope this paper will contribute to the task of rectifying this deficiency.

Although there are plenty of disputed issues in evolutionary biology today, the facts to which I shall appeal are fairly uncontentious. We can dismiss the claims of the burgeoning school of creation "scientists," who misleadingly, and to my mind dishonestly, contrive to present the biological controversies about the *structure* of the process of evolution as a controversy about whether the processes of evolution really occur at all. Evolution is being ably defended (see, e.g., Ruse 1982) and requires no additional assistance from me.

Biological thinking has had surprisingly little impact on the conduct of philosophical inquiry over the last century and a quarter. During this time two far-reaching and complementary innovations in biological thought have taken place: Charles Darwin's theory of the evolution of species by natural selection and the elucidation of the structure of DNA by James Watson and Francis Crick. These are not just the most significant discoveries in recent biological history; they are without question the most important discoveries in biology ever made. To show how these revolutionary innovations in biological thought relate to some old philosophical worries is one of the main aims of this paper. However, before taking up the biological story I will set the stage by introducing the philosophical problems which I claim that biology can illuminate.

In the first section of the paper I offer some reflections about the so-called problem of the meaning of life, both as it has traditionally been raised and also from the perspective of biology. Since biology is the discipline devoted to the study of life, we should expect it to illuminate these issues. Nevertheless, I do not believe that biology can silence all the worries which cluster around uses of the phrase *the meaning of life*, and some of the issues which biology fails to address are among the most important. However, I think that rehearsing a few solid biological platitudes can help show us how the search for significance should *not* be conducted, and that is an essential preliminary to any worthwhile inquiry. This is the task addressed in the second section. Finally in the last section of the paper I examine why the conclusions reached in the second section are so often found to be both unsatisfying and unsettling.

The Theistic Legacy

The questions raised by the serious use of the phrase *the meaning of life* are multiply ambiguous and vague. The phrase can be used to refer to the purpose or significance of life of any form, of human life in general, of the life of a group or a society, or of the life of an individual. Furthermore, the question can relate to the reason, purpose, significance, or cause of life in any of these senses. In this paper I shall concentrate on problems which concern the significance that individuals attach to their lives. The major source of confusion will emerge as a failure to clearly distinguish a teleological and a causal sense of the question "Why are we here?" The biological facts, I will argue, provide a solid foundation for a comprehensive causal answer to this question in a way that undercuts the need for a teleological account. There are no grounds for believing that there is purpose or significance in the world in any grand or cosmic sense—indeed there are good reasons for rejecting any such proposal.

A contrary view, expressed in the writings of some religious thinkers, is that human life could be worthwhile only if it were part of some divinely ordained cosmic scheme. On this view, our lives could be significant only if they had a role in such a divine plan. Thinkers who hold this view have often maintained, moreover, that if death is *really* the end of our existence, then life as a whole would be deprived of its significance: "If we are to believe that all our striving is without final consequence, life is meaningless . . . it scarcely matters how we live if all will end in dust and death" (Clark [1958] 1967, 467). Life would be meaningless on the view expressed here by C. H. D. Clark, either if it were not part of a divinely ordained scheme or if it were not eternal.

Clark here mistakenly supposes that if something comes to have no significance later, then it can have no significance now. Yet events in our lives are not deprived of significance purely as a result of their transience.[1] Conversely, Thomas Nagel (1971) has argued that if something lacks significance now, then it is hard to see how some *later* occurrence could invest it with significance. In general (at least in the long term) significance cannot be retrospectively added or removed from what happens in the world.[2] The claim that anything must be eternally significant to be significant at all is mistaken.

Both of Clark's claims, namely that if a life is significant it must be (a) part of some grand cosmic design and (b) eternal, must be rejected. Kurt Baier (1957, 102-10) has suggested, moreover, that a life whose sole point was to serve the purpose of another, even a divine other, would be a degrading life of serfdom.

The view that death deprives life of its significance has been vigorously challenged from a different direction by Bernard Williams (1973,

82), who has argued that it is mortality rather than immortality which would make our lives meaningless. Williams's argument is very different from the one advanced by some existentialists, who have suggested that death gives meaning to life because the *fear* of death gives meaning to life—a view suggested in some places, for examples, by Fyodor Dostoyevski. Williams's argument is based rather on the claim that the nature of human motivation and happiness are such that life without end would be intolerable. *Pace*, Dr. Johnson, eventually one tires even of London.

Against Clark's view that life can have point, and our existence a reason, only if it plays a role in some grand design, one can readily construct ironical arguments analogous to those which David Hume ([1779] 1948) directed against the argument from design, to the effect that the "grand scheme" looks more like the work of an exceedingly indifferent artisan than that of an omniscient and benevolent architect. I shall not pursue this line of polemic. I want to question such views from a different direction.

What seems to me to be especially dubious, and what I want to question, is the idea that the significance of life can be derived only from some externally imposed goal. If life can be said to have any significance, then that significance must be accounted for in some other way. The facts—in particular the biological facts—reveal that any cosmic design or purpose is quite superfluous. Although purposive answers are a common response to "Why?" questions, there is an important range of cases in which they are often invoked but in which they serve no explanatory purpose. In order to appreciate this point it is necessary to say something about the ascription of significance.

A familiar way of ascribing significance to artifacts and natural objects is in terms of the roles which they play, or might play, in our lives. Indeed the significance of a great many items is quite properly provided in functional terms in relation to our projects. Artifacts are usually shaped intelligently, precisely with a view to their role in our lives. This fact has undoubtedly led thinkers to suppose that the *only* way we can ascribe significance to items in the world, including human beings, is in terms of such functional significance.

Indeed the idea that functional aptness is incontrovertible evidence for intelligent design is so natural that it was long regarded as self-evident. To suppose that design or adaptive fitness could be the product not of intelligence but of quite impersonal forces was a disturbing idea implicit in Darwin's account of phylogeny. This possibility however had already been anticipated by Hume ([1779] 1948, Part V), who had pointed out that although it might appear that a ship, for example, was the outcome of ingenious design, it could as well have been produced by a long period of trial and error from generations of "stupid mechanics." Darwin conjectured, in effect, that the "mechanics" of evolution were nothing like divine architects but were entirely natural forces: the exquisite mechanisms of biology could be explained without appeal to an intelligent creator.[3]

The erroneous view that significance can only be ascribed to items in terms of their role in an intelligently designed project is an important source of the misconception that human life can have significance only if it is part of some grand, superhuman purpose or project. An analogous view operates in the other direction of the "Chain of Being" in the claim that animal species and the natural world are good only insofar as they (or it) are good for humans or their purposes (see Passmore 1974, Part I). This pattern of thought may then be uncritically extended to the question: "What are humans good for?" Such extrapolations are the product of weak analogies which are, of course, quite unwarranted. Whatever significance human life may have, it should *not* be modelled on the sort of functional significance associated with artifacts. It is this insight which lies at the heart of Baier's rejection of a theistic account of human value, mentioned above. We should also resist the instrumental model as the sole basis for the value of natural items—but that is another story (see Godfrey-Smith 1979).

The Revolution in Biology

The synthetic theory of evolution, that is, Darwin's theory of evolution combined with modern genetics, can provide a comprehensive answer to the question: "Why are we here?" Although an answer to the causal question does not exhaust the issues associated with the significance of life, it does serve to clarify them. From the point of view of biology, there is no reason to suppose that the question why human beings exist will be any different in kind from the question as to why whales or tigers or starfish exist. Why on earth should we expect a special answer for the existence of just one biological species? The reason for the existence of any biological species is, briefly, the differential survival rate of self-replicating molecules. A comprehensive understanding of the principles involved and the mechanisms of the process (though not of course all the detail) has been provided by modern genetic theory, and in particular by molecular biology.

Modern genetic theory has provided a solid foundation for evolutionary theory in that it has provided a mechanism for evolution. Darwin of course had an account of evolution: it was the result of what he called "natural selection."[4] However, he was compelled to present his argument for evolution by selection in abstract and metaphorical terms because he had no idea how to account for the cause of variations or how it was that favorable variations came to accumulate (Young 1971, 488). He was able to demonstrate conclusively *that* speciation occurred but was unable to explain *how*. In fact Darwin had no mechanism for evolution, only an analogy with artificial selection. In particular, he lacked a particulate theory of heredity, the distinction between somatic and germ cells, and the concept of dominance. Gregor Mendel had provided the basis for a conception of hereditary units as indivisible particles, but no one then realized the significance of his discovery, almost certainly not even Mendel (Dawkins 1976, 36).

In the absence of a particulate genetic theory it is hard to explain how any favorable variations are not diluted out of existence—the problem of "swamping." This was the basic objection wielded half a century earlier with great effect by Archdeacon William Paley to refute the evolutionary theory proposed by Darwin's grandfather Erasmus Darwin (Young 1971, 488).[5]

Lacking a satisfactory mechanism for the process of evolution, Darwin hedged and qualified his theory. "I am convinced," he wrote, "that natural selection has been the main but not the exclusive means of modification" (Young 1971, 493). The longer he lived the less exclusive was the role he assigned to natural selection. He was greatly troubled, for example, by the structure of the vertebrate eye. How could such an exquisitely refined mechanism possibly evolve through piecemeal incremental changes? It seemed to demand a purposive explanation for its development: "To suppose that the eye with all its inimitable contrivances for adjusting the focus to different distances, for admitting different amounts of light, and for the correction of spherical and chromatic aberration, could have been formed by natural selection seems, I confess, absurd in the highest degree" (Gould 1977, 103). This sort of objection to Darwin's theory is addressed in Stephen Jay Gould's paper "The Problem of Perfection." As Gould puts it, "the dung-mimicking insect is well-protected but can there be any adaptive advantage in looking only five per cent like a turd?" (Gould 1977, sec. 12). What, we may ask, is the adaptive advantage of 5 percent of an eye? The twentieth century has provided a vindication of the earlier and less compromising Darwin, and numerous Nobel prizes have been won by scientists who have uncovered the fine structure of mechanisms by which evolution can be explained by natural selection, and by natural selection alone (Young 1971, 497).

The most important post-Darwinian development in the theory was the elucidation of the structural basis of replicative invariance by Watson and Crick. According to Jacques Monod (1972, 103) this is without doubt the most important discovery ever made in the history of biology. Although some might award this laurel to Darwin, it is clear that the major honors must be divided between these two discoveries. Why is the unravelling of the structure of DNA a fact of such profound significance? The answer has considerable philosophical interest (Monod 1972, 98-113).

We can begin by observing two fundamentally opposed stances in the history of Western philosophy reaching back over two and a half millennia to (the usual) Greek origins. (Here I abandon analytic caution and paint with a broad brush.) On the one hand there is the conviction that truth and reality reside in stable and immutable forms—the tradition of Parmenides and Plato. On the other hand there is the conviction that reality is ceaseless flux, change, and decay—the tradition exemplified in the philosophy of Heraclitus. These opposed metaphysical conceptions have continued, after a fashion, to the present.

Many philosophers of distinction, especially those with an interest in the natural sciences, have favored the Platonic tradition. This perhaps is not surprising since science is, after all, an attempt to formulate theories about the world which take the form of immutable truths. The basic aim of science is to analyze phenomena by penetrating the mutable appearances and revealing the underlying invariants—that is, characteristics which do *not* change. The laws of physics, for example, specify invariant relations, and in general the fundamental statements of a science are expressed as conservation principles. In fact the analysis of *any* phenomenon is possible only if we analyze it in terms of some invariant which is conserved throughout the change. The formulation of the laws of kinetics by Sir Isaac Newton was a discovery which demanded the invention of differential equations, that is, a method of defining change in terms of something which remained unchanged. Without invariants, whether the subject be physics or economics of demography, it is impossible to formulate precise testable laws. Certainly descriptive inquiry is possible in the absence of law-like statements, but at this stage of their development an inquiry is inchoate and phenomenological; it is perhaps an essential preliminary, but only a preliminary, to the establishment of mature science, which is quantitative and not just descriptive.

It is of course much debated whether there *are* any absolute invariants in reality. One school of thought holds that invariants are not features of reality we discover but are fictions we invent and employ in our models; and while they may be indispensable tools for thinking about the world, there is no reason to suppose that they actually reflect the structure of reality. This is a crude characterization of an anti-realist position which could take the form of conceptualism, instrumentalism, or pragmatism. Whether the invariants are to be located in our heads (conceptualism), in our theories (instrumentalism), in our practices (pragmatism), or in the world (realism), any decent explanatory theory has to locate them *somewhere*. The debate between realists and anti-realists is one which we do not however need to take up here. The important point to note is that significant progress in theoretical understanding typically takes place as a result of the unification of a field of inquiry brought about by the discovery of some new invariant principle.

Biology is no exception. The global significance of Darwin's theory lay in its power to unify. Instead of a vast variety of immutable species whose ultimate origin and purpose God alone knew, it became possible to conceive that all species had developed by a slow process of incremental adaptation from a small number, perhaps a single variety, of organism. Like other great unifiers, Darwin of course stood on the shoulders of giants, among whom must be included Carolus Linnaeus, and the much-maligned Jean-Baptiste Lamarck. Linnaeus must be credited with much of the groundwork of tracing the continuities and patterns from the seemingly chaotic gestalt of the biological community. The patterns which emerged from the taxonomies of Linnaeus, however, were purely formal. It was Lamarck who proposed that the relations between species were causal, replacing the "Great Chain" by what has been called the "Escalator of Being." The revolutionary importance of Lamarck's suggestion has been unfortunately overshadowed by his discredited proposal that speciation occurred because individuals inherited favorable characteristics from their parents.

We can broadly characterize the development of biology over the last two hundred years as follows. The work of Linnaeus (in particular) in the late eighteenth century established taxonomies which suggested that there were underlying continuities or patterns which could be traced between different species. In the nineteenth century Darwin (and Lamarck) suggested that these patterns were not merely formal similarities but represented a systematic causal relatedness between species, with Darwin's account prevailing over Lamarck's. The discovery of *formal* patterns of continuity between species was not too upsetting to the prevailing religious orthodoxy: all this discovery showed, it was thought, was that God used a basic set of blueprints (or archetypes) when creating species. Darwin's suggestion however could not be accepted with equanimity. Finally, in the twentieth century Watson, Crick, and others elucidated the causal mechanism by which the process of evolution takes place.

Darwin's unified conception of the biological world naturally invites comparison with two other great triumphs of synthetic thought.[6] These are Euclid's demonstration that a huge body of incorrigible geometrical truths could be systematically generated from a small number of axioms and postulates, and Newton's synthesis which showed how the motions of material objects could be explained in terms of a few underlying physical principles. Euclid, Newton, and Darwin each showed how a huge body of not obviously related facts could be understood and explained in terms of a single unifying framework of principles. Each of these achievements has had a profound and exciting impact on subsequent generations of thinkers, who have attempted to emulate these exemplary models of systematic thought. To be called a "Darwin" or a "Newton" is the highest compliment that can be paid to a theoretical innovator. Thus, Karl Marx has been called the Darwin of the social sciences, Adam Smith the Newton of civil society.

Darwin's synthesis of the biological world has been progressively strengthened by advances in biochemistry over the past fifty years, which have revealed "the profound and strict unity, on the microscopic level, of the whole living world" (Monod 1972, 101). We now know that all organisms, from microscopic bacteria to blue whales, rely on chemical machinery which is the same in its structure and its function. This is a truly awesome discovery.

The chemistry of life as we know it is universal in its structure because all living beings without exception are made up from two principle classes of macromolecular components: proteins and nucleic acids. These, moreover, are made up from the same basic structural units: twenty amino acids for the proteins and four kinds of nucleotides for the nucleic acids. It is the same in function because the same sequences of reactions are used by all organisms for their essential chemical operations: the mobilization and storage of energy, and the biosynthesis of components. The unification of the biological world, which received its first solid theoretical foundation from Darwin, has been profoundly deepened and entrenched by the discovery by Watson and Crick of the fundamental biological invariant, DNA.

DNA is the uniquitous initiator of all biological replication, providing the coded instructions for generating the elaborate protein structures—the somatic tissues—which constitute the "life support systems" for the nucleic acids. This has led some sociobiologists to envisage individual organisms as mere instruments employed by genes as vehicles for their propagation. An enthusiastic exposition of this frequently overstated conception is presented in Richard Dawkins (1976).[7]

Changes in the DNA code are changes in the sequence of nucleotides. These changes—mutations—are due to various copying errors and scramblings. However, once they have occurred, they will be copied faithfully in subsequent generations, thanks to the organization effected through the laws of chemistry.

Changes in the genetic text are random occurrences in the sense that they occur with no preferred adaptive direction. In particular, there are no grounds for supposing that changes are the result either of divine intervention or of some striving toward perfection. The situation has recently changed with the advent of genetic engineering. It may be that not all *future* changes in the genetic text will be the exclusive product of chance; we have now acquired the capability to edit the genetic text deliberately. This is a possibility which some thinkers find very disturbing.[8] Putting that special case to one side, pure chance is otherwise the source of all variation. All evolutionary changes are accidental. It is through the imperfections of the copying mechanism that modifications have occurred, when the information in the code has become contaminated with fortuitous "noise" which, thanks to the structure of DNA, has been faithfully replicated along with what Monod (1972, 114) has called "all the music of the biosphere." The selection for change takes place at the phenotypic level and admits only acceptable mutations, which are those that do not lessen the coherence of an organism's somatic character. Continuing the musical metaphor, the only "noise" admitted is that which is in harmony with the music of the biosphere.

The synthetic theory has been tested by many disparate lines of criticism. It has been claimed that the theory is unfalsifiable, and thus really metaphysical rather than an empirical theory. This criticism is indeed sometimes justified, for the theory has frequently been carelessly presented in a tautological form, although careful expositions do not make this mistake (see Ruse 1973).

There are also a battery of empirical objections that have been directed at the theory. It has been claimed, for example, that speciation demands the heritability of acquired characteristics (Lamarckism); that the account of evolutionary change as gradual, serial, irreversible, and governed by selection of genes through the interaction of an organism with its environment is not sufficient; and that the account must be supplemented with instant "macromutations."

These and the many more objections which have been (and continue to be) raised have not proved fatal to the Darwinian research program. They have on the contrary led to its strengthening and improvement through modifications to and elaborations of the theory. In Kuhn's ([1962] 1969) terminology, they constitute anomalies or puzzles, the solution of which has led to the elaboration and articulation of the Darwinian paradigm. The articulation of the theory and the solving of its puzzles are by no means exhausted. The molecular reduction of genetics has transformed the synthetic theory into a truly formidable structure; it is not credible to envisage any large scale changes to the unifying framework which the theory provides for biology.

The theory of evolution and its molecular mechanisms provides a conclusive, comprehensive, and satisfying causal answer to the question: "Why are we here?"—which I said at the outset is one of the central questions which people have addressed in reflecting upon the meaning of life. The problem, from a biological perspective, undergoes a Wittgensteinian dissolution ("The riddle does not exist"). We are here as a result of chemical principles of organization, which provide the explanation for the existence of organisms. There is no reason, purpose, point, end, or externally employed goal at all which gives, or is needed to give, a reason for our existence. The search for an answer in terms of a role in some grand cosmic scheme is gratuitous and vain.[9]

The plain fact is that, thanks to the accumulation of fortuitous errors in the copying of self-replicating molecules, one species of organism—our species—has managed to acquire the intellectual capacity (through the development of a highly complex central nervous system) that has enabled it to ask—and to *answer*—the question, "Why are we here?" However, our existence is utterly contingent, an accident, or a long series of accidents of evolution. Appeals to teleological principles or cosmic purposes are gratuitous. We are here for the same reason that tigers and whales and starfish are here—because of our adaptive fitness to our biological circumstances. More precisely, it is because of our ancestors' adaptive fitness to *their* circumstances. It is not unreasonable to conjecture that our rapid and violent transformation of our surroundings may in fact be generating an environment for which human beings are not adaptively fit. The fact of species extinction demonstrates that adaptive fitness is not an eternal property.

We can still raise the question of how one should live in order that one's life should be meaningful. I said at the outset that this is an important question that the biological considerations raised thus far leave almost untouched. Nevertheless, I think that here too biology can be of assistance by clarifying the sort of life to which human beings are adapted (see Midgley 1978, 358). Biology provides constraints which exclude various modes of life as unfitting, or simply "inhuman." In particular our species has evolved (most recently) from social primates, and this contingent evolutionary history has provided us with a legacy of a particular motivational structure and an associated set of values.[10] We are not infinitely malleable; we must choose our lives from within a received set of biological constraints. It is however up to *us* to provide purpose or point to our lives from within these constraints; we cannot expect it to be dictated to us externally, either causally from biology or theistically from the dictates of a deity.

That purpose is dependent on *our* decisions is of course a thought that can be extracted elsewhere, in particular from various existentialist thinkers, although I will argue below that they exaggerate the amount of human autonomy in determining how a meaningful life is to be realized. Even if one could, with qualifications, agree with Jean-Paul Sartre that "All existing beings are born for no reason, continue through weakness, and die by accident," it does not follow, as Sartre thinks, that "It is meaningless that we are born; it is meaningless that we die" (Sartre 1956, 547). David Wiggins (1976) has also suggested, from rather different premises, that purpose must to a great extent be dependent on human invention.

The Uneasy Consequence

To have reached this point in our discussion, however, is unsatisfying. We can accept all of the biological story and still suspect that a problem of real significance has somehow slipped through our fingers. I want to pursue the problems further to try to determine whether this feeling that we have somehow missed a crucial point is well grounded. It is helpful here to compare the maturation of biology with that of some other sciences.

Typically, a science takes a significant step towards maturity when it dispenses with teleological explanations of natural phenomena. Physics came of age when Galileo and Newton showed that the motions of physical objects did not require an Aristotelian explanation in terms of goal-directedness. Rocks do not fall in order to achieve some sought after goal state: we can replace teleological statements which have the form "X does Y in order to . . ." with causal statements of the form "X does Y as a result of . . ." (e.g., rocks fall as a result of the law of gravity). In the case of physics, despite inevitable resistance, the elimination of the teleological story is profoundly satisfying. Likewise, primitive explanations of recurrent patterns of experience in other domains have given way to objective causal accounts; only in our more superstitious moments do we invest inanimate objects with an intention to thwart us.

The parallel development in biology, however, has proved to be profoundly disturbing. Darwin, in a sense which I shall qualify shortly, banished teleology from biology. However, his causal account of the biological world has quite evidently not produced the same universally agreeable and intellectually satisfying unification that the Newtonian research program provided. Why is banishing teleology from physics perfectly acceptable whereas banishing it from biology is so disturbing?

I suspect that one worry is that a world which is the product of random or accidental events seems to be a world devoid of objective purposes. Without objective purposes we are deprived of the most obvious ground for objective values, that is, the removal of cosmic purposes from the world removes (at least one attractive basis for) objective values.

Existentialists respond ambivalently to this evaporation of objective values. In some moods they seem to be prepared to accept the consequent absurdity of human existence but still advocate a continuation of our pursuit of subjective goals and desires in a spirit of heroic defiance, shaking our fists at the uncaring world, as it were. This romantic bravado is expressed by Albert Camus in various places (see Nagel 1971). In other moods, existentialists rise above despair and suggest that the values and significance which we create by our autonomous decisions are perfectly adequate to provide us with rewarding and meaningful lives (see Barnes 1967).

It is questionable in any case that a divinely organized cosmic plan could be able to provide our lives with significance independently of our aims and desires. Suppose that God wanted me to participate in His divine plan, but I lacked any inclination or desire to do so. God's purposes and desires, of themselves, would then surely be of no help in providing significance or purpose in my life: at best we would have an instantiation of Baier's conception of an abject life of serfdom. So for the world to be meaningful to me it will still require my own desires and aims (see Joske 1974, 95), and once we have these individual subjective desires and aims to appeal to, we might start wondering why the transcendent authority of an omniscient being is still necessary.

It is misleading, in any case, to suppose that Darwin abolished teleology from biology entirely: it would be more accurate to say that he relocated it. The course of evolution *itself* certainly has no object or end; but biological organisms most certainly do strive for goals and manifest preferences in their lives—even though only a privileged sub-class can express or articulate their preferences. Jacques Monod acknowledges this fact, although he attempts to dissociate himself from what he regards as the disreputable associations of "teleology" and speaks instead of the *teleonomic* nature of organisms. The final result of the biological story is to show how life and purpose can arise in a lifeless and purposeless world. The fact that life and purpose have emerged from, and are based upon, the exquisite and impersonal processes of chemistry in no way compromises the existence of purpose, and hence of value and significance. Existentialists, while rightly rejecting transcendent cosmic purposes, are wrong in supposing that the only source of significance and value is located in individual (human) choice and commitment. Their claim that human beings have no nature or are free to create their own nature through their autonomous decisions, is a significant mistake.

I suggested above that biology can help throw some light on the question of what sort of life we ought to pursue. Any acceptable answer will have to take into consideration important constraints that derive from the nature of human nature, and biological considerations can be of assistance here (as I suggested above) by helping to elucidate the sort of life to which the species homo sapiens is best adapted. This involves an Aristotelian supposition that organisms, including ourselves, have natural ends: teleology is in fact built into our nature. This innately programmed purposiveness, which is of the greatest importance for the project of establishing significance in our lives, is itself the product of the aimless processes (and "mistakes") of evolution. An Aristotelian account can not only provide the basis for a naturalized ethic for mankind; it can also provide the basis of an ethic for nature.[11]

This does not involve, as existentialist writers like Hazel Barnes (1967) suppose, an intolerable imposition of some straight jacket of conformity on human values and behavior: a unitary conception of human nature is quite compatible with an indefinitely rich variety of ways in which this nature can be expressed. The existentialist worry is as absurd as supposing that a piano's permitting the production of only a finite number of chords is a lamentable restriction of the expressive powers of the instrument. Quite the contrary! Without some conception of human nature we could not make sense of a wretched, impoverished, degraded—literally "inhuman"—existence. Of course, pontificating about what is and is not "natural" has clear and well-known dangers, but they are not as great as the dangers of denying that there are limits in the treatment of our fellows (human and animal) which it is intolerable to transgress.[12]

Yet I suspect that this biological story will not have silenced everyone's worries. It may still be felt that if there is no transcendent or cosmic purpose, human life and choice must appear trivial and inconsequential. This tiny speck of matter which we occupy for a fleeting moment of time is, from a grand cosmic point of view, so inconspicuous and insignificant. It is this process of stepping back and locating the here and now in the vastness of time and space which appears to reduce all to something quite trivial.

One surely legitimate response to this apparent dwarfing of human concerns from a grand cosmic perspective was advanced by Frank Ramsey in reply to Bertrand Russell. Ramsey (1931, 291) declared himself quite unimpressed by the stars, however overwhelming their size, as they were incapable of thinking or feeling love. (Ramsey, being distinctly more generously proportioned than Russell, was perhaps less inclined to be overawed by sheer physical bulk!)

The temporal dwarfing of value, to which I have already alluded, is a more commonly articulated worry. If time annihilates all that we do, what then is the point?[13] This worry is based on a confusion. As Thomas Nagel (1971) has pointed out, if nothing matters in a million years, then by the same token nothing that will matter in a million years matters now. In particular, the fact that in a million years nothing will matter, does not matter now. That is, the (alleged) future insignificance of the present entails the present insignificance of the future, and hence the present insignificance of the future insignificance of the present. Likewise, if nothing matters from a cosmic point of view, the fact that nothing matters from a cosmic point of view does not, from that point of view, matter. We cannot validly infer our cosmic insignificance but only our cosmic nonsignificance; that is, we can infer only the irrelevance of such a perspective for considerations of significance, and this does nothing to undermine the fact of significance from our more parochial temporal and spatial perspectives.[14]

NOTES

1. Plato seems to have held the same belief about "the good," to which Aristotle famously replied: "It [the good itself] will not be good any the more for being eternal; after all, that which lasts long is no whiter than that which perishes in a day" (*Nichomachean Ethics*, 1096b3). Likewise, significant events in our lives are not rendered the less so as a result of their transience.

2. I say "in general" because it seems plausible that in the short term the significance of events often *is* tied up with the causal outcomes which the event may set in motion. We might reasonably suppose, for example, that the assassination of Archduke Ferdinand may be an event of greater significance than the assassination of Lord Mountbatten, and it may be that this is a judgment which can only be made retrospectively. I will say some more about the way that temporal considerations can affect questions of significance in the final section of the paper.

3. One might still invoke an intelligent creator to explain the origins of the physical universe, and perhaps the laws which govern its development. There may, that is, still be room for a theistic response to ultimate questions about origins. Such a response, arguably, does not really resolve these ultimate questions but serves merely to relocate them. However the naturalistic story of the development of life, for which there is overwhelming evidence, certainly removes God's hand from the tiller of biological evolution.

4. The phrase *survival of the fittest* is not Darwin's; it was the coinage of Herbert Spencer. Also, Darwin seldom used the word *evolution*, preferring to speak of "descent with modification." This was partly because *evolution* was associated with Albrecht von Haller's untenable "homuncular" theory of embryology and partly because Darwin did not want to suggest that the modification of species implied any sense of progress. There is no sense of improvement apart from that of better adaptive fitness (see Gould 1977, sec. 3).

5. Paley appears to have provided a *locus classicus* for several notable lost causes: another major claim to fame of his derives from his statement of what is probably the most widely quoted version of the argument from design—some twenty-three years *after* Hume's ([1779] 1948) devastating refutation!

6. It appears likely that within the foreseeable future physicists will achieve another spectacular synthesis: the incorporation of the fundamental forces of nature into a single unified field theory (see Davies 1984). It can be argued that the theoretical unifications provided by physics and biology provide a basis for empathy with the nonhuman world (see Grey 1986).

7. The account is reminiscent of Arthur Schopenhauer's conception of a *species* imposing its will upon individuals, which is used as an instrument for the species' propagation (Schopenhauer [1818] 1958, vol. 2, chap. 44). For an attempt to temper the overstated claims of Richard Dawkins, Richard Alexander, et al., see Stephen Jay Gould

(1977, 267), Mary Midgley (1979), and Michael Ruse (1979). Notable attempts to articulate the sociobiological paradigm in the area of human affairs are Edward Wilson (1978) and Alexander (1979). Alexander ambitiously interprets human history and culture as the result of individuals attempting to maximally propagate their genes, either directly through parenthood or indirectly by supporting propagation of and by close relatives. Much of his energy is (not surprisingly) devoted to attempting to square his "inclusive fitness" model with a plethora of counterexamples. It is certainly quite credible that a sociobiological account can be provided for general behavioral tendencies, such as aggression, altruism, and sexual behavior. It is much less plausible to suppose that the more detailed manifestations of behavior, such as the creation of specific sex roles or the expression of emotions like depression and anger, can be explained in terms of an underlying genetic fine structure. Developing and articulating the sociobiological paradigm—and in the process tempering its more extravagant claims—is one of the most vital tasks on the agenda of social scientists.

8. A somewhat alarmist expression of these anxieties can be found in Jeremy Rifkin (1983).

9. To appropriate another gnomic utterance from Ludwig Wittgenstein ([1921] 1961, 6.521): "The solution to the problem of life is seen in the vanishing of the problem."

10. These values include (very schematically and neither exhaustively nor in order of importance) enthusiasm, curiosity and wonder, exaltation from problem solving and discovery, group identification, ethnic and national pride, satisfaction from triumph in competition (see Wilson 1978, chap. 9). Advocating even this modest claim invites the charge of biological determinism. This I reject. Many of the worst forms of social degradation are precisely impositions which offend against the sorts of life to which we are adapted. Biology does not dictate a unique form of life to which we should all conform: constraints are not straightjackets. Part of our nature is precisely to choose how to live, but our (biological) nature is a factor which conditions the satisfaction which we gain from our choices. (See Lorenz [1963] 1966, chap. 12.)

11. For Aristotle, it is a factual matter what conditions are required for an organism to flourish and thrive. If this manner of deriving an "ought" from an "is" involves the claim that it is a good thing to flourish and thrive, and right to act so as to assist others (or at least not to prevent them) in their flourishing and thriving, it is one that I at least am prepared to accept. The rather vacuous naturalistic maxim which suggests itself is that one ought to pursue the kind of life to which one is suited. However dubious Aristotelian physics may appear now, his moral philosophy (and his biologically based metaphysics) contains much of contemporary value, as Stuart Hampshire (1977) and Alisdair MacIntyre (1981) have argued.

12. A good deal of opposition to the sociobiological claim that we have a biologically based motivational structure is based on an analogous error (although some of the more extravagant pretensions of sociobiology are ill-founded: see n. 7). For an evolutionary account of the development of our motivational structure see George Pugh (1979).

13. Benedict Spinoza advocated viewing the world *sub specie aeternitatis* as a means of consolation, suggesting that the phenomenon of dwarfing can be exploited to positive advantage. On a suitably generous perspective, present pain, distress, and injustice can seem relatively insignificant. That is, when conceived as part of an indefinitely vast cosmic order, we transcend our relatively parochial human preoccupations. For Spinoza of course the vast cosmic order was itself something at which to marvel, indeed *the* thing at which to marvel; so turning from the temporal world did not in any way undermine questions of significance but rather provided a means by which they could be properly located.

14. Any alternative perspective which appears to show the unimportance of what we recognize to be important thereby rules itself out of order just because it fails to take account of what we know to be the case.

REFERENCES

Alexander, Richard D. 1979. *Darwinism and Human Affairs*. Seattle: Univ. of Washington Press.

Aristotle. 1941. *Nichomachean Ethics*. Trans. W. D. Ross. In *The Basic Works of Aristotle*, ed. R. McKeon, 927-1112. New York: Random House.

Baier, Kurt. [1957] 1981. "The Meaning of Life." In *The Meaning of Life*, ed. E. D. Klemke, 81-117. New York: Oxford Univ. Press.

Barnes, Hazel. [1967] 1980. "The Far Side of Despair." In *The Meaning of Life: Questions, Answers and Analysis*, ed. Steven Sanders and David R. Cheney, 105-11. Englewood Cliffs, N.J.: Prentice-Hall.

Clark, C. H. D. [1958] 1967. *Christianity and Bertrand Russell*. Quoted in Paul Edwards, "Meaning of Life." In *The Encyclopedia of Philosophy*, ed. P. Edwards, 4:467-77. New York: Macmillan.

Davies, Paul. 1984. *Superforce: The Search for a Grand Unified Theory of Nature*. London: Heinemann.

Dawkins, Richard. 1976. *The Selfish Gene*. Oxford: Oxford Univ. Press.

Godfrey-Smith, William. 1979. "The Value of Wilderness." *Environmental Ethics* 1:309-19.

Gould, Stephen J. 1977. *Ever Since Darwin*. New York: Norton.

Grey, William. 1986. "A Critique of Deep Ecology." *Journal of Applied Philosophy* 3:211-16.

Hampshire, Stuart. 1977. *Two Theories of Morality*. Oxford: Oxford Univ. Press.

Hume, David. [1779] 1948. *Dialogues Concerning Natural Religion*. Ed. Henry D. Aiken. London: Macmillan.

Joske, W. D. 1974. "Philosophy and the Meaning of Life." *Australasian Journal of Philosophy* 52:93-104.

Klemke, E. D., ed. 1981. *The Meaning of Life*. New York: Oxford Univ. Press.

Kuhn, T. S. [1962] 1969. *The Structure of Scientific Revolutions*. 2d ed. with Postscript. Chicago: Univ. of Chicago Press.

Lorenz, Konrad. [1963] 1966. *On Aggression*. Trans. Marjorie Latzke. London: Methuen.

MacIntyre, Alasdair. 1981. *After Virtue*. Notre Dame, Ind.: Notre Dame Univ. Press.

Midgley, Mary. 1978. *Beast and Man*. Ithaca, N.Y.: Cornell Univ. Press.

———. 1979. "Gene-juggling." *Philosophy* 54:439-58.

Monod, Jacques. 1972. *Chance and Necessity*. London: Collins.

Nagel, Thomas. [1971] 1981. "The Absurd." In *The Meaning of Life*, ed. E. D. Klemke, 151-61. New York: Oxford Univ. Press.

Passmore, John. 1974. *Man's Responsibility for Nature*. London: Duckworth.

Pugh, George E. 1979. "Values and the Theory of Motivation." *Zygon: Journal of Religion and Science* 14:53-82.

Ramsey, Frank. 1931. *The Foundations of Mathematics*. Ed. R. B. Braithwaite. London: Routledge & Kegan Paul.

Rifkin, Jeremy. 1983. *Algeny*. New York: Viking Press.

Routley, Richard and Nicholas Griffin. 1982. *Unravelling the Meanings of Life*. Canberra: Australian National Univ. Research School of Social Sciences.

Ruse, Michael. 1973. *The Philosophy of Biology*. London: Hutchinson.

———. 1979. *Sociobiology: Sense or Nonsense?* Dordrecht: Reidel.

———. 1982. *Darwinism Defended*. Reading, Mass.: Addison Wesley.

Sartre, Jean-Paul. [1943] 1956. *Being and Nothingness*. Trans. H. E. Barnes. New York: Philosophical Library.

Schopenhauer, Arthur. [1818] 1958. *The World as Will and Representation*. Trans. E. F. J. Payne. 2 vols. Indian Hills, Colo.: Falcon's Wing Press.

Smith, John Maynard. [1958] 1975. *The Theory of Evolution*. 3d ed. Harmondsworth: Penguin.

Wiggins, David. 1976. "Truth, Invention and the Meaning of Life." *Proceedings of the British Academy* 62:331-78.

Williams, Bernard. 1973. "The Makropulos Case." In *Problems of the Self*, 81-100. Cambridge: Cambridge Univ. Press.

Wilson, Edward O. 1978. *On Human Nature*. Cambridge, Mass.: Harvard Univ. Press.

Wittgenstein, Ludwig. [1921] 1961. *Tractatus Logico-Philosophicus*. Trans. D. F. Pears and B. F. McGuiness. London: Routledge & Kegan Paul.
Young, Robert M. 1971. "Darwin's Metaphor: Does Nature Select?" *The Monist* 55:442-503.

[24]

THE POSSIBILITY OF MEANING IN HUMAN EVOLUTION

by Barbara Forrest

Abstract. Science undermines the certitude of non-naturalistic answers to the question of whether human life has meaning. I explore whether evolution can provide a naturalistic basis for existential meaning. Using the work of philosopher Daniel Dennett and scientist Ursula Goodenough, I argue that evolution is the locus of the *possibility* of meaning because it has produced intentionality, the matrix of consciousness. I conclude that the question of the meaning of human life is an existentialist one: existential meaning is a product of the individual and collective tasks human beings undertake.

Keywords: biology; consciousness; emergence; emergent functions; evolution; existence; existentialism; intentionality; language; life; life forms; meaning; naturalism; organisms; philosophy; purpose; reductionism; religion; science; self-consciousness; significance; species; symbol; value; worldview.

The human species will one day be extinct. The impact of this realization upon the human psyche is jarring, yet science provides evidence for it: 99 percent of all species that have ever lived on earth are now extinct (Wilson 1992, 344). And if the past rate of extinction does not constitute a guarantee of our future, the death of the sun surely does: several billion years from now, our sun will go the inevitable way of all stars, depriving Earth of the only source of energy that makes human life possible (Friedman 1986, 229–35).

Barbara Forrest is Associate Professor of Philosophy in the Department of History and Political Science at Southeastern Louisiana University. Her address is SLU 10484, University Station, Hammond, LA 70402, and her e-mail address is bForrest@selu.edu. A shorter version of this paper was presented at the Conference on Science and Society hosted by the Russian Academy of Sciences and the Institute for History of Natural Sciences and Technology, St. Petersburg State University, St. Petersburg, Russia, 21–25 June 1999. It was also presented at a meeting of the South Place Ethical Society in London, England, 27 June 1999.

We have established scientifically some disquieting facts: (1) human beings have evolved from nonhuman life forms, meaning that (2) at one time we did not exist, and that (3) according to paleontological and astronomical evidence, at some time in the future we shall cease to exist. Furthermore, from a scientific standpoint, there is no discernible reason that we *had* to evolve in the first place, and there is no guarantee that we shall continue to evolve successfully; more hominid species have become extinct than have survived. The price of such knowledge has been the gnawing question of whether human existence has genuine meaning if it was constructed with cranes rather than supported by skyhooks, as Daniel Dennett says.[1]

The problem of meaning is easily resolved for those who embrace a preconstructed system of meaning such as religion.[2] However, religion cannot help us find meaning in any honest sense unless it can assimilate the truth about where human beings have come from, and the only real knowledge we have about where we came from we have acquired through science. Yet the journey from ignorance to knowledge about our origins has deposited us at a point that Philip Kitcher calls "painful enlightenment," a sometimes-experienced result of scientific inquiry in which "people acquire beliefs that have an impact on their values" and experience a loss of "psychological comfort" (Kitcher 1998, 52–53):

> The normal course of scientific inquiry may make our community better off in either (or both) of two ways. First, one of the items valued may be knowledge of some aspect of nature, and a new discovery may deliver that knowledge. Second, inquiry can expand the available strategies, making it possible for the community to pursue goals that previously seemed beyond reach or to proceed with greater efficiency and thus attain far more than it would otherwise have done. . . .
>
> Painful enlightenment is different. Even though the goal of knowing some aspect of nature may be achieved, the principal consequence of the advance in knowledge is constriction of the set of available strategies and/or destabilization of the scheme of values. The set of available strategies associated with a valued item may become empty, or the community may come to believe that there are no strategies for attaining that item, or the standard justifications for valuing the item may become untenable, or the community may come to believe that there is no possible way of justifying the value of that item. If the item has a high index of value, representing its importance and centrality to the lives of the community, then the dislocation will be severe. (Kitcher 1998, 53–54)

Human existence as inherently meaningful—one of the cornerstones of the religious worldview—is such an item.

If the human species is headed for extinction, which the evolutionary record and solar astronomy tell us it is, then the problem of meaning for many people becomes acute. Is the choice either to reject science, with its unhappy discoveries and implications, or to ignore it in favor of a more comfortable but less supportable worldview? If we accept what science shows us about ourselves, does human existence mean anything?

In this paper I explore the question of whether evolution can provide a basis of meaning for human existence. The word *meaning* here will conjointly denote purpose, value, and significance. I understand the *purpose* of human existence to refer to a plan or agenda advanced either during an individual life or through collective human existence. The *value* of human existence refers to whether it is of any merit or account, whether it is *important*. The notion of *significance* is roughly the same as that of value, but it may include the idea that human existence is indicative of something beyond itself. Meaning in the higher sense I specify includes these connotations and will be referred to as *existential meaning*. So the question of meaning is the question of whether human existence is endowed with purpose, whether it is important to something/someone, and whether it is meritorious in any sense. A related question is whether meaning is inherent in human existence or is an artifact or construct.

The implication of evolution is that the problem of existential meaning has not always existed, because it is a problem only for human beings, and they have not always existed, nor have they always had the ability to pose questions about meaning. The problem of meaning exists only because there are human beings, and there are human beings only because of evolution, which was taking place long before we appeared on its timeline and would have continued even if we had not showed up at all.

I argue that, although the phenomenon of human evolution itself endows human existence with no existential meaning, it *is* the origin of the *possibility* of *creating* such meaning, because our ability to pose the question of meaning is rooted in our existence as intentional beings, and intentionality is a product of evolution. Human intentionality is of a sufficiently high grade to enable us to make our existence one of the focal points of this intentionality. The human capacity to seek meaning in the existential sense is rooted in our capacity for meaning in the intentional sense: our ability to direct our thinking, first toward something external to ourselves and then to ourselves, and to be aware or conscious that we do this. Meaning in both senses is rooted in our ability to make connections—between our thoughts or words and what they represent and between ourselves and what we see as vital to human interests.

I also argue that meaning in the higher sense is an *existential* artifact, constructed out of capabilities we possess by virtue of the particular evolutionary path we have traveled. What science shows us about ourselves has seriously undermined—or at least forced changes in—the belief that human existence is either naturally or divinely endowed with predefined meaning. Humans must *achieve* meaning in the sense enunciated by various existentialist thinkers, and evolution has endowed us with the capacities essential for this achievement.

I base my position on the work of Dennett, a philosopher who constructs his views in large measure on the basis of evolutionary biology and cognitive science, and Ursula Goodenough, a cell biologist who has ventured onto the humanistic side of the academic field in an attempt to ground meaning in the continuity not only between human beings and other animals but between human life at the cellular level and at the conscious level.

LOOKING FOR MEANING

Humans were intentional beings before they became conscious ones, so we can refine our earlier understanding of meaning by viewing it as a continuum, with simple intentionality on the *lower*, or evolutionarily earlier, end, semantic or symbolic (representative) meaning on the ascent toward the *higher*, or evolutionarily more recent, end, and existential meaning on the *highest*, or most recent, end. Meaning in the highest sense is derived from whatever enables an individual to live with a sense of worth and importance—a belief system, a purpose, a mission. It may be understood by those who seek it as originating in an ultimate, transcendent dimension, or it may not be so understood, depending on the social and psychological needs and epistemic commitments of the meaning seeker.

Clearly our lives revolve around meaning. On the lower end of the meaning continuum, things shape our behavior by being objects we seek or avoid, acquiring value for us insofar as they serve our purposes. A berry or a fish meets a metabolic need and acquires survival value. A flat surface "means" we can use it as a table where we can cut the fish, and thus it acquires utilitarian value. When human beings began to wonder whether we ourselves serve a purpose—the ability to wonder about this having been made possible by the ability to examine our lives representatively through language—the search for meaning in the higher sense began.

The concept of meaning in the higher sense, understood as a product of evolution, is not a reassuring notion. Edward O. Wilson's explanation deromanticizes meaning even further by defining it in the lower sense, in terms of its neurobiological matrix: "What we call *meaning* is the linkage among the neural networks created by the spreading excitation that enlarges imagery and engages emotion" (Wilson 1998, 115). He is referring to semantic meaning as it is employed by semantic memory, the ability to connect "objects and ideas to other objects and ideas" (1998, 134). Wilson thus locates the origin of meaning in the ability not only to remember but to imagine, to symbolize, and to feel emotively, all of which in their most basic forms are neural activities. This is related to what I have specified as meaning on the lower end of the continuum.

The import of such neural activity for meaning in the higher sense is that if our brains had no capacity to produce imagery, we could not envision or create possibilities for ourselves—eliminating the ability to consider the meaning of our existence; if we could not symbolize possibilities, we could not communicate our visions to ourselves or to others; and finally, if imagining and symbolizing were not linked to emotive capability, then we would not *care* if life had meaning in any sense at all. Furthermore, it is important that these present capabilities are built on the (genetically and culturally) conserved capabilities of our evolutionary ancestors, whose successful adaptation made human existence possible.[3]

MEANING'S LOCUS IN INTENTIONALITY

Human beings are not so much meaning seekers as meaning makers, and human existence acquired the possibility of meaning when human persons became able to reflect on themselves and their situations. Our ability to make meanings is rooted in our nature as intentional systems, a nature we share with all living things, and our intentionality is a product of the evolutionary process.

Not all intentional systems are organisms—computers are also intentional systems—but all organisms are intentional systems. Because I am human, the fact that my thinking/mental state is *about* something, is focused or directed, marks me as an intentional being. The fact that I am *aware* of what it is about marks my uniquely human intentionality. The fact that my thinking is *about* whether my existence itself has meaning marks my relatively high level of human intentionality. Although we share intentionality with even the lowliest life forms, we are unique in that we alone are aware of our intentionality, and we can express both our intentionality and our awareness of it in words and other symbols.

An intentional system is one whose internal disposition or state or functioning is directed toward or linked to, aims at, or functions in conjunction with something external to itself. As Dennett puts it, an intentional state is "about" something. An intentional state is directed toward the consumption of food, for example; thus, the state has content. According to Dennett, "*Intentional systems* are, by definition, all and only those entities whose behavior is predictable/explicable from the intentional stance. Self-replicating macromolecules, thermostats, amoebas, plants, rats, bats, people, and chess-playing computers are all intentional systems" (Dennett 1996, 34). The "intentional stance," Dennett says, is "the strategy of interpreting the behavior of an entity (person, animal, artifact, whatever) by treating it *as if* it were a rational agent who governed its 'choice' of 'action' by a 'consideration' of its 'beliefs' and 'desires'" (1996, 27). In short, if the intentional stance means explaining the behavior or functioning of something as if that something could think about what it is doing, then intentional systems are those whose functioning or behavior exhibits *apparently* purposeful, goal-directed, rational activity, that is, the functioning *looks* purposeful even when the system is not the kind in which conscious purpose is present. At the very least, an intentional system's functioning or behavior is the kind for which there are *reasons,* even when the system is not truly rational. Human, and some animal, behavior is purposeful and goal-directed. The functioning or behavior of all other intentional systems is more properly referred to as quasi-purposeful, quasi-goal-directed, and quasi-rational. An intentional system is a teleological system whose telos is entirely natural, determined by its systemic configuration and the external availability of whatever answers to the requirements of its particular configuration.

Dennett explains intentionality as a trait of even the most primitive organic system such as the amoeba:

> Consider a simple organism—say, a planarian or an amoeba—moving nonrandomly across the bottom of a laboratory dish, always heading to the nutrient-rich end of the dish, or away from the toxic end. This organism is seeking the good, or shunning the bad—*its own* good and bad, not those of some human artifact-user. Seeking one's own good is a fundamental feature of any rational agent, but are these simple organisms seeking or just "seeking?" [*sic*] We don't need to answer that question. The organism is a predictable intentional system in either case. (Dennett 1996, 32)

The amoeba, of course, is not conscious of its behavior. That does not preclude our explaining its behavior from the intentional stance, but it does point out a contrast between human beings and other intentional systems. An amoeba, for example, can "track" or follow a food source and can discern the difference between a nutrient and a toxin; its intentionality lies in the possession of appropriate receptors. Human beings, however, can track an object or person over time and *reidentify* it as the same. Dennett points out that "The practice and projects of many creatures require them to track and reidentify individuals . . . but no evidence suggests they must appreciate that this is what they are doing. . . . Their intentionality never rises to the pitch of metaphysical particularity that ours can rise to" (Dennett 1996, 117). For example, a human being is capable not only of reidentifying another human being but of doing so with joyful anticipation of the reunion, resentful memories of past conflicts, or deliberately cultivated indifference.

For all its uniqueness, however, human intentionality is what Dennett calls "derived intentionality." It is not self-constitutive. The fact that human thought is *about* something implies a referent in which the thought is grounded. Dennett explains it this way:

> A shopping list written down on a piece of paper has only the derived intentionality it gets from the intentions of the agent who made it. . . . It is . . . an artifact created by your brain and means what it does because of its particular position in the ongoing economy of your brain's internal activities and their role in governing your body's complex in the real, surrounding world.
>
> . . . the brain is an artifact, and it gets whatever intentionality its parts have . . . from the intentions of its creator, Mother Nature (otherwise known as the process of evolution by natural selection).
>
> . . . the intentionality of brain states is derived from the intentionality of the system or process that designed them. (Dennett 1996, 51–53)

Dennett's explanation of human intentionality embeds it in a natural matrix from which the mind is not differentiated in any metaphysically significant way.

Until Darwin, saying that human intentionality is derived would have been tantamount to saying that human intentionality derives from divine intentionality. Insofar as human beings are rational, they share one of God's characteristics, that is, human thought is a derivative of divine thought, and thus human beings are blessed with their quasi-divine and uniquely human essence. However, modern biology, beginning with Darwin, continuing through current work in neuroscience, and supplemented by cognitive science, has undermined this essentialist idea by undermining the idea of divinely derived intentionality; at the very least, it has pushed the concept of the divine origin of human intentionality further back in the explanatory scheme.

We now know that we are one species among millions, all linked by the presence of DNA, the possession of which we share with the lowliest life forms. We have our set of 100,000 human genes only because having this genome turned out to be advantageous in making us the kind of creature whose technological prowess permits the successful occupancy of almost any ecological niche on the planet. There was nothing initially special about this; the specialness derives from the end result—successful adaptation and successful continuation. This is the kind of knowledge that results in the painful enlightenment of which Kitcher speaks.

Evolutionary biology, which now, unlike in Darwin's day, includes genetics, undercuts completely the Aristotelian view that the manifest properties of human beings are the constituents of an essence that eternally defines what being human means. It has made human rationality a historical phenomenon. Dennett aptly expresses this:

> In the beginning there were no reasons; there were only causes. Nothing had a purpose, nothing had so much as a function; there was no teleology in the world at all . . . [because] There was nothing that had interests. But after millennia there happened to emerge simple replicators. . . . *If* these simple replicators [were] to survive and replicate . . . their environment [had to] meet certain conditions . . . conducive to replication. . . .
>
> When an entity arrives on the scene capable of behavior that staves off, however primitively, its own dissolution and decomposition, it brings with it into the world its own "good" . . . it creates a point of view from which the world's events can be roughly partitioned into the favorable, the unfavorable, and the neutral. (Dennett 1991, 173–74)

The entity now has interests to pursue. There are specifiable reasons for what it does, although these reasons are certainly not known to it, as Dennett emphasizes: "The first reasons preexisted their own recognition. Indeed, the first problem faced by the first problem-facers was to learn how to recognize and act on the reasons that their very existence brought into existence" (Dennett 1991, 174).

This means that reasons themselves have an evolutionary history precisely coextensive with the evolution of life forms. The first self-replicating macromolecule was the first intentional system, and from this point in history there is an entity of which it makes sense to understand its behavior as having reasons, or having a rationale. Moreover, if life evolved from nonlife, we can infer that meaning (in both the lower and the higher senses) evolved from nonmeaning. So meaning originates in the ability to do something, that initial something being self-replication, and is therefore rooted in this primitive kind of agency. Dennett explains this:

> Through the microscope of molecular biology, we get to witness the birth of agency, in the first macromolecules that have enough complexity to "do things." This is not . . . intentional action, with the representation of reasons, deliberation, reflection, and conscious decision—but it is the only possible ground from which the seeds of intentional action could grow. . . .
>
> . . . An impersonal, unreflective, robotic, mindless little scrap of molecular machinery is the ultimate basis of all the agency, hence meaning, and hence consciousness, in the universe. (Dennett 1995, 202–3)
>
> There is no longer any serious informed debate about this: *we are the direct descendants of these self-replicating robots.* (Dennett 1996, 22)

In short, consciousness, the construction site of meaning in the higher sense, is rooted in intentionality, the locus of meaning in the lower sense. Intentionality in turn is rooted in primitive agency, beginning with the self-replication of the first macromolecules on earth. This realization points to the major thesis of this paper, also expressed by Dennett:

> This is the defining theme of existentialism in its various species: the only meaning there can be is the meaning you (somehow) create for yourself. . . . Darwinism does have some demystification to offer in its account of the process of meaning-creation . . . *importance itself,* like everything else that we treasure, gradually evolves from nothingness. (Dennett 1995, 184)

At this point the anguished cry of "Reductionism!" may be anticipated: If we are nothing but robots, then our lives are meaningless, worthless! To interpret reductionism so drastically, however, is to misunderstand what it truly is.

Let us accept, for the sake of argument, Dennett's assertion that the most basic physiological processes of which our higher capabilities are constructed are those occurring in the most basic organic system, the cell. A robot is a machine programmed with instructions to accomplish a certain task. That is exactly what a cell does—it accomplishes the tasks specified by its genome. The human brain, the locus of all human higher capabilities, including meaning making, is composed of cells, as is every part of the body. However, this kind of reductionism does not imply that there is no more worth to human existence than to a single cell or a sophisticated machine. The reduction of these human higher capabilities is merely the explanation of them. They still exist, functioning exactly the same way, endowing human experience with the same richness as before the explanation. A proper reductionism does not explain away our higher capabilities; rather, our understanding of them is simply grounded in a better understanding of their foundations.

Knowledge of the basic processes of consciousness is no threat to the value of consciousness itself unless, as Dennett says, our understanding of consciousness is "based all along on confusion or mistaken identity" (Dennett 1995, 82). The fact that we are "descended from robots"—the first self-replicating macromolecules—does not mean, according to Dennett, that we are robots ourselves when considered at the level of the full, particularized configuration of organs that constitutes a human being:

> Now, it certainly does not follow from the fact that we are descended from robots that we are robots ourselves. After all, we are also direct descendants of fish, and we are not fish. . . . But unless there is some extra ingredient in us (which is what dualists and vitalists used to think), we are *made* of robots—or, what comes to the same thing, we are each a collection of trillions of macromolecular machines. And all of these are ultimately descended from the original self-replicating macromolecules. So something made of robots *can* exhibit genuine consciousness, because you do if anything does. (Dennett 1996, 23–34)

According to Dennett, whatever we lose through better scientific understanding of human consciousness and intentionality is offset by the deeper, more accurate understanding we gain. This point is illustrated by the fact that water is no less enjoyable and essential to human well-being after we understand that it is composed of hydrogen and oxygen atoms. We also can draw upon a metaphor used by Goodenough to show that reductionism is not a threat to the value and appreciation of something at a higher level. Goodenough asks us to consider a Mozart sonata. Understanding the composition techniques and the notes diminishes neither its beauty nor our capacity to appreciate it (Goodenough 1998, 34). If one subsequently loses the capacity to enjoy the music, Dennett will say that this loss is caused not by one's understanding the circumstances under which the music was composed but by one's initially mistaken idea of its unearthly origins. A more realistic conception of Mozart's genius at the outset forestalls the disillusionment that might result from its demystification.

Now let us apply this reductive analysis to the question of the meaning of human existence. Evolutionary theory, by demonstrating that human life has evolved from nonhuman life, is thus accused of robbing human existence of meaning. But this accusation is well placed only if there is no other source of meaning in human existence. If the possibility of meaning is contingent upon the development of intentionality, and if intentionality is a product of evolution, making the possibility of meaning likewise a product of evolution, to deny the value of any meaning human beings themselves construct because of its roots in our evolutionary development is to commit the genetic fallacy of condemning or devaluing something because of its origin, which is irrelevant to value.

Moreover, if a reductionist explanation is a threat to the possibility of meaning in human existence, it may be that our concept of meaningful existence is unrealistic from the outset. Maybe the human search for existential meaning beyond existence itself is an inflation of our own importance in the cosmos. Why do our lives have to mean anything in any sense that goes beyond the span of each individual life or the life span of the species?

Although consciousness of our own intentionality is impressive, the value of its singularity should not be overestimated. As Dennett argues philosophically and Goodenough explains scientifically, consciousness is part of the continuum of life. Although on the "higher" end—the end at which we, the conscious, meaning-making organism, can puzzle over whether our existence has any meaning—it is firmly related to the "lower" end, which Dennett says is "the only possible ground from which the seeds of intentional action could grow" (Dennett 1995, 202). Our consciousness of our intentionality is the product of the human genome, which ultimately determines the fate of our species. Roger Masters puts human existence in a perspective that should serve to keep our hubris in check:

> Evolutionary biology does not permit such an exaggerated view of human nature. We are living beings, no more precious than any other living form except in our own eyes. Because we can eat or kill virtually all other animals in the environment, we are at the top of the food chain—what is technically called "top carnivores." But this does not mean that we are independent of natural necessity or in control of our evolutionary destiny. (Masters 1989, 122)

Being top carnivore is certainly an enviable position, but can we wring any meaning out of it?

The Continuity of Intentionality

I have tried to explain how the human capacity for meaningful existence is rooted in consciousness of our intentionality, the product of evolution. At this point I shall connect this philosophical view to Goodenough's scientific one, showing how it is supported and complemented by her scientific explanation of the origin of intentionality and meaning at a very basic organismic level.

In *The Sacred Depths of Nature* (1998), Goodenough says, "Reproductive success is governed by many variables, but key adaptations have included the evolution of awareness, valuation, and purpose. In order to continue, genomes must dictate organisms that are aware of their environmental circumstances, evaluate these inputs correctly, and respond with intentionality" (Goodenough 1998, 170–71). Given the fact of reproductive success in any species, Goodenough's remark implies that, despite being top carnivore, humanity is not the only life form capable of purposive and evaluative behavior. Although human consciousness is the highest manifestation of awareness and intentionality in the natural world, Goodenough shows that there are degrees of awareness and intentionality, beginning with the light-sensitive reactions of one-celled life forms and existing even in plants: "Indexical meaning systems are found throughout life: in plants, perception of red light by the seed's phytochrome system means that the seed should germinate" (Goodenough 1998, 111). This is intentionality in a very basic sense, but the fact that it is intentionality at all indicates that perhaps human beings have elevated their self-awareness to a metaphysical status it does not deserve. As stated earlier, learning more about our situation in the cosmos can be painful; Goodenough, too, has experienced the "existential shudder" such knowledge can produce:

> I've had a lot of trouble with the universe. It began soon after I was told about it in physics class . . . I was overwhelmed with terror. . . .
>
> - Our Sun . . . will die, frying the Earth to a crisp during its heat-death, spewing its bits and pieces out into the frigid nothingness of curved spacetime.
>
> . . . And when I later encountered the famous quote from physicist Steven Weinberg—"The more the universe seems comprehensible, the more it seems pointless"—I wallowed in its poignant nihilism. A bleak emptiness overtook me whenever I thought about what was really going on out in the cosmos or deep in the atom. So I did my best not to think about such things. (Goodenough 1998, 9–10). . . .
>
> We are told that life is so many manifestations of chemistry and we shudder, a long existential shudder. (Goodenough 1998, 33)

Goodenough appears to have experienced what Hilary Putnam refers to when he says, "Science is wonderful at destroying metaphysical answers, but incapable of providing substitute ones. Science takes away foundations without providing a replacement" (Putnam 1987, 29).

Although science has made some metaphysical answers untenable (the existence of mind as an ontologically independent substance, for example), I do not agree that it leaves us without foundations. Rather, it has *changed* those foundations; taking away some, it has given us others. Science gives us a platform that is not only all around us but—perhaps less vividly because less *tangibly* for us—in our genes. With respect to the need for a stable platform upon which to construct an epistemologically justifiable worldview, Goodenough's view is that "our scientific understanding of nature seems to me like a good place to begin since it at least tells us what it is that we're working with. It doesn't follow that this understanding becomes a blueprint, but rather a touchstone."[4] That is, science does not prescribe every aspect of any worldview but rather serves as a reference point by which to gauge the accuracy of the empirical claims underlying the belief systems that constitute a worldview.

In response to the problem of nihilism posed by Steven Weinberg's observation that the universe appears pointless, Goodenough offers the possibility of existential meaning rooted in the *continuity* of all life forms and in the human goal of *continuation* of our species. Continuity is present not only in the biological universality of DNA but in the presence of "meaning systems," which are "unique to biology" (Goodenough 1994, 608). The presence of meaning systems—intentionality—at all biological levels indicates various levels of awareness, though only human beings are *consciously* aware. Human consciousness—mystifying to most people—is continuous with the bacterium's awareness of its environment, although human consciousness is at the "high" end of the continuum. All awareness, and thus human consciousness, is the product of evolution—not only at the human level but, according to Goodenough, even at the bacterial level: "I see the whole enterprise, from bacteria to starfish to maples to humans, as operating on the same principles, as profoundly homologous" (1994, 604). Recognizing the religious implications of this view, Goodenough informs us that "Recent discoveries in biology tell us that concepts central to religious thought [meaning, valuation, and purpose], concepts that we have believed to be unique to human perceptions and concerns, are in fact operant throughout the biological world" (1994, 604). With respect to meaning, the beginning of the process of valuation, Goodenough asserts that "meaning . . . is in fact fully applicable to the perceptions of a bacterium or a starfish or a maple" (1994, 605). The question, then, concerns the nature of the intentional continuity between the bacterium and human beings. Human beings have the ability to represent meanings mentally through the use of symbols, but meaning is present at both the molecular level and the mental level. Goodenough first explains human meaning making, the most familiar and well-established form.

At the mental level—where we are aware of meaning making—the process begins, for example, with the visual perception of a chair; the perception elicits a physical response, or meaning, such as sitting. The brain can also respond to the *word* "chair" by producing a mental image or by summoning the concept of a chair, and the meaning becomes symbolic. If one spots a chair when one is tired, the perception elicits an affective meaning as well. The continuity of the mental level of meaning with the molecular level consists in the simultaneous biochemical reactions that occur at the cellular level, for example at the thought of the chair: "The purpose of the word *chair* . . . is to elicit the biochemistry necessary to call up the mental concept of the piece of furniture" (1994, 607). A biochemical process is also initiated, of course, with the visual perception of a chair.

There is a less intuitively recognizable but just as genuine production of meaning in human beings at the molecular level, for example, during the production of insulin. The presence of high blood sugar indicates, or "means," that insulin is needed, and the pancreas cells are accordingly stimulated to produce it. The insulin molecule then binds to the insulin receptor on a cell, "meaning" that sugar is present, and stimulates the absorption of the sugar by the cell. The sugar's indication of the need for the insulin constitutes a rudimentary yet genuine case of meaning. This molecular meaning making is the same thing that happens in mental perception when neurotransmitters, stimulated by the perception of the chair, bind to their receptors on brain cells (1994, 606–7). Mental activity and cellular activity are continuous by virtue of the common presence of such molecular meaning production.

In the bacterium, intentionality is evident—as in human beings at the molecular level—in the functioning of receptors, proteins that "serve as transducers of meaning." Bacterial receptors interact with molecules released by decaying organisms that indicate the presence and location of food—meaning in the most basic sense. This interaction stimulates a "cascade" of shape changes in the bacterium enabling it to move toward the food, a process known as *chemotaxis*. Bacteria "use receptors continuously to evaluate their circumstances," and in the presence of toxic molecules, a cascade of shape changes results in the bacterium's moving away from the toxin (1994, 605–7).

It is clear that Goodenough sees the functioning of receptors in bacteria and other life forms as genuine instances of meaning, not just analogues of human meaning making. There *is* something unique about meaning at the mental level in human beings: "The uniqueness of humans is that we know the meaning of the word *meaning*" (1994, 608). Yet Goodenough also says that "This ability, while an astonishing innovation, is only the most recent innovation in the evolution of receptors. Meaning and valuation systems, per se, prevail all the way down" (1994, 608). Evaluation, too, consequently, is engaged in by organisms at all levels of life, evaluative capability—the ability of an organism to respond to its environment—having further evolved out of simple awareness and intentionality:

> The evolution of awareness has spun off two important capabilities:
>
> - Organisms usually attach a *value* to the things they perceive—this is good, that is bad—which, in complex animals, is experienced via neural and hormonal emotional systems.
> - Organisms usually attribute a *meaning* to something they're aware of, an ability that has for us become manifest in our capacity to think and act symbolically.
>
> These capabilities have converged in human brains as our ability to symbolize ideas and emotions, integrate them, and present them to the working memory. (Goodenough 1998, 105)

In the ability to form symbolic representations Goodenough sees the uniquely human aspect of meaning making. Human beings have the capacity to infuse life with meaning at the cognitive level by means of one of the most vital instruments of meaning making, language, through which we articulate emotional and intellectual states that form the basis of culture. Goodenough sees all of our higher capabilities, including our ability to create morally normative meanings, as issuing from our ability to create symbols, which is an evolved capability.[5]

> It is, I believe, our capacity to apprehend the meaning and the emotion embedded in symbols that endows us with our capacity for empathy. . . . Once there is empathy, then there can be the feeling we call compassion. . . . And emergent from our sense of compassion, in mortal conflict with our insistent sense that we should win, is our haunting sense that things should be fair. (1998, 114–15)

The idea that moral norms emerged from our evolved cognitive and emotive capacity, if correct, shows that all of our higher capabilities are rooted in our evolved capacity to preserve meanings through symbols. However, while language is uniquely human—an aspect of conscious human intentionality—it is nevertheless continuous with what happens at the molecular level, not only in human beings themselves but in all other forms of life. Goodenough can explain human linguistic ability as she explains other increasingly ascendant capabilities of life forms for which the matrix is molecular activity: as "emergent" functions. Through the concept of emergence, the specter of reductionism ceases to threaten the value of higher human capabilities, because we can understand and appreciate the capabilities at one level while recognizing their origin at a lower level.

> Life can be explained by its underlying chemistry, just as chemistry can be explained by its underlying physics. But the life that emerges from the underlying chemistry of biomolecules is something more than the collection of molecules. . . . once these molecules came to reside within cells, they began to interact with one another to generate new processes, like motility and metabolism and perception, processes that are unique to living creatures, processes that have no counterpart at simpler levels. These new, life-specific functions are referred to as emergent functions. (1998, 28)

And just as life emerges from the underlying chemistry of biomolecules, so human consciousness—with all that it entails—emerges from intentionality, which in turn has emerged from the evolution-produced human brain in interaction with its environment.[6]

WHENCE MEANING? (THE *REALLY* IMPORTANT KIND)

Goodenough is doing what Darwin did, but more thoroughly, aided by the mountain of biological data accumulated since Darwin. She is explaining the human capacity for meaning naturalistically by locating it in a natural function. Human intentionality, she and Dennett tell us, is a product of evolution. Goodenough's explanation as a scientist—that meaning making at the conscious human level is continuous with the bacterium's ability to detect differences in the substances surrounding it—is the same as Dennett's explanation as a philosopher; both are based on science. However, she knows that her explanation is not likely to strike a positive chord in people who seek meaning in a transcendent source.

> For me, the existence of all this meaning and intent, and my ability to apprehend it, *is* the ultimate value. The continuation of life reaches around and grabs its own

tail, and forms a sacred circle that requires no further justification, no purpose other than that the continuation continue until the sun collapses or the final meteor collides. . . .

Very well. Such a statement, which we can call a credo of continuation, may or may not elicit emotional resonance. (Goodenough 1994, 612)

The plain fact is that human beings exist because earlier life forms constituted the biological ancestry that has produced us. But this is the *historical* reason we exist. It does not address the question of the *reason* for our existence when reason is understood as *purpose*. The purpose of human existence implies a future with a task to be accomplished or a plan to be fulfilled—but there is no evidence that human beings exist in order to accomplish a task or in order to fulfill a plan determined by anyone but ourselves. The tasks and intentions we understand ourselves to have are the result of existing in a cosmos that would—and did, and will—exist without us. The mere fact of existence, with its natural pressure toward continuation, confers tasks and suggests purposes that become our conscious intentions (understanding *intentions* in the usual sense), along with any other goals we adopt. Most fundamentally, the goal of biological continuation, Goodenough says, can "suggest principles and practices for the leading out of our lives" (1994, 612). In addition to the basic process of continuation, however, or perhaps as part of it, values and purposes are suggested by what we must do to further the process. Human beings have created social, intellectual, and spiritual structures, all of which have genuine, if not demonstrably transcendent, meaning.[7]

Finding meaning, then, is a task we can assume only because we have evolved to the stage of self-consciousness—consciousness of our own consciousness. Death consciousness also surely plays a significant role in the effort to find meaning in our existence; each individual's certainty of death exerts a more pronounced pressure to locate a source of meaning than does the prospect of species extinction. In an imaginary scenario in which human beings were naturally immortal, if we had all the time in the world, if there were no end toward which life moves biologically, there would arguably be little need for concern over its purpose. The question of meaning would likely cease to press us at all. In addition to individual death, the natural telos of the species—extinction—merely compounds the urgency of the question of existential meaning.

For Goodenough, the sufficiency of a naturalistic explanation and the fruitfulness of the search within nature for the meaning of existence is the product of a realization that one need not have ultimate answers in the traditional sense: "The realization that I needn't have answers to the Big Questions, needn't seek answers to the Big Questions, has served as an epiphany" (Goodenough 1998, 12). In this epiphany, she finds herself capable of acceptance, which, she says, can be disappointed and resentful; it can be passive and acquiescent; or it can be the active response we call assent. When my awe at how life works gives way to self-pity because it doesn't work the way I would like, I call on assent. . . . To give assent is to understand, incorporate, and then let go. With the letting go comes that deep sigh we call relief, and relief allows the joy-of-being-alive-at-all to come tumbling forth again. . . . Once [assent] is freely given, one can move fluidly within it." (1998, 47)

So can evolution be the locus of meaning in human existence? The answer is in one sense yes and in another sense no. It is yes if by *meaning* we understand intentionality and human consciousness, which make a meaningful existence possible, because the question itself is not possible without consciousness of our intentionality, a direct product of evolution. The answer is no if we expand meaning to include inherent purpose or value or significance. From an evolutionary standpoint, human beings have no more significance than any other organism. If we did, we would be special in the sense of being uniquely important to the planetary and cosmic scheme of things; however, the evolutionary history of all living things indicates that we are not so privileged. In *The Meaning of Evolution*, George Gaylord Simpson conveys an estimation of human existence that is not likely to strike a positive chord in most people: "Man is the result of a purposeless and natural process that did not have him in mind. He was not planned" (Simpson 1967, 345). So how can human existence have any meaning in the higher sense?

The answer is an existentialist one: We must constitute our own significance as existing beings, bearing the existential burden of choosing what we will become individually and collectively—a burden human beings inherited along with the particular spot we occupy on our particular branch of the evolutionary tree of life. Yet if we occupied any other branch of this tree, we would not be capable of even wondering what our choices are. Dennett endorses the distinctly existentialist implications of Stephen Jay Gould's assertion that "We are the offspring of history, and must establish our own paths in this most diverse and interesting of conceivable universes—one indifferent to our suffering, and therefore offering us maximal freedom to thrive, or to fail, in our own chosen way" (Dennett 1995, 311).

There are various responses to our inability to discern *intrinsic* meaning in human existence. I have noted Ursula Goodenough's response of acceptance in letting go of the need for answers to the "Big Questions." Another possibility, common since Darwin, is retreat into religious dogma, ignoring the findings of science. Of course, there is also the possibility of assimilating evolution into the religious worldview, a possibility not to be lightly dismissed provided certain conditions are met. Finally, there is the existentialist response, exemplified by Nietzsche and others.[8]

Eliminating outright the viability of a worldview based on the dogmatic religious rejection of science, and acknowledging the less-than-conclusive epistemological justification for any religious view that requires a supernatural source of meaning, the most well-founded choice is the existentialist view when it is grounded naturalistically in the scientific illumination of human existence.[9] This is essentially Goodenough's view, as well as Dennett's; it consists of a forward-looking acceptance (without existentialism's usual morbidity) of the task of creating meaning at both the individual and societal levels. If evolution is the source of intentionality, and conscious intentionality is the matrix of the possibility of existential meaning, then existence really does precede essence, as Sartre asserted—if we define essence as the kind of conscious intentionality that has evolved in human beings and accept the temporality and mutability—the *historicality*—of this kind of essence. Furthermore, there is no inherent reason why other higher animals cannot share this kind of intentionality to some degree, and such shared capability provides a sense of connection between ourselves and other intelligent creatures.

Once we accept the idea that there are no skyhooks, as Dennett asserts, then where must we look for meaning? Precisely where some are already looking: at the projects we choose for ourselves individually and collectively. This choice offers no ultimate solace, but Goodenough's description of the continuity of all life may diminish the threat of alienation in an existence without ultimate answers. It can offer the possibility of naturally grounded purpose if not the hope of an ultimate purpose to human life.

The project of the continuation of biological life that Goodenough proposes is a long-term one, however. In the short term, human beings confront the existentialist task of choosing projects. Both collective and individual projects require existential fortitude in the face of knowledge that even if the project of continuation is successfully undertaken, it, too, will be nullified by the one astronomical event with which human life is most vitally connected—the death of the sun—not to mention the statistical probability that we will be extinct long before our most important star becomes "a black rock, cold as the void of space" (Friedman 1986, 235).

The yearning for an ultimate answer to the problem of meaning may be too deep in most persons to be given up. Certainly we may credit the *desire* for ultimate answers as a stimulant to inquiry, but if it were possible to achieve them, there would no longer be such a stimulant. In any case, no answer to any fundamental question is truly an ultimate one; we can always raise further questions. So we must settle for penultimate answers, stopping when our existential yearnings cannot be satisfied from within the natural context out of which they arise. And while the enhanced understanding of human existence made possible by science is alone not a *sufficient* determinant of value and meaning, it can engender in us a "natural piety" before the universe and the process that has produced us.[10]

The locus of the *possibility* of meaning—human evolution, encompassing the development of intentionality—is determinate. The content and future of meaning are open-ended and indeterminate, however. Evolution means change, and the constancy of change means we will never have ultimate answers. This promises inexhaustible possibilities of meaning as human beings individually and collectively search for deeper natural knowledge and social understanding. Science by its nature cannot yield ultimate answers; as its history amply documents, neither can philosophy. Scientific knowledge increases incrementally in a permanently asymptotic relationship to the uniquely human goal of truth, at the pinnacle of which we stand as the initiators of the search. The fact that only human beings occupy this pinnacle makes it appropriate now to return to George Gaylord Simpson's remarks in *The Meaning of Evolution*, which, although beginning with the discomfiting observation that human beings are not an inevitable product of the evolutionary process, end with the assurance that we are indeed a unique one: "It is . . . a gross misrepresentation to say that [man] is just an accident or nothing but an animal. . . . man is unique . . . defined by qualities found nowhere else, not by those he has in common with apes, fishes, trees, fire, or anything other than himself" (Simpson 1967, 345).

NOTES

1. "Cranes," in Dennett's terminology, are the natural mechanisms that enhance the power of natural selection, whereas "skyhooks" are explanations of life forms that presuppose the need for a supernatural "mind" to account for their design (Dennett 1995, 73–80).

2. I call religion a preconstructed system of meaning because for almost everyone, except the few people in history who have been innovative enough to found new religions, religion comes to them in a form determined by its earlier adherents. Of course, there can be other sources of meaning—political worldviews such as Marxism, for example. But these have been less universal and less fundamental to the human search for meaning than religion has been.

3. "[O]ne of the basic principles of evolution is the conservation of previous gains in adaptation" (Donald 1991, 165).

4. Ursula Goodenough, personal communication, 6 March 1999. This comment was in response to my question of how one can avoid the dilemma of relativism.

5. Dennett says that language "is ultimately grounded in the rich earth of biological function" (Dennett, 1995, 402). Many nonhuman animals experience emotional states. Emotional experience is not dependent on the capacity to create symbols. Other animals, however, cannot reflect on these states and represent those reflections. For that, symbol-making capability is needed.

6. Emergence is not a new idea; nor is continuity. John Dewey discusses both in much the same form as Goodenough conceives them: "Continuity . . . means that rational operations *grow out of* organic activities, without being identical with that from which they emerge" (Dewey 1938, 19).

7. Robert Pennock provides a whole slate of such structures, none of which is undermined by our evolutionary origins: "Ask people what is most valuable in their lives, what gives their lives meaning, and you will get a wide range of answers. Certainly some people will cite their faith in God (though for most of these their faith does not depend upon whether or not the Genesis account is literally true). Many more will mention the pride and joy they feel for their children, the tenderness they feel for their lovers and friends, the sense of accomplishment they derive from their work, the pleasure they receive from music and art, or the deep satisfaction they feel in the struggle to build a better tomorrow. People find value in a well-crafted novel and a well-cooked meal, in vigorous athletic activity and in quiet moments of reflective contemplation. They find purpose in the building of a home, the furtherance of social justice, and the pursuit of scientific knowledge. How easy it is to extend such a list!" (Pennock 1996, 22)

8. "The sense that the meaning of the universe had evaporated was what seemed to escape those who welcomed Darwin as a benefactor of mankind. Nietzsche considered that evolution presented a correct picture of the world, but that it was a disastrous picture. His philosophy was an attempt to produce a new world-picture which took Darwinism into account but was not nullified by it" (Hollingdale 1965, 90).

9. Despite my reference to Nietzsche, my use of the concept of existentialism in this paper is generic, employing the concept central to virtually all existentialist thinkers: A fundamental aspect of the human condition is the search for meaning in life, and this search is of pivotal importance to individuals. My reference to Nietzsche is based on his recognition of the implications of evolution for existential meaning and his emphasis on the essentiality of an individual's establishing meaning and identity through chosen tasks. I could just as well have used Sartre to make the latter point. For this clarification, I am indebted to a question from Peter Derkx, director of the research program, "Humanism, Meanings of Life, Worldviews," at the University for Humanist Studies in Utrecht, the Netherlands.

10. "Natural piety" is Sidney Hook's term in *The Quest for Being*: "Man can live with a natural piety for the sources of his being. He can rely upon nature and himself without worshipping them. Man in fact relies only on his own natural and human resources even when he claims to rely on other resources" (Hook 1991, 208).

REFERENCES

Dennett, Daniel. 1991. *Consciousness Explained.* Boston: Little, Brown.

———. 1995. *Darwin's Dangerous Idea: Evolution and the Meanings of Life.* New York: Simon and Schuster.

———. 1996. *Kinds of Minds: Toward an Understanding of Consciousness.* New York: Basic Books. Science Masters Series.

Dewey, John. 1938. *Logic: The Theory of Inquiry.* New York: Henry Holt.

Donald, Merlin. 1991. *Origins of the Modern Mind: Three Stages in the Evolution of Culture and Cognition.* Cambridge: Harvard Univ. Press.

Friedman, Herbert. 1986. *Sun and Earth.* New York: Scientific American Library.

Goodenough, Ursula. 1994. "The Religious Dimensions of the Biological Narrative." *Zygon: Journal for Religion and Science* 29 (December): 603–18.

———. 1998. *The Sacred Depths of Nature.* New York: Oxford Univ. Press.

Gould, Stephen Jay. 1989. *Wonderful Life*, 323. New York: Norton. Quoted in Dennett 1995, 311.

Hollingdale, R. J. 1965. *Nietzsche: The Man and His Philosophy*, 90. London: Routledge and Kegan Paul. Quoted in Dennett 1995, 181.

Hook, Sidney. 1991. *Quest for Being.* Buffalo, N.Y.: Prometheus Books.

Kitcher, Philip. 1998. "Truth or Consequences?" *Proceedings and Addresses of the American Philosophical Association* 72:49–63.

Masters, Roger D. 1989. "Evolutionary Biology and Naturalism." *Interpretation* 17:111–26.

Pennock, Robert. 1996. "Naturalism, Creationism and the Meaning of Life: The Case of Philip Johnson Revisited." *Creation/Evolution* 16:10–30.

Putnam, Hilary. 1987. *The Many Faces of Realism.* LaSalle, Ill.: Open Court.

Simpson, George Gaylord. 1967. *The Meaning of Evolution: A Study of the History of Life and of Its Significance for Man.* New Haven: Yale Univ. Press.

Wilson, Edward O. 1992. *The Diversity of Life.* Cambridge: Harvard Univ. Press, Belknap Press.

———. 1998. *Consilience.* New York: Alfred A. Knopf.

[25]

Is the Spell Really Broken? Bio-psychological Explanations of Religion and Theistic Belief

JUSTIN L. BARRETT

Abstract *Recent advances in the evolutionary and cognitive sciences of religion have raised questions about whether the assumptions and findings of these fields as applied to religion conflict with belief in gods. Specifically, three scientific approaches to religion (Neurotheology, Group Selection, and Cognitive Science of Religion) are sketched, and five arguments against theistic belief arising from these approaches are discussed and evaluated. None of the five arguments prove formidable challenges for belief in gods.*

Key words: Belief; Cognitive science; Evolution; Religion; Theism

The long and adversarial relationship between Darwinism and theism continues. Not satisfied to explain the nature of living things, evolutionary scientists now aim to explain religion itself. Pascal Boyer's *Religion Explained* raised a few eyebrows with subtle suggestions that religion is "airy nothing" but a byproduct of evolved human minds.[1] Scott Atran's *In God's We Trust: The Evolutionary Landscape of Religion* even more directly affronted some religious sensibilities by purporting to explain theistic beliefs as irrational and counterfactual thought propped up by evolved mental capacities.[2] Repackaging these same types of arguments, Daniel Dennett has finally managed to gain considerable popular attention for the movement with his recently released *Breaking the Spell: Religion as a Natural Phenomenon*.[3] Dennett's book and its apparently lethal implications for religious belief have been discussed on multiple occasions on National Public Radio and in wide-circulation publications such as the New York Times.[4]

That this area of scholarship, captured under the umbrella *bio-psychological theories of religion*, appears to be growing in visibility and influence suggests a need to consider carefully the current and potential implications of the evolutionary and cognitive sciences of religion on religious belief. Will evolutionary sciences finally win the war and defeat theism once and for all? Some atheists within and outside these sciences hope so. Likewise, the fear that the bio-psychological theories of religion finally hold the right weapons to destroy theism torments some believers observing the field and even working in it. In this essay, I offer a brief overview of recent bio-psychological theories of religion and then present a number of arguments against theistic belief based on this scholarship.

Three subfields

The bio-psychological theories of religion come in three different varieties: *neurotheology*, *group selection*, and *cognitive science of religion*. All three subfields represent attempts to explain religion reductively by appealing to evolved properties of the mind-brain and their impact on behavior. That is, religious beliefs and practices arise because of the activity of evolved functional properties of human brains. Nevertheless, the three schools take importantly different approaches to explaining religion.

Neurotheology

Neurotheology primarily concerns identifying which components and dynamics of the brain underlie religious experiences and subsequent beliefs.[5] The suggestion is that religious phenomena can be identified as the (perhaps accidental) output of evolved neural circuitry. Evolved brains have components that have arisen because of their usefulness to survival that happen to interact in such a way as to generate religious experiences, including experiences of the presence of an invisible, disembodied being. These experiences are shared and may be codified into common religious beliefs, particularly in the existence of gods that account for these experiences.

The most dramatic and controversial element of the neurotheology school is work by Michael Persinger. Reportedly, Persinger has been able not only to identify brain regions responsible for certain types of religious experiences, but has also succeeded at artificially inducing such experiences using electromagnetic fields, experiences that appear similar to being in the presence of a supernatural being.[6] If such experiences may be artificially induced, could it be that god concepts arise from misfiring of these same brain regions?

Group selection

David Sloan Wilson's *Darwin's Cathedral* represents the *group selection* perspective. In short, Wilson argues that religious systems encourage pro-social behavior, and groups that exhibit pro-social behavior (cooperation, lack of cheating and stealing, etc.) will tend to out survive and reproduce groups that do not exhibit these traits. So, religious communities will tend to survive better than non-religious communities do.[7]

As religious communities will have stronger pro-social tendencies, they will cooperate better. As they cooperate better, they will survive and thrive better than competing communities. Hence, religious communities—and whatever genetic information accounts for their religiosity and pro-sociality—will tend to survive and expand at a greater rate than non-religious communities. Over time, then, people with the biological disposition to be religious will increasingly outnumber non-religious people. This selection process thereby accounts for the widespread existence of religious people.

By this account, religion does not persist because it makes any special truth claims or inject lives with meaning. Rather, religion exists because it helps people survive. Religion possesses utility.

Cognitive science of religion

The most developed bio-psychological field dealing with religion has come to be called the *cognitive science of religion.* This camp includes Atran, Boyer, and a growing number of others.[8] This work has been strongly associated with evolutionary psychology, particularly that of Atran and Boyer, but an evolutionary theoretical foundation is not strictly necessary.

Though research has concerned religious rituals, religious community morphology, god concepts, prayer, morality, after-life concepts, scripture use and a number of other areas, what unifies these projects are a number of fundamental commitments. The cognitive science of religion begins by acknowledging that:

(1) In general, the basic functional processes of human minds are the same regardless of cultural environments. That is, by virtue of a common human biology living in a remarkably uniform natural world, minds develop similarly everywhere as regarding, for instance, intuitive causal reasoning.
(2) Further, as the cognitive sciences have shown, human minds are not general-purpose information processing devices but are highly specialized conglomerates of many functional subsystems that solve particular problems.
(3) These subsystems importantly color or shape perception and cognition regarding the natural and social world. They do not receive passively and indiscriminately whatever is "out there."
(4) These contours of human minds inform and constrain recurrent patterns of human thought and action including religious thought and action.
(5) Hence, recurrent features of religious thought and action (e.g. belief in gods) can be explained (or predicted) by appealing to requisite conceptual structures. Particular thoughts and actions will occur more frequently among humans than other possible thoughts and actions by virtue of their foundation in the basic dynamics of human minds.

To illustrate, we can confidently predict that in no yet-to-be-studied religious tradition will people practice a ritual system that demands a sacrifice of sparrows 13 times the square-root of the number of days past since the last full moon. We can also predict that the gods of the system will not be five-dimensional amoeboids that experience time backwards, only know what is not true, and behave beneficently on every third day. Why not? Because such a religious system incorporates reasoning completely foreign to the naturally occurring preferences and capabilities of our minds. The difficulty does not lie in the complexity of the information required *per se*—primitive computers could easily calculate the appropriate sparrow sacrifice—but because our minds are receptive to a very narrow subset of theoretically possible information. Find out what information we thrive on consuming and producing, and you find out why religions (or any cultural phenomena) tend to look the way they do.

The cognitive science of religion also includes a number of scholars interested in the formation of children's religious beliefs. Young children's understanding of death and the possible persistence of persons after death,[9] children's affinity for intentional design explanations of the natural world,[10] and children's acquisition of various aspects of god concepts receive attention.[11] What these cognitive developmental psychologists offer to the more general cognitive science of religion is the observation that religious concepts for which children's minds seem to have a natural tendency to entertain and generate will tend to become more common. Showing, for instance, that children find creationist accounts of animals more compelling than evolutionary accounts regardless of parents' beliefs, helps explain why creationism is so resilient in the face of alternative accounts.[12]

The threat to theism

Though scholars in these fields have largely avoided explicit published treatments of what their science might mean for religious belief, popular reactions clearly demonstrate that believers fear behind titles such as *Religion Explained* and *Breaking the Spell* and *Is God an Accident?*, lurks a genuine assault on religious belief at the experienced hand of science. Authors in this area typically do not directly attack religious belief but leave plenty of reasons for believers to be suspicious. Psychologist Paul Bloom's Atlantic Monthly article serves as a fine case in point.[13]

In *Is God an Accident?*, Bloom elegantly summarizes some central observations from the cognitive science of religion including his own work concerning what appears to be a natural and early-developing tendency in people to be mind-body dualists.[14] Such a tacit belief makes acquisition of various life-after-death notions quite easy indeed. Bloom offers that the organization of human minds accounts for why religious beliefs are so persistent in the face of science. Nowhere in his article does Bloom offer an argument for how the bio-psychological approaches show religious belief to be false. Rather, Bloom gently asserts the incompatibility of science and religion as when he imagines what would happen if religious folk allowed for insights from science to permeate their belief system: "Scientific views would spread through religious communities. Supernatural beliefs would gradually disappear as the theologically correct version of a religion gradually became consistent with the secular world view."[15] In addition, characteristic of many authors in the cognitive science of religion, Bloom casually assumes the falsity of religious beliefs in his presentation of the scientific evidence. For instance, he explains that natural systems "go awry" and consequently give rise to religion, by "inferring goals and desires where none exist."[16] Here we have no direct argument or evidence that gods or other entities do not exist. This conclusion is assumed in the presentation. These subtle rhetorical moves give observers the impression that these new treatments of religion are fundamentally dangerous or even outright hostile to theism.

To make explicit just where the potential points of danger for theistic belief might lie, in the following, I discuss five possible arguments against theism on the basis of the bio-psychological explanations of religion. These arguments represent positions that I suspect (but cannot prove) some authors in the area hold based on my conversations with these authors and exposure to both their written and spoken ideas. These arguments also represent those that believers fear lie behind the texts based on the conversations I have had with theists both inside and outside the field.

Arguments against theism

Argument no. 1: neural substrate

Arising from advances in neurotheology, one implicit attack on religious belief could go this way:

> We have identified the regions of the brain responsible for religious experience and can artificially induce religious experience. Therefore, its causes are entirely natural and so, we have no need to appeal to supernatural to account for them. Hence, theistic belief is unjustified.

To get this argument off the ground we have to grant the claim that neurotheology has actually identified the brain regions and functions that give rise to those religious experiences that are fundamental to belief in gods (broadly construed). This supposition is certainly debatable, and its full achievement might prove impossible, for now, let us assume that neurotheology will eventually be successful at detailing all the biological functions at play when someone comes to believe in a god.

Even with this enormous concession, this argument remains a non-starter. To get from a perfect biological specification of a belief to the belief being wholly explained by natural processes, we need a premise that the supernatural does not regularly causally act upon the neural substrate. Otherwise, a god could be directing neuronal networks to experience it. I imagine some theists would resist granting such a premise, and certainly, science cannot marshal requisite evidence to the contrary, as it is notoriously bad at measuring supernatural activity.

However, suppose the theist grants the further premise that the supernatural does not causally act upon the neural substrate. Then does the argument manage to defeat this form of theism? Not yet. The accommodating theist could simply maintain that a god or gods put into place the natural order—including the organization of human brains via evolution—such that human brains naturally give rise to religious experiences under particular situations. That natural biological processes have been found to correspond to religious experiences and beliefs does not entail that gods do not exist.

I find no way to salvage this argument even with the concessions made. Without the concessions it is even more fragile. Its pivotal flaw seems to be a misunderstanding about explaining epistemic states in terms of biology. Suppose I believe I see a robin outside my window. You tell me you can exactly specify the neural pathways responsible for generating my belief. Does that mean the robin is not really there? Hardly. That religious experiences can be artificially manufactured is irrelevant in the same way. Neurologists have found that by stimulating the cortex they can create various perceptual experiences. No one wants to argue that if scientists can use electromagnetic fields to make me believe I see a robin that suddenly I am not justified under normal conditions in believing I see robins.

For some theologies that see minds as wholly separate from bodies (including brains) the findings of neurotheology may present difficulties for this view of humanness. That particular bodily activities correlate perfectly with particular experiences should give one pause to suppose that all mental life is causally separate from bodily processes. Nevertheless, aside from complete dualists, finding biological substrates bears little impact. How many contemporary theists doubt that their brains are active during religious thought or experiences? The only news is just which parts of the brains are active. That brains might be active during religious thought or experience is no affront to religious belief so why would the insight that this or that particular brain area might be active matter?

Argument no. 2: evolutionary byproduct

Another implicit argument appears to be that

> Selection pressures (operating on either groups or individuals) have led to various dispositions or propensities in human minds that happen to give rise to religious belief. Religious beliefs are, therefore, accidents or byproducts of evolution. As such, religious beliefs cannot be trusted. Belief in gods amount to cognitive illusion.

This argument, sometimes attributed to group selectionists but more in line with the cognitive science of religion, is rich in rhetoric but poor in substance. Words like "accident," "byproduct," and "illusion" sound damning, but what do they really mean in this context?

"Accident" and "byproduct" both capture the idea that evolution did not select for people to believe in gods. Rather, it selected for other human behaviors (and the genetic material that gives rise to the behaviors and any prerequisite beliefs), and religious beliefs just happened to spring out of the way humans are put together. Religious beliefs themselves did not confer any selective advantage; rather they ride on beliefs and behaviors that did confer advantages. Music, too, may be an accidental byproduct of evolution in this sense. Natural selection did not favor music lovers and producers over others, rather, other functional units of the brain that did confer selective advantage happen to promote music.

The religion as byproduct position is fairly common in cognitive science accounts of religion, but what follows from this observation, assuming it is true?

Theists generally do not believe in gods, nor do they justify their beliefs, on the grounds that such beliefs conferred a selective advantage in our evolutionary history. As no weight rests on this foundation, to remove it does no harm to these beliefs.

Further, many beliefs and values that the scientists of religion themselves hold dear likely would be weakened by the same argument if it applied to theistic commitments. Contemporary beliefs and behaviors bestowed by science and technology arose far too late in our history to have played a role in natural selection of humans. Evolution did not select for calculus, quantum theory, or natural selection. Are these beliefs then suspect for being "accidents" or "byproducts" of evolution? With this line of reasoning, Darwinism would face the ax alongside theism. Suggesting that when beliefs arise as byproducts of evolution they must be jettisoned would force abandonment of the premises regarding selection in the argument—more on this *Suicidal Tendency* of such arguments below.

A different variation of the Evolutionary Byproduct argument suggests that theistic beliefs are not simply accidents or free riders on adaptive systems, but byproducts in the way that perceptual illusions arise from properly functioning evolved perceptual systems. Accusing religion of being a "cognitive illusion" (or "delusion") fares no better as a serious attack but nonetheless raises interesting comparisons. Psychologists have documented numerous illusions in which our minds tend to tell us something is the case that, upon closer inspection, turns out to not be so. For instance, people tend to see random events (such as series of coin tosses) as more orderly than they actually are, and see correlations where none exist.[17] These count as conceptual illusions brought about by how our minds naturally function. Perceptual illusions similarly tell us that something in the world is different than it actually is. For instance, when viewing a rainbow, people typically perceive several bands of color. Science has demonstrated, however, that a rainbow does not display bands of color but a perfect continuum of light wavelengths. The illusion of bands occurs by our visual system's tendency to categorize stimuli into meaningful units (termed *categorical perception*). The cognitive capacities that give rise to conceptual and perceptual illusions conferred selective advantages but occasionally produce relatively harmless illusions or mistakes.

One might argue that religious beliefs have much in common with these other cognitive illusions. The evolved, natural functions of human minds operating in the ordinary natural world prompt the belief in and spread of religious ideas, much as the evolved, natural functions of human minds operating in the ordinary natural world prompt the belief in illusions such as illusory correlations and bands of color in rainbows. Again, this comparison amounts to a conventional claim in cognitive science of religion. What the term "illusion" adds is the evaluation that a perception or belief is in error. On the basis of closer inspection (often using scientific or statistical methods) we can discover that a correlation we thought existed does not actually exist, or a sequence of coin tosses that looks unlikely could in fact arise by chance, or a rainbow does not actually present bands of color. To be able to call genuinely religious beliefs "illusions" we need to be able to demonstrate that they too, upon further examination, are in error. However, this task is not aided by the evolutionary or cognitive sciences of religion. To determine that a theistic belief amounts to an "illusion" requires a metaphysical commitment. To call theism "cognitive illusion" is a premise and not a conclusion of this argument.

64 Theology and Science

Argument no. 3: religious utility

The third argument to consider is the flip side of the preceding argument. Instead of emphasizing religious beliefs as unnecessary free riders on natural selection, it emphasizes the value of religious beliefs and subsequent behaviors.

> Religious beliefs have utility (e.g. in social arrangements) leading to their natural selection. As religious beliefs are selected for their utility and not their truth, they should not be trusted in terms of truth-value.

Though it starkly diverges from the previous argument, it carries its own fatal flaw. Nevertheless, such an argument is valuable to consider as it expresses an important but often overlooked element of evolution by natural selection.

This argument, derivable from the group selection account of religion, begins with an interesting empirical claim about religious beliefs, that they have utility in the sense of conferring advantages in reproduction and survival. Obviously, this claim must be a general claim about belief in superhuman intentional beings with moral concerns (i.e. gods) and not a claim about any and every specific religious belief. (A belief that universal celibacy is the true path to salvation would likely lead to negative reproduction effects.)

However, even taken as a claim about religious beliefs generally, it is not clear how believing in gods—even those that promote pro-social behavior—can be encouraged through natural selection. Natural selection, after all, selects for behaviors and not beliefs or attitudes. A person that generously shares with neighbors and never cheats because she believes in a punitive god is treated the same by natural selection as a person that behaves identically with no particular beliefs (let alone theistic ones) surrounding the behaviors. As it is pro-social behavior that allegedly is favored by natural selection, the argument should be recast with *pro-social behavior* taking the place of *religious beliefs*.

> Pro-social behaviors have utility (e.g., in social arrangements) leading to their natural selection. As pro-social behaviors are selected for their utility and not their truth, they should not be trusted in terms of truth-value.

As behaviors do not have truth-value, clearly such an argument is confused.

However, suppose we fortify the argument in another way. As much as any particular cognitive structure typically gives rise to behaviors, natural selection may reward those behaviors and the underlying cognitive structures that give rise to the behaviors. Hence, indirectly, evolution may select for cognitive devices and any beliefs they might generate (that then motivate behaviors). Replacing *pro-social behaviors* with *cognitive devices that generate religious beliefs* in the argument, we get

> Cognitive devices that generate religious beliefs have utility (e.g. in social arrangements) leading to their natural selection. As cognitive devices that generate religious beliefs are selected for their utility and not their truth, they should not be trusted in terms of truth-value.

This version of the argument more closely resembles what a cognitive scientist of religion might suggest instead of what is claimed by Wilson's group selection account. Substantively, it does not differ from the Evolutionary Byproduct argument. Both claim religious beliefs free ride on naturally selected cognitive structures. The difference is only in the introduction of a premise concerning what should and should not be trusted for belief. The Evolutionary Byproduct argument suggests that if natural selection *did not select* for a particular belief, but only its underlying cognitive mechanisms, it is not to be trusted. (However, note that, strictly speaking natural selection does not select for any beliefs but only underlying cognitive mechanisms and these only indirectly.) The Religious Utility argument takes the other side and suggests that if natural selection *did select* for a cognitive device, in part because of the beliefs and subsequent behaviors it generates, then the beliefs cannot be trusted.

If a belief is an accident of evolution, it cannot be trusted. If it is a legitimate product of evolution, it still cannot be trusted. Why not? Here is an important but often neglected feature of evolution by natural selection: it only favors minds that generate survival behaviors and not necessarily true beliefs. As already discussed, natural selection does not care about beliefs. If true beliefs lead to the extermination of a gene pool, so be it. If systematic errors or a particular profound disconnect with reality encourage survival and reproduction, then those genes win. In fact, the illusions discussed above are part of the vast psychological literature demonstrating that human minds seem systematically to get things wrong for the sake of survival. When it comes to natural selection, Truth is expendable.

However, the twice-revised version of the Religious Utility argument still fails as a defeater for religious belief. Granted, just being a product of evolution surely does not give us confidence in the truth of religious beliefs. However, as discussed in relation to the Evolutionary Byproduct argument, to say that being the product of evolution undermines religious beliefs is overstating the opposition between natural selection and Truth-representing minds. Natural selection does not discriminate against Truth; it just does not attempt to preserve it.

Argument no. 4: inherited belief

At a conference I was asked if children simply acquire belief in gods (1) because such beliefs cannot be proven wrong (i.e. they are not falsifiable) and (2) because of the persuasive and coercive power of parents and communities. A cognitive scientist of religion answered the question by explaining that regardless of the coercive or persuasive techniques of adults, children will not believe just any claims that cannot be proven wrong. Religious ideas are readily acquired because of their particular fit with a large number of cognitive devices that normally develop in humans. Persuasion or coercion can only augment acquisition of this particular subset of beliefs that enjoy the support of cognitive devices.

66 **Theology and Science**

The questioner could have responded (but did not) with another proposed argument against belief in gods.

> People are credulous recipients of theistic beliefs (e.g. from parents). Natural selection provided people with the cognitive faculties that make us credulous recipients. As we now know why people so readily believe in gods, continuing to believe is irrational.

An example may illustrate the strength of this argument.

Suppose someone persuades you of some proposition *P* so that you believe that *P* is true. Later you discover that at the time someone convinced you that *P* was true, you were under the influence of a drug that makes you particularly gullible. Are you then still rationally justified to believe that *P* is true? At the very least, you should reevaluate *P* while sober. Perhaps *P* will still be rationally justifiable on other merits but perhaps not.

The difficulty with applying this illustration to the Inherited Belief argument against religion is that we are never "sober." That is, the same factors that made us vulnerable to theistic beliefs are still operating once we are aware that they are there and from where they have come. One might suggest that these, too, are grounds for rejecting *P* or theistic beliefs outright, just to be sure, we do not adopt an unmerited belief. After all, how could theistic beliefs be fairly evaluated if we have these evolved capacities pushing us toward belief? Theism will always seem plausible. Therefore, theism should be rejected.

However, when do we ever use the plausibility of a proposition as grounds *against* believing the proposition? Something clearly is not right in this line of reasoning.

One problem with the Inherited Belief argument as presented is that it tacitly assumes that theistic beliefs are *only* credible because of the operation of these evolved mechanisms. That is, it assumes that no alternative reasons for belief are available outside of the cognitive mechanisms specified by cognitive scientists of religion. As essentially all of the mechanisms cognitive scientists point to as supporting religion are intuitive or even implicit, explicit reasoning such as that done by philosophers and theologians falls outside the pail. Arguments and evidence gathered through these other avenues are safe from the disturbing influence of the cognitive science of religion.

Perhaps, then, a weaker version of the Inherited Belief argument is in order.

> People are credulous recipients of theistic beliefs (e.g. from parents). Natural selection provided people with the cognitive faculties that contribute to making us credulous recipients. As we now know why people so readily believe in gods through these faculties, their contribution toward belief should be discounted. Hence, fewer reasons exist for theistic belief.

While not defeating religious belief outright, this version purports to weaken theism by cutting away some (perhaps most) available reasons for belief. Some individuals might be able to marshal enough justification to continue to believe but many believers would likely no longer have enough rational ground on which to stand.

Before feeling too satisfied with this vicious blow against theism, let us consider the drug illustration anew. A revision to parallel the renewed Inherited Belief argument might be the following. Suppose someone persuades you of some proposition *P* so that you believe that *P* is true. Later you discover that at the time someone convinced you that *P* was true, you were under the influence of a drug that makes you particularly gullible regarding many reasons to believe *P*. You should then reevaluate *P* in light of this information and disregard the reasons for believing *P* accounted for by the drug. Belief in *P* should be weakened.

However, consider another alternative. Someone persuades you of some proposition *P* so that you believe that *P* is true. Later you discover that at the time someone convinced you that *P* was true, you were under the influence of a "smart drug" that makes you particularly clear-headed and likely to believe only true propositions. Does this information undermine your belief in *P*? Surely not. If anything, it bolsters your confidence in *P*. If, on the other hand, you learned that you had been under the influence of a "stupid drug" that not only made you gullible but also especially gullible concerning the most patently false propositions imaginable, you would be especially anxious to reevaluate and probably reject belief in *P*.

The cases of the smart drug and the stupid drug illustrate that the ability for the Inherited Belief argument to undermine *or encourage* religious belief lies in whether the evolved mechanisms that allegedly support theistic belief are prone to produce accurate beliefs or inaccurate ones. If they generally produce accurate beliefs, such an argument actually favors theism. Hence, by itself, the Inherited Belief argument fails. I now turn to a fifth and final argument that explicitly builds in the charge that these evolved capacities are error prone.

Argument no. 5: error-prone minds

> The cognitive science of religion demonstrates that cognitive mechanisms of our evolved minds provide most impetus for believing in gods. As these mechanisms are error-prone, they cannot be trusted to give us Truth. As we cannot trust these aspects of our minds to give us Truth, we cannot trust most of our impetus for believing in gods. Therefore, cognitive science of religion weakens belief in gods.

Having learned from the mistakes of its predecessors, this argument avoids trying to defeat theism outright but aims only to weaken theistic belief. Its opening premise is a reasonable inference from theoretical claims and relevant evidence from the cognitive science of religion. Certainly, if a belief arises largely because of evidence (or reasons) from sources known to be untrustworthy, the belief is in danger unless sufficient other evidence (or reasons) exist apart from the dubious source. Assuming that the cognitive science of religion is on the right track, the only question seems to be whether the cognitive mechanisms that support belief in gods are error-prone to an extent that they cannot be trusted.

To illustrate, one mechanism implicated in theistic belief has been dubbed HADD—the Hypersensitive Agency Detection Device.[18] This device scans the environment for intentional agents and their activity. It has been suggested, and experimental evidence seems to support, that this device has a tendency to "detect" agency given fairly ambiguous evidence. The reason given for this hypersensitivity is a "better safe than sorry" strategy.[19] As other intentional agents (including people and animals) represented our ancestors' greatest threat and greatest promise for survival and reproduction, those with an agency detection device that did not often miss agency would have been at a selective advantage. Better to assume that a noise in the brush is an agent than to miss the chance of detecting an agent and miss out on capturing prey, or miss detecting an agent and being eaten. Such a strategy would lead to relatively harmless false-detections. HADD's ability to find agency given scant evidence would make HADD a strong candidate for detecting the activity of ghosts, spirits, and gods and thus promote belief in their existence. If HADD is tuned to find agency a bit too easily, then HADD is an error-prone mechanism promoting belief in gods. Being error-prone, HADD cannot be trusted and so, the evidence HADD contributes to belief in gods should be discounted.

As HADD's role in religious belief illustrates, this argument carries some potential merit. Nevertheless, here I focus on two weaknesses of the general argument (including HADD's role), and not additional criticisms specific to HADD.

Though HADD might appear to be the epitome of an error-prone device, fairly judging the accuracy of the entire concert of cognitive mechanisms involved in promoting belief in gods is no mean feat. If HADD worked alone in determining when or where we discovered the existence of agents, we would never be able to tell definitively when it was wrong. It would be analogous to trying to determine how many colors are in a rainbow only by looking at it over and over again. We would likely get the same answer repeatedly. Fortunately, HADD does not work alone. Other cognitive mechanisms, including our abilities to consider evidence reflectively, can override HADD or any other single cognitive mechanism that tries to generate a belief. This realization helps us decide that HADD is indeed error-prone, but it creates another problem. As HADD and the other cognitive mechanisms that promote belief in god do not work alone to generate beliefs, their accuracy cannot be evaluated in isolation.

What evidence, then, do we have that our total system (including HADD) that generates belief in intentional agents (such as gods) is error-prone to the extent that it cannot be trusted? Here we might be able to show that when people swear that computers are malicious intentional agents, upon further examination, computers turn out to be simply complex machines without intentions (or malice). The believer in computer agency himself might recant. That would be a case of mistaken agency detection. Alternatively, a person might be convinced that an intruder is in the house and lock herself in her bathroom for an hour as a precaution only later to discover that the sinister noises came from a radio. That would be a case of having mistakenly detected an agent. These sorts of episodes could be tallied up against the cases of having accurately detected agency (e.g. I thought a person was in front of me and I was right; I thought this letter was sent to me deliberately and I was right), and we could see how accurate agency detection typically is. However, note that we still have the problem of not definitively knowing when agency has been accurately detected. Are people really agents? How do we know apart from the cognitive systems responsible for agent detection? Do we count detections of ghosts, spirits, and gods as mistakes or as accurate detection? If as mistakes, we beg the question at hand. The same problem confronts other hypothesized cognitive systems responsible for promoting belief in gods.

These observations concerning determining the accuracy of our cognitive mechanisms to give us accurate beliefs raise another problem with the Error-Prone Mind argument. I call this problem *Collateral Damage*.

Collateral damage is a euphemism for killing unintended civilians in the course of warfare. I am afraid that in this war with belief in gods, several of the arguments presented here including the Error-Prone Mind argument run the risk of causing severe collateral damage. If it can be successful at damaging belief in gods, it will likewise damage other, unintended, beliefs.

Suppose we can agree that our agency detection system is indeed so inaccurate that it cannot be trusted. It gets things wrong more often than right. Then, if it encourages us to believe in a supernatural agent, we have every reason to ignore its encouragement. Belief in gods is certainly weakened, but so is belief in any and all intentional agency.

On what basis do we believe that other people have minds? At least in large part by virtue of the same cognitive mechanisms responsible for encouraging belief in superhuman minds. HADD and other mechanisms converge on the intuition that other people are intentional agents with beliefs and desires and not merely automatons or machines inside of human shells. We have no direct, conclusive evidence to support the belief that people are intentional agents with minds. Minds cannot be directly observed. We have no empirical evidence for their existence.[20] If belief in gods is weakened by the Error-Prone Mind argument, so too is belief in all intentional agents. This collateral damage may be tolerable to some cognitive scientists, but few others.

The cognitive sciences (including experimental psychology) have given us plenty of reason to doubt the veracity of our minds on a number of fronts. Above I mentioned perceptual and conceptual illusions as examples. Some of these cognitive errors are persistent and systematic and influence how we perceive the world, think, reason, and conduct ourselves in everyday life. Suppose our agency detection system is profoundly error-prone. So too are a host of other cognitive systems.

If the Error-Prone Mind argument is well formed, it should allow for innumerable additional substitutions.

> The cognitive science of X demonstrates that cognitive mechanisms of our evolved minds provide most impetus for believing X. As these mechanisms are error-prone, they cannot be trusted to give us truth. As we cannot trust these aspects of our minds to give us truth, we cannot trust most of our impetus for believing X. Therefore, cognitive science X weakens belief X.

I regard such an argument as yielding not only intolerable *collateral damage*. In fact, it suffers from a *Suicidal Tendency*. Not only does it potentially weaken belief in a host of different domains, but it also casts doubt on its own foundations and presuppositions.

70 **Theology and Science**

As noted above, evolution by natural selection has no regard for Truth. We can only be confident that it has given us minds helpful for survival in our ancestral environments. Further, the cognitive sciences have given us evidence that our minds—for the sake of survival—can be systematically fallible, trading survival and reproduction for accurate representations of reality. On what basis then do we trust our minds at all? Our minds cannot be trusted to tell us that gods exist, that other human minds exist, that our memories are reliable, or that natural laws remain the same from moment to moment, or that cognitive science can produce accurate findings or that evolution is true. The Error-Prone Mind argument proves to be self-defeating—it has a Suicidal Tendency.

The way out of this problem is to reject the claim that our mind and its component mechanisms are too error-prone to be trusted. Do they make mistakes? Sure. But do they make enough mistakes to completely undercut them as sources of beliefs? I hope not. They are all we have; but on this point, the evolutionary and cognitive sciences do not offer any assurances. Natural selection offers no truth guarantees and the cognitive sciences seem to give us plenty of reasons for doubt. One will have to build an epistemological foundation using pre- or extra-scientific timber.

The theist may build such an epistemological foundation by appealing to the divine as a trustworthy source of Truth that has imparted the ability to conceive Truth (at least under some conditions) through cosmic fine-tuning or supernatural selection or supernaturally generated mutations that then were naturally selected to produce human minds. I leave the details and the coherence of such a theology up to the individual theist.

For the moment, it appears that theists have nothing to fear from the bio-psychological explanations of religion. Although these scientific endeavors may grant new insights into the mechanisms that play a role in shaping religious thought and action, whether or not belief in gods is rationally justified remains a question outside of science.

Acknowledgement

This work was supported by a grant from the John Templeton Foundation.

Endnotes

1 Pascal Boyer, *Religion Explained: The Evolutionary Origins of Religious Thought* (New York: Basic Books, 2001), 2, 4.

2 Scott Atran, *In Gods We Trust: The Evolutionary Landscape of Religion* (Oxford: Oxford University, 2002). In keeping with the comparative character of this area of scholarship, I use *theism* and *theistic* to refer to beliefs in non-natural intentional agents generally including God, gods, ghosts, and ancestor-spirits. Monotheism and polytheism are regarded as subclasses of theism in this sense. As the arguments here equally apply to belief in any of these intentional agents, I use the term *gods* as an inclusive umbrella. The Judeo-Christian God, Muslim Allah, Hindu Shiva and Vishnu, and any number of other gods are all objects of the arguments presented here.

3 Daniel Dennett, *Breaking the Spell: Religion as a Natural Phenomenon* (New York: Viking, 2006).
4 Leon Wieseltier, "The God Genome," *New York Times,* (February, 19, 2006).
5 Eugene G. D'Aquili and Andrew B. Newberg, *The Mystical Mind: Probing the Biology of Religious Belief* (Minneapolis: Fortress, 1999); Rhawn Joseph, ed., *NeuroTheology: Brain, Science, Spirituality, Religious Experience* (New York: University Press, 2002); Michael A. Persinger, *Neuropsychological Bases of God Beliefs* (New York: Praeger, 1987).
6 Michael A. Persinger, "The sensed presence within experimental settings: Implications for the male and female concept of self," *Journal of Psychology: Interdisciplinary and Applied,* 137:1 (2003): 5–16.
7 David Sloan Wilson, *Darwin's Cathedral: Evolution, Religion, and the Nature of Society* (Chicago: University of Chicago, 2002). Whitehouse suggests that a particular configuration of social morphology and ritual practices he terms the "imagistic mode of religiosity" might likewise produce group cohesion, pro-social behavior, and resultant competitive advantage over other groups; but Whitehouse's account does not appeal to genetic, adaptationist selection but cultural selection. Harvey Whitehouse, *Modes of Religiosity: A Cognitive Theory of Religious Transmission* (Walnut Creek, Calif.: AltaMira, 2004).
8 Justin L. Barrett, *Why Would Anyone Believe in God?* (Walnut Creek, Calif.: AltaMira, 2004); Brian Malley, *How the Bible Works: An Anthropological Study of Evangelical Biblicism* (Walnut Creek, Calif.: AltaMira, 2004); Robert N. McCauley and E. Thomas Lawson, *Bringing Ritual to Mind: Psychological Foundations of Cultural Forms* (Cambridge: Cambridge University Press, 2002); Ilkka Pyysiäinen, *Magic, Miracles and Religion: A Scientist's Perspective* (Walnut Creek, Calif.: AltaMira, 2004); D. Jason Slone, *Theological Incorrectness: Why Religious People Believe What They Shouldn't* (New York: Oxford University Press, 2004); Todd Tremlin, *Minds and Gods: The Cognitive Foundations of Religion* (New York: Oxford University Press, 2006); Whitehouse, *Modes of Religiosity.*
9 Jesse M. Bering, "Intuitive conceptions of dead agents' minds: The natural foundations of afterlife beliefs as phenomenological boundary," *Journal of Cognition & Culture,* 2:4 (2002): 263–308; Jesse M. Bering, Carlos Hernández-Blasi, David F. Bjorklund, "The development of 'afterlife' beliefs in secularly and religiously schooled children," *British Journal of Developmental Psychology,* 23:4 (2005): 587–607; Paul L. Harris, and Marta Gimenez, "Children's Acceptance of Conflicting Testimony: The Case of Death," *Journal of Cognition & Culture,* 5:2 (2005): 143–164.
10 E. Margaret Evans, "Cognitive and contextual factors in the emergence of diverse belief systems: Creation versus evolution," *Cognitive Psychology,* 42:3 (2001): 217–266; Deborah Kelemen, "Why are rocks pointy? Children's preference for teleological explanations of the natural world," *Developmental Psychology,* 35:6 (1999): 1440–1453; idem, "Functions, goals, and intentions: Children's teleological reasoning about objects," *Trends in Cognitive Sciences,* 3:12 (1999): 461–468; idem, "Are children 'intuitive theists?' Reasoning about purpose and design in nature," *Psychological Science,* 15:5 (2004): 295–301; Oliveria Petrovich, "Understanding of non-natural causality in children and adults: A case against artificialism," *Psyche en Geloof,* 8 (1997): 151–165; idem, "Preschool Children's Understanding of the Dichotomy Between the Natural and the Artificial," *Psychological Reports,* 84:1 (1999): 3–27.
11 Justin L. Barrett, Roxanne M. Newman, and Rebekah A. Richert, "When seeing does not lead to believing: Children's understanding of the importance of background knowledge for interpreting visual displays," *Journal of Cognition & Culture,* 3:1 (2003): 91–108; Justin L. Barrett and Rebekah A. Richert, "Anthropomorphism or preparedness? Exploring children's concept of God," *Review of Religious Research,* 44:3 (2003): 300–312; Justin L. Barrett, Rebekah A. Richert, and Amanda Driesenga, "God's beliefs versus mother's: The development of non-human agent concepts," *Child Development,* 72:1 (2001): 50–65.
12 Evans, "Cognitive and contextual factors in the emergence of diverse belief systems."

72 **Theology and Science**

13 Paul Bloom, "Is God An Accident?," *Atlantic Monthly*, (December 2005): 1–8.
14 Paul Bloom, *Descartes' Baby: How the Science of Child Development Explains What Makes Us Human* (New York: Basic Books, 2004).
15 Bloom, "Is God An Accident?," 8.
16 Ibid.
17 Thomas Gilovich, *How We Know What Isn't So: The Fallibility of Human Reason in Everyday Life* (New York: Free Press, 1991).
18 Stewart E. Guthrie, *Faces in the Clouds: A New Theory of Religion* (New York: Oxford University, 1993).
19 If we assume that people and sophisticated animals do qualify as agents, such a strategy of tabulating accurate versus inaccurate agency detections would likely lead us to the conclusion that our cognitive equipment is extremely reliable and not so error-prone after all. Such a finding would encourage rather than discourage confidence in theistic beliefs.
20 Alvin Plantinga, *God and Other Minds: A Study of the Rational Justification of Belief in God* (Ithaca, N.Y.: Cornell University Press, 1990).

Biographical Notes

Justin L. Barrett is Senior Researcher at the University of Oxford's Centre for Anthropology and Mind. He earned degrees in psychology from Calvin College (B.A.) and Cornell University (Ph.D.). He served on the psychology faculties of Calvin College and the University of Michigan (Ann Arbor). Dr. Barrett is an editor of the *Journal of Cognition & Culture* and is author of numerous articles and chapters concerning cognitive science of religion. His book *Why Would Anyone Believe in God?* (AltaMira, 2004) presents a scientific account for the prevalence of religious beliefs.

[26]

How Firm a Foundation? A Response to Justin L. Barrett's "Is the Spell Really Broken?"[1]

HOWARD J. VAN TILL

Abstract *In his essay, "Is the Spell Really Broken?," cognitive psychologist Justin L. Barrett evaluates several ways in which critics of theism might employ various bio-psychological theories of religion to discredit basic theistic beliefs. Although appreciative of the substance of his evaluations, I nonetheless find reasons for challenging both Barrett's wording of some anti-theistic objections, and his advice on how theists might proceed to build an epistemological foundation for theistic beliefs—a foundation that would withstand erosion by the new and noteworthy criticisms of cognitive science. If the phenomenon of religious belief in general is a fully natural outcome of human brain evolution, can the content of specific theistic beliefs still be convincingly warranted?*

Key words: Cognitive science; Evolution; Religious belief; Theism; Warrant

Off to an awkward start

Justin Barrett deals courageously with a sensitive topic in his essay, but he begins it with what I find to be a series of unfortunate word choices. "The long and adversarial relationship between *Darwinism* and *theism* continues," says Barrett in his opening line. And in the next paragraph he poses the rhetorical question, "Will evolutionary *sciences* finally win the war and defeat *theism* once and for all?"[2]

Speaking candidly, I must say that I have grown extremely weary of seeing the term *Darwinism*, without any qualification whatsoever, used as if it were a synonym for *atheism*. In spite of the efforts of the Intelligent Design movement and other religiously energized anti-evolution movements to make it so, it is not.

Furthermore, it is not *theism*, broadly conceived, that sees itself at war with evolutionary *science*. No, it is *episodic creationism*—with its inclination to reduce the concept of divine creative action to episodes of irruptive, form-imposing, supernatural intervention—that has declared war on the acceptability and/or adequacy of biological evolution.

Whether from the pen of a theist or an atheist, rhetoric that blurs the distinction between, on the one hand, a cluster of scientific theories regarding the formational history of living organisms, and on the other hand a comprehensive metaphysical system, is rhetoric more likely to amplify rather than to diminish the confusion that persists in the discussion of scientific and religious beliefs regarding the phenomenon of biological evolution. There is indeed a war of *worldviews* going on, as has long been the case; but identifying this as a battle between *evolutionary science* and *theism* encourages the propagation of a wearisome misunderstanding of the issues.

On to more interesting issues

A number of scholars cited by Barrett have called for the application of scientific investigation and analysis to the phenomenon of religious belief. Barrett himself is an active participant in this enterprise, and has been credited with playing a major role in founding the particular field of study now called the "cognitive science of religion."

These scientific approaches should, I believe, be welcomed by all parties. Even by religious "believers"? Sure, why not? Persons confident of their own worldview should welcome any open and competent investigation of its "bio-psychological" dimension. If the human capacity and proclivity for religious belief can be understood, at least in part, in terms of the mental tools that evolved in our formational history, why not pay respectful attention to this insight?

But theism's welcome mat is not always out, and self-appointed spokespersons for science sometimes fail to honor the limits of science's domain of competence. Mutual antagonism is likely to thrive whenever worldviews become the targets of reciprocal criticism. In the essay under review, Barrett offers his critical examination of five possible arguments against theistic belief that appeal to the results of recent scientific studies. Recognizing that some representatives of science do occasionally employ rhetoric that is dismissive of religious truth-claims, Barrett is especially eager to restore confidence to religious believers who have come to fear "that these new treatments of religion are fundamentally dangerous or even hostile to theism."[3] (For the record, I think it only fair to note in passing that some spokespersons for theism have also been known to speak dismissively of scientific truth-claims. But that's another story.)

Five arguments against theism, and Barrett's critiques of them

The wording of the five arguments is Barrett's, not that of any particular critic. Unfortunately, this opens Barrett to the criticism that these may be straw-man arguments. Indeed, some of them seem to be stated by Barrett in such a strident tone that they would be very difficult to defend, difficult even for an ardent critic of theism. Following are my thoughts on four of Barrett's responses to the arguments as he stated them.

The "evolutionary byproduct" argument, in Barrett's words

> Selection pressures (operating on either groups or individuals) have led to various dispositions or propensities in human minds that happen to give rise to religious

> belief. Religious beliefs are, therefore, accidents or byproducts of evolution. As such, religious beliefs cannot be trusted. Belief in gods amount to cognitive illusion.[4]

In response, Barrett points out that just because certain kinds of beliefs (or propensities for these beliefs) may be "accidents" or "byproducts" of evolution, that fact by itself tells us nothing about whether the beliefs are illusory or true. "To be able to call genuinely religious beliefs 'illusions' we need to be able to demonstrate that they ... upon further examination, are in error."[5]

Agreed, of course; but that sword cuts both ways. Accidental beliefs are not necessarily false, but neither are they necessarily true. In his book, *Why Would Anyone Believe in God?*,[6] Barrett argues ardently that belief in gods generally (and in the God of Abrahamic traditions in particular) can be well understood as the outcome of *natural* human mental tools operating without coercion and without unusual environmental requirements. But the *naturalness* of religious belief is no more a guarantor of *truth* than is its *accidentalness* a guarantor of *falsehood*.

What critics of theism need to do, if they wish to discredit some religious belief, is to construct a line of reasoning from substantial evidence to the conclusion that this particular belief is false, or at least *unwarranted*. Theists, on the other hand, must own the burden of demonstrating that the belief is true, or at least *warranted* by something other than loudly repeated assertions or cocksure private judgments—not Barrett's style, of course, but one that is familiar to us nonetheless. Hard *proof* may be out of reach for both, so that the best approach for advocate and critic alike is to examine the *warrant* (evidence, reasons) for either belief or disbelief. That being the case, perhaps the stridently worded criticism crafted by Barrett could profitably be restated in a more balanced form as follows:

> Selection pressures (operating on either groups or individuals) have led to various dispositions or propensities in human minds that naturally give rise to the holding of religious beliefs. Religious belief is a *natural* phenomenon. However, whether or not any particular religious belief is *warranted* is an entirely different matter that will have to be settled outside the court of scientific inquiry.[7]

Stated in this non-inflammatory way, perhaps this proposition could initiate a constructive conversation between persons of good will who, for whatever reasons of nature or nurture, of accident of biology or accident of cultural heritage, hold vastly differing worldviews.

The "religious utility" objection, as stated by Barrett

> Religious beliefs have utility (e.g. in social arrangements) leading to their natural selection. As religious beliefs are selected for their utility and not their truth, they should not be trusted in terms of their truth-value.[8]

The logic of this objection flows in the opposite direction relative to the previous one. Here, the argument is that if belief in gods *does* have survival value (and is *not* simply a byproduct or accident of evolution), then its truth-value must be doubted. As Barrett himself correctly notes, "When it comes to natural selection, Truth is expendable."[9]

But Barrett goes on to argue that this utility-based objection (whose specific wording he fine-tunes a couple of times) "still fails as a defeater for religious belief ... [T]o say that being the product of evolution undermines religious belief is overstating the opposition between natural selection and Truth-representing minds. Natural selection does not discriminate against Truth; it just does not attempt to preserve it."[10]

Point well taken, but the problem of "overstating the opposition" is, in the context of this essay, a problem entirely of Barrett's own making! The wording of the objection is Barrett's wording, not the wording of any named or quoted critic of theism (although I'm sure that Barrett could find examples in this style).

But my principal criticism of Barrett's rhetorical strategy here is centered on his use of the word, "defeater." If I understand Barrett correctly, he is saying that the religious utility argument fails to be able to deliver a knockout punch to religious belief. True, perhaps; but intellectual arguments, like boxing matches, are routinely won or lost on the cumulative effect of punches far less damaging than a smashing knockout. Critics of religious beliefs need not knock their opponents unconscious in order to win an argument. Suggesting otherwise, as Barrett here seems to be doing, is also a form of overstatement. Knockout or not, the critic of religious belief is still in a viable position to carry on with the contest.

To demonstrate a more even-handed approach, let me suggest the following restatement of this objection that avoids the problems of overstatement:

> Religious beliefs have utility (e.g. in social arrangements) leading to their natural selection. Because, in the context of evolutionary dynamics, religious beliefs are selected for their *utility* and not for their *veracity*, the question of whether or not any particular religious belief is *warranted* or *true* is an entirely different matter that will have to be settled outside the court of scientific inquiry. Let all participants understand, however, that this out-of-(scientific)-court settlement regarding *warrant* for the *content* of specific beliefs must be performed in the full awareness of what cognitive science has learned about the general *phenomenon* of religious belief.[11]

The "inherited belief" argument (weak version) as stated by Barrett

> People are credulous recipients of theistic beliefs (e.g. from parents). Natural selection provided people with the cognitive faculties that contribute to making us credulous recipients. As we now know why people so readily believe in gods through these faculties, their contribution toward belief should be discounted. Hence fewer reasons exist for theistic belief.[12]

Even in this softened version, edited by Barrett to avoid particular pitfalls built into his original version, numerous word choices ensure that it will elicit a hostile reaction from theists. Barrett himself characterizes it as a "vicious blow against theism."[13] Referring to theists as "credulous recipients" of theistic beliefs is clearly not designed to encourage a constructive conversation. Neither is the term "discounted" likely to evoke a sympathetic hearing or mutual respect.

My main criticism, however, concerns the minimal attention that Barrett pays in this essay to the actuality and effectiveness of the biases that are unavoidably introduced by an individual's membership of a religious community. Barrett's own work in the cognitive science of religion does an outstanding job of demonstrating the "naturalness of religion" thesis. The human inclination to posit the existence and action of superhuman Agents is the natural product of mental tools built into our evolved brains. Barrett has also convincingly argued that the more numerous and diverse the occasions on which some particular religious belief is reinforced by social experiences with other "believers," the more credible that belief becomes and the more likely it is to be incorporated into one's reflective belief system, *whether true or not.*[14]

Let me diverge briefly on a bit of relevant personal experience. Some years ago it dawned on me, in the wake of certain unpleasant interactions with an especially feisty and sometimes ill-mannered portion of the religious community, that I needed to re-examine the content of my religious belief system. As I reflect on that awakening experience, I continue to ask myself questions (some stated quite bluntly) like these: Why did it take me so long to realize that this re-evaluation was necessary? Why had I so long been content to remain bound to the stringent requirements of creeds written four centuries ago? Was I dull of mind? Was my brain turned off?

The work of Justin Barrett, Pascal Boyer and other scholars in the cognitive science of religion has been immensely helpful to me in my search for answers to these self-reflective questions.[15] I think I now understand my own actions and reactions (as well as those of critics and supporters alike) far better than I did two decades ago in the heat of turmoil. As it turns out, I think that my mind/brain was neither dull nor inoperative. I would even be so bold as to suggest that it was functioning quite normally in the natural manner that the cognitive sciences are now beginning to understand, *with one substantive qualification*: I would suggest that cognitive science needs to pay additional attention to the way in which membership in a tightly-knit religious community affects the *emotional security* of its individual members. In the context of a tight, religion-based coalition, tribal dynamics (which have immense power either to amplify or diminish an individual's emotional security, with its associated survival value) have a strong impact on a member's behavior and beliefs. With that caveat in mind, let me suggest the following as a modified version of the "inherited belief" argument that, in my judgment, deserves more rigorous examination:

> Natural selection has provided people with an inclination toward religious beliefs generally. It has also provided mental tools that enhance the credibility of particular beliefs when they are reinforced by diverse social experiences (e.g. affirmation by believing parents or by other members of one's belief-based coalition). Even reflective beliefs can be strongly biased by common social dynamics. Since cognitive science now understands the effect of biases introduced by these natural mechanisms, it is imperative that their contribution toward theistic belief be carefully scrutinized and that the warrant for particular theistic propositions be rigorously evaluated.[16]

The "error-prone mind" argument

> The cognitive science of religion demonstrates that cognitive mechanisms of our evolved minds provide most impetus for believing in gods. As these mechanisms

> are error-prone, they cannot be trusted to give us Truth. As we cannot trust these aspects of our minds to give us Truth, we cannot trust most of our impetus for believing in gods. Therefore cognitive science of religion weakens belief in gods.[17]

This objection deserves respectful attention because, as Barrett candidly notes, "Certainly, if a belief arises largely because of evidence (or reasons) from sources known to be untrustworthy, the belief is in danger unless other evidences (or reasons) exist apart from this dubious source."[18] Fair enough. The question, then, is this: How error-prone is the cognitive system that leads to belief in gods, or God?

One of the mind's cognitive tools that contributes to belief in gods is known as HADD (the hypersensitive agency detection device). HADD alerts a person to the possible presence and action of an intentional agent that may be either beneficial or dangerous. Some of the agents indicated by HADD are familiar *embodied* agents (animals, other humans) whose capabilities and existence in the natural world are well documented. Other agents postulated by HADD, however, are different in significant ways—superhuman *unembodied* agents (ancestral spirits, gods, God) whose existence and capabilities are the subject of a great deal of conjecture and debate. In either case, certain environmental signals are quickly (often *prematurely*) taken by HADD to be indicators of the presence or action of an intentional agent. The move from HADD-initiated non-reflective beliefs to stable reflective beliefs ordinarily requires a slower and more deliberate process.

As Barrett points out, HADD is tuned, for the sake of survival, to err in the direction of over-detection—the "better safe than sorry" strategy. The hypersensitivity-based objection that Barrett is here criticizing is this: "Being error-prone, HADD cannot be trusted; so, the evidence HADD contributes to belief in gods should be discounted."[19]

I concur with Barrett that a summary dismissal of HADD-initiated agency indicators would be as out of place as would be their uncritical acceptance. What needs more attention, it seems to me, is the critical issue of whether or not there is a systematic difference in the error-proneness of HADD's postulation of an *unembodied* superhuman Agent in comparison to its positing a familiar *embodied* agent. These are two remarkably different categories of agency. Errors in positing action by familiar embodied agents (animals, other humans) are relatively easy to detect. But the question of whether or not HADD errs in postulating the existence or activity of unembodied superhuman Agents (ancestral spirits, gods, God) is precisely the issue at hand, and its answer is far from self-evident.

Is belief in gods or God *necessarily* weakened by knowing how our cognitive mechanisms, including HADD, work? No; that would be another overstatement. But the more pertinent and difficult question on the table is this: When we do the deliberate reflection on and evaluation of our HADD-generated *non-reflective* belief in God, is there sufficient *warrant* to support our raising that intuition to the level of a *reflective* belief? Knowing how our cognitive mental tools operate does not necessarily weaken belief in God, but it certainly ought to alert us to the need to warrant that belief with something far more convincing than a simple assertion or the unthinking adoption of one's tribal tradition.

This is not a trivial challenge; in order to meet it, "One will have to build an epistemological foundation using pre- or extra-scientific timber" and, says Barrett, "... whether or not belief in gods is rationally justified remains a question outside of science."[20]

Once again, I find myself in agreement with Barrett. Theists must, I believe, bear the burden of demonstrating to critics that theistic beliefs are *warranted* by a substantial epistemological foundation. Furthermore, I would argue that that foundation-building project, although it necessarily extends far beyond the realm of the sciences, must now be carried out in full awareness of what the cognitive science of religion has so far accomplished. But I am not at all encouraged by Barrett's specific suggestion for the shape of that foundation:

> The theist may build such an epistemological foundation by appealing to the divine as a trustworthy source of Truth that has imparted the ability to conceive Truth (at least under some conditions) through cosmic fine-tuning or supernatural selection or supernaturally generated mutations that then were naturally selected to produce human minds. I leave the details and the coherence of such a theology up to the individual theist.[21]

In this statement, Barrett acknowledges the necessity of incorporating what we know of both cosmic and biological evolution into theistic thinking about how we got to be the way we are. So far, so good. But there is a serious inconsistency embedded in Barrett's wording that puzzles me. Barrett's positive reference to the concept of *cosmic fine-tuning* suggests that he welcomes the idea of crediting the success of cosmic evolution (the formational history of the physical universe) to a divine choice for the correct numerical values of the universe's fundamental constants. Cosmic fine-tuning is apparently taken to be a manifestation of divine provision. It would seem, then, that a key trait of divine creative action is the provision, at time-zero, of whatever would eventually be required for the evolution of cosmic structures *without need for irruptive, form-conferring interventions in the course of time.* That's what cosmic fine-tuning accomplishes.

But Barrett's treatment of divine action in relation to *biological* evolution takes the exact opposite tack. His references to "supernaturally generated mutations" and "supernatural selection" imply that certain episodes of supernatural intervention may have been essential to the formation of humans with Truth-conceiving minds. If so, then it would seem that the key trait of divine creative action in the biological realm is *not* the provision, at time-zero, of all that would eventually be needed (biological fine-tuning) for the evolution of life forms, but rather *the interjection of occasional form-conferring actions to compensate for what had not been provided in the first place.* In the absence of biological fine-tuning, form-conferring interventions become necessary.

I suppose one could posit that there is no *a priori* requirement that a Creator proceed in a consistent fashion for both cosmic and biological evolution, but this two-style approach (cosmic fine-tuning, yes; biological fine-tuning, no) has all the marks of an *ad hoc* hypothesis framed in the style of the old ruse, "heads I win, tails you lose."[22] This hybrid mechanism of *empirically accessible natural processes* episodically punctuated by a series of conjectured but *empirically invisible supernatural interventions* is straight from the rhetoric of the Intelligent Design movement, a movement whose claims I have evaluated for more than fifteen years, a movement that, in my judgment, has done nothing to strengthen the warrant for theistic belief.[23]

If, in the context of the important discoveries that Barrett and other cognitive scientists have made concerning the way in which we naturally build religious belief systems, this *ad hoc* and inconsistent approach is the best that theists can do to build a foundation of rational warrant for belief in God, then the cause seems hopeless to me. An epistemological foundation fabricated from little more than the recitation of restated tribal myths involving supernatural feats of form-imposing action may enhance the emotional security of persons who are already true believers, but I see nothing in these highly conjectural and *ad hoc* insertions of irruptive intervention that would be attractive to anyone else. Given the probative force of the *naturalness of religion thesis* that Barrett and other cognitive scientists have demonstrated, *ad hoc* conjectures will not provide the firm epistemological foundation that theism desires.

Endnotes

1 Justin L. Barrett, "Is the Spell Really Broken?," *Theology and Science*, vol. 5, no. 1 (2007): 57–72.
2 Ibid., 57, emphasis added.
3 Ibid., 60.
4 Ibid., 62.
5 Ibid., 63.
6 Justin L. Barrett, *Why Would Anyone Believe in God?* (Lanham, Md: AltaMira Press, 2004).
7 Compare to Barrett, "Is the Spell Really Broken?," 62.
8 Ibid., 64.
9 Ibid., 65.
10 Ibid.
11 Compare to ibid., 64.
12 Ibid., 66.
13 Ibid., 67.
14 As one example of these influential social experiences, consider the hearing of repeated proclamations from tribal religious leaders. "Those who work in the persuasion and propaganda fields, such as advertising, know the power of proclamation. When people hear similar claims repeatedly, even though they receive no evidence or proper justification for the claim, they tend to believe the claim. The more familiar the claim, the more intuitively true it seems. Thus, someone who hears the truth of the Divine Trinity affirmed weekly for years will tend to accept its truth, even if no strong evidence or justification has ever been offered" (Barrett, *Why Would Anyone Believe in God?*, 69).
15 In addition to Barrett's *Why Would Anyone Believe in God?*, see also Pascal Boyer, *Religion Explained: The Evolutionary Origins of Religious Thought* (New York: Basic Books, 2001) and Todd Tremlin, *Minds and Gods: The Cognitive Foundation of Religion* (New York: Oxford University Press, 2006).
16 Compare to Barrett, "Is the Spell Really Broken?," 65–67.
17 Ibid., 67.
18 Ibid.
19 Ibid., 68.

20 Ibid., 70.
21 Ibid.
22 A two-style approach similar to Barrett's was offered by Intelligent Design advocate Stephen C. Meyer in *Science and Christianity: Four Views*, ed. Richard F. Carlson (Downers Grove: Intervarsity Press, 2000), 127–174, with my response on 188–194.
23 For a sample of my evaluation of the ID movement and its specific claims for scientific support, see my essay, "Are Bacterial Flagella Intelligently Designed? Reflections on the Rhetoric of the Modern ID Movement," *Science and Christian Belief*, vol. 15, no. 2 (October, 2003): 117–140.

Biographical Notes

Howard J. Van Till is Professor Emeritus of Physics and Astronomy at Calvin College, located in Grand Rapids, Michigan. After graduating from Calvin College in 1960, he earned his Ph.D. in physics from Michigan State University in 1965. His research experience includes both solid-state physics and millimeter-wave astronomy. His current interests include the assessment of belief systems and the processes by which we construct them. Professor Van Till is a founding member of the International Society for Science and Religion, has served on the executive council of the American Scientific Affiliation and the advisory board of the John Templeton Foundation, as well as being a member of the editorial boards of both *Science and Christian Belief* and *Theology and Science*.

Name Index

Printed and bound by CPI Group (UK) Ltd, Croydon, CR0 4YY
17/10/2024
01775696-0001